T0331767

# Rock Fractures in Geological Processes

Rock fractures largely control many of the Earth's dynamic processes. Examples include plate-boundary formation and development, tectonic earthquakes, volcanic eruptions, and fluid transport in the crust. How rock fractures form and develop is of fundamental importance in many theoretical and applied fields of earth sciences and engineering, such as volcanology, seismology, hydrogeology, petroleum geology, natural hazards, and engineering geology. An understanding of rock fractures is essential for effective exploitation of many of the Earth's natural resources including ground water, geothermal water, and petroleum.

This book combines results from fracture mechanics, materials science, rock mechanics, structural geology, hydrogeology, and fluid mechanics to explore and explain fracture processes and fluid transport in the crust. Basic concepts are developed from first principles and are illustrated with numerous worked examples that link models of geological processes to real field observations and measurements. Calculations in the worked examples are presented in detail with simple steps that are easy to follow – providing the readers with the skills to formulate and quantitatively test their own models, and to practise their new skills using real data in a range of applications. Review questions and numerical exercises are given at the end of each chapter, and further homework problems are available at www.cambridge.org/gudmundsson. Solutions to all numerical exercises are available to instructors online.

*Rock Fractures in Geological Processes* is designed for courses at the advanced-undergraduate and beginning-graduate level, but also forms a vital resource for researchers and industry professionals concerned with fractures and fluid transport in the Earth's crust.

**Agust Gudmundsson** holds a University of London Chair of structural geology at Royal Holloway. He has a Ph.D. in Tectonophysics from the University of London and has previously held positions as research scientist at the University of Iceland, professor and Chair at the University of Bergen, Norway, and professor and Chair at the University of Göttingen, Germany. Professor Gudmundsson's research interests include volcanotectonics, seismotectonics, and fluid transport in rock fractures and reservoirs. He has published more than 130 research papers on these and related topics, is on the editorial boards of *Terra Nova*, *Tectonophysics*, *Journal of Geological Research*, and *Journal of Volcanology & Geothermal Research*, and is a fellow of the Iceland Academy of Sciences and Academia Europaea. The book draws on Professor Gudmundsson's extensive experience in field, analytical, and numerical studies of crustal fractures and of teaching undergraduate and graduate courses in structural geology, geodynamics, hydrogeology, rock mechanics, reservoir geoscience, seismotectonics, and volcanotectonics.

# Rock Fractures
# in Geological Processes

AGUST GUDMUNDSSON

**Royal Holloway University of London**

# CAMBRIDGE
## UNIVERSITY PRESS

University Printing House, Cambridge CB2 8BS, United Kingdom

Cambridge University Press is part of the University of Cambridge.

It furthers the University's mission by disseminating knowledge in the pursuit of education, learning and research at the highest international levels of excellence.

www.cambridge.org
Information on this title: www.cambridge.org/9780521863926

First published 2011
4th printing 2015

*A catalogue record for this publication is available from the British Library*

*Library of Congress Cataloguing in Publication data*
Gudmundsson, Agust.
Rock Fractures in Geological Processes / Agust Gudmundsson.
p.   cm
Includes bibliographical references and index.
ISBN 978-0-521-86392-6 (hardback)
1. Rocks–Fracture.   2. Rock mechanics.   3. Hydrogeology.   4. Geology, Structural.
5. Fluid mechanics.   I. Title.
TA706.G83 2011
624.1´5132–dc22        2010051429

ISBN 978-0-521-86392-6 Hardback

Additional resources for this publication at www.cambridge.org/gudmundsson

# Contents

# Preface

Many of the Earth's most fascinating natural processes are related to rock fractures. Volcanic eruptions, tectonic earthquakes, geysers, large landslides and the formation and development of mid-ocean ridges all depend on fracture formation and propagation. Rock fractures are also of fundamental importance in more applied fields such as those related to fluid-filled reservoirs, deep crustal drilling, tunnelling, road construction, dams, geological and geophysical mapping and field geology and geophysics.

There has been great progress in understanding fracture initiation and propagation over the past decades. The results of this progress are summarised in many papers, textbooks and monographs within the fields of fracture mechanics and materials science. Much of this improved knowledge of fracture development is of great relevance for understanding and modelling geological processes that relate to rock fractures.

The purpose of this book is to offer a modern treatment of rock fractures for earth scientists and engineers. The book is primarily aimed at, first, undergraduate or beginning graduate students in geology, geophysics and geochemistry and, second, scientists, engineers and other professionals who deal with rock fractures in their work. The book has been designed so that it can be used (1) for an independent study, (2) as a textbook for a course in rock fractures in geological processes and (3) as a supplementary text for courses in structural geology, seismology, volcanology, hydrogeology, geothermics, hazard studies, engineering geology, rock mechanics and petroleum geology.

Each chapter begins with an overview of aims and ends with a summary of the main topics discussed and a list of all the main symbols used in that chapter. There are many worked examples (solved problems) and exercises (supplementary problems) in each chapter. The worked examples serve to illustrate the theoretical principles and show how they may be applied to fracture-related processes in the crust. The examples and exercises are meant to provide a deeper understanding of the basic principles of rock fractures, so that the reader can use them with great confidence in solving rock-fracture problems.

I have taught much of the material in the book over the past 12 years to earth science students in Norway, Germany and England. The basic material has been used in undergraduate and graduate courses on such diverse topics as volcanotectonics, seismotectonics, structural geology, geodynamics, rock mechanics, rock-fracture mechanics, hydrogeology, petroleum geology and applied geology. While most of these students were educated in geology, many were educated in geophysics, geochemistry, physical geography and engineering. Based on this experience, almost all the material in the book should be suitable for students with a very modest knowledge of mathematics and physics. The only exceptions are parts of Sections 13.4, 13.5 and 14.6, where more advanced mathematics is used. All the necessary physics is explained in the book.

The book is also meant for professional scientists whose work involves rock fractures, in particular fluid transport in fractured rocks. These include geologists, geophysicist, geochemists, hydrogeologists, civil engineers, petroleum engineers and experts in related fields. Many of these may neither have the time nor inclination to read the entire book. I have therefore written the chapters, particularly those in the second half of the book, so as to make each of them comparatively independent of the other chapters. Thus it should generally be possible to read and understand the content of one chapter without having to read all the other chapters. For this reason, and also for pedagogical reasons, there is considerable repetition of various basic principles and results, particularly in the chapters that constitute the second half of the book. The repetition should help in effective learning of the main topics.

As regards referencing of the technical literature, I follow the common tradition of citing comparatively few references in the part of the text dealing with general solid mechanics in the first half of the book. The basic topics treated in this part are well established and are treated in numerous standard textbooks and monographs, many of which are included in the reference lists at the ends of the chapters. Many of the topics discussed in the second half of the book, however, are still the subject of very active research in the field of rock-fracture development and related fluid transport. In this part of the book there are thus many more references in the text, as well as extensive reference lists at the ends of the chapters. Although the reference lists cannot be exhaustive, they indicate the papers and books that were used when writing the chapters and may also serve as guides to the general literature on rock fractures.

Agust Gudmundsson

# Acknowledgements

Many colleagues and students have made contributions to this book, some through technical discussions over the years, which have helped in my formulating some of the ideas presented in the book. An exhaustive list of all these people is not possible, but below I mention some colleagues, students and friends who have been most directly involved with the book itself.

First, I would like to mention two colleagues and friends who are no longer with us. Both shared my interest in, and enthusiasm for, rock fractures. One, Neville J. Price, wrote the first monograph on rock fractures, Fault and Joint Development in Brittle and Semi-Brittle Rock, which had very great effects on the field of rock fractures. The other, Jacques Angelier, was the world's leading expert on palaeostresses and their relation to rock fractures – topics that are still at the forefront of fracture-related research.

Then, I would like to mention several colleagues who read and commented on the manuscript. Very helpful reviews of the manuscript were provided by Adelina Geyer, Shigekazu Kusumoto and Sonja L. Philipp. They read the entire manuscript and made many corrections and suggestions for improvement for which I am very gratefully. In addition, many of the numerical models in the book are from my collaboration with Adelina Geyer and Sonja L. Philipp, whereas some of the analytical parts in Chapter 13 are from my collaboration with Shigekazu Kusomoto. Additional numerical models were made in collaboration with Ruth E.B. Andrew, Otilie Gjesdal, Belinda Larsen, Ingrid F. Lotveit and Trine H. Simmenes. I also thank Ken Macdonald, Philip Meredith and Stephen Sparks for providing the very positive, and much appreciated, general comments on the back cover of the book.

Most of the illustrations in the book are either original or have been remade from various sources. All the sources are cited in the reference lists. For some of the illustrations, particularly those that are not much modified, if at all, and are from recent papers, there are also direct citations in the figure captions. Most of the previously published illustrations are from my own papers in various journals (see the note 'Illustrations' below). I thank the publishers for permission to use the illustrations in the book.

Many people have helped with the illustrations, most of which have been modified many times. Some were originally made by the technical staff at the University of Bergen, Norway, and by students and assistants at the University of Göttingen, Germany. Others were originally made by students and colleagues in France, Germany, Iceland, Italy, Norway and Spain. Any list of names would necessarily be incomplete, so I prefer to offer here a warm thank you to all those who have contributed to the illustrations in the book.

Several people provided photographs, as mentioned in the appropriate captions. These include Valerio Acocella, Jacques Angelier, Ines Galindo, Aevar Johannesson and

Sonja L. Philipp. I am particularly grateful to Valerio Acocella for the photograph on the front cover of the book.

While working on this book, I have received much, and greatly appreciated, help from Nahid Mohajeri. She has redrawn and modified most of the earlier illustrations and made many of the original illustrations in the book.

Although this book project has not received direct funding as such, many of the ideas and results presented have been obtained through many funded projects. In particular, some of the results presented here derive from various projects on seismic and volcanic risk funded by the European Union.

At Cambridge University Press, Laura Clark, Susan Francis, David Hemsley and Emma Walker have been very helpful and positive during the work on the book. In particular, Susan Francis has been very encouraging and patient during the time it took to complete the book. I take this opportunity to thank Cambridge University Press for a splendid collaboration.

Agust Gudmundsson

# Illustrations

Apart from the papers cited in the appropriate figure captions, the following papers are the main sources for the illustrations modified from scientific journals. I am an author on all the papers; the titles and other details are in the reference lists. *Earth-Science Reviews*, **79**, 1–31, 2006 (Figs. 2.10, 6.13, 6.14, 6.15, 6.16, 6.17, 6.19, 6.26, 6.27); *Tectonophysics*, **220**, 205–221, 1993 (Figs. 2.12, 14.13); *Journal of Structural Geology*, **9**, 61–69, 1987 (Fig. 3.13); *Journal of Structural Geology*, **32**, 1643–1655, 2010 (Figs. 6.3, 14.21, 14.22, 14.24, 14.25, 14.26); *Tectonophysics*, **336**, 183–197, 2001 (Figs. 12.2, 16.2); *Tectonophysics*, **139**, 295–308, 1987 (Fig. 13.4); *Bulletin of Volcanology*, **67**, 768–782, 2005 (Fig. 13.6); *Journal of Geophysical Research*, **103**, 7401–7412, 1998 (Fig. 13.7); *Hydrogeology Journal*, **11**, 84–99, 2003 (Figs. 13.25, 14.15); *Bulletin of Volcanology*, **65**, 606–619, 2003 (Figs. 13.29, 13.30); *Journal of Structural Geology*, **22**, 1221–1231, 2000 (Fig. 14.18); *Geophysical Research Letters*, **18**, 2993–2996, 2000 (Figs. 16.4, 16.7, 16.8); *Terra Nova*, **15**, 187–193, 2003 (Fig. 16.9); *Journal of Structural Geology*, **23**, 343–353, 2001 (Figs. 17.1, 17.2).

# 1 Introduction

## 1.1 Aims

Rock fractures occur in a variety of geological processes and range in size from plate boundaries at the scale of hundreds of kilometres to microcracks in crystals at the scale of a fraction of a millimetre. This chapter provides a definition of a rock fracture as well as of some of the mechanical concepts used in the analysis of fractures. Some of these definitions are preliminary and more accurate ones will be provided in later chapters. The primary aims of this chapter are to:

- Provide a definition of a rock fracture.
- Indicate some of the many earth-science topics and processes where fractures play a fundamental role.
- Provide definitions of stress, strain, constitutive equations, and material behaviour.
- Explain the one-dimensional Hooke's law.
- Summarise some basic fracture-related definitions in structural geology.
- Discuss the basic information needed to solve fracture problems.
- Define and explain boundary conditions, rock properties, and rock-failure criteria.
- Explain accuracy, significant figures, and rounding of numbers.
- Explain the basic units and prefixes used.

## 1.2 Rock fractures

A **rock fracture** is a mechanical break or discontinuity that separates a rock body into two or more parts (Fig. 1.1). The continuity or cohesion of the rock body is lost across a fracture. A fracture forms in response to stress. More specifically, the rock breaks and forms a fracture when the applied stress reaches a certain limit, namely the **rock strength**. The stress associated with the fracture formation may be normal or shear or both. An ideal **brittle fracture** shows no effects of plastic (or ductile) deformation on the fracture surface. Thus, the two halves of a solid body broken by an ideal brittle fracture can be fitted back together perfectly. For rocks and most solid materials, however, there is normally some, and sometimes considerable, plastic deformation associated with the propagation of a fracture, particularly at its tips.

**Fig. 1.1**    Tension fracture in the rift zone of North Iceland. View (looking) north, the fracture is located in a Holocene (less than 10 ka-old) pahoehoe (basaltic) lava flow. Behind the persons is a normal fault in the same lava flow. Tension fractures such as this one form when the tensile stress in the rock reaches its tensile strength (0.5–6 MPa). The somewhat irregular shape of the fracture is largely attributable to its path being partly along existing (mainly hexagonal) columnar (cooling) joints in the lava flow.

As a noun, the word 'fracture' denotes the structure, the **crack**, but as a verb it indicates the **process** by which the material breaks or the crack forms. Thus, we distinguish between a rock fracture (a noun), namely a structure, and to fracture a rock (a verb), which means the process responsible for the fracture formation. In the latter sense, the word fracture is often regarded as being synonymous with **rupture.**

Rock fractures are the most common outcrop-scale structures in the Earth's crust (Figs. 1.1–1.5). Their sizes vary from microcracks that dissect small grains or crystals to mid-ocean ridge segments and transform faults that form parts of the boundaries of the Earth's lithospheric plates.

The way by which fractures form and develop plays a vital role in many theoretical and applied fields of earth sciences and engineering. Fields within the earth sciences where rock fractures are of fundamental importance include structural geology, tectonics, volcanology, seismology, field geology, hydrogeology, geothermics, geological hazard studies, and petroleum geology and geophysics. Fields within engineering where rock fractures play an important role include general rock mechanics, tunnelling, civil engineering, petroleum engineering, and general engineering geology.

As an example of the importance of fractures in geological processes, consider volcanic eruptions. Most of the magma transported to the surface in volcanic eruptions such as the one in Fig. 1.2 is through rock fractures that are driven open by the magma pressure. When the magma in the fractures solidifies, they become inclined sheets (Fig. 1.2) or subvertical dykes (Fig. 1.3). Other examples of fluid transport through rock fractures are mineral veins

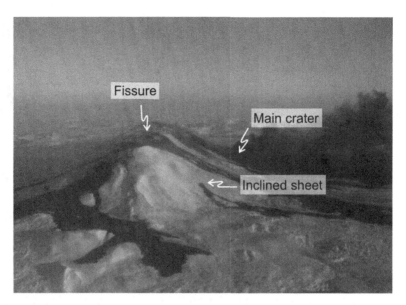

**Fig. 1.2**  Eruption of the volcano Hekla in South Iceland in January 1991 was partly through vertical dykes (the main volcanic fissure along the top of a volcano) and partly through inclined sheets (seen on the slopes of the volcano). Dykes and inclined sheets are hydrofractures or, more specifically, magma-driven fractures. Hydrofractures and tension fractures (Fig. 1.1) are both extension fractures.

(Fig. 1.4). The veins form as a result of transportation of hot water in geothermal fields, in volcanic areas as well as in sedimentary basins. Also, all tectonic earthquakes are related to fractures; more specifically, to faults (Fig. 1.5).

## 1.3  Notation and basic concepts

A rock fracture forms when certain mechanical conditions are satisfied. To solve mechanical problems associated with rock fractures, such as their initiation, propagation paths, and fluid transport, we must understand and use some basic concepts from the mechanics of solids and fluids.

Many of the concepts and techniques discussed in the following chapters are well known in such earth-science fields as hydrogeology, reservoir engineering, petrophysics, earthquake mechanics, rock mechanics, rock physics, and structural geology. However, many of the definitions used are primarily applicable to one or two of these special fields and are not easily translated into a useful conceptual framework for other fields. Furthermore, the notation used in the different fields varies widely and may lead to confusion. In this book a unified notation is adopted as far as is possible, and all the formulas and concepts are presented in such a way that a solution of a fracture problem in one field of earth sciences can be easily applied to another.

**Fig. 1.3** Vertical dyke exposed in a river canyon in North Iceland. View northeast, the dyke is the feeder of (it supplied magma to) the crater cone and associated lava flow. The dyke is of Holocene age, about 6 m thick in the lower part of the exposure, but about 13 m thick where it meets the crater cone (cf. Fig. 1.6; Gudmundsson *et al.*, 2008).

Clearly, however, some **symbols** must be used in more than one meaning. This follows for two reasons. First, there are more parameters and constants used in the equations than there are letters in the alphabets used. Second, the standard notation in each field is followed as much as possible. For example, the Greek letter $\mu$ (mu) is commonly used for dynamic viscosity in fluid mechanics and hydrogeology, and is thus used in the equations for fluid transport in rock fractures. But $\mu$ is also the standard symbol for an important concept in rock physics, namely the coefficient of internal friction. Even in solid mechanics the same symbols are used in different meanings. For example, the letter $G$ is commonly used for an elastic constant called shear modulus (the other commonly used symbol for that constant is $\mu$, which already has two other meanings). But $G$ is also a standard symbol in fracture mechanics for a concept referred to as the energy release rate during fracture propagation. Normally, the different meanings of the symbols are clear from the context. However, to make the meaning as clear as possible, and to make the chapters more readable and the mathematics easier to follow, most symbols used are defined where they first occur in each chapter, and commonly the definition is repeated in the chapter. In addition, a list of the main

Mineral veins seen in a horizontal section (at the surface) of a basaltic lava flow in North Iceland. Most mineral veins are extension fractures, that is, hydrofractures. The 15-cm-long pencil provides a scale.

symbols used is provided at the end of most chapters. While this makes the text somewhat longer, it should be of help to many readers and make it easier to read the chapters and sections of chapters, and to use the equations, independently of the rest of the book.

Many important words, concepts, facts, and definitions are printed in **bold**. **Figures** are referred to as follows: Figure 1.5 or (Fig. 1.5) indicates figure number 5 in Chapter 1, as is done above. In the photographs, the view indicated means the direction in which the photographer is looking. Thus, **view north** means that the photographer was looking (facing) north when the photograph was taken. All **formulas** and **equations** are referred to by numbers in parentheses. Thus, Eq. (1.2) means equation number 2 in Chapter 1. **Worked examples** at the ends of chapters are referred to by numbers. Thus, Example 5.1, or (Example 5.1), refers to details that are given in example number 1 in Chapter 5. Some examples are derivations of formulas, but most are calculation exercises where specific numerical data are given, commonly taken directly from the technical literature. In **Appendices A–E** at the end of the book there are data on crustal rock and fluid properties as well as lists of the base SI units (and many derived units), the SI prefixes (such as kilo, mega, and giga), the Greek alphabet, and some mathematical and physical constants. While the examples commonly use data that may differ somewhat from those in Appendices D and E (data on rocks and crustal fluids), the appendices should be useful in solving some of the **Exercises** at the ends of the chapters, as well as solving general problems related to rock fractures and associated fluid transport. Most exercises refer only to the material of the chapter to which they belong, but some assume knowledge of the earlier chapters as well.

The purpose of this section is to provide preliminary explanations and definitions of some of the main concepts used in the book. More accurate definitions and detailed discussions of many of these and related concepts are provided in the subsequent chapters. Analysis of

**Fig. 1.5**  Part of a normal fault in the rift zone of Southwest Iceland. View northeast, the normal fault, located in a Holocene pahoehoe lava flow, has a vertical displacement (throw) of about 40 m, and an opening (aperture) of as much as 60 m. The cars on the roundabout provide a scale.

rock fractures requires an understanding of stress, strain, and the stress–strain relations. A brief preliminary definition and clarification of these topics should thus be helpful.

### 1.3.1 Stress

**Stress** is a measure of the intensity of force per unit area. The unit of stress is the pascal, Pa, where $1\ \mathrm{Pa} = 1\ \mathrm{N\ m^{-2}}$. Using this simple definition, the formula for stress $\sigma$ is

$$\sigma = \frac{F}{A} \tag{1.1}$$

where $F$ is the force in newtons (N) and $A$ is the area in square metres ($\mathrm{m^2}$) on which the force acts. We use the symbol $\sigma$ when the force that generates the stress is normal to the area on which it acts. Thus, $\sigma$ is the **normal stress** (Fig. 1.6). When a force operates parallel with the plane of interest, such as a fault plane, it is a shear force and generates shear stress.

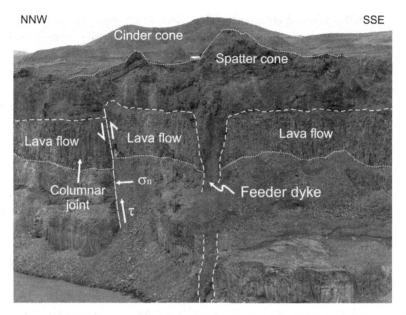

**Fig. 1.6**    Normal stress $\sigma_n$ and shear stress $\tau$ on a reverse fault close to the feeder dyke in Fig. 1.3. The fault is presumably an old normal fault that was reactivated as a reverse fault due to the horizontal stress generated by the pressure of the nearby feeder dyke. The normal stress on the reverse fault is denoted by $\sigma_n$ and the shear stress by $\tau$. The contacts (dotted lines) between the lava flows, as well as their columnar joints (primarily vertical fractures), are indicated (cf. Gudmundsson *et al.*, 2008).

We use the symbol $\tau$ for **shear stress**, so that

$$\tau = \frac{F}{A} \tag{1.2}$$

where $F$ is here the shear force (or force component) acting parallel with a plane whose area is parallel with the direction of the force is $A$.

Stress as defined in Eqs. (1.1) and (1.2) is really a **vector** and referred to as the **traction,** the **traction vector,** or the **stress vector.** To explain why it is a vector, recall that vector quantities have both size (or magnitude) and direction. The direction is the line of action of the quantity. By contrast, **scalar** quantities have only size (or magnitude); no other information is needed to specify them. For example, if a fault length is specified as 10 km, then 10 km is a scalar quantity. Similarly, temperature, mass, and time are scalars. The velocity of tectonic-plate movements, specified at a point as 10 mm per year due east, is a vector quantity. Similarly, a force of 100 N at an angle of 30° to a fault plane striking due north as well as a geothermal gradient (the rate of temperature increase with depth in the crust) are vector quantities.

The quantity in Eq. (1.1) is a vector because force is a vector and area $A$ is a scalar and so is its reciprocal, $A^{-1}$. The product of a vector and a scalar is a vector, so that $\sigma$ in Eq. (1.1) is a vector. As is explained in Chapter 2, the **state of stress at a point** is the collection of all stress vectors, for planes of all possible directions, at that point. This three-dimensional collection of stress vectors specifies the complete stress at the point and is not a vector but rather a second-order tensor. Thus, **stress is a tensor**, not a vector. But the **component** of the stress in any particular direction is a vector. In geology, materials science, and mechanical engineering it is common to refer to the stress in a given direction; for example, the stress on a fault plane or stress on a crystal slip plane. In this book, we shall not use the term traction (for the stress vector), but rather stress, both for the stress vector and the stress tensor.

In mechanical engineering, materials science, and the mathematical theory of elasticity, tensile stresses are much more common and important than compressive stresses. This follows because fracture propagation in, say, an aeroplane or a geothermal pipe at the surface is much more likely to be due to tensile stresses than compressive stresses. As a consequence, tensile stresses are considered positive and compressive stresses negative. In geology, geophysics, rock mechanics, and soil mechanics, however, compressive stresses dominate. Almost anywhere in the Earth's crust, there are compressive stresses. Only the uppermost parts of areas of active rifting, or the contact zones with fluid-filled reservoirs, are likely to develop outcrop-scale tensile stresses. Thus, in these scientific fields tensile stresses are commonly regarded as negative, whereas compressive stresses are positive.

It is generally recognised that it is more natural to derive many of the basic equations for stress, strain, and elasticity on the assumption of tensile stress being positive. However, Mohr's circles of stress (Chapter 2), for example, look awkward if a tension-positive convention is used in geology. Some authors define tensile stress and strain as positive while deriving the main equations and then as negative (and compressive stress and strain as positive) in geological and geophysical applications (Jaeger, 1969). Others use the definition that a positive displacement will always act in a negative coordinate direction (Davis and Selvadurai, 2002). And there are other variations on this theme. In this book **tensile stresses** (and **strains**) are generally regarded as **negative** and **compressive** stresses and strains as **positive**. The negative signs for tensile stresses and strains, however, are commonly omitted, since it is clear from the discussion that they are implied. Also, some stress and strain values are treated as absolute values, in which case the signs are also omitted. This should not cause any confusion since the signs are carefully explained in all the worked examples.

## 1.3.2 Strain

When loads are applied to a solid rock body, it deforms. The deformation of the body can occur in three main ways, namely through rigid-body translation, rigid-body rotation, and changes in its internal configuration. In this book, only the last type of deformation, that is, changes in the internal configuration of the body, is referred to as strain. Some authors use strain and deformation roughly as synonyms, but here strain is regarded as only one of three ways by which a solid body can respond to loads. The other two ways, rigid-body translation and rotation, are **deformation but not strain.**

More specifically, the fractional change in a dimension of a rock body subject to loads is **strain**. For an elongated body, such as a bar-shaped piece of rock, subject to tensile or compressive force, strain is the ratio of the change in length or extension of the body, $\Delta L$, to its original length $L$. Denoting strain by the symbol $\varepsilon$ we have, for one-dimensional tensile

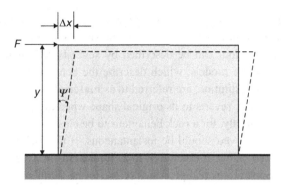

Fig. 1.7 Shear strain $\gamma$ defined. The deflection or movement of the upper face of the body relative to the lower face is $\Delta x$, $y$ is the vertical distance between the faces, and $\Psi$ is the angle generated by the deflection.

strain,

$$\varepsilon = \frac{\Delta L}{L} \tag{1.3}$$

Since Eq. (1.3) represents a length divided by a length, $\varepsilon$ is dimensionless, that is, a pure number, and has **no units**. Strain is often expressed as a percentage, that is,

$$\varepsilon\% = \frac{\Delta L}{L} \times 100 \tag{1.4}$$

Equation (1.3) gives one-dimensional **normal strain**. For a shear force $F$, the **shear strain** $\gamma$ is given by

$$\gamma = \frac{\Delta x}{y} = \tan \psi \tag{1.5}$$

where $\Delta x$ is the deflection or movement of the upper face of the body relative to the lower face, with $y$ being the vertical distance between the faces (the height of the body) and $\psi$ the angle generated by the deflection (Fig. 1.7).

### 1.3.3 Mechanical behaviour

Several concepts from solid mechanics should be introduced at this stage to make the subsequent discussion more precise. These concepts relate to the general mechanical behaviour of materials; here the focus is on rocks. Many of these concepts are given a more rigorous definition in subsequent chapters. These and the following definitions from structural geology (Section 1.4) are presented in a brief, summary fashion. You may want to skip them on first reading, and come back to them when needed while reading other chapters.

By the **mechanical behaviour** of a rock we mean how the rock responds to forces or loads: more specifically, what relation between force and displacement or between stress and strain the rock shows when loaded. In solid mechanics, the word **load** normally means the forces, stresses, or pressures applied to the body and external to its material. In geology, loading can also refer to displacements applied to a rock body. Under load, the rock may deform elastically or fracture or flow. The way the rock deforms under load varies depending on the mechanical properties of the rock, the strain rate, the temperature, and the state of stress. Rock behaviour can be described by several ideal models. In mechanics and materials science these models, which describe the behaviour of ideal materials based upon their internal constitution, are referred to as **material equations** or **constitutive equations**.

If the rock reverts to its original shape when the load is removed, the rock behaviour is **elastic.** Strictly, for a rock behaviour to be elastic, its return to the original shape once the load is removed should be instantaneous. If the rock behaviour follows Hooke's law, as described below, it is referred to as **linear elastic**. Under short-term loads, particularly at low temperatures and pressures (close to the Earth's surface), most solid rocks behave as approximately linear elastic so long as the strain is less than about 1%. Many rocks, because they contain pores that are partly or totally filled with fluids, show time-dependent shape recovery. Such a time-dependent elastic behaviour is called **viscoelastic** (also poroelastic). If the rock does not return to its original shape when the applied load is removed, its behaviour is **inelastic**.

Some rocks, when a critical stress is reached, behave as **plastic**. Plastic deformation is permanent; when the load is removed, the rock does not return to its original shape. Plasticity means that the rock is permanently deformed by a load without developing a fracture. For plastic deformation or flow to occur in a rock body, a certain stress level must be reached or exceeded. This stress level, at which plastic deformation occurs, is known as the **yield stress**. In some materials, plastic deformation occurs as soon as the yield stress is reached. Many rocks, however, show time-dependent plastic deformation. Such a time-dependent plastic deformation, which is common at high temperatures and pressures in the lower crust and in the mantle, is called **creep**.

Commonly, there is some plastic deformation in a rock before it fractures. If the plastic deformation before fracture is minor and essentially limited to a region close to the tip of the fracture, then the rock behaviour and the fracture are said to be **brittle**. Thus, the brittleness of a rock is indicated by a fracture developing without appreciable prior plastic deformation. If, however, there is extensive plastic deformation in the rock before fracture, the fracture is said to be **ductile** and the rock is said to be **tough**. Fractures may thus be classified as brittle or ductile according to the strain in the solid at which they occur. With brittle fracture, there is little if any plastic deformation and change of body shape before fracture. Many rocks, particularly close to the Earth's surface, fail through the development of brittle fractures. When ductile fractures form, the plastic deformation that begins once the yield stress is reached reduces the cross-sectional area of a part of the body, say the central part of a bar under axial tension, resulting in **necking**. The fracture that eventually develops in the necked region is the result of shear stresses; that is, the ductile fracture occurs when the material shears at 45° to the direction of the axial tension. Some rock fractures, particularly those formed at high temperatures and pressures, may be described as ductile.

A tough material absorbs a great deal of energy before it fractures. **Toughness** is a fundamental concept in fracture mechanics in general and in the mechanics of rock-fracture formation in particular. If no fracture develops in the rock when loaded but rather large-scale plastic deformation, the rock is said to behave in a ductile manner or, simply, to be **ductile**. Thus, the rock is said to be ductile when it deforms plastically without fracture.

We have already defined the term **fracture** both as the structure (the crack) and as the process of breaking. A fracture forms in a rock when a critical stress is reached; this stress is the **strength** of the rock. Depending on the loading conditions, the strength that must be reached for a brittle rock to fail may be the **tensile strength**, the **shear strength**, or the **compressive strength**. There are also other measures of rock or crustal strength, as discussed in later chapters. Rock fractures in the crust form when either the tensile strength or the shear strength is reached. Fracture formation normally occurs as soon as either the tensile or shear strength is reached. Some fractures, however, form and propagate at stress levels that are lower than the tensile or shear strengths. These fractures are normally generated by repeated loading at regular intervals, such as the stresses due to the Earth's tides. This process of fracture formation is referred to as **fatigue**.

### 1.3.4  Hooke's law and elasticity

For brittle rock fractures, the most important constitutive or material equation is Hooke's law. Originally, it stated that, within the limit of proportionality, the extension or displacement of a bar or spring is proportional to the applied force (Fig. 1.8). In modern terms, **Hooke's law** states that, within the limits of proportionality, the strain in a body is directly proportional to the stress producing the strain (Fig. 1.9).

One reason for the use of stress and strain instead of force and extension is that the ratio of stress $\sigma$ to strain $\varepsilon$ is the same regardless of the size and shape of the body. This ratio is called the modulus of elasticity, or **Young's modulus**, and is denoted by the symbol $E$. Thus, the **one-dimensional Hooke's law** may be given as

$$\sigma = E\varepsilon \tag{1.6}$$

Young's modulus, that is, the ratio $E = \sigma/\varepsilon$, is a material property. Since strain has no units, Young's modulus has the same units as stress, namely newtons per square metre, or pascals.

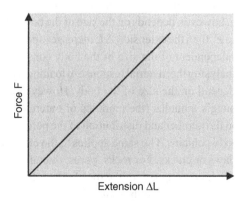

**Fig. 1.8**   The stiffness $k$ of a rock is indicated by the slope of its force–extension ($F - \Delta L$) curve.

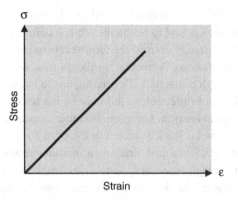

$\sigma$

Stress

Strain

$\varepsilon$

**Fig. 1.9**    The Young's modulus $E$ of a rock is indicated by the slope of its stress–strain ($\sigma - \varepsilon$) curve. The steeper the slope, the higher the Young's modulus and the stiffer the rock.

The value of the modulus can be determined from the slope or gradient of the straight-line stress–strain graph in Fig. 1.9.

Hooke's law is a constitutive equation; it describes the behaviour of a linear-elastic material. In other words, a linear-elastic material is one that obeys Hooke's law. Once the stress (load) is removed, a linear-elastic body returns to its original shape and the strain disappears.

Young's modulus is a measure of the material's **stiffness**, that is, the material resistance to strain, such as is generated through stretching or bending. A material with a large value of Young's modulus has a high material stiffness or is simply stiff. Strictly, stiffness and Young's modulus are not the same, however, since stiffness $k$ is the ratio of force $F$ to extension $\Delta L$, namely

$$k = \frac{F}{\Delta L} \tag{1.7}$$

Thus, stiffness $k$ has units of $\mathrm{N\,m^{-1}}$, whereas Young's modulus has units of $\mathrm{N\,m^{-2}}$. The value of the stiffness is indicated by the slope or gradient of the force–extension graph (Fig. 1.8). This value, however, depends on the size of the body considered. If the size is increased for a given force $F$, then the extension $\Delta L$ increases and the measured stiffness $k$ decreases. Since stress is independent of the size of the body considered, it is generally more suitable than force for analysing the material response to loading. Generally, $E$ as measured in laboratories does not depend on the size of the body. However, as discussed in Chapter 4, the field or **in-situ** Young's modulus (the modulus of outcrop rocks or, generally, of rocks in nature) depends on the number and distribution of the pores, fractures, and other discontinuities that the rock body contains. The same applies to the effective Young's modulus of any solid that contains flaws or cracks. For rocks, pores, fractures, and discontinuities tend to increase in number with increasing size of the rock body. It follows that, up to a certain limit, for rocks in the upper (brittle) part of the crust, Young's modulus generally decreases with increasing size of the rock body being considered.

In solid mechanics and elasticity, a **bar** is an elongated solid of uniform cross-sectional area, the shape of which is normally circular, rectangular, or hexagonal. A bar is commonly used when discussing the response to loading in one direction (one-dimensional loading) and deriving the related stress, strain, and elasticity equations. Consider a bar of cross-sectional area $A$ and length $L$. Young's modulus and stiffness are related in the following way (Example 1.3):

$$E = \frac{kL}{A} \tag{1.8}$$

It follows that a material with a large Young's modulus is stiff or inflexible, and a material with a small Young's modulus is flexible, floppy, or compliant. Thus, **compliance** is the reciprocal of stiffness and means the ease by which a body can be deformed elastically. Since **softness** also indicates a tendency of a body to deform easily, although not necessarily elastically, it is common to use the word soft as a reciprocal of stiff in the geological literature. In this book **soft** normally means that the material is elastic (unless stated otherwise) but with a low Young's modulus.

The value of Young's modulus is commonly large. Thus, for example, for many igneous and metamorphic rocks the value is between $10^9$ and $10^{11}$ Pa. Consequently, the values of Young's modulus for rocks and many other solid materials are given in **gigapascals (GPa)**, where 1 GPa = $10^9$ Pa (Appendix A.3). From Eq. (1.6) we can understand why the value of Young's modulus is commonly so large. If the strain is 100%, so that the length of, say, a solid bar is doubled, then we have $\varepsilon = 1.0$ and Eq. (1.6) gives $\sigma = E$. Thus, the value of Young's modulus is equal to the stress needed to double the length of the bar. Normally, a brittle, solid bar would fracture long before it could double its length. If, however, we consider a bar of rock or steel and assume that it could be extended so as to double its length, we can imagine the enormous stress that would be needed for this to happen and, thereby, understand why the value of Young's modulus for many solids is so large.

## 1.4  Some definitions from structural geology

The education of scientists dealing with rock fractures varies widely. Although most are presumably educated as geologists, in particular as structural geologists, many are trained as geophysicists or geochemists. Also, an increasing number of scientists working on this topic have their basic training in applied mathematics, physics, solid mechanics, civil engineering, and materials science. Most non-geologists are unfamiliar with many of the terms on rock fractures and associated structures that are used in structural geology, tectonics, and related fields. Although most of the terms related to field studies of rock fractures are defined in the chapters where they are used, it may be helpful to summarise the definitions of some of the main terms here, particularly for readers who are unfamiliar with structural geology. These basic definitions follow the standard terminology in structural geology. Further information on how they are used, together with illustrations, are provided in textbooks on structural geology and tectonics as well as in the subsequent chapters of this book. To make them easier to understand, some of the terms defined below are illustrated by photographs or line drawings.

1. A **fracture** is any significant mechanical break in the rock (Figs. 1.1–1.5). Some authors define a fracture so that the break should not be parallel with a visible fabric in the rock. (Here fabric means the geometrical arrangement of structures or component features in a rock.) Since most fractures originate at stress concentrations, including fabric elements (such as grains), the word fracture will be used as defined above without any restriction as to its relation to fabric.

2. The **attitude** or orientation of a fracture is defined by its strike and dip (Fig. 1.10). **Strike** is the trend of a horizontal line in the fracture plane. Alternatively, strike is the trend of the intersection (a line) between the fracture plane and a horizontal plane. **Dip** is the acute angle between the fracture plane and a horizontal plane. Dip is measured in an imaginary vertical plane trending at 90° to the fracture plane. **True dip** is the maximum dip of the fracture plane; other (less steep) dips are referred to as **apparent dips**.

3. **Crack** is essentially a synonym for fracture, but is here used primarily in fracture models. Thus, we refer to three ideal shapes of fractures as through cracks, part-through cracks, and interior cracks (Chapter 9). And we refer to the ideal modes of displacement of the fracture surfaces during loading as mode I, mode II, and mode III cracks (Chapter 9). We also use the term crack when describing microcracks such as occur at the tips (ends) of fractures.

4. An **extension fracture** is formed by normal stress or fluid pressure and opens in a direction parallel with the maximum tensile (minimum compressive) principal stress. There is thus no shear stress parallel with the fracture and no fracture-parallel movement of the walls, only opening perpendicular to the fracture plane. There are two main types: one is a **tension** fracture, which forms when the stress perpendicular to the fracture is negative (Fig. 1.1), the other is a **hydrofracture**, which is driven open by internal fluid pressure (Figs. 1.2–1.4, and 1.6). Sometimes the words **absolute tension** and **relative tension** are used in connection with extension fractures. Absolute tension is used when the stress perpendicular to the extension fracture, the minimum principal stress, is negative, in which case the extension fracture is a tension fracture. Relative tension is used when the minimum principal compressive stress is reduced, such as is common prior to rifting episodes – with dyke emplacement, normal faulting, or both – at divergent plate boundaries, but remains compressive. Relative tension is much more common in the crust than absolute tension which, on an outcrop scale, can only occur close to or at the surface in areas undergoing extension. The term **outcrop scale** means rock exposures with linear dimensions reaching metres, tens of metres, or more.

5. A **shear fracture** forms by shear stress and shows clear evidence of fracture-parallel movement of the fracture walls or surfaces. Shear fractures are normally referred to as faults (Figs. 1.5–1.6, and 1.10).

6. A **discontinuity** is any significant mechanical break or fracture of negligible tensile strength. The terms does thus include fractures, but also many contacts between rock bodies (Figs. 1.3 and 1.6) and layers (beds), and many bedding planes.

7. A **bedding plane** or bedding coincides with the surface of deposition. A bedding plane is formed by a change in mineralogy, chemistry, grain orientation, grain size, and colour in any combination. The bedding plane may form a discontinuity, but commonly does not. In many sedimentary rocks, for example shales, the rock splits along bedding planes which are thus discontinuities. However, in many sandstones where the bedding is marked by changes in colour or grain size there is no preferred splitting along bedding planes, which, therefore, do not form discontinuities.

8. An **interface** is a layer across which there is an abrupt change in some parameters from the rock on one side to the rock on the other side of the interface. The layers on either side may be of the same or a different rock type (Fig. 1.6). The surfaces between rock layers are commonly referred to as contacts. Strictly, a **contact** is two-dimensional. An interface is more general than either a surface or a contact in that an interface is three-dimensional, that is, it has thickness across which the various properties change.

9. A **joint** is a fracture with a slight fracture-normal displacement (opening) but no visible fracture-parallel movement (Figs. 1.3 and 1.6). A joint is thus, by definition, primarily an extension fracture. Most joints studied in the field are presumably extension fractures. However, some may have a fracture-parallel displacement too small to be seen in the field, in which case they are part extension and part shear (hybrid or mixed-mode) fractures (Chapters 9 and 11).

10. A **fault** is a shear fracture where there is a clear sign of fracture-parallel displacement of the walls or fracture surfaces (Figs. 1.5–1.6). Evidence of faulting includes offset marker layers and horizons such as beds, contacts, dykes, and mineral veins, as well as fault breccia and slickensides. The main ideal types are a **strike-slip fault** (Fig. 1.11), where the displacement is parallel with the fault strike, and a **dip-slip fault** (Fig. 1.10), where the displacement is parallel with the fault dip. For a dip-slip fault, the vertical displacement is called **throw** and the horizontal displacement **heave**. The displacement is most easily measured if the rock contains an easily recognisable **marker layer** (or a marker horizon). Many faults, however, have an **oblique slip**. A strike-slip fault is **sinistral** or left-lateral if the displacement of the opposite fault wall (the wall on the far side of the fault plane) is to the left, but **dextral** or right-lateral if the displacement of the opposite fault wall is to the right (Chapter 9).

11. For dip-slip faults, the wall above the inclined fault plane is referred to as a **hanging wall**, whereas the wall below the fault plane is referred to as the **footwall** (Fig. 1.10). A dip-slip fault is named a **normal fault** if the hanging wall is displaced down relative to the footwall. If the hanging wall is displaced up relative to the footwall, the fault is named a **reverse fault** if the fault dip exceeds 45°, but **thrust fault** or **thrust** if the fault dip is less than 45°.

12. A **slickenside** is a polished and smoothly striated surface resulting from frictional sliding and abrasion along a fault plane. Slicken (an English dialect word) means smooth, so that 'polished surface' is a synonym for slickenside. The polished striations themselves are also commonly referred to as slickensides. They indicate the direction of the (primarily most recent) displacement on the fault.

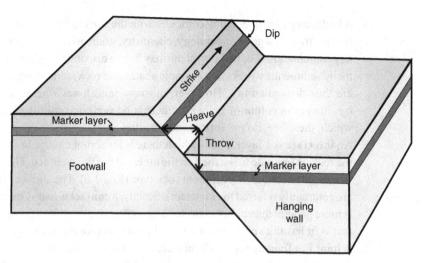

**Fig. 1.10**    The attitude of a rock fracture, such as the dip-slip (normal) fault seen here, is defined by the strike and dip of the fracture plane. For a dip-slip fault, the part of the fault above the fault plane is known as the hanging wall, and the part below the fault plane is known as the footwall; the vertical displacement is called throw and the horizontal displacement heave. The displacements are most easily measured if the rock contains a marker layer, that is, an easily identified layer that can be found on either side of the fault.

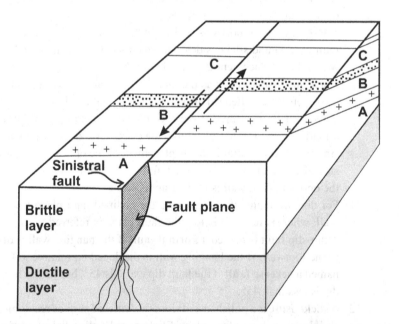

**Fig. 1.11**    Sinistral (left-lateral) strike-slip fault dissecting and displacing several gently dipping marker layers (A–C). A large strike-slip fault, with a length of tens or hundreds of kilometres, normally cuts through the entire brittle part or layer of the lithosphere. This is the part where brittle deformation takes place and earthquakes are generated, and is also called the **seismogenic layer**.

## 1.5  How to solve fracture problems

As a very practical example of what is needed to solve a fracture problem, consider the conditions for rupture of a fluid-filled reservoir or chamber (Fig. 1.12). The fluid in the reservoir need not be specified; it could be ground water, geothermal water, oil, gas, or magma (as in Fig. 1.12). To decide whether the reservoir or magma chamber is likely to rupture so that a hydrofracture fracture initiates and propagates into the host rock, we need some basic information. These include the following:

1. The **geometric boundary conditions**. These must be specified with information as regards the following items:

    (a) The overall shape or geometry of the reservoir. Has it, for example, approximately the shape of an oblate ellipsoid, a sphere, a cylinder, or perhaps a flat disk (that is, sill-like as it would be called in volcanology)?

    (b) The depth of the reservoir. Is the reservoir located at the depth of several hundred metres below the Earth's surface, or perhaps at the depth of several or many kilometres?

    (c) The size of the reservoir, particularly in relation to its depth below the surface. This is particularly important in obtaining analytical solutions since it determines whether the stresses around the reservoir are affected by the free surface (surface in contact with fluid, such as air, water or magma) of the Earth. If the reservoir is close to the surface in relation to its horizontal dimensions, the greatest stress at the surface or periphery of the reservoir will be at locations and reach magnitudes that differ from those of a reservoir of otherwise similar geometry, but at so great a depth that, in comparison, its horizontal dimensions are small.

2. The **loading boundary conditions**. In solid mechanics the word 'load', as indicated above, normally refers to the forces, stresses, or pressures applied to (acting on) a body

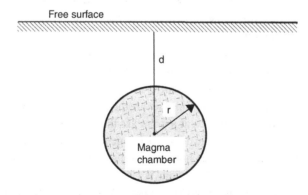

**Fig. 1.12**  Fluid-filled reservoir, here a magma chamber, with a radius *r* and located at a depth *d* below the free surface of the Earth. The reservoir has a circular cross-section and may be either a horizontal cylinder or a sphere depending on its third dimension (not shown here). Reservoirs have many different shapes and occur at various depths below the Earth's surface; both factors have strong effects on the local stresses and fracture formation around the reservoirs.

and external to its material. We follow this usage, so that loading boundary conditions refer to the stresses or fluid pressures applied to the rock body and causing its deformation. The loading conditions may also, but rarely in this text, refer to displacements. To solve a problem involving rock fractures where the crustal segment is subject to a specific stress field, we must know the crustal stresses. There are many methods used to calculate or infer the **state of stress in a rock body**, including the following:

(a) **Focal mechanisms of earthquakes**. The waves from earthquake sources can be used to infer, crudely, the state of stress in the crust where they occur. Focal mechanisms, which are also called fault-plane solutions, provide a general overview of the regional stress field in an area.

(b) **Hydraulic fracturing**. This is a method that gives the local stress field in the vicinity of a drill hole (Chapter 6). Pressured water is used to generate an extension fracture which propagates from the hole. The fracture orientation and opening (aperture) provide an indication of the local stress field.

(c) **Local engineering methods**. These are mainly used to estimate stresses in areas surrounding tunnels or in small rock masses for engineering purposes. They include techniques such as **overcoring** and **flat jack**. Both of these are based on making (drilling) a cut in the rock and measuring the deformation or opening displacement of the cut (that is, how it tends to open or close). The deformation of the cut is an indication of the local stress field around the cut.

(d) **Palaeostresses**. Many rock structures, such as dykes and stria and other features on smooth fault surfaces (slickensides), can be used to infer the local stress field at the time of the formation of these structures. Such stress fields are referred to as palaeostresses since they do not necessarily reflect the present-day stress field but rather the fields that operated in the past when the observed structures were formed. Systematic studies of palaeostress indicators provide reliable information about the stress history of the area within which the structures occur. Comparison of palaeostresses and present-day stresses, determined through methods (a)–(c) above, often yields very useful indications as to the tectonic evolution of a region.

3. **Rock properties**. Rocks vary in their response to loading. Some behave as ductile or plastic, others as quasi-brittle, and still others as brittle materials. The behaviour of a rock subject to loading depends also on the rock anisotropy and heterogeneity. If the properties vary from point to point within the rock it is **heterogeneous**. If the properties have different values in different directions at a given point, the rock is **anisotropic**. By contrast, if the rock properties are the same from point to point, the rock is **homogeneous;** and if the properties at a point are the same in all directions, the rock is **isotropic**. All rocks are to a certain degree heterogeneous and anisotropic, but their behaviour can often be approximated as homogeneous and isotropic. For brittle rock, it is then commonly assumed that its behaviour is elastic, or more specifically, linear elastic, until failure. For a rock that behaves as linear elastic and is also approximately homogeneous and isotropic, there are five elastic constants or moduli. Only two of these are independent, however, and those most commonly used are **Young's modulus** and **Poisson's ratio**. All the other moduli can be derived from these two. The moduli for a rock mass can be estimated in various ways, such as the following:

(a) Field (site) or in-situ **seismic measurements**. The results provide so-called **dynamic moduli**, which are normally higher than the **static moduli**. The static modulus is the ratio of stress to strain under static conditions. For most modelling situations we use the static moduli because the rate of deformation is slow in comparison with the velocities of the seismic waves (from which the dynamic moduli are determined). For example, a magma-driven fracture such as a dyke (Figs. 1.2–1.3, and 1.6) may propagate at an average rate of less than one metre per second. For comparison, seismic waves have typical velocities of several kilometres per second. Thus, the rate of rock deformation during dyke propagation (mainly compression of the dyke-fracture walls as the dyke extends) is slow in comparison with the deformation generated during the propagation of seismic waves. It follows that for modelling dyke emplacement we use the static elastic moduli. By contrast, when modelling earthquake fractures we normally use dynamic elastic moduli.

(b) **Laboratory measurements** are made on small rock samples (normally, in the order of centimetres in length and width). These results must be scaled if they are to be made suitable for modelling and describing the mechanical behaviour of a rock mass in an outcrop or a reservoir. Generally, a rock mass contains many pores, fractures, contacts, and other discontinuities. As indicated, the effective or in-situ moduli of a rock mass containing discontinuities, particularly at shallow crustal depths (and comparatively low pressure), is normally much lower than that of discontinuity-free small laboratory samples.

4. **Rock failure**. To assess whether a reservoir or a chamber (Fig. 1.12) is likely to fail, we must specify the appropriate rock-failure criterion. The terms failure criterion, yield criterion, and flow criterion are often used as essentially synonyms. There is some difference, however, in the meaning of these concepts. A **failure criterion** is the most general of the terms. It refers to the condition of permanent deformation in brittle, quasi-brittle, and ductile solids. Thus, the Griffith criterion describes the conditions of brittle failure through fracture propagation, whereas the Treca criterion describes the conditions for plastic yield. A **yield criterion** applies to ductile and brittle materials and states the conditions under which a material reaches its yield stress (or yield strength) whereby non-elastic (plastic) permanent deformation starts. A **flow criterion** normally refers to the stress (flow stress) at the onset of plastic flow in an already deformed material. There are several rock-failure criteria used. The best known are the following (Chapter 7):

(a) The **Coulomb criterion**. This criterion states that shear failure occurs when the shear stress on the potential fracture plane reaches a certain magnitude.

(b) The **Mohr's criterion**. This criterion states that shear failure occurs on a plane where the shear stress and the normal stress have a certain relationship which is characteristic of the rock. The Coulomb criterion may be regarded as a special (linear) case of the Mohr's criterion.

(c) The **Griffith criterion**. This criterion derives from the Griffith theory of small cracks or flaws that start to propagate, and thus link up, when, for a given crack orientation, the local stresses reach a certain magnitude. In contrast to the Coulomb and the Mohr criteria, both of which are empirical, the Griffith criterion is based on a physical theory, namely the Griffith theory of fracture. The Griffith theory may be applied both to a predominantly compressive regime and, more successfully, to a predominately tensile regime. For the failure of a fluid-filled reservoir or a magma chamber, the most likely fractures to form are extension fractures, more specifically hydrofractures (such as dykes), driven open by the fluid pressure in the reservoir. For such fractures, the Griffith criterion is the most appropriate.

(d) The **Tresca** and **von Mises criteria** are used to describe the stress condition for plastic deformation or yield in the rock. These criteria are particularly used to describe ductile deformation in the lower crust and upper mantle. In a typical fault zone that extends from the surface of the brittle crust and into the ductile lower crust (where it is properly referred to as a shear zone) the most appropriate failure criteria are as follows (with increasing crustal depth): (1) the Griffith criterion, (2) the Coulomb criterion, and (3) the Tresca criterion or the von Mises criterion (Chapter 7).

## 1.6 Accuracy, significant figures, and rounding

Many of the data used in rock-fracture calculations are not well constrained. The laboratory and field data may be as precise as we want them to be, but it does not follow that they are accurate in the sense of reflecting the actual values at the time of fracture formation. This applies particularly to rock properties such as the elastic constants. For example, the values of Young's modulus (Eq. 1.6) for solid rocks range over several orders of magnitude. And the modulus measured on a small specimen in the laboratory may be a poor indication of the in-situ value from which the specimen was taken, either today or at the time of fracture formation. While the data **precision**, which is a measure of how similar the results of repeated measurements are (under constant conditions), may be very high, the data **accuracy**, which is a measure of how close the data we use are to their true values, may be, and commonly is, much less so. This lack of high accuracy in many fields of earth sciences (such as solid-earth geophysics, rock mechanics, hydrogeology, and tectonics) is related to the problem of scaling up the results from laboratory measurements to in-situ or field conditions (Chapter 4).

**Significant figures** are counted from the first non-zero numeral in the number, starting from its left. This means that in a number that is less than one, none of the zeros before the first non-zero digit are significant. As an example, 0.000 067 89 is given to four significant figures. Also, the number 8.4534 to two significant figures is 8.5, and to three significant figures is 8.45. Similarly, 0.001 463 to two significant figures is 0.0015. The most significant numeral or digit is the leftmost non-zero digit; the least significant digit is the rightmost significant digit. For example, in the number 385.6 the digit 3 has the value of 300, the digit 8 the value of 80, the digit 5 the value of 5, and the digit 6 the value of 0.6. Thus, 3 is the most significant digit, whereas 6 is the least significant. If the number was 385.60, then the least significant digit would be 0.

The standard **SI units** (Système Internationale d'Unités; International System of Units) are used in this book (Appendix A). These have seven base units (time, mass, length, current, temperature, amount of substance, and intensity) and many derived units (such as force, pressure, stress, energy, speed, and acceleration). Time is given in seconds (s), mass in kilograms (kg), and length in metres (m). In the SI system, temperature is measured on the Kelvin scale and expressed as K; however, in many examples in the book the temperature is given using the Celsius scale, a derived scale in the SI system expressed as °C (widely called centigrade) and still very common in the scientific literature. Temperature on the Celsius scale is equal to temperature on the Kelvin scale minus 273.15.

Equations such as Eq. (6.1) equate quantities that have not only units and magnitudes (values, sizes), but also dimensions. Independently of the magnitude and units of a quantity, it has **dimensions** which indicate its type. Thus, the measured values for fracture openings (apertures) and lengths may vary widely in magnitude (5 km, 300 m, and 1 m for length, and 10 m, 5 cm, and 1 mm for opening, for example), but they all have the dimension of length. Spreading rates at a mid-ocean ridge may, for example, be 1 cm per year or 17 cm per year, but the dimensions in both cases are the same, namely those of velocity (length per unit time). The fundamental dimensions are length (L), mass (M), and time (T). Spreading rates thus have the dimensions of $LT^{-1}$, force (mass $\times$ acceleration) the dimensions of $MLT^{-2}$, and stress and pressure (force per unit area) the dimensions of $ML^{-1}T^{-2}$.

**Scientific notation** presents all numbers as values between 0 and 10 multiplied by the number 10 raised to a suitable power. This notation is used throughout the book. For numbers given in scientific notation, all the figures are significant. For example, the viscosity of water may be given as $1.55 \times 10^{-3}$ Pa s (pascal·second), with three significant figures. If the viscosity of a fluid were given as $1.123 \times 10^2$, there would be four significant figures. Commonly, we use **prefixes**, such as kilo (e.g. kilometre and kiloannum for thousand metres and thousand years, respectively). The most common multiplies for the prefixes are in steps of $10^3$. The most commonly used prefixes in the book, with the multiplies in parentheses, are: milli ($10^{-3}$), kilo ($10^3$), mega ($10^6$), and giga ($10^9$). Examples include cracks in grains and crystals with the lengths of millimetres (mm), forces in kilonewtons (kN), tensile and shear strengths of rocks in megapascals (MPa), and the values of elastic constants such as Young's modulus in gigapascals (GPa). When discussing the age of formation of a structure, such as a lava flow, the word 'ago' is commonly regarded as being implied and is thus not stated. For example, in the phrase 'a lava flow formed (at) 9 ka' would mean 'a lava flow formed nine thousand years ago'.

The accuracy of a decimal number used in calculations is indicated by its number of significant digits. Thus the numbers 12.5 and 0.000 156 are both accurate to three significant figures. The precision of a decimal number is indicated by the number of decimal places it has. For example, the number 5.65 is precise to the nearest hundredth or, alternatively, precise to two decimal places.

It follows from these considerations that the use of very precise values for elastic constants and other data in rock-fracture calculations is normally not appropriate and can, in fact, be misleading as to the accuracy of the results. Consequently, in the examples in the book the standard methods of **rounding** to the appropriate significant figures are used when necessary. The **rules of rounding** are as follows:

- If the least significant digit is less than 5, it is dropped and the remaining digits do not change. This is called rounding down.

- If the least significant digit is greater than 5, then the next significant digit (to the left in the number) is increased by 1. This is called rounding up.
- If the least significant digit is exactly 5, then the following odd-add rule is used: the next significant figure is increased by 1 if it is odd, but remains unchanged if it is even. Thus, the number 0.3475 is rounded up to 0.348 to three significant figures (since 7 is odd), whereas the number 0.4645 is rounded to 0.464 to three significant figures (since 4 is even).

In the calculations in many examples in the book, as in geology and geophysics in general, the values of constants and parameters are often known only to within a factor of 2–10. Thus, it serves little purpose to show many digits to the right of the decimal point. In many of the examples the results are therefore rounded so that they show the same number of digits or, at most, one more digit to the right of the decimal point, as in the least precise numerical quantity used in the calculations. Thus, for example, if one constant has a value of 50 and another of 0.25, as would be common in calculations using Young's modulus and Poisson's ratio, the answer would normally be given with, at most, one digit to the right of the decimal point. For a stress magnitude in megapascals the results would therefore be given as, for example, 10 MPa or, possibly, as 10.6 MPa, but not as, for example, 10.634 MPa.

## 1.7 Summary

- Fractures are mechanical breaks that separate a rock body into two or more parts. They form when stresses in the rock reach the rock strength. A fracture is brittle when little or no plastic deformation takes place prior to fracture, but ductile if considerable plastic deformation occurs prior to fracture. Mechanically, brittle fractures are of two main types: extension fractures and shear fractures. Extension fractures are either tension fractures or hydrofractures. Shear fractures are primarily faults.
- Stress is force per unit area; strain is the fractional change in a dimension of a body subject to stress. Stress has units of newtons per square metre or pascal. Normal stress acts perpendicular to a plane (such as a fault plane), whereas shear stress acts parallel with the plane. Strain is change in dimension divided by the original dimension and thus has no units (is a pure number).
- Stress and strain are related through material or constitutive equations, one of which is Hooke's law. This law states that there is a linear relationship between the stress and strain of an elastic material and that the ratio between stress and strain is a material property referred to as Young's modulus, which is thus the slope of the stress–strain curve. Rocks with a large Young's modulus are commonly called stiff, whereas those with a small Young's modulus are called compliant or soft. A linear-elastic material, by definition, obeys Hooke's law. Once the stress is removed, the material returns immediately to its original shape and the strain disappears. Many solid rocks, particularly at shallow crustal depths, show approximately linear-elastic behaviour up to strains of about one percent.
- To solve rock-fracture problems, such as the condition for rupture of a fluid-filled reservoir, we need certain basic information. First, we need to know the geometric boundary

conditions. These include the geometry, size, and depth of the reservoir. Second, we need to know the loading (normally stress or pressure) boundary conditions. The state of stress in the crust, in particular, is important and can be obtained from various types of stress measurements. Third, we need to know the rock properties, in particular whether the rock behaviour is brittle or ductile (plastic). The rock properties can be obtained from in-situ and laboratory measurements. Fourth, we need to know the appropriate criterion for rock failure. The main criteria used are the Coulomb, Mohr, and Griffith criteria, for essentially brittle rock behaviour, and the Tresca and von Mises criteria, for essentially ductile or plastic rock behaviour.

• The precision of data is a measure of how similar the results of repeated measurements are (for unchanging conditions). The accuracy of data is a measure of how close they are to the true values. Many earth-science data in general, and those for rock fractures in particular, are quite precise but, at the same time, not very accurate. Significant figures are counted from the first non-zero numeral in the number, starting from its left. In the examples in the book, the standard methods of rounding the values of the physical parameters and the results of the calculations to the appropriate significant figures are used when necessary. The SI system of units and scientific notation are used in the book, commonly using prefixes such as in megapascals and gigapascals.

## 1.8 Main symbols used

| | |
|---|---|
| $A$ | area |
| $E$ | Young's modulus |
| $F$ | force |
| $L$ | length |
| $\Delta L$ | extension, change in length |
| $k$ | stiffness |
| $\gamma$ | shear strain |
| $\varepsilon$ | normal strain |
| $\sigma$ | normal stress |
| $\tau$ | shear stress |
| $\psi$ | angle |

## 1.9 Worked examples

### Example 1.1

Problem

A rock bar with a cross-sectional area of 1000 mm$^2$ is subject to an axial tensile force of 2 kN in a laboratory experiment. Calculate the stress in the bar. Would the bar be able to sustain the stress without failure?

Solution

We always give the results in SI units. Thus, we must first change all the units given in the example into SI units. We have:

The tensile force is 2 kN = $2 \times 10^3$ N

The cross-sectional area is 1000 mm$^2$ = $1 \times 10^{-3}$ m$^2$

From Eq. (1.1) we have

$$\sigma = \frac{F}{A} = \frac{2 \times 10^3 \, \text{N}}{1 \times 10^{-3} \, \text{m}^2} = 2 \times 10^6 \, \text{N m}^{-2} = 2 \times 10^6 \, \text{Pa} = 2 \, \text{MPa}$$

Since the rock bar is subject to uniaxial tension, it would be most likely to fail, if at all, through brittle tension fracture. The field or in-situ tensile strength of common solid rocks is 0.5–6 MPa (Appendix E). However, the laboratory tensile strength of solid rocks is as much as 10–20 MPa. Thus, the rock specimen would be likely to tolerate a tensile stress of 2 MPa without failure.

## Example 1.2

Problem

A rock bar is subject to axial compressive stress and, as a consequence, contracts by 3 mm. If the original length of the bar is 1.5 m, calculate the compressive strain in the bar due to the loading.

Solution

The original length of the bar is 1.5 m

The shortening or contraction is 3 mm = 0.003 m

From Eq. (1.3) we have

$$\varepsilon = \frac{\Delta L}{L} = \frac{3 \times 10^{-3} \, \text{m}}{1.5 \, \text{m}} = 0.002$$

Thus the strain is 0.002.

Often the strain is given in percentage. From Eq. (1.4) and the calculated strain we have

$$\varepsilon\% = \frac{\Delta L}{L} \times 100 = 0.002 \times 100 = 0.2\%$$

Thus, the compressive strain, which is positive in geology, is 0.2%. This is a comparatively small compressive strain and can be sustained without failure by most solid rocks.

## Example 1.3

Problem

Derive Eq. (1.8).

Solution

From Fig. 1.8 we know that stiffness $k$ is force $F$ divided by extension $\Delta L$, or

$$k = \frac{F}{\Delta L}$$

From Eqs. (1.1), (1.3), and (1.6) we also know that

$$\sigma = \frac{F}{A}$$

$$\varepsilon = \frac{\Delta L}{L}$$

$$\sigma = E\varepsilon$$

Thus we have

$$E = \frac{\sigma}{\varepsilon} = \frac{F/A}{\Delta L/L} = \frac{FL}{A\Delta L} = \frac{F}{\Delta L} \times \frac{L}{A} = \frac{kL}{A}$$

---

## Example 1.4

### Problem

A cylindrical rock specimen of basalt is subject to a compressive axial stress of 100 MPa. The measured compressive strain is 0.2%. Find the Young's modulus of the basalt.

### Solution

When making calculations using results where the strain is given as percentage, we should always change the strain into a proper fraction. Thus in this case, the strain of 0.2% corresponds to a fractional compressive strain of $0.2/100 = 0.002$. Similarly, when the stress is given in megapascals, we change it into pascals before we make our calculations, so as to use the proper SI units. Thus, a compressive stress of 100 MPa is equal to $1 \times 10^8$ Pa. From Eq. (1.6) we have

$$E = \frac{\sigma}{\varepsilon} = \frac{1 \times 10^8\,\text{Pa}}{0.002} = 5 \times 10^{10}\,\text{Pa} = 50\,\text{GPa}$$

This is a reasonable laboratory Young's modulus for basalt (Appendix D).

We can give the results either in pascals or megapascals. Pascal is a very small unit: it corresponds to the pressure generated by a water film which depth is only 0.0001 m or 0.1 mm. For this reason gigapascals are normally used when discussing the value of Young's modulus and megapascals when discussing the magnitudes of stresses or fluid pressures. However, it is best to do the calculations using pascals.

# 1.10 Exercises

1.1 Explain what is meant by the general term rock fracture. What is a brittle fracture? What is a ductile fracture?

1.2 What is the main difference between an extension fracture and a shear fracture? Give examples of extension fractures and shear fractures.

1.3 Give examples of geological processes where rock fractures play a fundamental role.

1.4 Define stress and give its units. What is the difference between normal stress and shear stress and what are their symbols? Why is tensile stress commonly regarded as negative in geology and related fields but positive in physics and engineering?

1.5 Define strain and give its units. What is the difference between deformation and strain?

1.6 What is a constitutive (material) equation? Give an example of such an equation.

1.7 Define elastic and plastic rock behaviour. What is the basic difference as regards deformation between elastic and plastic behaviour?

1.8 Define the one-dimensional Hooke's law; name and explain all the symbols used and give their units.

1.9 Generally, Young's modulus and stiffness are used as synonyms and used interchangeably. Strictly, however, there is a difference in the definition of these terms. What is the basic difference?

1.10 What basic information is needed to solve rock-fracture problems?

1.11 What basic methods are used to infer the state of stress in a rock body?

1.12 When is a rock said to be (a) homogeneous, (b) heterogeneous, (c) isotropic, and (d) anisotropic?

1.13 What is the basic difference between a static modulus and a dynamic modulus (these are elastic constants)? How are they measured? Give examples of their appropriate use in modelling rock fractures.

1.14 What are the main rock-failure criteria?

1.15 What do the terms data precision, data accuracy, and significant figures mean in calculations?

1.16 The compressive axial stress at the end of a cylindrical rock specimen with a cross-sectional area of 4 cm$^2$ is 1 MPa. Calculate the compressive axial force acting on the specimen.

1.17 A rock bar is subject to a compressive strain of 0.4%. If the original length of the bar is 0.2 m, calculate its decrease in length or shortening.

1.18 A one-metre-long cylindrical rock specimen of andesite is subject to compressive axial stress of 60 MPa. If the measured axial compressive strain is 0.12%, calculate Young's modulus of the andesite.

# References and suggested reading

Adams, S. and Allday, J., 2000. *Advanced Physics*. Oxford: Oxford University Press.

Amadei, B. and Stephansson, O., 1997. *Rock Stress and its Measurement*. London: Chapman & Hall.

Ashby, M. F. and Jones, D. R. H., 2005. *Engineering Materials*, 3rd edn. Amsterdam: Elsevier.

Benenson, W., Harris, J. W., Stocker, H., and Lutz, H. (eds.), 2002. *Handbook of Physics*. Berlin: Springer-Verlag.

Bird, J. and Ross, C., 2002. *Mechanical Engineering Principles*. Oxford: Newnes.

Caddell, R. M., 1980. *Deformation and Fracture of Solids*. Upper Saddle River, NJ: Prentice-Hall.

Callister, W. D., 2007. *Materials Science and Engineering: An Introduction*. New York: Wiley.

Davis, R. O. and Selvadurai, A. P. S., 2002. *Plasticity and Geomechanics*. Cambridge: Cambridge University Press.

Eberhart, M. E., 2003. *Why Things Break*. New York: Three Rivers Press.

Fenna, D., 2002. *A Dictionary of Weights, Measures, and Units*. Oxford: Oxford University Press.

Gordon, J. E., 1978. *The New Science of Strong Materials*. Princeton, NJ: Princeton University Press.

Gordon, J. E., 1978. *Structures*. London: Penguin.

Gottfried, B. S., 1979. *Introduction to Engineering Calculations*. New York: McGraw-Hill.

Gudmundsson, A., Friese, N., Galindo, I., and Philipp, S. L., 2008. Dike-induced reverse faulting in a graben. *Geology*, **36**, 123–126.

Hosford, W. F., 2005. *Mechanical Behaviour of Materials*. Cambridge: Cambridge University Press.

Huntley, H. E., 1967. *Dimensional Analysis*. New York: Dover.

Jaeger, J. C., 1969. *Elasticity, Fracture, and Flow: with Engineering and Geological Applications*, 3rd edn. London: Methuen.

Johnson, A. and Sherwin, K., 1996. *Foundations of Mechanical Engineering*. London: Chapman & Hall.

Klein, H.A., 1988. *The Science of Measurement. A Historical Survey*. New York: Dover.

Means, W. D., 1976. *Stress and Strain: Basic Concepts of Continuum Mechanics for Geologists*. Berlin: Springer-Verlag.

Meyers, M. A. and Chawla, K. K., 2009. *Mechanical Behavior of Materials*. Cambridge: Cambridge University Press.

Shackelford, J. F., 2005. *Introduction to Materials Science for Engineers*, 6th edn. Upper Saddle River, NJ: Pearson Prentice-Hall.

Woan, G., 2003. *The Cambridge Handbook of Physics Formulas*. Cambridge: Cambridge University Press.

# Stress

## 2.1 Aims

Fractures form when the stress in the rock becomes so high that the rock breaks. Stress is thus of fundamental importance for understanding rock fractures. Stress has already been defined (Chapter 1). Here the focus is on those aspects of stress analysis that are most useful for understanding rock fractures. The primary aims of this chapter are to:

- Explain some of the concepts from stress analysis, particularly those relevant to rock fractures.
- Clarify the difference between a stress vector (traction vector) and a stress tensor and the relationship between these concepts.
- Define the principal stresses and principal stress axes and planes.
- Explain the stress ellipsoid.
- Discuss stresses on an arbitrary plane and Mohr's circles of stress.
- Define mean stress and deviatoric stress.
- Indicate and explain the use of some special stress states.
- Define stress fields and stress trajectories and explain their use in stress analysis.

## 2.2 Some basic definitions

Stress at a **point** is a tensor of the second rank, whereas the stress on a **plane** is a vector, that is, the stress vector or traction (Chapter 1). The magnitude of a stress, given as, say, 10 MPa, is a scalar. Before we discuss the concept of stress in more detail, it is necessary to recall some basic definitions that are assumed to be known. This section, of necessity, must be brief. More details on tensors, vectors, and scalars are provided in some of the books listed at the end of this chapter.

- A **scalar** is a non-directional physical quantity that has only a magnitude and is given by a single number. A scalar is a tensor of the order or rank zero; it has one component. Examples include temperature, mass, density, and stress magnitude (size or value). The density of cold water, for instance, is 1000 kg m$^{-3}$, and the eruption temperature of basaltic magma is commonly around 1150 °C. Similarly, the magnitude of a certain stress vector may be 5 MPa. All these quantities are scalars.

- A **vector** is a physical quantity that has both magnitude and direction. For its specification, two numbers are needed. A vector is a tensor of the first rank; it has three components. Examples include force, geothermal gradient, and stress vector. A common vertical geothermal gradient in continental areas is $25°C \, km^{-1}$. Similarly, a stress vector acting upon a fault plane might be specified as trending N30°E and having a magnitude of 10 MPa. Thus, for both the geothermal gradient and the stress vector the magnitudes (25°C, 10 MPa) and the directions (vertical, N30°E) are specified.
- A **tensor** (of the second rank) is a physical quantity that remains unchanged (invariant) when the frame of reference (the coordinate system) used to define the quantity itself is changed. A tensor of the second rank has nine components. In this book, the frame of reference is Cartesian and the tensors, as well, are Cartesian and of the second rank. So the word 'tensor' here means a second-order Cartesian tensor. A more specific definition of a tensor of the second rank is given below (Section 2.3.2).

We may summarise these definitions as follows: (1) if the value of a physical quantity in space can be specified by one number, the quantity has one component and is a scalar or a tensor of rank zero; (2) if the value must be specified by three numbers, the quantity has three components and is a vector or a tensor of rank one; and (3) if the value must be specified by nine numbers, the quantity has nine components and is a tensor of rank two. In a three-dimensional physical space, the number of components of a tensor is given by $3^N$, where $N$ is the rank. Thus, a scalar has $3^0 = 1$ component, a vector has $3^1 = 3$ components, and a tensor has $3^2 = 9$ components.

Continuum mechanics in general, and rock physics (including rock-fracture mechanics) in particular, deals with physical quantities that are independent of any particular coordinate systems that may be used to specify the quantities. The quantities include stress and strain. For a complete description of such quantities, we need tensors of the second rank. The main physical laws used in continuum mechanics, such as the generalised Hooke's law in elasticity, are expressed by tensor equations. These laws do not change, that is, they are invariant, when the coordinate system is changed (through coordinate transformation), and this is one principal reason for the use of tensors in continuum mechanics and rock physics.

## 2.3 The concept of stress

Recall that in continuum mechanics, stress has two basic meanings. One is the stress on a given plane, which is referred to as the **stress vector** or **traction**. The other is the state of stress at a point, which is referred to as the **stress tensor**. Thus, when referring to a specific plane, stress is a vector quantity with three components. But when referring to a general state of stress at some location in the crust, stress is a second-rank tensor quantity with nine components. Both, however, define stress as force per unit area, with the units of $Nm^{-2}$ or pascals (Pa).

In engineering mechanics, materials science, rock physics, geophysics, and geology the term stress is often used loosely with several meanings. One is the stress vector, another is

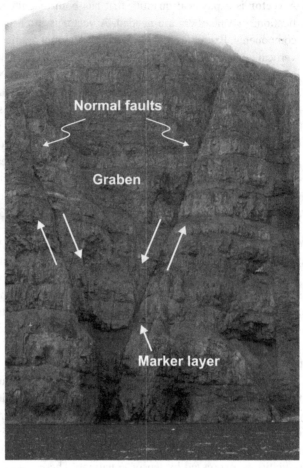

**Fig. 2.1**
Two normal faults forming a graben in the Tertiary lava pile in North Iceland. The throw (vertical displacement) on the faults is about 3 m (see the marker layer). The depth of erosion at sea level, below the initial surface of the Tertiary rift zone, is about 1500 m. View southwest, the cliff is about 120 m high.

the stress tensor, the third is a component of the stress tensor, and the fourth is the magnitude of the stress vector. The intended meaning is usually clear from the context. Let us now clarify these concepts.

Consider a crustal segment, such as a sea cliff with two normal faults dipping towards each other and thus forming **a graben** (Fig. 2.1). Today these faults are exposed in eroded sea cliffs but were at roughly 1500 m depth below the surface of the rift zone as it was when they formed. They are probably inactive today, but are very similar to dip-slip faults that are presently active in the rift zone of Iceland (Chapter 1). In general, we need to know the stresses, and the forces that give rise to those stresses, in the host rock so as to be able to say if and when faults are going to slip again.

There are two types of forces that act on a plane in a rock: body forces and surface forces. **Body forces** act at a distance and on all elements of a volume of the rock. Thus, body forces act directly on the particles of the rock and are given as force per unit mass or force per unit volume; usually the latter. A body force therefore does not result from direct contact between rock bodies but rather is associated with the rock mass and distributed throughout its volume. Examples include inertial, magnetic, and, most importantly for rock physics, gravitational forces.

**Surface forces** act across surfaces. The surfaces are of two types. One type is a physical surface such as that associated with fractures and faults (Figs. 1.6 and 2.1). Then the forces result from physical contact between the rock bodies on either side of the surface. The other type is an imaginary surface within the rock body. The imaginary surface can be of any orientation, and the surface force on it is as if it really existed as a physical surface. Such imaginary surfaces are highly relevant to rock fractures because they allow us to calculate the forces and stresses on any potential fracture plane. Thus, surface forces act directly on the particles at any physical surface, but indirectly on imaginary surfaces. For the imaginary surfaces, the surface force is transmitted along the atomic bonds inside the body. Surface forces are given as force per unit area, that is, in pascals.

Consider a force across a surface such as a fracture or fault plane (Figs. 1.1, 1.6, and 2.1). In general, the force is **compressive** when the particles on either side of the plane are pushed closer together so that the fault aperture (opening) decreases, but **tensile** when the particles on either side of the fracture plane are pulled farther apart so that the fracture aperture increases. Here, **compressive** forces and stresses and are regarded as **positive**, whereas **tensile** forces and stresses are **negative** (Chapter 1).

## 2.3.1 The stress vector

We have already given a general definition of stress (Chapter 1). For analysing many rock-fracture problems, however, a more complete definition is needed. In many books on advanced strength of materials and elementary elasticity theory, the definition of stress is as given in Example 2.8. This definition is simpler and may be easier to understand than the one given below. Thus, you may prefer to study first the definition in Example 2.8. The definition that follows is the one given in many standard books on continuum mechanics. It is more accurate than the one given in Example 2.8, but also somewhat more complex in that it uses more concepts to arrive at the definition. Because stress is one of the fundamental concepts used in analysing rock fractures, this definition is presented here for the sake of completeness.

Consider a part $\Delta A$ of an area $A$ at the surface of a much larger potential fracture plane, for example a potential fault plane, at depth in the crust (Fig. 2.2). The fault plane passes through an infinitesimally small point $P$. Surrounding $\Delta A$ and $P$ is a certain volume of rock, $V$, on which various body forces (such as gravity) and surface forces act (assumed but not shown in Fig. 2.2). The portion of the rock volume on one side of the area $A$ is denoted by I, that on the other side by II. Since the parts I and II are in contact across the plane there will be internal forces transmitted across $\Delta A$ which, at point $P$, give rise to a resultant force $\Delta F_i$ and a moment $\Delta M_i$. Here the subscript $i$ stands for the coordinate directions or axes $x$, $y$, and $z$ or, in a different notation, $x_1, x_2$, and $x_3$. Since $\Delta F_i$ and $\Delta M_i$ have one subscript, they are defined by three components and are thus vectors. The moment of a force acting

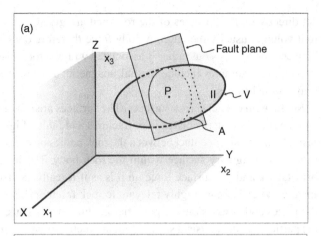

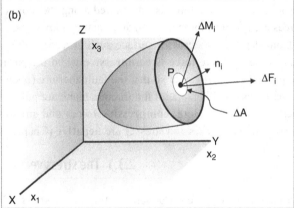

**Fig. 2.2** Definition of stress. (a) A continuous rock body of volume $V$ is located at depth in the crust. The body is dissected by a potential or actual fault plane, with the portion of the rock volume on the left side of the plane denoted by I and the portion on the right side of the plane by II. The part of the fault-plane surface located within the rock volume $V$ is referred to as area $A$. (b) The part of $A$ surrounding the point $P$ is denoted by $\Delta A$. The force $\Delta F_i$, the moment $\Delta M_i$, and the normal $n_i$ to the point $P$ are indicated. Two symbols for the coordinate axes are given: a Cartesian notation, $x$, $y$, $z$, and a 'tensor' or suffix notation, $x_1, x_2, x_3$.

about the point $P$ is the turning effect of that force; the moment is measured as the force times the perpendicular distance between $P$ and the line of action of the force.

To define the orientation of the area $\Delta A$ at the point $P$ we take $n_i$ as the outward unit normal of the area at $P$, that is, the line trending perpendicular to the area at the point $P$. Notice that $\Delta F_i$ is a force element, namely that part of the total force that acts across the small-element surface $\Delta A$. Let the area $\Delta A$ now shrink to zero with $P$ remaining inside it. Then the moment $\Delta M_i$ tends towards zero and we obtain the following definition of the **stress vector**:

$$\vec{\sigma} = \lim_{\Delta A \to 0} \frac{\Delta F_i}{\Delta A} = \frac{\mathrm{d}F_i}{\mathrm{d}A} \tag{2.1}$$

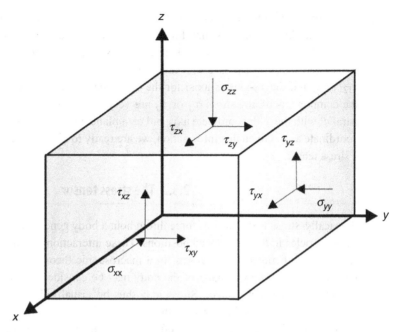

**Fig. 2.3** State of stress at a point may be presented by the nine components of the stress tensor, representing three stress vectors. The normal-stress components (or normal stresses) are here denoted by $\sigma$ and the shear-stress components or shear stresses by $\tau$. When discussing normal stresses on fault planes and when using the Mohr's stress circles the normal stress is traditionally denoted by $\sigma_n$.

Note that the stress vector in Eq. (2.1) depends on the particular surface element $\Delta A$ and its orientation, as defined by the unit normal $n_i$. Thus, if a surface through the point $P$ has a different normal, then the stress vector on it will be different. This follows because the stress on a plane depends on its attitude (strike and dip) in relation to the force generating the stress. When a force acts on a rock body dissected by fractures and faults with different attitudes, the stress vector on each fracture will therefore be different (Section 2.6).

Like any other vector, the stress vector on a plane can be resolved into three components (Fig. 2.3). Two of these components are tangential to, or lie in, the plane and one is normal (perpendicular) to the plane. The three components are perpendicular to each other. The normal component is referred to as normal stress on that plane, and is commonly given the symbol $\sigma$ and called the **normal-stress component** or simply **normal stress**. The tangential components are commonly given the symbol $\tau$ and called the **shear-stress component** or simply **shear stress**. However, there are many other symbols that are used for these components.

All the stress components have subscripts depending on the coordinate system used as reference system and the attitude of the components. If the coordinates are denoted by $x_1, x_2$, and $x_3$ or $x$, $y$, $z$ (Figs. 2.2 and 2.3), then the normal stress on the plane perpendicular to the $x_1$-axis ($x$-axis) would be given as $\sigma_{11}(\sigma_{xx})$, whereas the shear stresses on that plane would be $\tau_{12}$ and $\tau_{13}$ ($\tau_{xy}$ and $\tau_{xz}$). The first index or subscript indicates the coordinate axis perpendicular to the plane: here the $x_1$-axis (the $x$-axis), so that the first index is 1 $(x)$. The second index indicates that direction or coordinate axis in which the stress is acting (in which the arrow is trending): here the $x_1$-axis ($x$-axis) for the normal stress (so that the second index of $\sigma_{11}$ is 1 $(x)$) and the $x_2$-axis ($y$-axis) for one of the shear stresses, $\tau_{12}(\tau_{xy})$, and the $x_3$-axis ($z$-axis) for the other shear stress, $\tau_{13}$ $(\tau_{xz})$. Thus, the rules for the components of any stress $\sigma_{ij}$ or $\tau_{ij}$ are very simple: the component of the stress acts parallel with the $j$-th coordinate axis and on a plane whose normal is parallel with the $i$-th coordinate axis. Given this information, we are ready to discuss and explain the concept of a stress tensor.

## 2.3.2 The stress tensor

Physically, stress is the state of force throughout a body generated by the interaction of the body particles in their displaced positions. These interactions occur over distances similar to the sizes of molecules or atoms. In a macroscopic theory of stress, as used here, the interactions between two parts of the body may be considered as a surface force applied across their common contact. Stress may thus be visualised as a distribution of surface forces between successive layers of the body.

The state of stress at a point is defined by all the stress vectors associated with all the (infinite number of) planes, of all possible attitudes, that pass through that point. To define the stress at a point, consider the surface forces on the planes or faces of a small cube of material around that point (Fig. 2.3). Each face of the cube is parallel with two of the coordinate directions or axes, and perpendicular to the third axis. We then give the stress vector on each of these three mutually perpendicular faces and resolve each vector into components. From the discussion above it follows that the component perpendicular to a particular plane is the normal stress, the components parallel with the plane are the shear stresses. There is one normal component on each plane and two shear-stress components, each one parallel with one coordinate axis. Thus, the **state of stress at a point** is defined by nine stress components, namely

$$\sigma = \begin{array}{ccc} \sigma_{xx} & \tau_{xy} & \tau_{xz} \\ \tau_{yx} & \sigma_{yy} & \tau_{zz} \\ \tau_{zx} & \tau_{zy} & \sigma_{zz} \end{array} \qquad (2.2)$$

These are the components of a **stress tensor**. They are nine because they represent **three vectors**. The symbols used here are $\tau$ for the shear-stress components or shear stresses and $\sigma$ for the normal-stress components or normal stresses. For each stress component, the first subscript or suffix refers to the normal to the plane or face on which the component is acting, whereas the second subscript refers to the direction in which the component acts. From the discussion above it follows that, for example, the normal to the plane on which the component $\tau_{xy}$ acts is in the direction of the $x$-axis, whereas the component itself acts in a direction parallel with the $y$-axis. The component $\tau_{xz}$ acts on the same plane, so that the normal to it is parallel with the $x$-axis, hence the first subscript. However, the component itself acts in a direction parallel with the $z$-axis, hence the second subscript. For the normal components, the normal to the plane coincides with the direction of action. The component

$\sigma_{yy}$, for instance, acts in a direction parallel with the $y$-axis, and that is also the direction of the normal to the plane, hence the subscript $yy$. Because the subscript is always repeated for normal stresses, it is common to omit the second subscript and write the components as $\sigma_x, \sigma_y, \sigma_z$.

The cube (Fig. 2.3) is, by definition, in equilibrium, that is, it is not moving or rotating. Clearly, some of the shear components, for example $\tau_{zy}$ and $\tau_{yz}$, operate in opposite directions. If either shear component was larger than the other, the cube would rotate about the $x$-axis, but because the cube is in equilibrium this cannot happen. It therefore follows that these shear components must be equal, that is, $\tau_{zy} = \tau_{yz}$. Using the same logic, we conclude also that $\tau_{xz} = \tau_{zx}$, and $\tau_{yx} = \tau_{xy}$, reducing the number of independent stress components in Eq. (2.2) to six. Since compressive stress components are considered to be positive, the positive normal-stress components in Fig. 2.3 show that the cube is under compression.

## 2.4 Principal stresses

For any stress state at a point, there are three mutually perpendicular planes or faces free of shear stress. These are referred to as principal stress planes or **principal planes of stress**. The normal stress components or normal stresses that act on these planes are known as the **principal stresses** and they act along the **principal axes**.

The principal stresses are denoted by $\sigma_1$, $\sigma_2$, and $\sigma_3$. Their arrangement is so that $\sigma_1$ is the maximum principal compressive stress, $\sigma_2$ is the intermediate principal compressive stress, and $\sigma_3$ is the minimum principal compressive stress, that is,

$$\sigma_1 \geq \sigma_2 \geq \sigma_3$$

The principal stresses are at right angles to each other, that is, they are mutually orthogonal. Because compressive stress is positive and tensile stress is negative, then under geological conditions we have as follows:

- $\sigma_1$ is always positive. Even at the surface of the Earth, the maximum principal compressive stress is positive and equal to the atmospheric pressure (1 bar or about 0.1 MPa).
- $\sigma_2$ is usually positive. Under some temporary conditions, at very shallow depths or at the surface, the intermediate principal compressive stress may become tensile, that is, negative. These conditions may be satisfied during doming, such as above shallow magma chambers or salt diapirs or during rapid erosion and uplift. Since $\sigma_3$ is then, by definition, also tensile, the result is biaxial tension.
- $\sigma_3$ can be either positive (the minimum principal compressive stress) or negative (the maximum principal tensile stress). In the absence of fluids, $\sigma_3$ can be negative (absolute tension) on an outcrop scale only close to or at the Earth's surface. As we shall see (e.g. Chapters 7 and 8), outcrop-scale tensile stresses are unlikely to exist at crustal depths exceeding one kilometre.

The principal stresses form the basis for analysing rock fractures. For example, the general classification of shear fractures or faults refers to the orientation of the principal stresses. Similarly, all extension fractures form in a direction that is perpendicular to $\sigma_3$ and, since the principal stresses are mutually orthogonal, parallel with $\sigma_1$ and $\sigma_2$. Generally, close to the Earth's surface outside areas with an Alpine landscape, one of the principal stresses is vertical and the other two horizontal.

## 2.5 Stresses on an arbitrary plane

Commonly, the stresses that we measure or infer in the Earth's crust are the principal stresses. In comparatively flat areas, the principal stresses are normally assumed to be arranged so as to be parallel and perpendicular to the Earth's surface. This presumed arrangement is used as a basis for fault classification into normal, reverse, and strike-slip in Anderson's (1951) widely used theory of faulting which is based on the Coulomb criterion of brittle failure (Chapter 7). This configuration of the principal stresses, however, assumes that the crust is homogeneous and isotropic and the absence of any fluid-filled reservoirs or magma chambers (plutons if solidified) that may significantly affect the orientation of the principal stresses. Assuming that one of the principal stresses is vertical at the surface, in which case, by definition, the other two must be horizontal, it is convenient when analysing stresses related to rock fractures to define a coordinate system that coincides with the directions of the principal stresses. The principal axes can then be used as the coordinate axes.

In many geological (and engineering) situations, only two of the principal stresses need to be considered; the third one may be regarded as zero. This means that there are neither normal stresses nor shear stresses acting on one of the faces of the cube in Fig. 2.3. Generally, such an assumption may be appropriate when two of the dimensions of the body under consideration are much larger than the third dimension. For example, when we analyse the stresses and displacements in the major lithospheric plates on Earth, it is relevant that their thicknesses are much smaller than their lateral dimensions. A typical plate thickness is around one hundred kilometres, whereas the lateral dimensions are thousands of kilometres. For instance, when studying the effects of ridge push on stresses and displacements, the plate may be regarded as being only loaded by the lateral stresses, so that a **plane-stress** or two-dimensional stress formulation is appropriate (see Chapter 4).

Consider an inclined plane in a crustal segment. This plane could, for example, be a fault plane as in Figs. 2.1 and 2.4. The normal stress on the plane is denoted by $\sigma_n$ and the shear stress by $\tau$ (Figs. 1.6 and 2.4). The normal stress makes an angle $\theta$ with the principal $\sigma_1$-axis. For convenience we take the distance between A and B as a unit length, that is, one side of a square with a unit area (in three dimensions). It can then be shown (Example 2.5) that the normal stress on the plane is given by

$$\sigma_n = \sigma_1 \cos^2 \theta + \sigma_3 \sin^2 \theta \tag{2.3}$$

Fig. 2.4 Normal fault cutting through layers of sediments (conglomerates) and basaltic lava flows close to the Holocene rift zone of Southwest Iceland. View northeast, the fault changes its dip abruptly from 73° in the upper lava flow (where a person is standing) to 58° in the upper conglomerate layer. The change in dip may be partly because of different coefficients of internal friction but is likely to be primarily due to the rotation of the local stresses at the contact between these mechanically dissimilar layers (Chapter 13). The regional maximum ($\sigma_1$) and the minimum ($\sigma_3$) principal stresses during the fault formation are indicated, as are the normal ($\sigma_n$) and shear ($\tau$) stresses on the fault plane.

Similarly, the shear stress on the plane is given by

$$\tau = (\sigma_1 - \sigma_3)(\sin\theta\cos\theta) \tag{2.4}$$

Using well-known trigonometric identities (Example 2.6), Eq. (2.3) can be rewritten in the useful form

$$\sigma_n = \frac{\sigma_1 + \sigma_3}{2} + \frac{\sigma_1 - \sigma_3}{2}\cos 2\theta \tag{2.5}$$

Similarly, Eq. (2.4) can be rewritten in the form

$$\tau = \frac{\sigma_1 - \sigma_3}{2}\sin 2\theta \tag{2.6}$$

Equations (2.5) and (2.6) are very useful and will be applied below for the analysis of normal and shear stresses on fault planes. In fact, the calculated shear stress is one measure of the shear stress that acts on an actual or a potential fault plane before an earthquake. As is discussed in Example 2.7, $\theta$ is the angle between the normal to the fault plane and the direction of the maximum principal compressive stress, $\sigma_1$. However, Eqs. (2.5) and (2.6) can also be presented in terms of the angle $\alpha$ between the fault plane and $\sigma_1$ (Example 2.7). This is particularly useful because when we consider faulting in terms of Anderson's (1951) theory (Chapter 7), the fault planes make a certain angle with $\sigma_1$, so that this presentation is easily visualised. Using the trigonometric relations in Example 2.7 it can be shown that when $\alpha$ is used instead of $\theta$, Eqs. (2.5) and (2.6) become

$$\sigma_n = \frac{\sigma_1 + \sigma_3}{2} - \frac{\sigma_1 - \sigma_3}{2} \cos 2\alpha \qquad (2.7)$$

$$\tau = \frac{\sigma_1 - \sigma_3}{2} \sin 2\alpha \qquad (2.8)$$

## 2.6 Mohr's circles

Equations (2.5)–(2.8) can be presented in a visually compelling way on a diagram referred to as **Mohr's stress circle** or Mohr's circle (sometimes written Mohr stress circle or referred to as Mohr's construction or Mohr's diagram). The main purpose in using these circles, and their construction, may be summarised as follows (Figs. 2.6 and 2.7):

- What we want to find are the shear and normal stresses on a plane at a certain depth in the crust. This plane may be an actual fault plane (Figs. 2.1 and 2.4), but it does not have to be. It may be a potential fault plane, a contact of an inclined sheet and its host rock, a contact between lava flows, or any other real or imaginary plane of interest to us.
- We start the Mohr's diagram by constructing a two-dimensional coordinate system where the horizontal axis (the abscissa) is the normal stress, $\sigma_n$, and the vertical axis (the

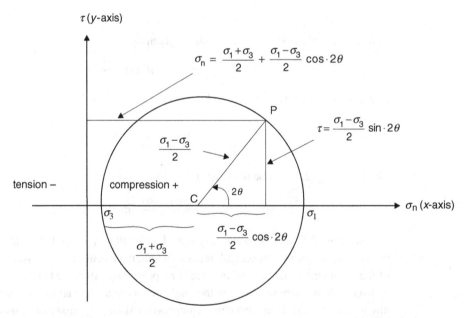

**Fig. 2.5**  Mohr's stress circle or diagram showing the normal stress by $\sigma_n$ and shear stress $\tau$ on a plane, defined by point P. In this construction, the angle $\theta$ between the normal to the fault plane and the maximum principal compressive stress $\sigma_1$ is used.

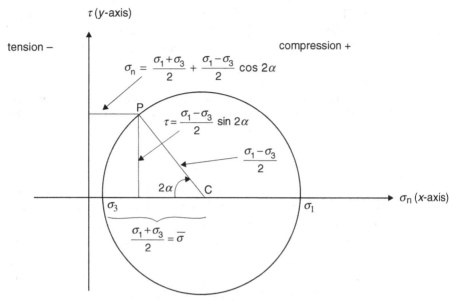

$\tau\,(y\text{-axis})$

tension −                                                        compression +

$$\sigma_n = \frac{\sigma_1 + \sigma_3}{2} + \frac{\sigma_1 - \sigma_3}{2}\cos 2\alpha$$

P

$$\tau = \frac{\sigma_1 - \sigma_3}{2}\sin 2\alpha$$

$$\frac{\sigma_1 - \sigma_3}{2}$$

$2\alpha$    C

$\sigma_3$                                                $\sigma_1$                  $\sigma_n\,(x\text{-axis})$

$$\frac{\sigma_1 + \sigma_3}{2} = \bar{\sigma}$$

**Fig. 2.6** Mohr's stress circle for the normal stress $\sigma_n$ and shear stress $\tau$ on a plane, defined by point P. In this construction, the angle $\alpha$ between the fault plane and the maximum principal compressive stress $\sigma_1$ is used.

ordinate) is the shear stress, $\tau$. We mark the stresses along these axes either in pascals or, more commonly, megapascals.

- Then we mark the magnitude of the minimum and the maximum principal stresses $\sigma_3$ and $\sigma_1$, respectively, on the $\sigma_n$-axis. Since the magnitude of the compressive stress increases when moving to the right on the $\sigma_n$-axis, $\sigma_1$ is always to the right of $\sigma_3$.
- Then we draw a circle (or, more commonly, a semicircle) through $\sigma_3$ and $\sigma_1$, so that the circle/semicircle has a diameter equal to $\sigma_1 - \sigma_3$.
- We can use this diagram to obtain the values of the normal stress $\sigma_n$ and the shear stress $\tau$ on any crustal plane of interest. Each plane is marked by a point P on the circle/semicircle.
- The point P is connected to the centre of the circle C through a line that makes an angle of $2\theta$ with the positive (right) part of the $\sigma_n$-axis, that is, the segment to the right of the centre of the circle. Here $\theta$ is the angle between $\sigma_n$ and $\sigma_1$ (Eqs. 2.2–2.6; Example 2.7). Alternatively, we can use $\alpha$, the angle between the fault plane and $\sigma_1$ (Eqs. 2.7, 2.8; Example 2.7).
- The coordinates of P are then $\sigma_n$ and $\tau$, as given by Eqs. (2.5) and (2.6) or Eqs. (2.7) and (2.8). Thus, the circle consists of an infinite number of points that show the stresses on planes with all possible values of $\alpha$ or $\theta$.
- We make the following observations:
  (a) The distance of the centre of the circle C from the origin equals $(\sigma_1 + \sigma_3)/2$.
  (b) The radius of the circle is equal to $(\sigma_1 - \sigma_3)/2$, which is also the maximum shear stress.

(c) The shear stress reaches it maximum on planes oriented at $\theta$ (or $\alpha$) $= 45°$. This result follows from Eqs. (2.6) and (2.8) with $\sin 2\theta = \sin 2\alpha = 1$.

(d) Compressive stress is to the right, and tensile stress to the left, of the $\tau$-axis.

The centre of the circle $(\sigma_1 + \sigma_3)/2$ is the mean normal stress or simply the **mean stress**. In three dimensions, the mean stress is defined as the arithmetic mean of the normal stresses. When the normal stresses are the principal stresses, the mean stress is defined as

$$\bar{\sigma} = \frac{\sigma_1 + \sigma_2 + \sigma_3}{3} \tag{2.9}$$

The mean stress is isotropic and is also referred to as the **spherical stress**. A well-known example of a mean stress of this type is the stress in a fluid at rest; it is called the **hydrostatic stress** and is equal to the static fluid pressure. Thus, the mean stress corresponds to **hydrostatic pressure**. Mean stress produces only **dilation**, that is, a change in volume. When there is expansion, say a rock body expands, the mean stress must be negative (because of the sign convention where compression is positive). By contrast, when there is contraction, say, a rock body contracts or shrinks, the mean stress is positive.

The mean stress gives the centre of the Mohr's circle but not its diameter. To find the diameter of a Mohr's circle, we must know the difference between the principal stresses, such as $\sigma_1 - \sigma_3$. This difference is a direct measure of the shear stress on the particular plane and indicates also by how much the normal stresses on that plane deviate from the mean stress on the plane. The difference between a normal stress on a plane and the mean stress on that plane is named stress deviator or **deviatoric stress**. The magnitudes of the principal deviatoric stresses $(\sigma_1', \sigma_2', \sigma_3')$ are equal to the differences between the magnitudes of the corresponding principal stresses and that of the mean stress, thus

$$\begin{aligned} \sigma_1' &= \sigma_1 - \bar{\sigma} \\ \sigma_2' &= \sigma_2 - \bar{\sigma} \\ \sigma_3' &= \sigma_3 - \bar{\sigma} \end{aligned} \tag{2.10}$$

Deviatoric stress produces **distortion**. Shear stress depends only on the deviatoric stress, not on the mean stress. Any state of stress can be resolved into two parts: **mean stress** and **deviatoric stress**. In two dimensions, the deviatoric stress is **pure shear**, that is, $\sigma_1 = -\sigma_3$, and the mean stress is either isotropic tension or, more commonly, isotropic compression. Thus, any two-dimensional state of stress can be generated by a pure shear followed by additional mean stress. The pure shear provides a Mohr's circle of the correct diameter, but with a centre at the origin of the coordinate system. When the appropriate mean stress is added, however, the centre of the Mohr's circle moves (or the circle becomes shifted) to the appropriate position along the $\sigma_n$-axis. If the added mean stress is compressive, the centre moves to the right along the $\sigma_n$-axis, whereas if the added mean stress is tensile, the centre moves to the left.

It is important to recognise the different effects that the deviatoric stress and the mean stress have on the brittle deformation of rocks. The mean stress, by definition, increases with depth in the crust. But the mean stress is isotropic and as such does not generate fractures and faults. Experiments show that rocks deform permanently only in response to the deviatoric stress. However, the mechanism by which the deformation takes place, such as brittle fracture or ductile deformation, depends partly on the mean (isotropic) stress. The formation of and slip on all faults depends on the deviatoric stress. Thus, if there is no deviatoric stress, there is also no shear stress to generate faults. Generally, faulting and, in fact, all brittle deformation in the crust seeks to **reduce the deviatoric stress**. In other words, faulting and other brittle deformation is generally in such a direction as to reduce the stress difference, and thus the deviatoric stress, in the crust. When there is no stress difference, there is no tendency to brittle deformation, and the crust is in a **lithostatic** state of stress (Section 2.7; Chapter 4).

Mohr's circles can also be used to represent a three-dimensional state of stress. One example is given in Fig. 2.7. In this case, each Mohr's semicircle represents the family of potential fault planes oriented parallel with one of the principal stress directions. Again, we mark the magnitudes of the principal stresses on the $\sigma_n$-axis. But instead of marking only $\sigma_3$ and $\sigma_1$, we also mark the intermediate principal compressive stress, $\sigma_2$. Because $\sigma_1 \geqslant \sigma_2 \geqslant \sigma_3$ it follows that $\sigma_1$ will be at the greatest distance to the right of the origin, the other principal stresses being closer to the origin. As indicated above, in case $\sigma_3$ is an absolute tension, its magnitude is marked to the left of the origin.

In the example given (Fig. 2.7), the three Mohr's semicircles are interpreted as follows:

- The largest circle, with a diameter $\sigma_1 - \sigma_3$, represents normal and shear stresses on all possible planes oriented perpendicular to the $\sigma_1/\sigma_3$-plane. In this example, the stress difference or diameter $\sigma_1 - \sigma_3 = 40 - 5 = 35$ MPa.
- The intermediate circle, with a diameter $\sigma_1 - \sigma_2$, represents normal and shear stresses on all the possible planes oriented perpendicular to the $\sigma_1/\sigma_2$-plane. Here the diameter $\sigma_1 - \sigma_2 = 40 - 20 = 20$ MPa.
- The smallest circle, with the diameter $\sigma_2 - \sigma_3$, represents normal and shear stresses on all the possible planes oriented perpendicular to the $\sigma_2/\sigma_3$-plane. In this example, the diameter $\sigma_2 - \sigma_3 = 20 - 5 = 15$ MPa.

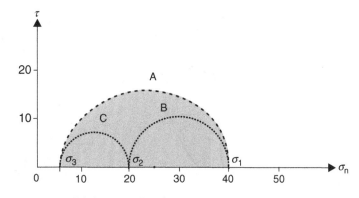

**Fig. 2.7**   Three-dimensional stress on a Mohr's diagram. Here the maximum principal compressive stress $\sigma_1 = 40$ MPa, the intermediate principal compressive stress $\sigma_2 = 20$ MPa, and the minimum principal compressive stress $\sigma_3 = 5$ MPa.

## 2.7 Special stress states

The stress vectors around a point in a rock body define a **stress ellipsoid**. The ellipsoid is defined by the mutually orthogonal principal stress directions (Fig. 2.8). More specifically, the surface of the ellipsoid is defined by the tips of the stress vectors associated with all possible planes through that point, in the case of tensile stress being positive, or by the tails of all the stress vectors, in case compressive stress is positive, as it is here. The principal axes of the ellipsoid are equal to the magnitudes of the principal stresses.

Based on its shape, the stress ellipsoid can be used to classify the various special states of stress at a point. Some of the terms applied to three-dimensional stress states, as can be represented by stress ellipsoids, include the following:

- **Polyaxial** (general) **stress**. All the principal stresses are unequal, that is, $\sigma_1 \neq \sigma_2 \neq \sigma_3$.
- **Triaxial stress**. All the three principal stresses, $\sigma_1, \sigma_2, \sigma_3$, have non-zero values and two of the stresses are equal. In rock-physics experiments (triaxial compression) this means that $\sigma_1 > \sigma_2 = \sigma_3$. In triaxial extension $\sigma_1 = \sigma_2 > \sigma_3$.
- **Hydrostatic stress**. All the principal stresses are equal, that is, $\sigma_1 = \sigma_2 = \sigma_3$. In this book, we shall commonly refer to this as **isotropic stress**, to avoid confusion with **hydrostatic pressure** (or stress) in crustal fluids. Since all the principal stresses are equal, the stress ellipsoid is a **sphere**.

(a) (b)

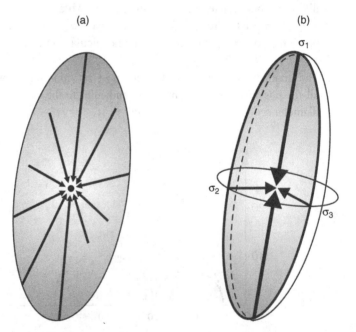

**Fig. 2.8** Stress ellipsoid. (a) The ellipsoid is defined by tails of all the arrows representing the stress vectors. (b) The principal stresses coincide with the principal axes of the ellipsoid. A two-dimensional version of the ellipsoid is a stress ellipse.

- **Axial stress**. Two of the principal stresses are equal, so that the stress is symmetric about one of the principal axes (axisymmetric). For axial extension we have $\sigma_1 = \sigma_2 > \sigma_3$ and the stress ellipsoid is an oblate spheroid (a flat, sill-like body). By contrast, for axial compression we have $\sigma_1 > \sigma_2 = \sigma_3$ and the stress ellipsoid is a prolate spheroid. It is important to remember that the stress ellipsoid has nothing to do with the shape of the body being loaded but rather shows the magnitudes of the principal stresses (the stress vectors). Thus, while a spherical body (such as a magma chamber) under axial compression parallel with the $\sigma_1$-axis would tend to become shorter in that direction, and thus approach the shape of an oblate spheroid, the magnitude of the compressive stress is largest parallel with the $\sigma_1$-axis, hence the largest axis of the stress ellipsoid is parallel with the $\sigma_1$-axis and its shape is a prolate ellipsoid.

There are many special states of stress referred to in the literature, some of which are commonly used in connection with rock fractures. Here we define some of the more commonly discussed states of stress. Several of these states of stress are presented by Mohr's circles in Fig. 2.9.

- **Uniaxial tension**. Here only one principal stress, namely $\sigma_3$, is non-zero, its value being negative (absolute tension). Consequently, the diameter of the Mohr's circle is equal to the magnitude of $\sigma_3$ and the maximum shear stress is equal to $\sigma_3/2$. Uniaxial tension is commonly used as a model when analysing fracture formation in areas undergoing extension, such as rift zones and sedimentary basins (Fig. 2.10). This is often justified, particularly in two-dimensional models. For example, at the surface of a rift zone, such as at a divergent plate boundary, the absolute tensile stress during tension-fracture formation is commonly 1–6 MPa and horizontal and equal to the magnitude of $\sigma_3$. The maximum compressive principal stress $\sigma_1$ at the surface during rifting is vertical and equal to the atmospheric pressure, namely 0.1 MPa. The intermediate principal stress $\sigma_2$ may be close to zero. Since the absolute value of the tensile stress $\sigma_3$ is 10–60 times $\sigma_1$, and $\sigma_2$ is close to zero, in modelling the rifting events close to or at the surface, the stress field may be approximated by uniaxial tension with $-\sigma_3 = 1 - 6\,\mathrm{MPa}$ and $\sigma_1 = \sigma_2 = 0$. Similar approximations may often be made at the surface of active magma chambers during rifting episodes and dyke injection.
- **Uniaxial compression**. One principal stress, namely $\sigma_1$, is non-zero and positive. The diameter of the Mohr's circle is equal to the magnitude of $\sigma_1$ and the maximum shear stress is equal to $\sigma_1/2$. This state of stress is often used in modelling, for example in analogue models on faulting and folding. Then it is assumed that one of the horizontal compressive stresses, $\sigma_1$, completely drives the deformation, so that the other principal stresses can be taken as zero, that is, $\sigma_2 = \sigma_3 = 0$. The principal stress $\sigma_1$ may, for example, be assumed to coincide with the convergent vector in continental-continental collision at convergent plate boundaries. In the crust, this state of stress may occur, but only close to or at the surface.
- **Biaxial tension**. Both $\sigma_2$ and $\sigma_3$ are nonzero and negative, whereas $\sigma_1$ is zero. The maximum shear stress is half the difference between the two tensile principal stresses. In engineering, biaxial tension is common, for example, in spherical containers of fluids under pressure. Such a state of stress exists also, for example, at the surface of an air-filled

| Special stress states | Mohr representation |
|---|---|
| 1. uniaxial tension | 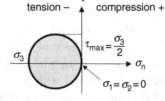 |
| 2. uniaxial compression | 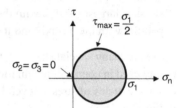 |
| 3. biaxial tension | 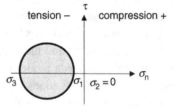 |
| 4. biaxial compression | 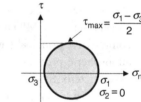 |
| 5. pure shear | 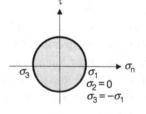 |
| 6. general state of stress | 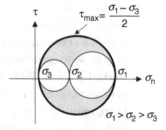 |

**Fig. 2.9**     Some typical states of stress as represented by Mohr's circles.

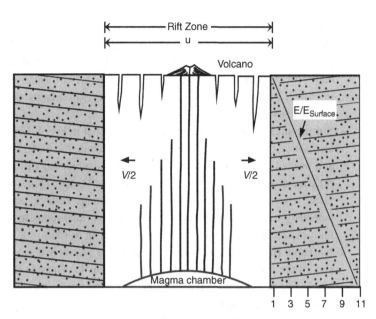

Two-dimensional presentation of a rift zone. Here it is assumed that the third dimension, into the page, is infinite. For such a conceptual model, uniaxial tension may often be considered an appropriate loading. Here $V/2$ is half the spreading rate, $u$ is the width of the rift zone, and $E/E_{\text{surface}}$ shows the ratio between Young's modulus at any particular depth and the surface modulus.

balloon. In geology, biaxial tensile stresses may occur at the surface of a magma chamber subject to internal fluid excess pressure, that is, pressure in excess of the lithostatic stress at the depth of the chamber. Such stresses may also occur at and close to the surface of various types of rock domes. Examples of possible biaxial tension include crustal domes above rising salt diapirs or expanding magma chambers, as well as crustal segments subject to rebound effects following deglaciation.

- **Biaxial compression**. Both $\sigma_1$ and $\sigma_2$ are non-zero and positive, whereas $\sigma_3$ is zero. The maximum shear stress is half the difference between the two compressive principal stresses (Figs. 2.5 and 2.6). This is a commonly assumed stress field for analysing faulting in the crust. Strictly, it is unlikely to apply except close to the Earth's surface, primarily in areas of horizontal compression such as at convergent plate boundaries. Many examples in the book use a two-dimensional Mohr's circle for analysing faulting and fracturing with only $\sigma_1$ and $\sigma_3$ specified. Then, however, it is not implied that $\sigma_2 = 0$, but rather that the intermediate principal stress does not affect the simple Coulomb criterion for shear failure (Chapter 7). A three-dimensional extension of this stress state is referred to as **general compression** or **triaxial compression**.
- **Pure shear**. Here $\sigma_1 = -\sigma_3$ and both are non-zero, whereas $\sigma_2 = 0$. Thus, the material (the rock) is simultaneously subject to tension in one direction and compression, of equal magnitude to the tension, at right angles to the tension. The Mohr's circle is centred on the origin of the coordinate system, so that the normal stress components within the material add up to zero. The shear stress is thus pure in the sense that it has non-zero

components whereas the normal stresses add up to zero. A well-known example of pure shear occurs at the surface of a solid circular shaft or a thin-walled tube subject to torque or torsional loading, that is, one end of the tube is rotated slightly with respect to the other end. This state of stress is sometimes assumed for deformation in the crust, particularly in connection with fault zones and is used in experiments on fault-rock development, such as clay smear.

- **Simple shear**. Here the loading is similar to that in Fig. 1.7 but with compressive stress also acting perpendicular to the top surface. Thus, the material is compressed vertically and pushed or pulled horizontally so that it is subject to simultaneous tensile and compressive distortion in the direction of movement (slip). This is a common model for fault zones where the fault rocks (the core and damage zone, Chapter 14) are subject simultaneously to fault-perpendicular normal stress (analogous to the vertical pressure) and fault-parallel shear stress (analogous to the push or pull).

- **Isotropic (hydrostatic) tension**. All possible planes are subject to equal tensile stress and all the principal stresses are equal and tensile. There is thus no shear stress on any plane. The stress state is represented by a point on the $\sigma_n$-axis. This would very rarely be a state of stress in the crust.

- **Isotropic (hydrostatic) compression**. All possible planes are subject to equal compressive stress and all the principal stresses are equal and compressive. There is thus no shear stress on any plane. This state of stress, represented by a point on the $\sigma_n$-axis, may be quite common at great depths in the crust. In rock mechanics it is sometimes referred to as **Heim's rule**, meaning that all compressive stresses tend to become equal at great depths in the crust. When applied to crustal stresses, however, it is most commonly referred to as **lithostatic stress** (Chapter 4). Lithostatic stress depends only on the overburden pressure, so that it increases (here as vertical stress $\sigma_v$) with depth according to the equation (Chapter 4)

$$\sigma_v = \rho_r g z \tag{2.11}$$

where $\rho_r$ is the average density of the crustal rock column to a depth of $z$ below the surface and $g$ is the acceleration due to gravity at or close to the Earth's surface.

- **General stress**. This is the polyaxial stress discussed above with $\sigma_1 > \sigma_2 > \sigma_3$. Except close to the surface of areas undergoing extension, and close to fluid-filled reservoirs, this is probably the most common state of stress in the brittle part of the crust, with all the principal stresses positive (compressive).

## 2.8 Stress fields

A **stress field** describes the variation of stresses through a rock body. If the stress field of a body is known, it implies that the state of stress is defined at every point in the body. The field is most commonly used to show the spatial variation in the principal stresses within a rock body at an instant. The field can, however, also be used to show the stress variation through the body with time. Well-known fields in physics are electric, magnetic,

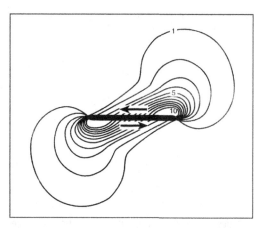

**Fig. 2.11** Stress contours showing the magnitudes of the principal stress, here the maximum tensile stress (minimum compressive stress) $\sigma_3$. The results are from a numerical model showing the tensile stresses around a sinistral strike-slip fault just before slip. Modified from Gudmundsson (2000).

and gravitational fields, which are all vector fields. The other two types of fields in physics are scalar fields and tensor fields, the name depending on whether the value of the field at every point is a scalar or a tensor. A stress field is a **tensor field**.

A common visual presentation of a field is a set of curves. For a stress field, there are several types of curves. However, in this book we use mainly two types: stress contours and stress trajectories. **Stress contours** give the magnitude of the stress at any point along the contour (Fig. 2.11). These can either show shear stress or normal stress. Commonly, the contours present the magnitudes of the principal stresses and are then sometimes referred to as **isobars**. Bar is an old unit of stress and pressure: 1 bar is equal to 0.1 megapascal.

**Stress trajectories** are curves whose directions at any point are the directions of the principal stress axes at that point. They can be either three-dimensional or two-dimensional (Fig. 2.12). They are also referred to as **isostatics**. They are sometimes thought of as '**lines of force**'. Since force per unit area is a measure of the stress intensity, it follows that the greater the number of 'lines of force' that pass through a unit area of rock, the greater will be the stress intensity. The spacing of the stress trajectories in a region is thus a measure of the stress intensity in that region. That is to say, the closer the spacing of the 'lines of force' in a given region, the greater the **stress concentration** in that region (Fig. 2.12, Chapter 6).

Because stress trajectories represent the directions of the principal stresses, they are, at any location, oriented perpendicular to principal stress planes, that is, planes on which there are no shear stresses. Absence of shear stress is also one of the defining characteristics of a **free surface**. Such a surface is normally in contact with a fluid (gas or liquid) since, by definition, a fluid cannot sustain any long-term shear stress. The Earth's surface is a free surface, since there the rock is in contact with air (on dry land) or water (in lakes or in the sea). The walls or boundaries of many **fluid-filled reservoirs**, such as many magma chambers, are also free surfaces because the rock walls are in contact with a fluid. However, this applies only if the reservoir is totally fluid (not just with fluid in its pores) at the contacts with the walls.

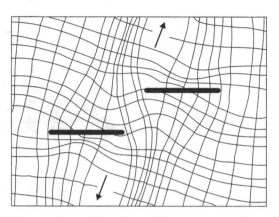

Fig. 2.12 Stress trajectories, as continuous curves, showing the stress field generated by uniaxial tension (indicated by arrows) oblique to two offset fractures. The curves subparallel with the tensile loading represent $\sigma_3$, whereas those subparallel with the trends of the fractures represent $\sigma_1$. Extension fractures such as these follow the $\sigma_1$-trajectories so that, in this model, the propagating nearby tips of the fractures would tend to curve towards each other. The reduced spacing between the $\sigma_3$-trajectories in the region between the nearby ends of the fractures indicates a high tensile stress concentration in that area (Chapter 6).

It follows from these definitions that stress trajectories must be either **perpendicular** to or **parallel** with a free surface. For example, in a rift zone the trajectories of $\sigma_1$ are normally at right angles to the Earth's surface, whereas those of $\sigma_2$ and $\sigma_3$ are parallel with the surface (but at mutually right angles). Similarly, for a totally fluid-filled reservoir under lithostatic or higher pressure (excess fluid pressure, Chapter 6), $\sigma_1$ would normally be at right angles to the walls or boundaries (Fig. 2.13) of the reservoir, whereas $\sigma_2$ and $\sigma_3$ would be parallel with the walls (Chapter 6).

Stress trajectories are either presented by continuous curves, or by short, straight line segments, referred to as ticks. The ticks show the orientations of the principal stresses at individual points, commonly the points where solutions have been obtained in numerical models. The solution points, hence the ticks, can theoretically be made as dense as we wish. However, for a clear presentation of the results, the ticks must have a reasonable minimum spacing (Fig. 2.13), from which the orientations of the principal stresses at other points in the field can be inferred.

The stress field may be either **homogeneous** or **heterogeneous** (inhomogeneous). In a homogeneous field, all the components of the stress tensor are the same at each point in the rock body. In a heterogeneous stress field, the components of the stress tensor differ between points in the body. As elaborated in later chapters, a strictly homogeneous stress field cannot exist in the crust. The reasons are many and include the following:

- Crustal rocks contain numerous cavities and discontinuities and inclusions with elastic properties that differ from those of the host rock and, commonly, from other flaws in the rock. Cavities concentrate stresses and discontinuities (such as contacts, faults, and fractures) modify the stress field, often abruptly.

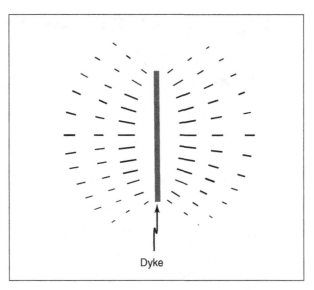

Dyke

**Fig. 2.13** Stress trajectories, as ticks, showing the orientation of $\sigma_1$ around a fracture (here a dyke) subject to internal overpressure (pressure in excess of $\sigma_3$) as the only loading. The results show, as is confirmed by field observations during dyke emplacement, that the fluid overpressure increases the horizontal compressive stress in the vicinity of the dyke. For a dyke emplacement in a rift zone this means that, temporarily, following the dyke emplacement the horizontal compressive principal stress perpendicular to the dyke need not be $\sigma_3$, as is the normal state of stress in a rift zone, but rather $\sigma_1$ (or $\sigma_2$). The implication is that, until the horizontal compressive stress generated by the dyke is relaxed (through plate movements, for example), renewed dyke emplacement close to this dyke is unlikely.

- Many rocks are layered. The layers commonly have different mechanical properties, in particular in igneous rocks such as lava flows and pyroclastic layers. Thus, even if the applied loading is uniform, the stress field will vary as a function of different rock properties.
- Body forces, such as gravity, and stress sources, such as fluid-filled pressured reservoirs, generate stress gradients.

Stress fields in rock bodies are thus, as a rule, heterogeneous. In subsequent chapters, there are many illustrations showing various stress fields and a detailed discussion of their importance for fracture formation and development. An example of variation in a local stress field, as indicated by the fracture (joint) pattern, is given in Fig. 2.14.

## 2.9 Summary

- Stress on a plane is a vector; stress at a point is a tensor. Both have the units of force per unit area, that is, pascals. In rocks, stresses are generated by two types of forces: body forces, such as gravity, and surface forces, such as the tectonic forces associated with plate movements. Body forces act on all the particles in the rock; surface forces

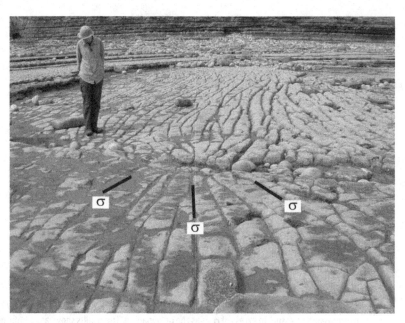

**Fig. 2.14** Joints in a limestone layer in Wales (the Bristol Channel). Presumably, most joints are extension fractures and thus propagate parallel with the $\sigma$-trajectories (Fig. 2.12), here either $\sigma_1$ or $\sigma_2$. On the assumption that these joints are extension fractures, they map out the curved $\sigma$-trajectories in this area. The curved geometry of the trajectories is here partly due to the joints being associated with a small cavity or inclusion (Chapter 6), a region of stress concentration, associated with a nearby fault.

act on real surfaces (such as fault planes, contacts, and joints) and imaginary surfaces. Compressive stress is positive, tensile stress is negative. When no confusion can arise, however, it is sometimes convenient to treat the tensile stresses as absolute values and omit the negative signs.

- The stress vector can be resolved into three mutually orthogonal components. One of the components is normal to the plane on which the stress vector acts and is called normal stress ($\sigma$), the other two components lie in the plane and are called shear stresses ($\tau$). The stresses have subscripts or indexes, such as $\tau_{xy}$. The first index (here $x$) refers to the coordinate axis that is perpendicular to the plane on which the stress vector acts, whereas the second index (here $y$) refers to the coordinate axis parallel with which the stress component is acting.

- The state of stress at a point may be defined by three mutually perpendicular stress vectors acting on a very small cube (representing the point). Since each stress vector has three components, the three vectors together have nine components. These nine components define the stress tensor. Only six of the stress components of the tensor, however, are independent. At the point, there are three mutually perpendicular planes with no shear stresses acting upon them. These are the principal planes of stress; the normal stresses acting on them are the principal stresses $\sigma_1, \sigma_2$, and $\sigma_3$, where $\sigma_1 \geqslant \sigma_2 \geqslant \sigma_3$. Their orientation is the basis for understanding extension fractures and shear fractures and, in particular, the formation and classification of the main types of faults (normal, reverse, and strike-slip). In many geological situations, particularly close to or at the Earth's surface (particularly if it is comparatively flat), one of the principal stresses is vertical and, by definition, the other two are horizontal and thus parallel with the surface.

- Commonly, when analysing rock fractures, only two of the principal stresses need be considered. For example, when studying a fault in a vertical outcrop (Figs. 2.1 and 2.4), the focus may be on the two dimensions of the wall or cliff where the fault is exposed (the vertical section) so that the dimension into the wall is ignored (Chapter 9). The normal and shear stress on a fault plane can be presented visually (also in three dimensions) using Mohr's stress circle. Mohr's diagram consists of a two-dimensional coordinate system with normal stress (usually in megapascals) on the horizontal axis and shear stress on the vertical axis. We mark the minimum and maximum principal stresses, $\sigma_1$ and $\sigma_3$, on the horizontal axis, and draw a circle or semicircle through them. From this circle we can read the values of the normal and shear stresses on any plane in the crust, including all fault planes. Each plane is marked by a point on the circle. The centre of the circle is the mean (average) stress at the location of the point of interest on the plane. The difference between the normal stress on the plane and the mean stress, at the point of interest, is the deviatoric stress. Mean stress produces expansion or contraction, whereas deviatoric stress produces distortion. Any state of stress is composed of two parts, namely mean stress and deviatoric stress. Shear stress is entirely due to deviatoric stress.

- Brittle deformation in the crust in general seeks to reduce deviatoric stress and make all the principal stresses equal, an isotropic compression. When the isotropic compression depends only on depth or overburden pressure, the state of stress is referred to as litho-static. There are many other special stress states, such as uniaxial and biaxial tension and compression, and simple and pure shear. However, probably the most common state of stress in the upper part of the crust is general stress, that is, polyaxial stress where $\sigma_1 > \sigma_2 > \sigma_3$, whereas in the lower part, and close to the fluid-filled reservoirs, the state of stress is more likely to be lithostatic.

- The variation of stress through a rock body gives rise to a stress field. The stress field can be presented visually in various ways. When dealing with rock fractures, the most compelling presentation of the stress field are stress contours and stress trajectories. The contours give the magnitude (usually in megapascals) of the stress at any point within the rock body. Stress trajectories give the directions of the principal stresses at any point. The trajectories are either presented as continuous curves or as short lines, ticks. When the trajectories meet a free surface, a surface free of shear stress – such as the Earth's surface – they must, by definition, be either perpendicular to or parallel with the surface.

## 2.10  Main symbols used

$A$        area
$\Delta A$        small part of an area, small area
$\Delta F_i$        resultant force

| | |
|---|---|
| $g$ | acceleration due to gravity |
| $\Delta M_i$ | momentum |
| $n_i$ | unit normal |
| $P$ | point |
| $L$ | length |
| $\Delta L$ | extension, change in length |
| $z$ | depth below the Earth's surface (crustal depth) |
| $\alpha$ | angle between a fault plane and $\sigma_1$ |
| $\gamma$ | shear strain |
| $\varepsilon$ | normal strain |
| $\rho_r$ | average rock or crustal density |
| $\sigma$ | stress tensor |
| $\sigma_{ii}$ | normal-stress component |
| $\vec{\sigma}$ | stress vector |
| $\bar{\sigma}$ | mean stress |
| $\sigma_1$ | maximum compressive principal stress |
| $\sigma_2$ | intermediate compressive principal stress |
| $\sigma_3$ | minimum compressive (maximum tensile) principal stress |
| $\sigma'_1$ | deviatoric stresses |
| $\sigma_n$ | normal stress; this symbol is normally used in connection with Mohr's circles |
| $\sigma_v$ | vertical stress, lithostatic stress |
| $\tau$ | shear stress |
| $\tau_{ij}$ | components of shear stress |
| $\theta$ | angle between normal to a fault plane and $\sigma_1$ |

## 2.11 Worked examples

### Example 2.1

#### Problem

(a) If the uppermost part of the crust has an average density of 2500 kg m$^{-3}$, at what depth would the vertical stress reach 25 MPa?

(b) If the state of stress is lithostatic, what are the magnitudes of the principal stresses at this depth?

#### Solution

(a) The vertical stress is given by Eq. (2.11) as

$$\sigma_v = \rho_r g z$$

The average crustal density $\rho_r$ is 2500 kg m$^{-3}$ and the acceleration (of free fall) due to gravity $g$ is 9.81 m s$^{-2}$. We want to find the depth at which $\sigma_v = 25$ MPa $= 2.5 \times 10^7$

Pa. We rewrite Eq. (2.11) and solve for the depth $z$ thus:

$$z = \frac{\sigma_v}{\rho_r g} = \frac{2.5 \times 10^7 \text{ Pa}}{2500 \text{ kg m}^{-3} \times 9.81 \text{ m s}^{-2}} = 1019 \text{ m}$$

This means that, for a typical crustal density in the upper part of the crust, the vertical stress increases by roughly 25 MPa per thousand metres. This a reasonable rule of thumb, namely:

Vertical stress **increases** by about **25 MPa for every kilometre** in the uppermost part of the crust.

(b) A state of stress is known as lithostatic when all the principal stresses are equal (an isotropic stress) and equal to the overburden pressure, that is, to the vertical stress. Thus, in this case all the principal stresses would be the same and equal to 25 MPa.

---

### Example 2.2

#### Problem

The sea cliff in Fig. 2.1 is about 120 m high and is composed of basaltic lava flows with thin layers of soil in between the lava flows. If the average density of the rock layers that constitute the cliff is 2600 kg m$^{-3}$ and the average dip of the faults is 73°, find:

(a) the vertical stress in the cliff at sea level
(b) the normal and shear stresses on the normal faults at a depth of 100 m, that is, 20 m above sea level (where there is an unusually thick and clear soil layer between the lava flows). Assume that the vertical stress is the maximum principal compressive stress and that the minimum principal compressive stress is one-third of the vertical stress, as is sometimes the case (Chapter 4).

#### Solution

(a) The vertical stress follows directly from Eq. (2.11) and is as follows:

$$\sigma_v = \rho_r g z = 2600 \text{ kg m}^{-3} \times 9.81 \text{ m s}^{-2} \times 120 \text{ m} = 3.1 \times 10^6 \text{ Pa} = 3.1 \text{ MPa}$$

This compressive stress is much less than the compressive strength of in-situ rocks (normally many tens of megapascals) so that compressive failure would be unlikely.

(b) The fault dip is 73°. Since $\sigma_1$ is vertical, the fault planes make an angle of $90 - 73 = 17°$ to $\sigma_1$, which is the angle $\tau$ in Eqs. (2.7) and (2.8). We must first calculate the vertical stress, that is, $\sigma_1$. Proceeding as in (a) we have (Eq. 2.11)

$$\sigma_1 = \rho_r g z = 2600 \text{ kg m}^{-3} \times 9.81 \text{ m s}^{-2} \times 100 \text{ m} = 2.6 \times 10^6 \text{ Pa} = 2.6 \text{ MPa}$$

From this follows that the minimum principal stress is

$$\sigma_3 = \frac{\sigma_1}{3} = \frac{2.6 \text{ MPa}}{3} = 0.9 \text{ MPa}$$

We can now use either Eqs. (2.5) and (2.6) or Eqs. (2.7) and (2.8) to calculate the normal and shear stresses on the normal fault of the graben. Since we already have $\alpha = 17°$, we shall here use Eqs. (2.7) and (2.8). The normal stress on the faults is then

$$\sigma_n = \frac{\sigma_1 + \sigma_3}{2} - \frac{\sigma_1 - \sigma_3}{2} \cos 2\alpha$$

$$= \frac{2.6 \text{ MPa} + 0.9 \text{ MPa}}{2} - \frac{2.6 \text{ MPa} - 0.9 \text{ MPa}}{2} \cos 34°$$

$$= 1.8 - 0.7 = 1.1 \text{ MPa}$$

Similarly, the shear stress on the fault is

$$\tau = \frac{\sigma_1 - \sigma_3}{2} \sin 2\alpha = \frac{2.6 \text{ MPa} - 0.9 \text{ MPa}}{2} \sin 34° = 0.5 \text{ MPa}$$

Both the normal and the shear stress are small, which is as expected because the loading is only attributable to the overburden pressure from the cliff above. Although driving shear stress of less than 0.5 MPa may occasionally generate fault slip (Chapter 9), such a low shear stress would normally indicate that fault slip is unlikely.

## Example 2.3

### Problem

The normal fault in Fig. 2.4 changes its dip abruptly on passing from the basaltic lava flow at the top of the section to the sedimentary layer (a conglomerate) below. In the lava flow, the fault dip is 73° but as low as 58° in the conglomerate layer in the central part of the photograph. The rock where the fault is exposed has been subject to glacial erosion, so that the depth of the present exposure below the top of the rift zone at the time of fault formation may be around 400 m. The density of the part of the pile eroded away, mostly lava flows and basaltic breccias (hyaloclastite) and sediments formed in subglacial eruptions, is estimated at 2300 kg m$^{-3}$. Since the normal fault formed in the flat rift zone in Iceland (the Holocene rift zone is only 20 km east of the exposed fault), the vertical stress may be taken as the maximum principal compressive stress $\sigma_1$. The minimum principal stress, $\sigma_3$, is then horizontal. The throw is 4 m and may have formed in a single slip during a rifting episode. The depth of the exposure below the top of the rift zone at the time of fault formation, estimated at 400 m, is similar to the maximum depth of absolute tension in rift zones (Chapters 7 and 8). It is thus reasonable to assume that $\sigma_3$ was either somewhat negative or close to zero at the time of fault slip. For the present purpose, we take $\sigma_3$ as zero. Using this information, calculate the normal and shear stresses on the fault plane, at its time of formation or slip, where it cuts through (a) the upper lava flow and (b) the upper conglomerate layer.

### Solution

We need first to calculate the vertical stress, $\sigma_1$. From Eq. (2.11) $\sigma_1$ is obtained as follows:

$$\sigma_1 = \rho_r g z = 2300 \text{ kg m}^{-3} \times 9.81 \text{ m s}^{-2} \times 400 \text{ m} = 9 \times 10^6 \text{ Pa} = 9 \text{ MPa}$$

The minimum principal compressive stress $\sigma_3 = 0$.

Where the fault dissects the lava flow, the fault dip is $73°$ so that $\alpha = 17°$. Thus, in the lava flow the normal and shear stresses on the fault plane are, from Eqs. (2.7) and (2.8), as follows:

$$\sigma_n = \frac{\sigma_1 + \sigma_3}{2} - \frac{\sigma_1 - \sigma_3}{2} \cos 2\alpha$$
$$= \frac{9 \text{ MPa} + 0 \text{ MPa}}{2} - \frac{9 \text{ MPa} - 0 \text{ MPa}}{2} \cos 34°$$
$$= 4.5 - 3.7 = 0.8 \text{ MPa}$$
$$\tau = \frac{\sigma_1 - \sigma_3}{2} \sin 2\alpha = \frac{9 \text{ MPa} - 0 \text{ MPa}}{2} \sin 34° = 2.5 \text{ MPa}$$

Where the fault dissects the conglomerate layer, the fault dip is $58°$ so that $\alpha = 90 - 58 = 32°$. Proceeding as before, we get the normal and shear stresses on the fault plane from Eqs. (2.7) and (2.8) as follows:

$$\sigma_n = \frac{\sigma_1 + \sigma_3}{2} - \frac{\sigma_1 - \sigma_3}{2} \cos 2\alpha$$
$$= \frac{9 \text{ MPa} + 0 \text{ MPa}}{2} - \frac{9 \text{ MPa} - 0 \text{ MPa}}{2} \cos 64°$$
$$= 4.5 - 2.0 = 2.5 \text{ MPa}$$
$$\tau = \frac{\sigma_1 - \sigma_3}{2} \sin 2\alpha = \frac{9 \text{ MPa} - 0 \text{ MPa}}{2} \sin 64° = 4 \text{ MPa}$$

The results show that both the shear and the normal stresses are much higher on the gently dipping part of the fault plane than on the steeply dipping part. This is as expected. In particular, the shear stress on the fault plane increases and reaches its maximum for a plane that makes an angle of $45°$ to the principal stresses. Because of fluid pressure, the effective normal stress on a fault plane during slip is presumably commonly close to zero (Chapters 7 and 9). Thus, it is primarily the shear stress that determines whether or not fault slip takes place.

The present results show that where a fault dissects a layered crustal segment, as is very common, the fault attitude, primarily the dip, may change abruptly and, consequently, the local stresses on the fault plane may change abruptly as well. Thus, the conditions for fault slip (Chapters 7 and 9) may be satisfied on one segment or part of a fault plane, but not on adjacent segments. This is one reason why most fault slips are limited to parts of faults, and that the rupture of entire fault zones is rare. It follows that large slips and earthquakes are also rare.

### Example 2.4

#### Problem

A normal fault is reactivated in a stress field where the maximum principal compressive stress $\sigma_1$ trends E–W and is horizontal and the minimum principal compressive stress $\sigma_3$ is vertical. The fault strikes north and dips $80°E$. Field measurements show that the stress magnitudes are as follows: $\sigma_1 = 30 \text{ MPa}$ and $\sigma_3 = 10 \text{ MPa}$.

(a) Illustrate the problem by a drawing.
(b) Calculate the normal and shear stresses on the fault plane and show the results using a Mohr's circle (or semicircle).

**Solution**

(a) An illustration of the problem is given in Fig. 2.15.
(b) Here we use Eqs. (2.5) and (2.6). The dip of the fault is 80°, so that the normal stress on the fault plane, $\sigma_n$, makes an angle of 10° to the plane. From Eqs. (2.5) and (2.6), the normal and shear stress on the fault plane are thus

$$\sigma_n = \frac{\sigma_1 + \sigma_3}{2} + \frac{\sigma_1 - \sigma_3}{2} \cos 2\theta$$

$$= \frac{30 \text{ MPa} + 10 \text{ MPa}}{2} + \frac{30 \text{ MPa} - 10 \text{ MPa}}{2} \cos 20°$$

$$= 20 \text{ MPa} + 9.4 \text{ MPa} = 29.4 \text{ MPa}$$

$$\tau = \frac{\sigma_1 - \sigma_3}{2} \sin 2\theta = \frac{30 \text{ MPa} - 10 \text{ MPa}}{2} \sin 20° = 3.4 \text{ MPa}$$

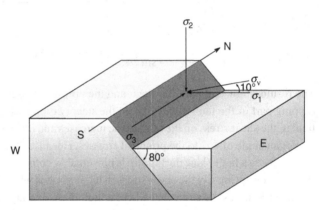

**Fig. 2.15**  Illustration of Example 2.4.

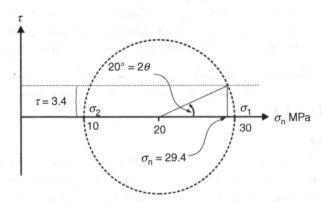

**Fig. 2.16**  Mohr's stress circle for Example 2.4.

The Mohr's circle is shown in Fig. 2.16. Recall that the construction is as follows. We mark the principal stresses $\sigma_1 = 30$ MPa and $\sigma_3 = 10$ MPa as points on the $\sigma_n$-axis, draw a (semi)circle through these points (the circle diameter being $\sigma_1 - \sigma_3 = 30 - 10 = 20$ MPa). From the centre of the circle we mark the angle $2\theta = 20°$, anticlockwise from the $\sigma_n$-axis, and find the point on the Mohr's circle corresponding to the fault plane. The stresses on that plane can then be read from the diagram as the coordinates of that point. That is to say, the horizontal coordinate of the point is the normal stress $\sigma_n$ on the plane and the vertical coordinate is the shear stress $\tau$ on the plane.

## Example 2.5

Problem

Derive Eqs. (2.3) and (2.4).

Solution

We shall derive these equations with reference to an arbitrary inclined plane marked by AB (Fig. 2.17). The axes of the coordinate system coincide with the direction of the principal stresses, $\sigma_1$ and $\sigma_3$, and the normal stress $\sigma_n$ makes an angle $\theta$ to $\sigma_1$. The length of line AB is one unit, and it is a part (one side) of a square of unit area. Then from trigonometry we have

$$\frac{OA}{1} = \sin\theta \Rightarrow OA = \sin\theta$$

$$\frac{OB}{1} = \cos\theta \Rightarrow OB = \cos\theta$$

We know that stress is force per unit area, so that force is stress times unit area. The force acting along OB is thus the stress ($\sigma_1$) along OB times the unit area perpendicular to that stress, namely $\cos\theta$, and similarly for the force along OB.

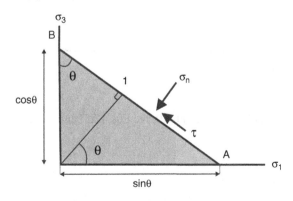

Fig. 2.17      Normal and shear stresses on an inclined plane, AB, of a prism of unit area.

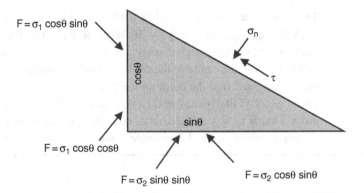

Fig. 2.18 Forces acting on the prism in Fig. 2.17.

The force along OA is therefore $\sigma_1 \cos \theta$ (the perpendicular component being $\sigma_1 \sin \theta$) and the force along OB is $\sigma_3 \sin \theta$ (the perpendicular component being $\sigma_3 \cos \theta$). The prism is in equilibrium so that the forces acting in one direction are balanced by forces acting in the opposite direction (Fig. 2.18). Remembering that the unit areas are implied (so that, for example, $\tau = \tau \times$ unit area), we have

$$\tau + \sigma_3 \cos \theta \times \sin \theta = \sigma_1 \sin \theta \times \cos \theta$$

Solving for $\tau$ we get

$$\tau = \sigma_1 \sin \theta \cos \theta - \sigma_3 \cos \theta \sin \theta = (\sigma_1 - \sigma_3)(\sin \theta \cos \theta)$$

which is Eq. (2.4). To find $\sigma_n$, recall that the prism is in equilibrium so that the force trying to force the prism down (namely, $\sigma_n \times$ unit area) is equal to the opposite force, so that

$$\sigma_n = \sigma_1 \cos \theta \cos \theta + \sigma_3 \sin \theta \sin \theta = \sigma_1 \cos^2 \theta + \sigma_3 \sin^2 \theta$$

which is Eq. (2.3).

## Example 2.6

### Problem
Use Eqs. (2.3) and (2.4) to derive Eqs. (2.5) and (2.6).

### Solution
Equation (2.3) is

$$\sigma_n = \sigma_1 \cos^2 \theta + \sigma_3 \sin^2 \theta$$

From trigonometry we have

$$\cos^2 \theta = \frac{1}{2}(1 + \cos 2\theta)$$

$$\sin^2 \theta = \frac{1}{2}(1 - \cos 2\theta)$$

$$\sin \theta \cos \theta = \frac{1}{2} \sin 2\theta$$

Using the first two identities, Eq. (2.3) becomes

$$\sigma_n = \sigma_1 \left[ \frac{1}{2}(1 + \cos 2\theta) \right] + \sigma_3 \left[ \frac{1}{2}(1 - \cos 2\theta) \right]$$

$$= \frac{1}{2}\sigma_1 + \frac{1}{2}\sigma_1 \cos 2\theta + \frac{1}{2}\sigma_3 - \frac{1}{2}\sigma_3 \cos 2\theta$$

Or

$$\sigma_n = \frac{\sigma_1 + \sigma_3}{2} + \frac{\sigma_1 - \sigma_3}{2} \cos 2\theta$$

which is Eq. (2.5).

Equation (2.4) is

$$\tau = (\sigma_1 - \sigma_3)(\sin \theta \cos \theta)$$

Using the third trigonometric identity above, Eq. (2.4) becomes

$$\tau = (\sigma_1 - \sigma_3)(\tfrac{1}{2} \sin 2\theta) = \frac{\sigma_1 - \sigma_3}{2} \sin 2\theta$$

which is Eq. (2.6).

---

### Example 2.7

Problem

Show how Eqs. (2.7) and (2.8) follow from Eqs. (2.5) and (2.6).

Solution

The relation between the angles $\alpha$ and $\theta$ is shown in Fig. 2.19. From trigonometry we have

$$\alpha + \theta = 90° \Rightarrow \theta = 90° - \alpha \Rightarrow 2\theta = 180° - 2\alpha$$

And also

$$\sin(180° - 2\alpha) = \sin 2\alpha$$

$$\cos(180° - 2\alpha) = -\cos 2\alpha$$

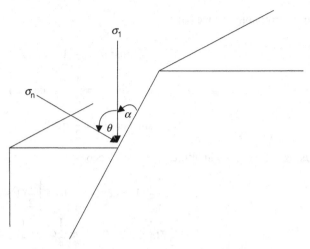

**Fig. 2.19** Relationship between the angles $\alpha$ and $\theta$ and $\sigma_1$ as used in deriving the equations for normal and shear stresses on a plane and constructing Mohr's circles.

Substituting these results into Eqs. (2.5) and (2.6), namely

$$\sigma_n = \frac{\sigma_1 + \sigma_3}{2} + \frac{\sigma_1 - \sigma_3}{2} \cos 2\theta$$

$$\tau = \frac{\sigma_1 - \sigma_3}{2} \sin 2\theta$$

we get Eqs. (2.7) and (2.8), namely

$$\sigma_n = \frac{\sigma_1 + \sigma_3}{2} - \frac{\sigma_1 - \sigma_3}{2} \cos 2\alpha$$

$$\tau = \frac{\sigma_1 - \sigma_3}{2} \sin 2\alpha$$

### Example 2.8

#### Problem

Provide a simplified version of the concept of stress as presented in Eq. (2.1).

#### Solution

We consider a solid subject to some surface forces $P_1$–$P_4$ on its boundary (Fig. 2.20). The surface forces are kept in equilibrium by the force exerted by part I of the body on part II across a circular cutting plane. The force by part I on part II is distributed over the entire cutting plane, so that any small area $\Delta A$ on the cutting plane is acted upon by a force $\Delta F$. The average force $\bar{P}$ per unit area of the cutting plane is thus

$$\bar{P} = \frac{\Delta F}{\Delta A}$$

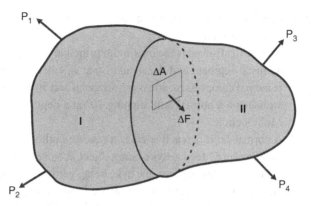

Fig. 2.20
External surface force on a solid body, denoted by $P_1$–$P_4$, through which a circular cutting plane runs. The part of the body to the left of the cut is denoted by I, the part to the right of the cut by II. $\Delta A$ is a small area on the surface of the cutting plane and is subject to the force $\Delta F$. In this very schematic illustration, the surface forces are shown as absolute values and thus without signs.

At a point located within the area $\Delta A$, the stress vector $\vec{\sigma}$ is defined as the limiting value of $\bar{P}$ as $\Delta A$ approaches zero, thus

$$\vec{\sigma} = \lim_{\Delta A \to 0} \frac{\Delta F}{\Delta A} = \frac{dF}{dA}$$

which is similar to Eq. (2.1).

## 2.12 Exercises

2.1 Define a scalar, a vector, and a tensor. Give geological examples of these physical quantities.

2.2 Explain the difference between a stress vector and a stress tensor. Give examples of their use in geology.

2.3 Define principal stresses and principal stress planes. Give the symbols of the principal stresses and rank them according to magnitude.

2.4 Define mean stress, deviatoric stress, and lithostatic stress.

2.5 Explain the concept of a stress ellipsoid.

2.6 Define triaxial stress, polyaxial stress, and hydrostatic stress.

2.7 Use Mohr's circles to show uniaxial tension, uniaxial compression, biaxial tension, biaxial compression, pure shear, isotropic tension, and isotropic compression.

2.8 Explain the concepts of stress fields, stress contours, and stress trajectories. Why must stress trajectories be either perpendicular to or parallel with the Earth's surface?

2.9 In a volcanic rift zone the average density of the uppermost 3 km of the crust is $2600 \text{ kg m}^{-3}$. Calculate the vertical stress at the depth of 3 km in the rift zone.

2.10 For an average crustal density of 2700 kg m$^{-3}$, at what depth would the vertical stress reach 100 MPa?

2.11 The vertical stress is the maximum principal compressive stress at a depth of 3 km in a crustal segment and the same as that calculated in Exercise 2.9, and the minimum principal compressive stress is horizontal and 50 MPa. Calculate the normal and shear stresses on a normal fault dipping 70° at a depth of 3 km and show the results using Mohr's circle.

2.12 A normal fault is reactivated as a reverse fault at a crustal depth of 100 m due to an overpressured feeder-dyke being injected close to the fault (Fig. 1.6). The fault and the dyke strike northeast, the dyke being vertical, whereas the fault dips 75° southeast. Because of the magma overpressure in the dyke, $\sigma_1$ becomes horizontal and equal to 20 MPa, whereas $\sigma_3$ becomes vertical and equal to the overburden pressure. The density of the uppermost 100 m of the crust is 2500 kg m$^{-3}$. Find the normal and shear stresses on the fault plane and show the results using a Mohr's circle.

# References and suggested reading

Anderson, E. M., 1951. *The Dynamics of Faulting and Dyke Formation with Applications to Britain*, 2nd edn. Edinburgh: Oliver and Boyd.

Caddell, R. M., 1980. *Deformation and Fracture of Solids*. Upper Saddle River, NJ: Prentice-Hall.

Chou, P. C. and Pagano, N. J., 1992. *Elasticity. Tensor, Dyadic, and Engineering Approaches*. New York: Dover.

Cottrell, A. H., 1964. *The Mechanical Properties of Matter*. New York: Wiley.

Holzapfel, G. A., 2000. *Nonlinear Solid Mechanics*. New York: Wiley.

Jaeger, J. C., Cook, N. G. W., and Zimmerman, R. W., 2007. *Fundamentals of Rock Mechanics*, 4th edn. Oxford: Blackwell.

Malvern, L. E., 1969. *Introduction to the Mechanics of a Continuous Medium*. Upper Saddle River, NJ: Prentice-Hall.

Mase, G. E., 1970. *Continuum Mechanics*. New York: McGraw-Hill.

Mase, G. T. and Mase, G. E., 1999. *Continuum Mechanics for Engineers*, 2nd edn. London: CRC Press.

Means, W. D., 1976. *Stress and Strain: Basic Concepts of Continuum Mechanics for Geologists*. Berlin: Springer-Verlag.

Niklas, K. J., 1992. *Plant Biomechanics*. Chicago, IL: The University of Chicago Press.

Nye, J. F., 1984. *Physical Properties of Crystals: Their Representation by Tensors and Matrices*. Oxford: Oxford University Press.

Oertel, G., 1996. *Stress and Deformation: A Handbook on Tensors in Geology*. Oxford: Oxford University Press.

Saada, A. S., 2009. *Elasticity Theory and Applications*. London: Roundhouse.

Verhoogen, J., Turner, F. J., Weiss, L. E., Wahrhaftig, C., and Fyfe, W. S., 1970. *The Earth*. New York: Holt, Rinehart and Winston.

# Displacement and strain

## 3.1 Aims

This chapter gives an overview of the concepts of displacement and strain. In textbooks on structural geology, strain is normally treated in great detail. The reason is that strain, particularly finite strain, is of fundamental importance for understanding the kinematics of ductile deformation, such as occurs during folding. Here, however, the focus is on those aspects of strain that are useful when analysing stress–strain relations for brittle rocks, which to a first approximation behave as linearly elastic and are subject to small (usually less than 1%) strain. The main aims of this chapter are to:

- Explain the difference between displacement and strain.
- Provide examples of the use of each concept in geology.
- Explain the difference between deformation and strain.
- Discuss general aspects of infinitesimal strain.
- Provide some basic formulas for measuring strain.
- Discuss strain rates and how they are measured.

## 3.2 Basic definitions

**Displacement** refers to the change in position of a particle. When the displacement is very small, it is referred to as infinitesimal; when large, it is referred to as finite. **Strain** is also related to the displacement of particles from their original position to a new position. Strain and displacement are thus closely related. There is, however, a significant difference. Strain always involves changes in the internal configuration of the body. During strain, the distances between the particles that constitute the strained body change.

Consider, for example, the basaltic lava flow in Fig. 3.1. When this lava flow becomes strained, such as during loading through plate pull or, as seen here, dyke injection, there will be some changes in the distances between the vesicles (the pores). That is, the internal configuration of the rock changes and it becomes strained. This is partly due to the magma overpressure in the dyke (Fig. 2.13). The basaltic lava flows outside a rift zone such as in Iceland (Fig. 2.10) are continuously being displaced as a result of divergent plate movements. However, far from the rift zone the vesicles or amygdales within the lava flows do not change their relative positions during the displacement. Thus, the lava flows of the

Fig. 3.1 Basaltic dyke emplaced in a basaltic lava flow (a part of the lava pile in East Iceland). The dyke is driven by magma overpressure, that is, pressure in excess of the minimum principal compressive stress in the host rock, $\sigma_3$. For a nearly vertical dyke such as this one, $\sigma_3$ is horizontal. The magma overpressure generates horizontal compressive stress in the wall rock, so that, following the dyke emplacement, the horizontal stress may temporarily change from $\sigma_3$ to $\sigma_2$ or $\sigma_1$. The horizontal stress generates strain in the host rock. In this section, the dyke is a non-feeder, as indicated by the exposed tip, that is, this dyke segment became arrested at about 1000 m below the surface of the rift zone at the time of dyke emplacement.

old parts of Iceland, or on the ocean floor far away from ridges and subduction zones, are subject to displacement that involves essentially no strain. These lava flows are thus carried away from the divergent plate boundaries as, effectively, a part of a rigid body.

Rigid displacement of a rock body, such as a lava flow, where there is no change in the internal configuration of the body, is **deformation** but not strain. Such ideal displacements of a body are referred to as **rigid-body translation** and **rigid-body rotation** (Fig. 3.2). There can thus be displacement of rock particles without strain, but there cannot be strain without displacement of particles. We can define these concepts as follows:

- Rigid-body translation. There is no internal distortion. The displacement vectors are equal. Approximate examples include non-rotational plate movements, ideal movements of fault blocks, and ideal rise and subsidence of crustal segments.
- Rigid-body rotation. There is no internal distortion. The displacement vectors are unequal. Approximate examples include ideal tilting of strata, ideal book-shelf faulting, and tilting of buildings during subsidence or seismogenic fault slip (Fig. 3.3).

When a rigid body is displaced, there is no change in its internal configuration: the distances between its particles remain the same. By contrast, when a body is strained the distance between its particles change: the particles move by different amounts and, commonly, in different directions. A rigid body strictly has infinite stiffness, that is, its Young's modulus

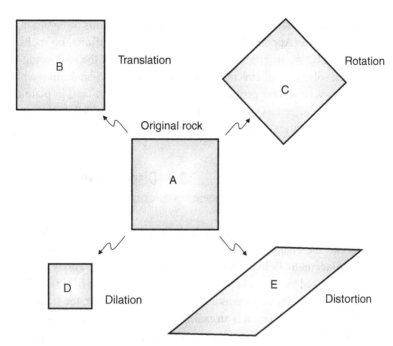

**Fig. 3.2**   Rock body A can deform in several ways. When the body is modelled as rigid, it may be translated (moved) B or rotated C without any change in its internal configuration. When, however, the internal configuration of the body changes, it becomes strained. The two types of strain are dilation D and distortion E.

**Fig. 3.3**   Example of an essentially rigid-body rotation. Slip on a reverse fault, generating a large earthquake in Taiwan in 1999, which runs under a house. The house was so well built that it suffered comparatively little damage (strain) during the earthquake slip and responded to the fault slip primarily through rigid-body rotation. Photograph: Jacques Angelier.

is infinite (Chapter 4). Because the Young's modulus of common, solid rocks is generally 1–100 GPa (Appendix D), which is clearly not infinite, no rocks (and, in fact, no known materials) are strictly rigid. But the assumption of the rock being rigid is sometimes made when solving problems in geology; for example, when dealing with some aspects of hydrogeology, structural geology, and plate tectonics. In fact, in plate tectonics the plates were initially defined as rigid except at their boundaries. While this assumption is still made in plate kinematics, it has gradually been relaxed when dealing with plate dynamics.

## 3.3 Displacement

Displacement is commonly shown by arrows, indicating **displacement vectors**. Each vector shows the change in the position of a rock particle. An array of such arrows defines a **displacement field**. When all the particles in a rock body show the same displacement, represented by parallel arrows of the same length, the field is said to be homogeneous. If, by contrast, the displacements among the particles differ, the field is said to be heterogeneous. Rigid-body translation is an example of a homogeneous displacement field. For rock bodies and crustal segments, no displacement field is strictly homogeneous. However, along parts of a rock body, for instance, a crustal segment, the field may be regarded as approximately homogeneous. For example, some displacement fields, as represented by the spreading vectors associated with ocean-ridge segments but at certain distances from the active rift zones themselves, are approximately homogeneous.

Within the active rift zones, however, the short-term displacement fields are generally heterogeneous. This follows because much of the deformation in the rift zones is related to faulting and dyke emplacement during particular (discontinuous) rifting events. The crust shows approximately elastic and thus effectively instantaneous response to the stress changes associated with these processes. As an example, consider the displacements generated as a result of feeder-dyke emplacement in a rift zone such as in Iceland (Figs. 3.4–3.7). The displacements are large close to the dyke, and decrease away from the dyke (cf. Fig. 2.13). Commonly, the displacements on either side of the dyke are small at distances similar to the strike dimension (horizontal length) of the dyke at the surface. This follows partly from a well-known principle in elasticity, namely **Saint-Venant's principle**. Effectively, this principle states that stress changes due to loading of a small part of a body are local and negligible at distances that are large in comparison with the dimension of the loaded part. For a feeder dyke, the dimension of the loaded part of the crust is primarily the strike dimension of the dyke.

While the example in Fig. 2.13 is hypothetical, very similar displacement fields have been observed around dykes in rift zones worldwide. For example, the Krafla rifting episode in North Iceland in 1975–1984 (Figs. 3.4–3.7) resulted in multiple dyke injections, producing a total maximum feeder-dyke thickness of 9 m (Fig. 3.5). Only part of the dyke reached the surface as a feeder, but the variation in opening displacement or dilation along the fissure swarm (Fig. 3.6) indicate that the dyke is segmented and the segments are not in line, not collinear, but rather offset (Gudmundsson, 1995; Sigmundsson, 2006). The effect of offset

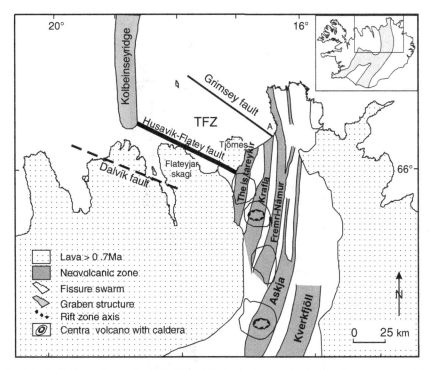

**Fig. 3.4** Location of the Krafla Fissure Swarm (the Krafla Volcanic System) where a major dyke emplacement occurred through tens of dyke injections in the Krafla Fires, 1975–1984. Nine of the injections reached the surface (Fig. 3.5). The Grimsey fault, the Husavik-Flatey fault, and the Dalvik fault form a part of an oceanic transform fault, the Tjornes Fracture Zone (TFZ). The Krafla Swarm is 80–100 km long and the entire swarm was active during the Krafla Fires.

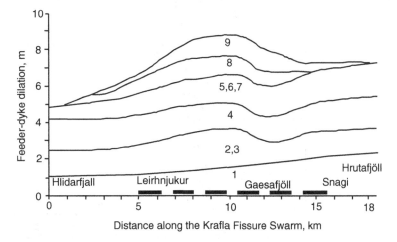

**Fig. 3.5** Nine feeder-dyke injections in the Krafla Swarm generated a dyke with a maximum surface thickness of about 9 m. The final length of the dyke at the surface (composed of offset segments) is indicated by the broken line from Leirhnjukur to Snagi and is about 11 km. Unpublished data from Eysteinn Tryggvason (cf. Gudmundsson, 1995; Sigmundsson, 2006).

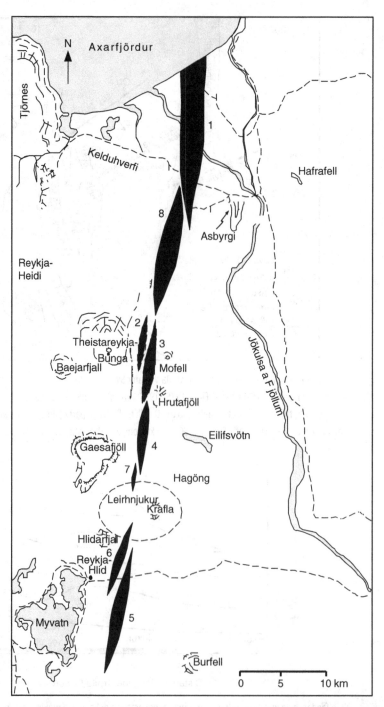

**Fig. 3.6**   Much of the surface extension (dilation) during the Krafla Fires was associated with opening of tension fractures (Fig. 1.1) and normal faults at the surface, and subsurface dyke emplacement. The main surface fracturing and extension (black areas) during eight rifting events (1–8) up to 1978 is indicated here. Data from Sigurdsson (1977).

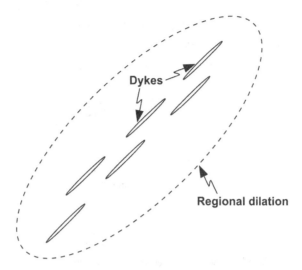

**Fig. 3.7**     Several offset dyke injections (Fig. 3.6), particularly if the segments are close in relation to their strike dimension, generate a crustal extension similar to that from a single dyke of a length equal to the total length of the segmented dyke (Chapter 13). The regional extension, and associated strain and stress, associated with the dyke emplacement is mostly limited to an elliptical zone of a dimension similar to that of the dyke.

segments of a dyke, or any other rock fracture, on the associated displacement field is commonly as if the fracture was single and of a strike dimension approximately equal to the distance between far ends or tips of the offset fracture segments. This is particularly so if the distances between the offset segments are small in comparison with their strike dimensions (Fig. 3.7). The dyke segments generate a dilation, or, in plate-tectonic terms, spreading, out to a certain distance from the segments (Fig. 3.7).

The overpressure (driving pressure) associated with the dyke in Fig. 3.6 generated an 'excess spreading rate' out to a distance of about 100 km, which is similar to the total length of the emplaced dyke (Bjornsson, 1985; Sigmundsson, 2006). Overpressure of emplaced dykes is one reason for the so-called **ridge-push** at divergent plate boundaries. Similar effects and displacement fields have been observed around large-scale dyke emplacement in the East African Rift (Hamling *et al.*, 2009). The displacement fields around feeder dykes are clearly heterogeneous since they vary in magnitude and trend within the plate segments affected by the intrusion.

Other examples of displacement fields measured in the horizontal plane are those associated with deformation of the Earth's surface above active magma chambers (Fig. 3.8). The ideal displacement field at the surface above a spherical chamber is radial from the projection of the centre of the chamber onto the surface. That is, from a point on the surface right above the centre of the chamber. This ideal field, however, is rarely seen in volcanoes during periods of unrest. The observed field is normally somewhat asymmetric. This is partly because few chambers are exactly axisymmetric. And even if many chambers are close to being ellipsoidal during parts of their lifetimes, particularly during their end stages (Gudmundsson, 1990), there are normally some irregularities in the geometry of an active

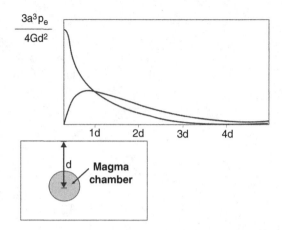

Fig. 3.8 Vertical and horizontal displacement around a spherical magma chamber (a point source) according to the Mogi model in volcanology (Mogi, 1958). Here $G$ is the shear modulus, $d$ is the depth of the magma chamber below the free surface, $a$ is the radius of the chamber, and $p_e$ is its excess magma pressure.

chamber. These irregularities include, for example, faulted parts of the roof, intrusions of various sizes and shapes associated with the chamber, as well as general chamber-shape variations, all of which result in local stresses and displacement that deviate from the ideal (Chapter 6). Such irregularities, however, do not much affect the surface displacement field if the diameter of the irregularity is less than the depth to the cavity (Saint-Venant's principle). But if the irregularity is large in comparison with its depth, the surface displacement can be much affected.

Host-rock heterogeneities may also have a large effect on the displacement field, such as during unrest in a volcano. During magma-chamber inflation (expansion) or deflation (contraction), the stiff rocks in the volcano, for example gabbro bodies or other mafic intrusions, become highly stressed but experience comparatively small displacements. By contrast, the compliant rocks, such as many pyroclastic and sedimentary layers, deform more easily and therefore show larger displacements (for a given magma pressure). Even if the magma chamber is close to axisymmetric, the associated surface displacement field may for this reason be somewhat asymmetric.

Many dip-slip surface faults illustrate well variations in the vertical displacement field. Generally, the vertical displacement or throw varies from the centre of the fault to its lateral tips (Chapter 9). It follows that the displacement vectors vary along the strike dimension of the fault at the surface. In fact, this applies to any rock fracture: each fracture shows a variation in its displacement along the strike dimension (as well as along its dip dimension) (Chapter 9).

As an example, consider the main fault in Fig. 3.9. This is a normal fault dissecting a Holocene lava flow in the Krafla Fissure Swarm in North Iceland (Fig. 3.4). A remarkable feature of this fault is that its vertical displacement decreases rapidly on approaching the tip of one of its segments. In fact, the displacement decreases from 32 m to 0 m in just over 300 m, a rate of reduction in displacement of about 1 m for every 10 m along strike. The

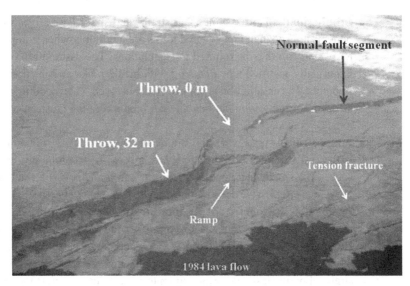

**Fig. 3.9** Abrupt variation in vertical displacement, throw, along one of the boundary normal faults of the Krafla Fissure Swarm (Fig. 3.4). View northwest, the fault is located in a Holocene pahoehoe lava flow (black lava flow, erupted in 1984, is seen at the lower margin of the photograph). The vertical displacement of the fault changes from 32 m to 0 m in just over 300 m along the length of the fault. The fault is composed of offset segments. A sloping surface between normal-fault segments, referred to as a relay ramp, is indicated.

displacement field is thus clearly heterogeneous. The rapid change in displacement over a short distance, as well as the mechanical interaction between the fault segments (Fig. 2.12), has resulted in variation in the slope or curvature of the surface close to the fault and, therefore, in changes in the internal configurations of the particles within the Holocene lava flow, that is, in strain.

# 3.4 Strain

When the particles of a rock body change position relative to one another, so that some points move closer together and others move further apart, the body is strained. More specifically, strain is the fractional change in a dimension of a non-rigid rock body when subject to loads. Some basic characteristics of strain may be summarised as follows:

- **Strain** is a pure **geometric concept**: it is the ratio of the change in configuration of a body to its initial configuration. In contrast to stress, which specifies the conditions in a body at an instant, strain compares the configurations of the body at two specified instants. There are two types of strain: dilation and distortion.
- **Dilation** denotes the change in size of the strained body; the body either increases in size (expands) or decreases in size (shrinks). **Distortion** denotes the change in shape of the strained body.

- **Infinitesimal strain** assumes that strains and space derivatives of displacement components are so small that squared terms are negligible when compared with first-power terms and can thus be omitted in calculations. The assumption of infinitesimal strains is normally justified when dealing with brittle rocks and in most applications to rock fractures. The assumption is not justified, however, when dealing with large strains, such as occur during folding and ductile deformation in the crust, in which case finite strain must be considered.
- **Finite strain** is used when the strains and displacements are comparatively large. This is the end strain, as observed today, in a rock. It is commonly brought about through successive small, incremental strains. It can sometimes be thought of as the sum of numerous progressive deformations, each of which involves a small, infinitesimal, strain. Finite strain is also referred to as **total strain**. The primary difference between finite and infinitesimal strain is thus the size or magnitude of the strain.
- **Engineering** or **nominal strain** are terms commonly used for small strains. These are based on the theory of infinitesimal strains and are thus applicable only to very small strains. For an elastic deformation up to rock failure (fracture), engineering strains give reasonably accurate answers and are used in this book unless stated otherwise.
- **Natural strain**, **logarithmic strain**, or **true strain**. This is a more useful and accurate measure of strain for the large strains, such as are common in geology, than engineering strains. Up to strains of about 0.1 (10%), engineering and true strains differ only by a small amount. At strains greater than about 0.1, however, the difference between the two types of strains becomes gradually larger. The difference between these strains is as follows:

$$\text{true strain} = \ln(1 + \text{engineering strain})$$

For strain, like stress, there are normal and shear components. **Normal strain** is the ratio of the change in length of a line to its original length. For an elongated body, say a bar-shaped piece of rock, the extension for a given load such as constant tensile stress $\sigma$ depends on the Young's modulus (stiffness) of the rock and on the bar's length. If the initial length of the bar is $L$, then the extension will be $\Delta L$ (Fig. 3.10). If the original bar is twice as long, that is, the original length is $2L$, and of the same rock and subject to the same load, then the elongation will be $2\Delta L$. Similarly, if the initial bar length is $3L$, the extension will be $3\Delta L$, and so on. Thus, the extension is proportional to the initial length of the extended linear structure. However, the ratio of the extension to the initial length is the same for all three bar lengths. That is, the normal or direct strain is

$$\varepsilon = \frac{\Delta L}{L} = \frac{2\Delta L}{2L} = \frac{3\Delta L}{3L} \tag{3.1}$$

Thus, the normal strain depends only on mechanical properties of the rock and the loading (here the tensile stress). Because Eq. (3.1) represents a length divided by a length, strain is a pure number and has **no units**. In geology, tensile strain is commonly regarded as negative and compressive strain as positive, the convention following the same rule as for stress.

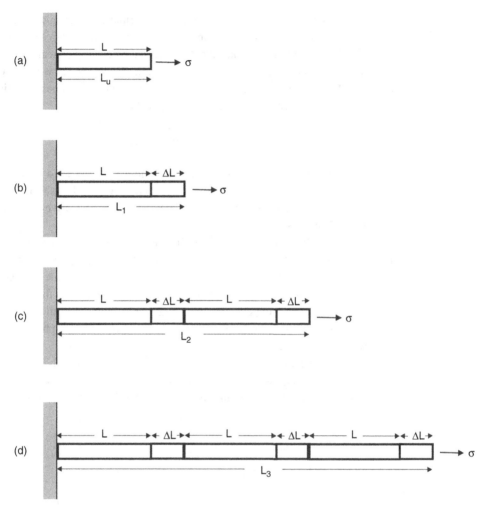

**Fig. 3.10** Bar of a rectangular cross-sectional shape extended by a tensile stress $\sigma$. The notation above the bar refers to Eq. (3.1), the notation below the bar to Eq. (3.17).

Using that convention, strains in Eq. (3.1) are negative, but the signs are omitted here to simplify the presentation of the basic concept.

Most strains in the brittle part of the crust, in contrast to those in the ductile part of the crust, are quite small. Commonly, therefore, we multiply the strains by 100, and give them as percentage (Chapter 1). Thus a normal strain of 0.001 would be equal to 0.1% strain. In calculations, however, we use the ratio itself, that is, the direct strain presented as the ratio of the extension or contraction to the initial length. Normal stresses thus produce changes in the linear dimensions of a rock body. If the normal stress is tensile, then there is extension; if the normal stress is compressive, then there is contraction.

The **shear strain**, denoted by $\gamma$, is the change in angle between two initially perpendicular lines (Fig. 1.7). If there were initially a grid of orthogonal lines (that is, the sets meeting at 90°) in the rectangle, then after the angular deformation the angles are $(90° - \gamma)$ and $(90° + \gamma)$. Here $\gamma$ is used for the change in angle from the initial 90°. The shear strain may thus be regarded as the displacement $\Delta x$ caused by a certain shear stress $\tau$ divided by the distance $y$ from the origin. From trigonometry, however, we know that $\Delta x / y = \tan \psi$, where $\psi$ is the angle that the deflected line of the grid makes with the initial line. Thus, the shear strain $\gamma$, measured in radians, may be defined as

$$\gamma = \frac{\Delta x}{y} = \tan \psi \tag{3.2}$$

which is Eq. (1.2). Thus, shear strain is, as it should be, the ratio between an angular change, that is, the displacement $\Delta x$, and the original length, $y$. The shear strain is regarded as positive when it is generated by positive shear stress, and negative when it is generated by negative shear stress.

The strain can also be analysed in three dimensions. Consider first normal strains. When a rock bar is stretched in the $x$-direction, it becomes elongated in that direction, by an amount $\Delta L$ (Eq. 3.1). However, as a result of the stretching in the $y$-direction, the bar becomes thinner. This is most noticeable when we stretch materials that tolerate unusually high strains while remaining elastic, that is, without failing in a brittle (rupture) or ductile (plastic) manner. One such material in daily use is rubber, which can be strained up to 3, or 300%, without rupturing.

During the elongation of the bar by $\Delta L$, there will thus also be a contraction in its transverse dimensions. If the original width of a rectangular bar is $W$ and its original height $H$, then the width decreases by an amount $\Delta W$ and the height by $\Delta H$ when the bar is

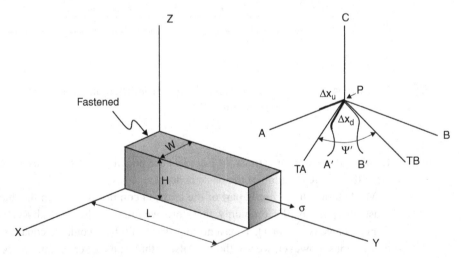

**Fig. 3.11**  Bar, of original dimensions $L$, $H$, and $W$, subject to tensile stress $\sigma$ is used for a general definition of normal strains (Eqs. 3.3–3.5) in the diagram on the left. A more formal definition of normal strain (Eqs. 3.7 and 3.8) and shear strain (Eqs. 3.9 and 3.10) is obtained from the diagram to the right. There TA and TB denote tangents to the deformed lines PA′ and PB′. The angle between the tangents is $\psi$, and $\Delta x_u$ refers to the undeformed (unstrained) state and $\Delta x_d$ to the deformed state.

stretched. Let the length of the bar be parallel with the $y$-axis of the coordinate system (Fig. 3.11), the width parallel with the $x$-axis, and the height parallel with the $z$-axis. We have then the following normal strains (again, the signs are omitted):

$$\varepsilon_{xx} = \frac{\Delta W}{W} \tag{3.3}$$

$$\varepsilon_{yy} = \frac{\Delta L}{L} \tag{3.4}$$

$$\varepsilon_{zz} = \frac{\Delta H}{H} \tag{3.5}$$

Similarly, the shear stresses in three dimensions follow from Eq. (3.2). In a similar way as we can define stress at a point, we can define **strain at a point** as the matrix or the **strain tensor**

$$\varepsilon = \begin{pmatrix} \varepsilon_{xx} & \gamma_{xy} & \gamma_{xz} \\ \gamma_{yx} & \varepsilon_{yy} & \gamma_{yz} \\ \gamma_{zx} & \gamma_{zy} & \varepsilon_{zz} \end{pmatrix} \tag{3.6}$$

There is also a symmetry among the shear strains, similar to the symmetry among the shear stresses, so that $\gamma_{xy} = \gamma_{yx}$, $\gamma_{zx} = \gamma_{xz}$, and $\gamma_{zy} = \gamma_{yz}$. It follows that there are only three independent shear strains. Thus, there are only six independent strain components, namely the three normal strains and three shear strains.

A more formal definition of strain can be given, as for stress, by taking the limit. Imagine three initially orthogonal material lines A, B, and C in a rock body that are initially parallel with the coordinate axes (Fig. 3.11). The lines meet at a point P. When the rock body becomes strained, the line A, for example, changes its length and shape to A′ (and line B changes to B′). Let the length of a part, a segment, of line A in the unstrained or undeformed body be denoted by $\Delta x_u$ and the length in the strained body be denoted by $\Delta x_d$. The subscript u refers to undeformed or unstrained state, and the subscript d to the deformed or strained state. The mean normal strain of the segment is then given by

$$\frac{\Delta x_d - \Delta x_u}{\Delta x} = \frac{\Delta u}{\Delta x} \tag{3.7}$$

Here $\Delta u$ denotes the change in length of the segment. The normal strain $\varepsilon_{xx}$ (and similarly for $\varepsilon_{yy}$ and $\varepsilon_{zz}$) can then be formally defined at point P as

$$\varepsilon_{xx} = \lim_{\Delta x \to 0} \frac{\Delta u}{\Delta x} = \frac{\partial u}{\partial x} \tag{3.8}$$

Similarly, the shear strain may now be defined as changes in the angles between the tangents to the lines at point P. For example, for lines A and B, the angle between the tangents at P after the deformation or strain is $\psi$ and the shear strain $\gamma_{xy}$ may be defined (and similarly for the two other independent shear strains) as (Fig. 3.11)

$$\gamma_{xy} = \tan\left(90° - \psi\right) \tag{3.9}$$

When $\psi$ is given in radians, Eq. (3.9) may be written as

$$\gamma_{xy} = \tan \psi \tag{3.10}$$

It should be noted that the notation $\gamma$ for the shear-strain components differs from the more general shear-strain tensor components by a factor of 2. In tensor notation, $\varepsilon_{xy}$, $\varepsilon_{yz}$, and $\varepsilon_{xz}$, or with other suffices, are used so that $\gamma_{xy} = 2\varepsilon_{xy}$, and so forth.

## 3.5  Deformation of a rock body

We have now described the deformation of a rock body as consisting either of rigid-body movement, where there is no change in the internal configuration of the body, or strain, where the internal configuration changes (Fig. 3.2). Here we briefly summarise the aspects of rock-body deformation that are referred to as homogeneous and heterogeneous.

- **Homogeneous deformation** implies that the amount of strain in all parts of the rock body is equal. It follows that all the lines that were parallel before the body was deformed remain parallel after its deformation. Also implied is that lines that were straight before deformation remain straight after deformation.
- **Heterogeneous deformation** (inhomogeneous deformation) implies that the amount of strain in different parts of the body is unequal. It follows that some (or all) lines that were parallel before body was deformed become non-parallel after its deformation. Also implied is that lines that were straight before deformation become curved after deformation (Fig. 3.11).

Rock deformation as reflected in finite strain, that is, the end stage of strain, is generally heterogeneous. The mathematical theory of heterogeneous strain is very complex. As a consequence, most finite-strain analyses in geology use the theory of finite homogeneous strain. Various markers are then used to try to determine the finite strain, on the assumption that it is homogeneous. Common structural markers include the following:

- oolites
- spherulites
- pebbles
- fossils
- worm tubes
- sun cracks
- cross bedding.

Unfortunately, the analysis faces several difficulties, many of which are well known and discussed by Ramsay (1967), Verhoogen *et al.* (1970), Means (1976), Davis and Reynolds (1996), and Ramsay and Lisle (2000). For example, there are, strictly, no rocks that undergo homogeneous, finite strain. There are always variations in the mechanical properties of the rock that result in different strains between different parts of the rock, even if the stresses responsible for the strains are uniform. Also, the initial shapes of the bodies used as markers for strain estimates are commonly poorly known. For example, we may assume an oblate-ellipsoidal, strained body to have been initially spherical. However, many oblate bodies, such as many pebbles, were initially oblate ellipsoids. Also, because the mechanical properties of the strained bodies, such as fossils, are generally very different from those of the host rock, the strain in the fossil can be very different from, and thus a poor indication of, the strain in the host rock.

In rock-fracture mechanics the strain considered is normally infinitesimal. Thus, finite strain, and the problems associated with assessing finite strain in various rock layers, is generally of little or no concern when analysing rock fractures. For the sake of completeness, however, some of the main formulas for measuring strain are listed in the next section.

## 3.6 Measuring strain

### 3.6.1 Elongation

Elongation is the ratio of the change in length of a material line or a rock bar, as a result of loading, to its initial length. This is also referred to as longitudinal strain or, more commonly, normal strain. From Eqs. (3.1) and (3.4), we may write the equation for elongation as follows:

$$\varepsilon = \frac{\Delta L}{L_u} = \frac{L_d - L_u}{L_u} \tag{3.11}$$

where now, in accordance with the notation in Eq. (3.7), we introduce the subscripts u and d to emphasise that $L_u$ refers to the undeformed original length and $L_d$ to the deformed length. In the extension-negative convention, the strains calculated from Eq. (3.11) have negative signs (as Eq. 3.1). We make the following observations:

- The strain is given as a fraction, say 0.01. If we want the result in percentage, the fraction must be multiplied by 100 (Chapter 1).
- If there is a complete compression of a material line, then $L_d = 0$. It follows from Eq. (3.11) that, in this case, $\varepsilon = -1$. In the extension-negative convention, however, the result $\varepsilon = -1$ should be multiplied by $-1$ and thus yield $\varepsilon = 1$.
- If there is complete extension, that is, the deformed material line becomes infinitely long, then $L_d = \infty$ and, from Eq. (3.11), $\varepsilon = \infty$ (or $-\infty$ for extension negative).
- From these considerations it follows that the strain range is from minus one to plus infinite, or in symbols: $-1 \leq \varepsilon \leq \infty$ for extension positive and for extension negative (somewhat awkwardly) $-\infty \leq \varepsilon \leq 1$.

### 3.6.2 Stretch

Stretch is the ratio of the deformed line length to the undeformed line length. Using $L_u$ for the undeformed line length and $L_d$ for the deformed length, then the stretch, $S$, is given by (Example 3.4)

$$S = \frac{L_d}{L_u} = (1 + \varepsilon) \tag{3.12}$$

We make the following observations:

- Stretch is always positive, even if the deformed material line length is smaller than the initial line length.
- If there is no change in the material line length, that is, no strain and $L_d = L_u$, then $S = 1$.
- From this, and Eq. (3.12), it follows that the range of the stretch is: $0 \leq S \leq \infty$.

### 3.6.3 Quadratic elongation

Quadratic elongation $\lambda$ is simply the square of the stretch, that is,

$$\lambda = \left(\frac{L_d}{L_u}\right)^2 = (1 + \varepsilon)^2 \tag{3.13}$$

We make the following observations:

- The range of the quadratic elongation is the same as that for the stretch, that is: $0 \leq \lambda \leq \infty$.
- Clearly, $\lambda$, $S$, and $\varepsilon$ are all related and thus not independent. If one of these strain measures is known, then the other two can be calculated. They all measure relative changes in material line lengths.

### 3.6.4 Dilation and deviatoric strain

**Dilation** (sometimes named dilatation) gives the ratio of the volume change to the initial, undeformed volume of the rock body. It is also referred to as volume strain. Dilation, denoted by $\Delta$, is thus a measure of size change, that is, fractional volume change, and is given by

$$\Delta = \frac{V_d - V_u}{V_u} = \frac{\Delta V}{V_u} = \varepsilon_{xx} + \varepsilon_{yy} + \varepsilon_{zz} \tag{3.14}$$

Consider the case when all the normal strains are equal, that is, the strain field is **isotropic** (or hydrostatic or spherical) and the shear strains of the rock body are zero. Since all the sides of the body change their dimensions in the same proportion, there is no distortion of the material lines of the rock body. And because the shear strains are zero, there are also no angular distortions of the body. It follows that during straining the rock body does not change its shape, only its volume. The change in volume is then due to the three normal strains in Eq. (3.14), all of magnitude $\frac{1}{3}\Delta$.

Any total strain can, in fact, be split into such a **volume change** or dilation (spherical strain) and **deviatoric strain**. The deviatoric strain is obtained by subtracting the isotropic strain field from the general strain field represented by Eq. (3.6) so as to get the deviatoric strain field matrix

$$\begin{matrix} \varepsilon_{xx} - \frac{1}{3}\Delta & \gamma_{xy} & \gamma_{xz} \\ \gamma_{yx} & \varepsilon_{yy} - \frac{1}{3}\Delta & \gamma_{yz} \\ \gamma_{zx} & \gamma_{zy} & \varepsilon_{zz} - \frac{1}{3}\Delta \end{matrix} \tag{3.15}$$

When the body is oriented along the principal strain axes, then matrix (3.15) reduces to

$$
\begin{matrix}
\varepsilon_{xx} - \frac{1}{3}\Delta & 0 & 0 \\
0 & \varepsilon_{yy} - \frac{1}{3}\Delta & 0 \\
0 & 0 & \varepsilon_{zz} - \frac{1}{3}\Delta
\end{matrix}
\tag{3.16}
$$

### 3.6.5 Natural strain

In the theory of infinitesimal strain, it is assumed that the rock deformation and the resulting strains are small. In other words, it is assumed that the strains are elastic. This is normally the case for most applications to rock-fracture mechanics, until the rock fails, and is a good first approximation in many analyses. However, when the deformation becomes ductile and the strains large, the assumption of infinitesimal strains is no longer tenable. Large strains involving ductile deformation are very common in geology. Most of the processes involved, however, are of little concern for rock-fracture mechanics. But they affect many rocks, so it is important to have an idea of how strain is defined for such large deformations.

The limitation of the definition used for infinitesimal strain, as given by Eqs. (3.1) and (3.3)–(3.5), when applied to large strains can be illustrated as follows. Consider a rock bar of initial length $L_u$. Let the bar be loaded in tension, that is, stretched until its length is doubled. For an elastic rock so large a stretch would not be possible, because the rock would fail, fracture, when the strain reached about 1–2%. However, for rocks that behave as ductile an elongation of this size, that is, extensional strain of 1.0 (100%), is entirely possible. If the same rock bar were then to be compressed by 100%, its length would have to be reduced to zero thickness. The strain in this shortening, according to the definition in Eqs. (3.1) and (3.3)–(3.5), would still be 1.0 (100%). However, a shortening of this magnitude would obviously alter the internal structure of the rock bar far more than extension of the same magnitude. Thus, even if the magnitude of the strain for the shortening and the extension is the same, the real strain is different.

Consider, for example, the bar in Fig. 3.10. Let's now think of the bar as a single one of initial length $L_u$. During extension, the normal strain is then calculated with reference to the initial length, $L_u$. However, if the initial length of the rock bar was $L_3$ and it was subject to compression, then the strain would be referred not to the length $L_u$ but rather to the length $L_3$. The absolute magnitude of the strain during extension from $L_u$ to $L_3$ would then be different from that calculated during compression from $L_3$ to $L_u$. But we would prefer these strains, while with opposite signs, to be of equal magnitude. To avoid this difference, we use the concept of **natural strain**, which is also referred to as logarithmic or true strain.

Natural strain is used for progressive or incremental strain, and primarily to analyse large strains. It is a measure of elongation that can be considered a series of small-strain increments. The incremental elongation is then divided by the length of the material line at that time (the instantaneous length). The natural strain $\varepsilon_N$ can be regarded as increments of engineering (infinitesimal) strain, namely (Fig. 3.10)

$$
\varepsilon_N = \left( \frac{L_1 - L_u}{L_u} \right) + \left( \frac{L_2 - L_1}{L_1} \right) + \left( \frac{L_3 - L_2}{L_2} \right)
\tag{3.17}
$$

where each term on the right-hand side of the equation is an incremental strain. For a very small (infinitesimally small) increment, we have $\Delta L = dL$ and can write

$$\varepsilon_N = \sum_{L=L_u}^{L=L_d} \frac{dL}{L} = \int_{L_u}^{L_d} \frac{dL}{L} = \ln\left(\frac{L_d}{L_u}\right) \tag{3.18}$$

Where $L_u$ and $L_d$ are the undeformed (initial) and deformed (final) lengths of the rock bar. On comparing Eq. (3.18) and Eqs. (3.11)–(3.13), we see that

$$\varepsilon_N = \ln(1 + \varepsilon) = \ln S = \tfrac{1}{2}\ln \lambda \tag{3.19}$$

To see how the definition of natural strain avoids the difference in magnitude between compression and extension, consider first the case of a rock bar extended to double its initial length so that $L_d = 2L_u$. Then it follows from Eq. (3.18) that $\varepsilon_N = \ln 2 = +0.693$. Consider next the case of a rock bar compressed to half its initial length, so that $L_d = \tfrac{1}{2}L_u$. Then, from Eq. (3.18), we have $\varepsilon_N = \ln 0.5 = -0.693$. Thus, as we prefer, the magnitude of the strain is the same in both cases.

We make the following further observations about natural strain and its relation to engineering (infinitesimal) strain:

- The range is from $\ln(0)$ to $\ln(\infty)$, that is: $-\infty < \lambda < \infty$.
- For homogeneous, isotropic material there is a linear relation between the natural strain and the applied stress.
- Incremental natural strains, Eq. (3.17), are tensor quantities regardless of the final strain.
- When the strain is less than about 0.1 (10%), the difference between engineering and natural strain is generally negligible.

## 3.6.6 Strain rates

Strain rate refers to the change in strain per unit time. The time unit used is the second. For normal strain, the rate is given as

$$\dot{\varepsilon} = \frac{d\varepsilon}{dt} \tag{3.20}$$

where the dot above $\varepsilon$ denotes a derivative with respect to time, as is standard in physics. Similarly, for shear strain the rate is given as

$$\dot{\gamma} = \frac{d\gamma}{dt} \tag{3.21}$$

We make the following further observations about strain rates:

- The unit of measurement of a strain rate is strain over time, that is, $t^{-1}$.
- A typical geological strain rate is $10^{-14}\,\mathrm{s}^{-1}$ (1 cm/year). This is, for example, a typical strain rate associated with spreading at mid-ocean ridges.
- The behaviour of a rock body depends on the strain rate to which it is subject. If the strain rate is high, the rock behaviour becomes more brittle, whereas if the strain rate is low, the rock behaviour becomes more ductile.

# 3.7 Summary

- Displacements, the change in position of rock particles, are usually indicated by arrows, called displacement vectors. A set of arrows defines a displacement field. Displacement fields show, for example, the velocity and direction of the spreading vectors at divergent plate boundaries. When all the particles in a rigid body move by the same amount as the rock body as a whole, the displacement is homogeneous and the displacement vectors are equal. This is rigid-body translation. When the displacement vectors in a rigid body are unequal, the result is rigid-body rotation. If the displacement vectors differ in magnitude and orientation, the displacement field is heterogeneous. Examples of heterogeneous and non-rigid local displacement fields are those generated by dyke emplacement and faulting in rift zones. Such fields give rise to strain.
- Strain means change in the distances between particles in a body, namely a fractional change in its dimensions, and has no dimensions (is a pure number). It is a pure geometric (kinematic) concept, of which there are two main types. One is dilation, whereby the body either expands or shrinks (changes its size). The other is distortion, whereby the shape of the body changes. There are many measures of strain, such as elongation, stretch, quadratic elongation, dilation, deviatoric strain, and natural strain.
- Strain rate is a measure of the change in strain during time. The time used is the second. The unit of measurement is $t^{-1}$, that is, the reciprocal of time. Rock behaviour is strongly dependent on the strain rate to which the rock is subject. If the strain rate is high, the rock may behave as brittle, whereas if the strain rate is low, the rock may behave as ductile (plastic). A typical geological strain rate, such as the rate of spreading at divergent plate boundaries, is $1 \times 10^{-14} \, \text{s}^{-1}$.

# 3.8 Main symbols used

| | |
|---|---|
| $H$ | height of a rectangular bar |
| $\Delta H$ | change in height of a rectangular bar |
| $L$ | length, length of a rectangular bar |
| $\Delta L$ | extension, change in length of a rectangular bar |
| $L_u$ | undeformed original length |
| $L_d$ | deformed length |
| $L_{1,2,3}$ | lengths (of a bar) |
| $S$ | stretch (a measure of strain) |
| $W$ | width (thickness) of a rectangular bar |
| $\Delta W$ | change in width (thickness) of a rectangular bar |
| $\gamma$ | shear strain |
| $\dot{\gamma}$ | shear-strain rate |
| $\gamma_{ij}$ | components of shear strain |
| $\Delta$ | dilation, dilatation (a measure of strain), change in a quantity |

$\varepsilon$    strain tensor
$\varepsilon$    normal strain
$\dot{\varepsilon}$    normal-strain rate
$\varepsilon_{ii}$   components of normal strain, normal strains
$\varepsilon_N$   natural strain
$\lambda$    quadratic elongation (a measure of strain)
$\psi$    angle

# 3.9  Worked examples

## Example 3.1

### Problem

The rod in Fig. 3.12 is extended 6 mm by a tensile force of 100 kN. Find the stress and strain in the rod.

### Solution

A force of 100 kN = $1 \times 10^5$ N. The rod has a circular cross-section with a radius $R$ of 12.5 mm. The rod has a cross-sectional area $A$ of

$$A = \pi R^2 = \pi (12.5 \text{ mm})^2 = 491 \text{ mm}^2 \doteq 4.91 \times 10^{-4} \text{ m}^2$$

From Eq. (1.1) the tensile stress in the rod is

$$\sigma = \frac{F}{A} = \frac{1 \times 10^5 \text{ N}}{4.91 \times 10^{-4} \text{ m}^2} = 2.04 \times 10^8 \text{ N m}^{-2} = 204 \text{ MN m}^{-2} = 204 \text{ MPa}$$

The original length of the rod is $L = 6$ m, whereas the increase in length because of the tensile force is $\Delta L = 6$ mm $= 6 \times 10^{-3}$ m. From Eqs. (3.1) and (3.11) the strain in the rod is

$$\varepsilon = \frac{\Delta L}{L} = \frac{6 \times 10^{-3} \text{ m}}{6 \text{ m}} = 0.001 = 0.1\%$$

While this strain, 0.1%, would in theory be tolerated by many rocks without failure (fracture), no rock rod could sustain a uniaxial tensile stress of 204 MPa without brittle failure. So this high stress could not be reached in a rod made of rock, although many metals and other materials may tolerate tensile stresses that are several hundred megapascals or more (Ashby and Jones, 2005).

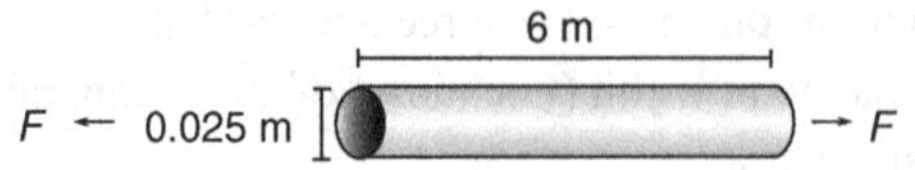

**Fig. 3.12**   Information for Example 3.1

---

### Example 3.2

Problem

A linear fossil has a stretch of $S = 0.8$. Find its:

(a) elongation
(b) quadratic elongation
(c) natural strain.

Solution

(a) From Eq. (3.12) the stretch $S$ is given by

$$S = \frac{L_d}{L_u} = (1 + \varepsilon)$$

It follows that the elongation $\varepsilon$ is

$$\varepsilon = S - 1 = 0.8 - 1 = -0.2$$

Thus, elongation is strictly negative because tensile stress and strain is negative, although commonly we omit the signs and just consider the magnitudes (absolute values) of the stresses and strains.

(b) From Eq. (3.13) the quadratic elongation $\lambda$ is given by

$$\lambda = \left(\frac{L_d}{L_u}\right)^2 = (1 + \varepsilon)^2 = S^2 = (0.8)^2 = 0.64$$

(c) From Eq. (3.19) the natural strain $\varepsilon_N$ is given by

$$\varepsilon_N = \ln(1 + \varepsilon) = \ln S = \tfrac{1}{2} \ln \lambda = \tfrac{1}{2} \ln(0.64) = -0.45$$

---

### Example 3.3

Problem

The Thingvellir Fissure Swarm forms a part of the Holocene rift zone in Southwest Iceland. A map of the swarm is provided in Fig. 3.13, and part of it is seen in the photograph in Fig. 3.14. The swarm consists of normal faults and tension fractures located in a Holocene (about 10 000 year-old) pahoehoe lava flow. The total elongation due to faults and tension fractures in the profile in Fig. 3.14 is 100 m. (In plate-tectonic context, such an elongation across a divergent plate boundary or a rift zone is commonly referred to as dilation although it is measured only in profiles (scan lines) at the surface and is thus not a three-dimensional dilation as defined in Eq. 3.14). Before the fractures formed with the resulting dilation the Holocene lava flow hosting the fissure swarm was 8 km wide so that its width has increased by 100 m due to the dilation. For the Holocene lava flow, find:

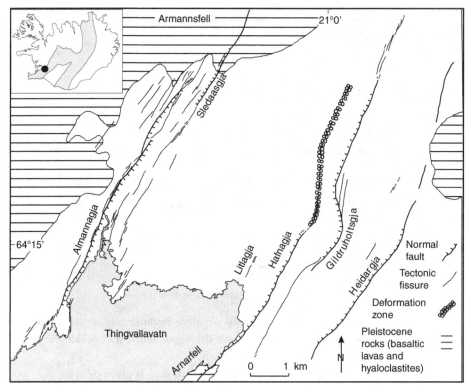

**Fig. 3.13**    Tectonic and geological background for Example 3.3. The Thingvellir Fissure Swarm, of Holocene age, is about 8 km wide and dissected by more than a hundred tension fractures and normal faults. Part of the graben is occupied by Lake Thingvallavatn.

**Fig. 3.14**    Aerial view of the Thingvellir Fissure Swarm, as seen in winter (with snow cover). View north, the main boundary faults of the graben are indicated. The profile where the extension was measured is close to the lake.

(a) the stretch
(b) the elongation
(c) the strain rate
(d) the time it will take part of the rift-zone crustal segment hosting the fissure swarm to double its width (to 16 km) at the strain rate obtained in (c).

Solution

(a) The undeformed width of the swarm is 8000 m ($L_u$), the deformed width is $8000 + 100 = 8100$ m ($L_d$). From Eq. (3.12) the stretch is

$$S = \frac{L_d}{L_u} = \frac{8100 \text{ m}}{8000 \text{ m}} = 1.01$$

(b) From Eq. (3.11) the elongation is

$$\varepsilon = \frac{\Delta L}{L_u} = \frac{L_d - L_u}{L_u} = \frac{8100 \text{ m} - 8000 \text{ m}}{8000 \text{ m}} = 0.01$$

We see immediately that the elongation could also be calculated directly from the stretch as $S - 1 = 1.01 - 1 = 0.01$.

(c) Strain rate is the strain that has occurred in the rock (here the extension in the profile across the fissure swarm) divided by the time, in seconds, that it took the strain to occur. The number of seconds in a year is about $3.15 \times 10^7$, so that the number of seconds in 10 000 years is about $3.15 \times 10^{11}$. Since the elongation (extension) is 0.01, we get the strain rate from Eq. (3.20) as

$$\dot{\varepsilon} = \frac{d\varepsilon}{dt} = \frac{0.01}{3.15 \times 10^{11} \text{s}} = 3.2 \times 10^{-14} \text{ s}^{-1}$$

This is a typical geological strain rate. If we translate it into perhaps more easily grasped measures, then we could say that this rate corresponds to about 0.01 metres per year or 1 centimetre per year. These are common spreading rates for very slow spreading ridges. This is in agreement with the observations that the rift zone of Iceland is a sort of slow-spreading ridge. In addition, in South Iceland it is split in two zones (Fig. 3.13, inset), each of which would get roughly half the long-term spreading. Geodetic measurements over the past few decades indicate that the current spreading rate in the Thingvellir Fissure Swarm is somewhat less than this (Sigmundsson, 2006), but a rifting episode with dyke injections could dramatically change the short-term local spreading rate, as happened in North Iceland during the Krafla Fires (Figs. 3.4–3.6).

(d) If the total width of the crustal segment hosting the fissure swarm is to double in width at the spreading rate calculated in (c), then from (b) the extension or elongation would have to be

$$\varepsilon = \frac{\Delta L}{L_u} = \frac{L_d - L_u}{L_u} = \frac{16\,000 \text{ m} - 8000 \text{ m}}{8000 \text{ m}} = 1.0 = 100\%$$

From (c) we have the spreading rate which is equal to the reciprocal of time, that is, $t^{-1}$. Using this information and Eq. (3.20) we have

$$\dot{\varepsilon} = \frac{d\varepsilon}{dt} = \frac{1}{t} = 3.2 \times 10^{-14} \text{ s}^{-1}$$

Solving for the time $t$ we get

$$t = \frac{1}{3.2 \times 10^{-14}\,\text{s}^{-1}} = 3.1 \times 10^{13}\,E = 984\,000 \text{ years}$$

Thus, it will take close to a million years to double the width of this rift-zone segment in Southwest Iceland at the rate of spreading estimated from open Holocene fractures.

## Example 3.4

### Problem

Show that the stretch $S$ can be expressed as $(1 + \varepsilon)$ as in Eq. (3.12).

### Solution

From the first part of Eq. (3.12) we have

$$S = \frac{L_d}{L_u}$$

And from the second part of Eq. (3.11) we have

$$\varepsilon = \frac{L_d - L_u}{L_u} \Rightarrow L_d - L_u = \varepsilon \times L_u \Rightarrow L_d = \varepsilon \times L_u + L_u$$

Using this expression for $L_u$ in Eq. (3.12) we get

$$S = \frac{L_d}{L_u} = \frac{(\varepsilon \times L_u + L_u)}{L_u} = \frac{L_u}{L_u} + \varepsilon \frac{L_u}{L_u} = (1 + \varepsilon)$$

## Example 3.5

### Problem

Show that the natural strain can be expressed through the equation $\varepsilon_N = \frac{1}{2} \ln \lambda$, as in Eq. (3.19).

### Solution

From Eq. (3.18) we have

$$\varepsilon_N = \sum_{L=L_u}^{L=L_d} \frac{dL}{L} = \int_{L_u}^{L_d} \frac{dL}{L} = \ln\left(\frac{L_d}{L_u}\right) = \ln S$$

From Example 3.4 we also know that $S = (1 + \varepsilon)$. Therefore

$$\varepsilon_N = \ln S = \ln(1 + \varepsilon)$$

From Eq. (3.13) we also have

$$\lambda = \left(\frac{L_d}{L_u}\right)^2 = (1 + \varepsilon)^2 \Rightarrow (1 + \varepsilon) = \lambda^{1/2}$$

Therefore

$$\varepsilon_N = \ln \lambda^{1/2} = \tfrac{1}{2} \ln \lambda$$

## 3.10 Exercises

3.1  What is the main difference between displacement and strain? Are rigid-body rotation and translation displacement or strain? Give geological examples where rigid-body rotation and translation would be appropriate descriptions of the deformation.

3.2  Define displacement field and make a schematic illustration of the surface displacement field around a typical feeder dyke (a volcanic fissure). Assume that the feeder dyke dissects an essentially homogeneous, isotropic lava flow at the surface.

3.3  What is Saint-Venant's principle? How does this principle relate to the deformation indicated in your illustration for the feeder-dyke in Exercise 3.2?

3.4  Define the following concepts: dilation, distortion, infinitesimal strain, finite strain, and natural strain.

3.5  Define normal strain, shear strain, and strain tensor.

3.6  Define elongation, stretch, dilation, deviatoric strain, and strain rate.

3.7  A rock rod when subject to axial tensile stress extends by 0.1 cm. If the corresponding elongation is 0.002 what is the original length of the rod (before extension)?

3.8  If the tensile stress causing the extension of the rock in Exercise 3.7 is 10 MPa, find the Young's modulus of the rock.

3.9  A linear fossil has a stretch of 1.4. If the undeformed length of the fossil was 0.1 m, what is the deformed length? Also, find the elongation, the quadratic elongation, and the natural strain of the fossil.

3.10  Geodetic measurements show that the distance between two measuring points on either side of an active volcanic rift zone has increased by 50 cm in the past 20 years. The rift zone forms a part of a divergent plate boundary and is 20 km wide. Find the stretch, the elongation, and the strain rate associated with the rift zone during this period. What other term is used for the strain rate across a rift zone of this type?

## References and suggested reading

Ashby, M. F. and Jones, D. R. H., 2005. *Engineering Materials*, 3rd edn. Amsterdam: Elsevier.

Bjornsson, A., 1985. Dynamics of crustal rifting in NE Iceland. *Journal of Geophysical Research*, **90**, 10 151–10 162.

Caddell, R. M., 1980. *Deformation and Fracture of Solids*. Upper Saddle River, NJ: Prentice-Hall.

Chou, P. C. and Pagano, N. J., 1992. *Elasticity. Tensor, Dyadic, and Engineering Approaches*. New York: Dover.

Cottrell, A. H., 1964. *The Mechanical Properties of Matter*. New York: Wiley.

Davis, G. H. and Reynolds, S. J., 1996. *Structural Geology of Rocks and Regions*, 2nd edn. New York: Wiley.

Geshi, N., Kusumoto, S., and Gudmundsson, A., 2010. Geometric difference between non-feeders and feeder dikes. *Geology*, **38**, 195–198.

Green, D.J., 1998. *An Introduction to the Mechanical Properties of Ceramics*. Cambridge: Cambridge University Press.

Gudmundsson, A., 1990. Emplacement of dikes, sills, and magma chambers at divergent plate boundaries. *Tectonophysics*, **176**, 257–275.

Gudmundsson, A., 1995. The geometry and growth of dykes. In: G. Baer and A. Heimann (eds.), *Physics and Chemistry of Dykes*. Rotterdam: Balkema, pp. 23–34.

Hamling, I. J., Ayle, A., Bennati, L., *et al.*, 2009. Geodetic observations of the ongoing Dabbahu rifting episode: new dyke intrusions in 2006 and 2007. *Geophysical Journal International*, **178**, 989–1003.

Heinbockel, J. H., 2001. *Introduction to Tensor Calculus and Continuum Mechanics*. Victoria (Canada): Trafford Publishing.

Malvern, L. E., 1969. *Introduction to the Mechanics of a Continuous Medium*. Upper Saddle River, NJ: Prentice-Hall.

Mase, G. E., 1970. *Continuum Mechanics*. New York: McGraw-Hill.

Mase, G. T. and Mase, G. E., 1999. *Continuum Mechanics for Engineers*, 2nd edn. London: CRC Press.

Means, W. D., 1976. *Stress and Strain: Basic Concepts of Continuum Mechanics for Geologists*. Berlin: Springer-Verlag.

Mogi, K., 1958. Relations between eruptions of various volcanoes and the deformations of the ground surfaces around them. *Bulletin of the Earthquake Research Institute, University of Tokyo*, **36**, 99–134.

Nye, J. F., 1984. *Physical Properties of Crystals: Their Representation by Tensors and Matrices*. Oxford: Oxford University Press.

Oertel, G., 1996. *Stress and Deformation: A Handbook on Tensors in Geology*. Oxford: Oxford University Press.

Ramsay, J. G., 1967. *Folding and Fracturing of Rocks*. New York: McGraw-Hill.

Ramsay, J. G. and Lisle, R. J., 2000. *The Techniques of Modern Structural Geology 3: Applications of Continuum Mechanics in Structural Geology*. New York: Academic Press.

Saada, A. S., 2009. *Elasticity Theory and Applications*. London: Roundhouse.

Sigmundsson, F., 2006. *Iceland Geodynamics: Crustal Deformation and Divergent Plate Tectonics*. Berlin: Springer-Verlag.

Sigurdsson, O., 1977. Volcano-tectonic activity in the Thingeyjarthing county (II) 1976–1978. *Tyli*, **7**, 41–56 (in Icelandic with English summary).

Solecki, R. and Conant, R. J., 2003. *Advanced Mechanics of Materials*. Oxford: Oxford University Press.

Verhoogen, J., Turner, F. J., Weiss, L. E., Wahrhaftig, C., and Fyfe, W. S., 1970. *The Earth*. New York: Holt, Rinehart and Winston.

# Relation between stress and strain

## 4.1 Aims

In this chapter we discuss the fundamental relationship between stress and strain, Hooke's law. This law describes approximately the stress–strain behaviour of many solid materials before failure, such as many metals, ceramics, and rocks. Elastic behaviour implies that the deformation is recoverable: a body that deforms elastically when loaded reverts to its original shape immediately when the load is removed. Most solid rocks behave as approximately elastic at low temperatures and pressures (that is, at shallow depths), up to strains of about 1%. Such strains are common in the crust before failure, hence the importance of Hooke's law for understanding processes leading to rock fractures. The main aims of this chapter are to:

- Explain the one-dimensional Hooke's law, as well as its extension to three dimensions.
- Discuss the elastic constants, the relations among them, and their physical meaning.
- Show how to estimate the vertical and horizontal stress in the Earth's crust.
- Provide information on the general state of stress in the Earth's crust.
- Discuss and explain the use of some reference states of stress in the crust.
- Explain the concept of elastic strain energy.

## 4.2 One-dimensional Hooke's law

Consider the comparatively isolated part of rock (a part of a basaltic dyke) in Fig. 4.1. This rock segment or 'bar' may serve as an illustration of the effects of the one-dimensional Hooke's law. If this rock bar is loaded by stress $\sigma$ at its end, then the strain $\varepsilon$ in the rock is related to the stress through the equation

$$\varepsilon = \frac{\sigma}{E} \qquad (4.1)$$

where $E$ is Young's modulus. This presentation of the one-dimensional Hooke's law is just another way of writing Eq. (1.6). When using Eq. (4.1) for rocks several assumptions are implied, including the following:

Basaltic dyke at the coast of a fjord in Iceland forms a wall that stands above its eroded surroundings. Some dykes are more resistant to erosion than the surrounding rocks, and thus form walls like this one. Other dykes, however, are less resistant to erosion than the host rock and thus form depressions, such as gullies. See person for scale.

(a) The rock (or, in general, the material) is **homogeneous**. This means that its properties do not vary with location within the rock body. In other words, the mechanical properties of the rock are the same throughout the rock body and are thus independent of position within that body. A material body that is not homogeneous is inhomogeneous or **heterogeneous**, in which case the material properties change with position within the body.

(b) The rock is **isotropic**. This means that its properties are the same regardless of direction of measurement. In other words, the mechanical properties of the rock body are the same in all directions at a point. A material body that is not isotropic is **anisotropic**, in which case the material properties are directional, that is, the properties have different values in different directions at a given point within the body.

(c) The relation between stress and strain in the rock is **linear** (Fig. 4.2). This is a basic assumption for a Hookean material. An elastic material that does not show a linear behaviour is **non-linear** (Fig. 4.2). Many elastic materials show non-linear behaviour; those that show linear behaviour are Hookean. If the behaviour of the body is not elastic, either linear or non-linear, then it is referred to as **inelastic**.

(d) The strains are **infinitesimal**. This means that the strains are so small that the second and higher-order powers of strain may be neglected. For some materials this assumption may be valid up to about 10% strain.

(e) When the rock body becomes loaded (stressed) it **instantaneously** becomes strained. When the load is removed, there is an instantaneous disappearance of the strain, and all the stored strain energy (Fig. 4.3; Section 4.7) is released. By instantaneous it is meant

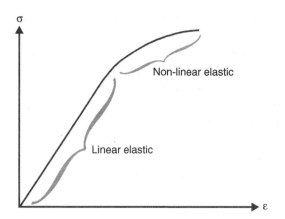

**Fig. 4.2** Material that follows Hooke's law shows a linear relationship between stress and strain. It is referred to as linear elastic. An elastic material where the stress–strain relation is not linear is referred to as non-linear.

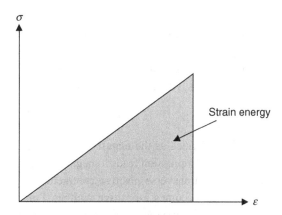

**Fig. 4.3** Elastic material subject to external forces or stresses stores elastic strain energy. The strain energy per unit volume of the material is equal to the area (shaded) under the stress–strain curve.

that there is no time lag between the loading or unloading of the body and its straining or strain recovery (strain disappearance).

Below, and in subsequent chapters, we discuss further how well real rock behaviour meets these assumptions and criteria. Here it is sufficient to say that the assumption of a linear elastic or Hookean behaviour of crustal rocks, particularly during geologically rapid loadings, has been very helpful in improving the understanding of many processes, such as in active fault zones, in volcanoes, and at plate boundaries.

Equation (4.1) presents only the longitudinal or axial strain due to an applied stress (stress vector) in the direction of the stress. When stress is applied to any three-dimensional body, such as the dykes in Figs. 1.3 and 1.5, there will also be lateral strains in directions perpendicular to the direction of the applied stress. If the applied stress is tensile, then the strike dimension or length will increase by a certain amount $\Delta L$ and, at the same time, the width or thickness will decrease by $\Delta W$ and the dip dimension or depth or height will decrease by $\Delta H$ (Fig. 3.11). By contrast, if the applied stress is compressive, then the length will decrease by a certain amount $\Delta L$, the width will increase by $\Delta W$, and the height will increase by $\Delta H$. We assume here, for the sake of the argument, that the rock body is free to contract and expand, and to simplify we ignore the different signs (and possibly different magnitudes) of the expansion and contraction depending on the tensile or compressive loading. Since strain is defined as change in dimension over the original dimension, it is clear that the stress in Fig. 3.11 not only gives rise to strain along the direction parallel but also in the directions perpendicular to the applied stress vector. From Eqs. (1.3) and (3.3)–(3.5) we know that the strains are

$$\varepsilon_{xx} = \frac{\Delta W}{W}$$

$$\varepsilon_{yy} = \frac{\Delta L}{L}$$

$$\varepsilon_{zz} = \frac{\Delta H}{H}$$

Because the rock is assumed to be isotropic, we know that the two transverse or lateral strains are equal. The lateral strains are proportional to the longitudinal or axial strain, the proportionality constant being referred to as **Poisson's ratio** and has the symbol $\nu$ (the Greek letter nu). Thus, Hooke's law implies that

$$-\nu \frac{\Delta L}{L} = \frac{\Delta W}{W} = \frac{\Delta H}{H} \tag{4.2}$$

Poisson's ratio is defined as the **negative** transverse strain divided by the longitudinal or axial strain. Axial compression leads to negative transverse expansion and axial extension (negative) leads to transverse positive contraction. Thus, by defining Poisson's ratio as negative, its value becomes a positive number. For rocks, Poisson's ratio is commonly between 0.15 and 0.35, the most typical value being about 0.25 (Appendix D.1). Poisson's ratio is thus a measure of the negative ratio of the lateral contraction or expansion to the axial expansion (elongation) or contraction. Equations (4.1) and (4.2) make it now possible to discuss the three-dimensional version of Hooke's law.

## 4.3 Three-dimensional Hooke's law

Consider a cube of rock in a three-dimensional (triaxial) stress field (Fig. 4.4). Assume that the sides of the cube are parallel with the principal stress axes. We assume that the rock cube is isotropic and that its coordinate axes coincide with the principal stress directions. For rocks that are isotropic, the axes of principal strain and stress coincide. This is a very convenient way of making the coordinate system in geology, since the principal stresses are generally of main concern when analysing fractures. Furthermore, the Earth's surface is a free surface, and one of the principal stresses is generally perpendicular to it; the other two being parallel with it (Chapter 2). Consequently, the principal stress directions, which also

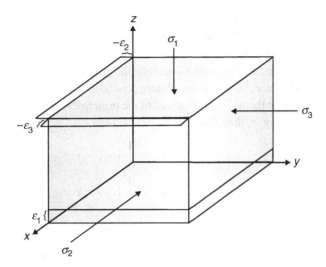

**Fig. 4.4** Rock parallelepiped or cube at depth in the Earth's crust subject to a triaxial stress field, represented by the principal stresses $\sigma_1, \sigma_2$, and $\sigma_3$. The principal compressive stress $\sigma_1$ gives rise to compressive strain $\varepsilon_1$, but also to tensile strains $-\varepsilon_2$ and $-\varepsilon_3$.

coincide with the principal strain directions, are a very appropriate framework for many rock-fracture studies.

It follows from Eq. (4.2) that the compressive principal stress $\sigma_1$ gives rise to compressive strain $\varepsilon_1$ parallel with $\sigma_1$, but also transverse tensile strains, that is, $-\varepsilon_2$ and $-\varepsilon_3$. Since the rock is isotropic $\varepsilon_2 = \varepsilon_3$. By analogy with Eq. (4.2), the principal compressive stress $\sigma_1$ thus gives rise to the following strains:

$$\varepsilon_1 = \frac{\sigma_1}{E} \quad \varepsilon_2 = -v\frac{\sigma_1}{E} \quad \varepsilon_3 = -v\frac{\sigma_1}{E} \tag{4.3}$$

Since all the equations are linear, the final strain in each principal stress direction of the coordinate system is the sum of the compressive strain due to the compressive stress in that direction, such as $\varepsilon_1$ in Eq. (4.3), and the tensile strains due to the compressive stresses $\sigma_2$, $\sigma_3$ in the two other principal directions. Thus the final strain in the $\sigma_1$ direction is

$$\varepsilon_1 = \frac{1}{E}\left[\sigma_1 - v(\sigma_2 + \sigma_3)\right] \tag{4.4}$$

For the other two principal strains, the procedure of finding the final strain is exactly the same, so that the **three-dimensional Hooke's law** for an isotropic rock body for the **principal strains in terms of principal stresses** is

$$\varepsilon_1 = \frac{1}{E}\left[\sigma_1 - v(\sigma_2 + \sigma_3)\right] \tag{4.5}$$

$$\varepsilon_2 = \frac{1}{E}\left[\sigma_2 - v(\sigma_3 + \sigma_1)\right] \tag{4.6}$$

$$\varepsilon_3 = \frac{1}{E}\left[\sigma_3 - v(\sigma_1 + \sigma_2)\right] \tag{4.7}$$

Clearly, if there is uniaxial compression, in which case $\sigma_1$ is not zero but both $\sigma_2$ and $\sigma_3$ are, then Eq. (4.4) reduces to the one-dimensional Hooke's law in Eqs. (1.6) and (4.1).

In Eqs. (4.5)–(4.7) we assumed that the loading is due to the principal stresses $\sigma_1$, $\sigma_2$, $\sigma_3$ and that the coordinates are along the principal axes. The relations represented by these equations are, however, completely general and independent of the coordinate system used to describe them. Thus, if instead of the principal stress axis we use the arbitrary coordinate system $x$, $y$, $z$, then Eqs. (4.5)–(4.7) become

$$\varepsilon_{xx} = \frac{1}{E}\left[\sigma_{xx} - v(\sigma_{yy} + \sigma_{zz})\right] \tag{4.8}$$

$$\varepsilon_{yy} = \frac{1}{E}\left[\sigma_{yy} - v(\sigma_{xx} + \sigma_{zz})\right] \tag{4.9}$$

$$\varepsilon_{zz} = \frac{1}{E}\left[\sigma_{zz} - v(\sigma_{xx} + \sigma_{yy})\right] \tag{4.10}$$

The results presented in Eqs. (4.5)–(4.10) show that, for an isotropic rock, the normal strains are independent of the shear-stress components.

When the sides of the rock cube in Fig. 4.4 are subject to shear stresses ($\tau_{xy}$, $\tau_{yz}$, $\tau_{zx}$) the resulting shear strains ($\gamma_{xy}$, $\gamma_{yz}$, $\gamma_{zx}$) are linearly related to the shear stresses through the equations

$$\gamma_{xy} = \frac{1+v}{E}\tau_{xy} = \frac{\tau_{xy}}{G} \tag{4.11}$$

$$\gamma_{yz} = \frac{1+v}{E}\tau_{yz} = \frac{\tau_{yz}}{G} \tag{4.12}$$

$$\gamma_{zx} = \frac{1+v}{E}\tau_{zx} = \frac{\tau_{zx}}{G} \tag{4.13}$$

Equations (4.11)–(4.13) show that shear stresses influence only the associated shear strains. In Eqs. (4.11)–(4.13) $G$ is the shear modulus, which is related to Young's modulus and Poisson's ratio through the equation

$$G = \frac{E}{2(1+v)} \tag{4.14}$$

and discussed in more detail in Section 4.4.

The equations above all show strains in terms of stresses. This presentation is very useful if we know the stresses and want to calculate the associated strains. Commonly, however, we know the strains or displacements from measurements and we want to calculate the stresses that generated them. Then it is useful to rewrite the above formulas in terms of stresses. When the principal stress directions are used as the axes of the coordinate system, Eqs. (4.5)–(4.7) may be rewritten so as to give the **principal stresses in terms of principal strains** thus:

$$\sigma_1 = \lambda\Delta + 2G\varepsilon_1 \tag{4.15}$$

$$\sigma_2 = \lambda\Delta + 2G\varepsilon_2 \tag{4.16}$$

$$\sigma_3 = \lambda\Delta + 2G\varepsilon_3 \tag{4.17}$$

where $\Delta$ is the dilation or volume strain (Eq. 3.14), and $\lambda$ (the Greek letter lambda) is Lamé's constant, which is related to Young's modulus and Poisson's ratio through the equation

$$\lambda = \frac{\nu E}{(1 + \nu)(1 - 2\nu)} \tag{4.18}$$

and discussed in Section 4.4. Thus, if instead of the principal stress axes we use the arbitrary coordinate system $x$, $y$, $z$, then Eqs. (4.8)–(4.13) may be rewritten for stresses in terms of strains as

$$\sigma_{xx} = \lambda\Delta + 2G\varepsilon_{xx} \tag{4.19}$$

$$\sigma_{yy} = \lambda\Delta + 2G\varepsilon_{yy} \tag{4.20}$$

$$\sigma_{zz} = \lambda\Delta + 2G\varepsilon_{zz} \tag{4.21}$$

$$\tau_{xy} = G\gamma_{xy} \tag{4.22}$$

$$\tau_{yz} = G\gamma_{yz} \tag{4.23}$$

$$\tau_{zx} = G\gamma_{zx} \tag{4.24}$$

All the equations relating stress and strain for a linear, elastic material can be presented in various forms. These forms depend partly on which elastic moduli or constants are used.

## 4.4  Elastic constants

### 4.4.1  Relationships

For isotropic elastic materials there are two independent elastic constants or moduli. However, several constants are used for convenience in various relations between stress and strain and for formulating and solving different types of problems. We usually refer to five constants, namely Young's modulus, Poisson's ratio, shear modulus, bulk modulus, and Lamé's constant. We have already used most of these in the equations above, but here we summarise the basic definitions of and relations between these and related moduli.

- **Young's modulus** is denoted by $E$ and also referred to as the modulus of elasticity. It relates the normal stress and strain and has the units of stress. It is the slope of the linear part of the stress–strain curve (Figs. 1.9 and 4.5). Young's modulus is a measure of stiffness, and commonly referred to as stiffness. Thus, rocks with a high or large Young's modulus are referred to as stiff; those with a low or small Young's modulus as compliant, floppy, or soft.
- **Poisson's ratio** is denoted by $\nu$ (nu) and is the negative of the ratio of transverse strain to longitudinal strain. For example, during axial extension in a uniaxial test, Poisson's ratio is the ratio of the lateral contraction to the axial extension. Poisson's ratio has no units; it is a pure number.

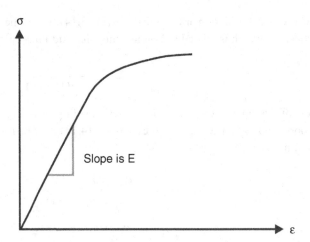

Fig. 4.5 Young's modulus or modulus of elasticity is the slope of the linear part of the stress–strain curve of an elastic material.

- **Shear modulus** is denoted by $G$ and also referred to as the modulus of rigidity. It relates shear stress and shear strain and has the units of stress. It is a measure of the resistance to shape changes.
- **Bulk modulus** is denoted by $K$ and also referred to as incompressibility. It is a measure of the change in rock volume, the dilation (at a constant shape), when the rock is subject to a given mean or spherical stress. It has the units of stress. Bulk modulus is the ratio of the mean stress $\bar{\sigma}$ to the dilation or volume strain $\Delta$, namely

$$K = \frac{\bar{\sigma}}{\Delta} \qquad (4.25)$$

- **Lamé's constant** is denoted by $\lambda$ (lambda) and has the units of stress. It has no simple physical meaning but is rather a mathematical convenience, defined by Eq. (4.18). Its main use is to make many equations such as Eqs. (4.19)–(4.21) more elegant and easier to handle.

The elastic constants are of fundamental importance for analysing and understanding rock fractures. To clarify their meaning and use, the following additional comments may be made:

1. Since only two of them are independent, it follows that two constants are sufficient to solve problems regarding fractures in isotropic elastic rock, or to describe the general behaviour of that type of rock. Many rocks are, of course, anisotropic because of the preferred orientation of crystals. Others, however, may be regarded as approximately isotropic. This follows because the grains in many sedimentary rocks and of crystals in many igneous rocks (and metamorphic rocks) are essentially randomly oriented so that their individual anisotropies cancel out.
2. The most commonly used elastic constants are Young's modulus and Poisson's ratio.

3. Together, Lamé's constant $\lambda$ and shear modulus $G$ are often referred to as Lamé's constants. Since

$$G = \frac{\lambda(1 - 2\nu)}{2\nu} \qquad (4.26)$$

it follows that when $\nu = 0.25$, as is common for many solid rocks, then $G = \lambda$. For this case, referred to as the **Poisson's relation**, the above five constants for isotropic rocks reduce to four.

4. The reciprocals of some of the five constants are commonly used. These include the **Poisson's number**, $m$, defined as the reciprocal of Poisson's ratio, namely

$$m = \frac{1}{\nu} \qquad (4.27)$$

Poisson's number is used in some formulas since it makes them simpler, as we shall see below. Poisson's number, as the name implies, is a pure number with no units. For a Poisson's ratio of 0.25, common for many solid rocks, Poisson's number is 4.0.

Another reciprocal, namely that of bulk modulus or incompressibility, $K$, is referred to as **compressibility** and denoted by $\beta$. It is defined as

$$\beta = \frac{1}{K} \qquad (4.28)$$

Compressibility has the units of 1/stress, that is, $Pa^{-1}$. It is particularly important in problems involving fluid-filled reservoirs. This is because the relative compressibilities of the fluid and the host rock (the matrix) in a porous aquifer or reservoir partly determine (1) how much fluid the reservoir can contain before rupturing, and (2) when it ruptures and fluid flows out how the reservoir responds mechanically to a reduced fluid volume. Fluid-volume reduction occurs, for example, as a result of pumping of a petroleum reservoir or a ground-water aquifer and through magma flowing out of a magma chamber during dyke injection. Compressibility also affects the response of the reservoir when fluid is added to it. Thus, reservoir expansion, when fluid is added to the reservoir, and contraction, when fluid is subtracted from the reservoir, are partly determined by the fluid and matrix compressibilities.

Appendix D.2 shows the common relations between the five elastic constants. Some general conclusions follow from these relationships as to the **restriction on the values of the constants**, that is, their ranges.

Consider first Young's modulus $E$ and shear modulus $G$. From the one-dimensional Hooke's law (Eqs. 1.6 and 4.1) we know that uniaxial tensile stress produces extension, and uniaxial compressive stress produces contraction or shortening. It follows that $E$ must be positive; if it were not, uniaxial tensile stress could produce contraction which is contrary to experience. We also know that positive shear stress produces positive shear strain (Eqs. 4.22–4.24) so that, by the same argument, the shear modulus $G$ must be positive. Because both $E$ and $G$ are regarded as constants, for a given material under given conditions, they must always be positive. Also, neither stress nor strain can be infinite, so that $E$ and $G$ must be finite.

Consider next the bulk modulus $K$ and Poisson's ratio $\nu$. When a rock body is subject to a compressive isotropic mean stress $\bar{\sigma}$ we expect its volume will decrease by a certain finite amount. From Eq. (4.25) it then follows that $K$ must be positive and finite. From Appendix D.2 we have the relationships

$$E = 3K(1 - 2v) \tag{4.29}$$

$$E = 2G(1 + v) \tag{4.30}$$

It follows from Eq. (4.29) that since $E$ and $K$ are positive then Poisson's ratio $v$ must be less than 0.5. For if $v = 0.5$, then $1 - 2v = 0$ and $E = 0$, which contradicts the results that $E$ must be positive and thus larger than zero for solid materials. The lower value of $v$ follows from Eq. (4.30). Since $G$ is also positive, then it follows from Eq. (4.30) that $v$ must be larger than –1. For if $v = -1$, then $1 + v = 0$, in which case $E = 0$. Thus, we have established the range of the values of Poisson's ratio as greater than –1 and less than 0.5.

Poisson's ratio values of 0.5 imply that the rock is **incompressible**. That is to say, there will be no decrease in the volume of the rock regardless of the mean stress or pressure applied to it. No rock bodies, or other materials for that matter, are truly incompressible, but rubber is close to being so. Negative values of Poisson's ratio imply that when a bar is subject to uniaxial tension (Fig. 3.11), the material does not contract in a transverse direction but rather expands. For natural materials such as rocks, and most engineering materials, this does not happen; by contrast, there is contraction in the transverse direction. Cork has a Poisson's ratio close to, but not quite, zero. While some materials with negative Poisson's ratios have been developed (Lakes, 1987), most solid materials and all rocks have Poisson's ratios larger than 0 and less than 0.5.

Using this information about the restriction of values of Poisson's ratio and Young's modulus for common materials, including rocks, we can estimate the restrictions on Lamé's constant $\lambda$ using the (Eqs. 4.18, 4.30; Appendix D.2) relation

$$\lambda = \frac{2Gv}{1 - 2v} \tag{4.31}$$

Since $E$ and $v$ are both larger than zero, and $v$ is also less than 0.5, it follows from Eq. (4.31) that $\lambda$ must be larger than zero and thus positive. Also, since $E$ is finite, then so must be $\lambda$. (In principle $\lambda$ may have negative values, but for natural materials it is always positive.)

We have thus established restrictions on the possible values that the elastic constants for common materials, including rocks, can take. These **ranges** of values for solid materials, including rocks, may be summarised as follows:

$$0 < E < \infty \tag{4.32}$$

$$0 < G < \infty \tag{4.33}$$

$$0 < K < \infty \tag{4.34}$$

$$0 < K < \infty \tag{4.35}$$

$$-1 < v < \frac{1}{2} \quad \text{and mostly} \quad 0 < v < \frac{1}{2} \tag{4.36}$$

We can now make some general observations as to the implications of these results for material behaviour in general, and that of rocks in particular. Here the focus is on the two most important elastic moduli used in this book, namely Young's modulus and Poisson's ratio.

1. **Young's modulus**
   - For a linear-elastic material, Young's modulus is the slope of the stress–strain diagram. Many materials, including rocks, however, are somewhat **non-linear** (that is, normal stresses and strains are not linearly proportional), in which case the tangent modulus may be used. It represents the instantaneous ratio of stress to strain, as given by the slope of the tangent to the stress-stain curve. For such a non-linear material, however, the (tangent) Young's modulus changes within the elastic range, that is, along the curve, since the slope of the curve is changing.
   - Young's modulus for **fluids** is zero. This is useful when making numerical models: a fluid-filled cavity, such a totally molten magma cavity, can then be modelled as an empty cavity (in three dimensions) or a hole (in two dimensions).
   - From Appendix D.2,

$$E = \frac{9KG}{3K + G} = \frac{3G}{1 + \frac{G}{3K}} \tag{4.37}$$

   For most isotropic solids, $3K \gg G$. For example, typical rocks have $v = 0.25$, $3K = 5G$. It follows that Eq. (4.37) can be written approximately as

$$E \approx 3G \left(1 - \frac{G}{3K}\right) \tag{4.38}$$

   - Formula (4.38) shows that the Young's modulus $E$ of a rock is largely controlled by the shear modulus $G$ and only slightly by the bulk modulus $K$. This means that the ratio of normal stress to normal strain in rocks is primarily controlled by the ability of the rock to resist shear stress rather than by its ability to change volume, that is, its dilation. This is partly because in uniaxial loading of a rock bar (Fig. 3.11), as is the main method for determining Young's modulus in the laboratory, the stress-free sides of the bar are free to contract or expand. Consequently, the resistance to volume change, that is, the resistance to dilation, is tested to a much lesser degree than the distortion.
   - The values of Young's moduli for rocks show a very large range. A common range is from **1 GPa to 100 GPa** (Appendix D.1). However, in the extreme, very compliant rocks such as some pyroclastic rocks and breccias and clays, may have Young's moduli as low as 0.01 GPa, and for unconsolidated sand and gravel and clay as low as 0.003–0.08 GPa. By contrast, the stiffest rocks reach Young's moduli of 150–200 GPa (Carmichael, 1989; Afrouz, 1992; Bell, 2000; Myrvang, 2001; Schön, 2004). Thus, at a typical plate boundary, the Young's moduli of the rocks may easily vary by two or three orders of magnitude and, in extreme cases, by as much as four orders of magnitude.

2. **Poisson's ratio**
   - The range in the values of Poisson's ratio is much more limited than the range of values of Young's modulus. When Poisson's ratio reaches 0.5, then the material behaves as a **fluid** at rest, that is, there are no shear stresses in the material. This follows from

Eq. (4.26) because with $\nu = 0.5$, then $G$ must be zero. Some plant tissues approach this value, and so does rubber. By contrast, if uniaxial extension produces no lateral contraction, then Poisson's ratio is 0: a well-known material approaching this value is cork. For most solid rocks, however, Poisson's ratios fall within a very narrow range, between 0.10 and 0.35, with most values between 0.2 and 0.3 (Appendix D.2). Thus, for typical solid rocks Poisson's ratios are unlikely to vary by more than a factor of three, and usually by a factor of two or less.

- This limited range of Poisson's ratios of rocks, in comparison with the range of Young's moduli, has important implications for modelling rock behaviour in general, and that of rock fractures in particular. For example, when modelling layered rocks, Poisson's ratio can often be taken as uniform, whereas Young's modulus varies by two orders of magnitude between layers.

- For a body subject to loading, Poisson's ratio measures the body's tendency to change its volume in proportion to its tendency to change its shape. As Poisson's ratio increases and approaches 0.5, the body increasingly responds to the loading through change in shape rather than change in volume. When Poisson's ratio reaches the theoretical value of 0.5, there will be no change in volume, regardless of loading. Such a material has a very high bulk modulus $K$ and is referred to as **incompressible**. Soft rubber is close to being incompressible, with a typical Poisson's ratio of 0.49. Thus, under load the shape changes of a soft rubber are large but the volume changes are minimal. Fluids such as water have Poisson's ratio of about 0.5, and are therefore often assumed incompressible. However, no materials are strictly incompressible, and the compressibility of fluids must, for example, be taken into account when modelling fluid-filled reservoirs such as ground-water aquifers and magma chambers.

- Alternatively, we can say that Poisson's ratio measures the relative tendency of the loaded body to respond through dilation and shear or distortion. A small Poisson's ratio indicates that the body is comparatively resistant to shear and distortion but less resistant to dilation.

### 4.4.2  Atomic view of elasticity and its constants

The details of the atomic basis for elastic behaviour and the elastic constants are quite complex and beyond the scope of the book. However, elastic behaviour and the elastic constants can be explained, in an elementary way, in terms of the atomic structure of matter. For this to be possible, a brief review of the atomic structure of matter is needed.

Atoms are made up of **protons, neutrons, and electrons**. The protons (positive charge) and neutrons (mass, no charge) constitute the atomic **nucleus**, which is surrounded by negatively charged electrons. A nucleus with electrons that together form an electrically neutral unit constitutes an **atom**. The electron is regarded as a fundamental particle, whereas the neutrons and protons are composed of still more fundamental particles known as **quarks**. The atomic number equals the number of protons in the atom. Material composed of atoms all of which have the same atomic number is referred to as an **element**, such as, for example, oxygen, O. All the atoms of a particular element have the same number of protons, but their number of neutrons may differ. Elements with different numbers of neutrons are known as **isotopes**.

On the Earth, matter is rarely in the form of pure elements but rather in a form composed of many types of atoms. In forming matter, the atoms are held together by electrical interactions between the outer electrons of each atom. These interactions are known as chemical bonds or simply **bonds**. Two or more atoms bonded together through electrical interactions form a **molecule**. The number of atoms in a molecule range from two or three (such as $O_2$, $H_2O$, $CO_2$) to several thousand.

For the purpose of understanding the basis of elasticity and elastic constants in terms of atomic structure and interactions, we now give a brief overview of the bonds that hold together atoms and molecules in solids such as minerals, crystals and rocks. The view presented here is much simplified and does not consider the quantum-mechanical view of the chemical bonding. For the purpose of this presentation, the words atoms, molecules, and particles are used interchangeably.

Each electron has a negative electric charge of the same magnitude as the positive charge of the proton. Opposite charges attract, whereas like charges repel. Thus, the negatively charged electrons are held to the positively charged (because of its protons) nucleus by electrostatic forces. An atom in its normal state is electronically neutral and thus with equal number of protons and electrons. In solids such as rocks, the atoms are held together by the electrostatic or **electric forces**. The electric force bonds between the atoms belong to one of two main groups: primary bonds and secondary bonds.

**Primary bonds** are strong and are classified as metallic, ionic, and covalent bonds. **Secondary bonds** are weak bonds and are classified as van der Waals and hydrogen bonds. Rock minerals and other ceramics as well as metals are held together by primary bonds. In metals, the main bond is the metallic bond; in ceramics, the main bonds are the ionic and covalent bonds. Because the primary bonds are strong, ceramics and metals tend to be strong (in shear, tension, and compression) and stiff (with high Young's moduli). They also have high melting points. By contrast, polymers (plastics) are held together by secondary bonds, so that they are more compliant and have melting points that are comparatively low. **Silicates** are the most abundant constituents of crustal rocks. The bonding between Si (silica) and O (oxygen) are hybrids: they are part ionic and part covalent. By contrast, in **carbonates**, a very important mineral of sedimentary rocks, the bonding is close to being purely covalent.

In solids such as rocks and minerals and crystals, the atoms are in close proximity to one another (Fig. 4.6). There are two types of electric forces between the atoms: **repulsive** and **attractive** (Fig. 4.7). Because a solid has an essentially fixed shape and does not spontaneously change its shape, there must be no resultant electric forces acting on its atoms. To keep the atoms together in the solid, there must be attractive forces. These are the forces that make it difficult to expand a solid (Fig. 4.7). For example, extending a rock bar requires large tensile forces or stresses, part of the reason being that the atoms in the crystals and minerals that constitute the rock bar are held together through attractive electric forces.

Similarly, when a solid is compressed or squeezed, it resists changes in its shape and size. This resistance is due to the repulsive electric forces. These are the forces that make

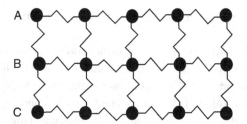

**Fig. 4.6**   Schematic presentation of the atomic structure of a solid. The atoms or particles may be thought of as connected by elastic springs and forming interconnected chains.

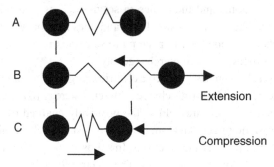

**Fig. 4.7**   When the solid in Fig. 4.6 is subject to extension that pulls the atoms further apart, the springs develop attractive forces between the atoms that try to pull them back to their normal, equilibrium positions. Similarly, when the solid is subject to compression that pushes the atoms closer together, the springs develop repulsive forces between the atoms that try to push then back to their normal positions.

it difficult to compress solids. For example, compressing solid-rock particles and minerals requires very large compressive forces or stresses. One reason that large compressive stresses are needed for any significant compressive strain in a solid rock is that the forces of repulsion oppose the atoms in the minerals coming closer together (Fig. 4.7).

A combination of repulsive and attractive forces yields a resultant force that varies with atomic separation. Both the forces of attraction and repulsion decrease as the separation between the particles (atoms, molecules) increases (Fig. 4.8). However, the force of repulsion decreases at a faster rate than the force of attraction. Consequently, in an unloaded solid there is a certain equilibrium separation between the atoms $a_0$ at which repulsive and attractive forces cancel. This separation corresponds to the position of the atoms where the resultant net force is zero so that they are in a stable equilibrium (Fig. 4.9). If the atoms are moved apart, so that the separation increases, the resultant (net) force acts to pull them back to the equilibrium separation. Conversely, if the separation decreases, so that the atoms come closer, then the resultant force acts to push them apart until they reach the equilibrium separation (Fig. 4.7).

Consider the case where a tensile force, $F_t$, is applied to a crystal (Fig. 4.9). The atoms then move apart until the attractive force balances the tensile force, resulting in a new separation $a_1$ between the atoms. When, however, the force is removed, the resultant attractive force

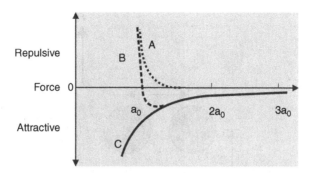

**Fig. 4.8**  Forces between particles or atoms in a solid. The separation or distance between the atoms is measured along the horizontal axis as $a_0$, $2a_0$, $3a_0$, and so forth. Curve A is the repulsive force, curve C the attractive force, and curve B the resultant force. The equilibrium separation between the atoms is $a_0$ and occurs when the repulsive and attractive forces are equal and the net or resultant force zero. Consequently, the equilibrium separation or distance between the atoms is where curve B intersects the horizontal axis (cf. Fig. 4.9).

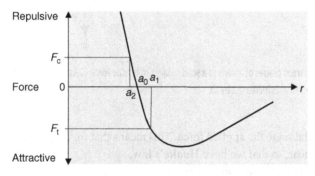

**Fig. 4.9**  Net force between atoms, represented by the curve, is zero at the equilibrium separation $a_0$ between the atoms (cf. Fig. 4.8). If a tensile force $F_t$ increases the separation of the atoms to a distance $a_1$, the net force becomes attractive and tries to pull the atoms back to position $a_0$. If a compressive force $F_c$ decreases the separation of the atoms to a distance $a_2$, the net force becomes repulsive and tries to push the atoms back to position $a_0$ (cf. Fig. 4.7).

pulls the atoms back (Fig. 4.7) to the original equilibrium position $a_0$ (Fig. 4.9). Similarly, when a compressive force $F_c$ is applied to the crystal, the separation between the atoms decreases until the repulsive force balances the compressive force (Fig. 4.7), resulting in a separation $a_2$ between the atoms (Fig. 4.9). Thus, as long as the forces are sufficiently small, the deformation is always reversible, that is, elastic. Now since the bulk elastic behaviour of a large solid body is just the aggregate effects of the deformation of individual bonds between the atoms, the overall behaviour of a body composed of atoms with bonds that behave in this way is perfectly **elastic**.

A closer examination of Fig. 4.9 also explains why the strain is proportional to the stress in the solid. Within the atom separation limits of $a_1$ and $a_2$, the force–separation curve is approximately linear. This implies that, within these limits, the separation shows a linear

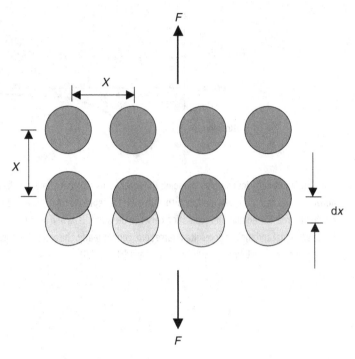

Fig. 4.10
Adjacent planes of atoms in a solid subject to tensile force $F$. As a consequence, the original distance between the planes, $x$, is increased by $dx$.

relation to the applied force. This means that for a larger solid the stress–strain curve is also linear, so that we have **Hooke's law**.

**Young's modulus** can then be derived from the force–separation relation in several ways. Consider first two adjacent planes of atoms (Fig. 4.10). If the equilibrium (original) separation of the atom pairs or planes is $x$, then each atom occupies a certain space with an area of $x^2$. If the area of a particular atomic plane is $A$, then the total number of atoms in that plane is the area of the plane divided by the area occupied by each atom, or $A/x^2$. If a tensile force pulls the atomic planes further apart by a distance of $dx$, then the attractive force between each pair of atoms will be $dF$. Since there are a total of $A/x^2$ atoms in each plane, the total attractive force resisting the moving apart of the planes is $(A/x^2)dF$. Since stress is force per unit area, the stress generated by this total force is $dF/x^2$. The corresponding strain, given as elongation, is $dx/x$. We therefore have the following theoretical formula for Young's modulus $E$:

$$E = \frac{\sigma}{\varepsilon} = \frac{1}{x}\frac{dF}{dx} \tag{4.39}$$

which has units of $N\,m^{-2}$, that is, pascals. Thus, this simple relation shows that the magnitude of Young's modulus is a measure of the resistance to an increased (out of equilibrium) separation of adjacent planes of atoms. Young's modulus is thus a **measure of the interatomic bonding forces**. More specifically, Eq. (4.39) shows that Young's modulus is inversely proportional to the separation of the atomic planes and directly **proportional to the slope or gradient**, $dF/dx$, of the force–separation curve so long as the atomic planes are, after the forced extra separation, still close to the equilibrium separation.

We can also relate the elastic constants to the potential or **bond energy** of the atoms. If we change the separation of two atoms from their initial equilibrium separation (Figs. 4.7, 4.9, and 4.10), work must be done (work is force times distance moved in the direction of the force). Consider an atomic pair (Fig. 4.7). The further the atoms in the pair are moved apart, the more work is needed. The atoms then gain **potential energy** because of the attractive forces between them. This is entirely analogous to a solid body gaining potential energy when it is moved to a higher elevation above the ground in a gravitational field. By definition, when the atoms are an infinite distance apart, the potential energy of each of them is regarded as zero. This is because at an infinite distance, the atoms have no attraction effects on each other. From this definition it follows that when there is an attractive force between the atoms, the potential energy is negative. In this view, the potential energy increases as the separation of the atoms approaches infinity since the energy become less negative (approaches zero).

The potential energy between atoms as a function of their separation can be derived from the force–separation curve (Fig. 4.8). This is because the potential energy at a particular separation of atoms is equal to the area under the force–separation (or displacement) curve (Fig. 1.8; Section 4.7). The general form of the potential energy curve as a function of separation between atoms is as shown in Fig. 4.11. Clearly, the equilibrium separation coincides with the potential energy minimum. This follows because at equilibrium there is no resultant attractive force between the atoms. Also, the deeper the potential energy minimum, the greater will be the energy needed to separate the atoms.

We can now use the potential energy curve (Fig. 4.11) to derive either Young's modulus or bulk modulus directly from atomic considerations. If $x$ is the separation between the atoms and $U$ is the potential or bond energy, then at the equilibrium position $a_0$ the slope is zero, so that

$$\frac{dU}{dx} = 0 \tag{4.40}$$

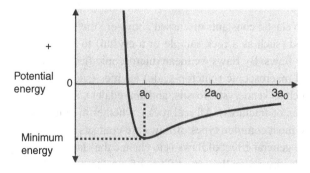

**Fig. 4.11**  Potential energy curve as a function of separation or distance, $a_0$, $2a_0$, $3a_0$, between particles or atoms in a solid. The shape of the curve can be used to derive elastic constants such as Young's modulus or bulk modulus.

Consider first the bulk modulus $K$. If we compress the solid containing the atoms so that the atomic separation becomes less than $a_0$, then the energy needed to reduce the separation is related to the slope of the potential energy curve (Fig. 4.11). The steeper the curve the greater the energy needed to reduce the distance, that is, the harder it is to compress and reduce the volume of the solid. The first derivative gives the gradient or the slope of the curve, so that the bulk modulus or incompressibility $K$ is related to the first derivative of $dU/dx$, that is, to the second derivative (the curvature) of the potential energy $U$. Thus, we have

$$K \propto \frac{d}{dx}\left(\frac{dU}{dx}\right) = \frac{d^2U}{dx^2} \tag{4.41}$$

In terms of stress and strain, the curvature of the potential energy curve is the bulk modulus of the solid containing the atoms. The same type of expression can be derived for Young's modulus. The following general conclusions can thus be made:

- The curvature of the potential or bond energy curve at the equilibrium separation is an elastic constant (Young's modulus or bulk modulus) of the solid.
- The radius of curvature of the energy curve at equilibrium separation of atoms is a measure of the magnitude or value of the elastic constant.
- If the radius of curvature is small, that is, the curve is sharp, then the elastic constant (stiffness) will be high or large; if the radius of curvature is large, the elastic constant will be low or small.

We have thus explained two of the elastic constants in terms of curvature of the potential energy curve at the equilibrium separation between atoms. Since for isotropic solids only two constants are independent, the others can be derived from these two. However, the constants derived from atomic considerations are considerably higher than those found in most rocks and other solids in nature or in small laboratory samples. This follows because all rocks and other natural solids in general contain flaws, at various scales, that reduce the constants. For rocks on an outcrop scale, these flaws include pores and fractures and give rise to effective elastic constants.

## 4.4.3 Effective elastic constants

The elastic constants discussed above are the ideal theoretical constants. They assume the solid (such as a rock sample or a crystal) to be homogeneous and isotropic and without any flaws. By flaws we mean microscopic flaws (such as dislocations and Griffith cracks) and macroscopic (outcrop-scale) cavities (such as pores and vesicles), elastic inclusions (such as xenocrysts, fossils, and amygdales), and solid interfaces (such as contacts and all types of fractures). All real rocks, whether in outcrops or in cores, contain flaws. Among the most common types of flaws are contacts and fractures, as well as pores and vesicles. The general effect of flaws is to change the elastic constants, so that the constants of in-situ rocks are normally very different from those obtained from theoretical considerations or those from small-sample experiments in the laboratory. The procedure of calculating or otherwise inferring **effective elastic constants** and other physical rock properties (such as permeability) from laboratory results is referred to as **upscaling**. Some general comments may be made on upscaling:

- It is one of the most fundamental problems to be solved in any study related to the behaviour of fluid-filled reservoirs.
- There are no known simple principles that can be used to obtain correct effective constants. In particular, for a fractured rock body it is often very difficult to reach even reasonably accurate estimates of its effective properties such as Young's modulus.
- The most difficult property to scale up and estimate accurately for in-situ conditions is the permeability of fractured rocks. The permeabilities of fractured rocks at a particular site may vary by 4–6 orders of magnitude, and in-situ crustal values in general range by as much as 12 orders of magnitude.
- Crude estimates of the effective elastic constants can, however, often be obtained from fracture frequencies using theoretical models. These models are primarily derived from micromechanics.

Of the two constants that are most commonly used for isotropic rocks, Poisson's ratio is the easier one to scale up. It does depend on the rock type under consideration, but its range is generally small compared with that of Young's modulus. For example, rocks as widely different as soft tuffs and basaltic lava flows may have the same Poisson's ratio, 0.25. Some fractured in-situ rocks may have Poisson's ratio as high as 0.35, but most values, in situ and in the laboratory, are between 0.2 and 0.3 (Appendix D.1).

The situation is very different for Young's modulus. For common elastic materials Young's modulus can vary by more than six orders of magnitude. Even if the range of variation for solid rocks is much less, Young's modulus may still vary by several orders of magnitude. For example, in standard laboratory tests on solid igneous, sedimentary and metamorphic rocks, Young's modulus may range by a factor of more than 100 (Appendix D.1). Even for a single rock type, such as sandstone, Young's modulus measured on laboratory samples may vary by several orders of magnitude (Appendix D.1).

Young's modulus of rocks is also different depending on the method of testing and on the geomechanical rating of the rock (e.g. Brady and Brown, 1993). Consider first the effects of testing. The two principal methods are dynamic and static (Goodman, 1989). The static tests are of two main types: laboratory tests and field tests. The laboratory static tests are mainly compression and bending tests. The static field tests are primarily loading using flat jacks or plate-bearing instruments and borehole-expansion measurements.

The dynamic methods are also of two types: laboratory tests and field tests. The laboratory tests are based on sending a stress wave through a small specimen and from the wave velocity calculating its Young's modulus. The dynamic field methods are primarily based on swinging a sledge hammer against an outcrop of the rock for which Young's modulus is to be determined and then measuring the velocity of the resulting seismic waves, commonly at distances as great as 50 m from the outcrop. Another dynamic field method is to use boreholes spaced at distances of 50–100 m and measure the velocity of a wave, generated by a hammer or an explosion, that travels from a point in one borehole to a point in the other.

There are several general statements that can be made for Young's modulus of an in-situ rock mass (cf. Link, 1968; Jumikis, 1979; Cheng and Johnston, 1981; Goodman, 1989; Brady and Brown, 1993; Myrvang, 2001; Jaeger *et al.*, 2007). These include the following:

- For a given rock, the dynamic modulus, particularly at shallow depths in tectonically active areas, is much higher than the static modulus. The dynamic modulus may be as much as 13 times higher than the static modulus. This difference is particularly noticeable when the rock is highly fractured. In general, fracture sets may largely alter the effective Young's modulus of a rock mass.
- Laboratory measurements on small specimens, dynamic and static, yield values that are commonly 1.5–5 times greater than those of the in-situ (field) modulus of the same rock. For igneous and metamorphic rocks the laboratory modulus is commonly 3 times the in-situ modulus.
- Increasing mean stress (and thus increasing crustal depth) increases Young's modulus. However, variations in Young's modulus depend greatly on the rock types. Thus, Young's modulus commonly decreases locally with depth when passing from a stiff layer to a soft layer or unit, but the general trend, for rocks of the same or similar type, is that Young's modulus increases with depth.
- Increasing temperature, increasing porosity, and increasing water content all decrease the Young's modulus of a rock of a given type.
- The most important effect on the magnitude or value of the field Young's modulus, particularly at comparatively shallow depths in tectonically active areas, is the fracture frequency in the rock mass.
- Young's modulus of a rock mass is normally less than that of laboratory samples of the same type of rock. This difference is mainly attributed to fractures and pores in the rock mass which do not occur in the small laboratory samples.
- With increasing number of fractures, in particular in a direction perpendicular to the loading, the ratio $E_{is}/E_{la}$ ($E$ in situ /$E$ laboratory) first shows a rapid decay. Thus, increasing fracturing normally decreases Young's modulus, and so do gouge and breccia in the core and damage zone of an active fault zone.
- When modelling crustal processes, the **static Young's modulus** should be used for processes that are slow in comparison with the velocity of propagation of the seismic waves, that is, much slower than several kilometres per second. For example, for modelling dyke emplacement, with an average rate of propagation of the magma-driven fracture of about 1 m s$^{-1}$ or less, the static modulus should be used. By contrast, when modelling an earthquake rupture, the **dynamic Young's modulus** should be used.

For elastic rocks, the principal effects of fracture sets is to decrease the effective Young's modulus of the rock mass. This applies in particular to rocks containing open fractures, pores, and other cavities, but also to rocks containing closed fractures, provided the fractures respond to the loading by slip, that is, provided they are or become shear cracks or faults. Highly fractured rocks, such as those that occur in the damage zones and fault cores of active fault zones (Chapters 10 and 14), do thus normally have smaller Young's moduli than non-fractured rocks of the same type. Also, for a given loading, the stresses in layers with low Young's moduli are normally less than those in layers with high Young's moduli. In other words, stiff layers take up more of the loading, and tend to magnify or raise (concentrate) the stresses, whereas soft layers take up little of the loading, and tend to suppress the stresses. Because the Young's moduli of layered and jointed rock masses may easily vary by an order of magnitude, the stresses associated with propagation of fractures and, in particular, the displacements of faults, depend much on variations in Young's modulus (Chapters 9 and 14).

A detailed theoretical discussion of the effects of flaws on the elastic constants is beyond the scope of the book. The topic is treated in detail by Nemat-Nasser and Hori (1999) and other micromechanics texts (Mura, 1987; Qu and Cherkaoui, 2006). However, some simple formulas showing the effects of fracture frequency on effective Young's modulus are given here.

Consider first a single set of fractures. The rock body is a cube with a side length $x$ dissected by fractures (or other discontinuities such as contacts) of an average spacing $\bar{s}$, so that the total number of discontinuities is $x/\bar{s}$. The unfractured rock body (that is, without any discontinuities) is homogeneous and isotropic and with Young's modulus $E$. Each fracture or discontinuity has a stiffness $k$, which has units of Pa m$^{-1}$ (Chapter 1). As the normal compressive stress increases from zero to a value of $\sigma_x$, there will be contraction $\Delta x$ in the rock, primarily because of compression or closing of the discontinuities. The equivalent or **effective Young's modulus** of the rock mass, $E_e$, is then given by (Priest, 1993)

$$E_e = \sigma_x \frac{x}{\Delta x} = \left( \frac{1}{E} + \frac{1}{\bar{s}k} \right)^{-1} \tag{4.42}$$

A rock body with many discontinuities is not strictly elastic because the deformation does not completely disappear once the load is removed. Thus, the concept of a normal Young's modulus, such as for an intact rock, may not be strictly applicable. Hence the use of the concept of 'effective' Young's modulus of the rock mass.

We can rewrite Eq. (4.42) so as to show how much the original Young's modulus of the intact rock is reduced because of the discontinuities in the rock mass. The Young's **modulus reduction factor** $M_r$ is given by

$$M_r = \left( 1 + \frac{E}{\bar{s}k} \right)^{-1} \tag{4.43}$$

These results can be extended to more complex models. For example, there exists a model for calculating the effective Young's modulus and Poisson's ratio of a rock body containing three sets of orthogonal fractures or discontinuities such as in some joint sets (Chapter 11). The results are generally similar to those for the effective modulus of a single set (Eq. 4.42). Formulas can also be developed for calculating the effective elastic constants for randomly oriented discontinuities in a rock mass (Priest, 1993; Nemat-Nasser and Hori, 1999). The results indicate that the reduction in Young's modulus is **much less** when the fractures have a **random orientation** than when they form a set of parallel discontinuities that are oriented in a direction perpendicular to the loading.

# 4.5 Rock stress

## 4.5.1 Vertical stress

Because the Earth's surface is free of shear stress, is a **free surface**, it is common to assume that one of the principal stresses is vertical close to and at the surface. As discussed earlier, this assumption is generally justified in comparatively flat regions, that is, regions with non-Alpine landscapes, as is supported by stress measurements (Chapter 1). When stresses are measured in a rock body (Chapter 1), it is usually the horizontal stresses that are measured; the vertical stress $\sigma_v$ is simply calculated from the following formula:

$$\sigma_v = \int_0^z \rho_r(z) g \, dz \tag{4.44}$$

where $z$ is the vertical coordinate axis, $g$ is the acceleration due to gravity ($9.81 \text{ m s}^{-2}$), and $\rho_r(z)$ is the density of the rock layers as a function of depth (Fig. 2.1). If the density of each layer is known, then Eq. (4.44) can be used to obtain the vertical stress at any particular depth $z$. Commonly, however, the detailed density distribution of the rock layers is not know, so that an average density of the layers is used. Then Eq. (4.44) simplifies to

$$\sigma_v = \rho_r g z \tag{4.45}$$

where $\rho_r$ is the average density of the crustal rock column to a depth of $z$ below the surface. This equation, identical to Eq. (2.11), can be clarified as follows. From the definition of stress, given in Eqs. (1.1) and (2.1), we have $\sigma = F/A$. Since force is mass times acceleration, and the acceleration at and close to the Earth's surface is $g$, it follows that we have $F = mg$. Now mass is equal to density times volume, and because the rock column is assumed to have a unit area, then its volume is the unit area times the depth, $z$. Thus, the volume is $1 \times z = z$. We can therefore write

$$\sigma_v = \frac{F}{A} = \frac{mg}{A} = \frac{\rho_r z g}{1} = \rho_r g z \tag{4.46}$$

The vertical stress $\sigma_v$ is sometimes referred to as **geostatic stress** or as **overburden pressure**. However, if the horizontal stresses are equal to the vertical stress as defined by Eqs. (4.44)–(4.46), then the state of stress is referred to as lithostatic.

We can thus define **lithostatic stress** state or field as follows. It is an isotropic (hydrostatic or spherical) state of stress (all the principal stresses are equal) where the stress magnitude increases proportionally with depth in the crust. The rate of increase of the stress magnitude with depth is determined by the density of the crustal rock, according to Eqs. (4.44)–(4.46).

Lithostatic state of stress or, simply, lithostatic stress, is very important when analysing rock fractures, particularly hydrofractures (Chapters 8, 11, and 13). Many such fractures are injected from fluid-filled crustal reservoirs. It is a common assumption that, during much of the active time (lifetime) of such a reservoir, it is in lithostatic equilibrium with its host rock. Thus, we commonly assume that the normal state of stress at the margin of the reservoir is lithostatic. This assumption is, for example, normally made for magma chambers. The implication is then that it is only during periods of unrest, usually associated with magma-chamber inflation or deflation, that the fluid-pressure change in the chamber results in stress concentrations that modify the state of stress around the chamber to non-lithostatic (Chapter 6). Many petroleum reservoirs are also in close to lithostatic equilibrium with their surroundings (Chilingar *et al.*, 2002).

## 4.5.2 Horizontal stress

Horizontal stress is commonly measured in drill holes (Chapter 6). Here, however, the focus is on the theoretical horizontal stress in relation to the vertical stress. This theoretical relationship is commonly used in geology and related fields to predict the horizontal stress in the absence of direct measurements. We shall first derive the theoretical horizontal stress and then explain the assumptions and limitations of this approach.

An isolated cube of rock subject to vertical compressive stress $\sigma_v$ will, according to Poisson's effect, expand laterally (Fig. 4.4). In the crust, however, there are no isolated rock cubes because rock is a continuum. A crustal segment may thus be visualised as consisting of numerous rock cubes, all of which are in contact with other cubes. It follows that when a crustal segment is subject to vertical stress $\sigma_v$, all the rock cubes tend to expand laterally. The net strain of all these expansions may be zero.

Let us assume that we are dealing with an area of extension so that the maximum principal compressive stress is vertical and thus $\sigma_1 = \sigma_v$. If both the horizontal principal stresses are equal, then we have

$$\sigma_2 = \sigma_3 = \sigma_h = \sigma_H \tag{4.47}$$

where $\sigma_h$ and $\sigma_H$ are the minimum and maximum horizontal stresses which, in this case, are equal. If the horizontal strains are zero, then we have $\varepsilon_h = \varepsilon_2 = \varepsilon_3 = 0$. It follows then from Eq. (4.6) that

$$\varepsilon_2 = 0 = \frac{1}{E}[\sigma_2 - \nu(\sigma_3 + \sigma_1)] \tag{4.48}$$

For solids, Young's modulus cannot be zero, nor its reciprocal, so that $E^{-1} \neq 0$. Then we have from Eqs. (4.47) and (4.48) that

$$\sigma_h - \nu\sigma_h - \nu\sigma_v = 0 \tag{4.49}$$

or

$$\sigma_h(1 - \nu) - \nu\sigma_v = 0 \tag{4.50}$$

or

$$\sigma_h = \frac{\nu\sigma_v}{1 - \nu} \tag{4.51}$$

Recalling the definition of Poisson's number (Eq. 4.27) as $m = 1/\nu$, it follows that

$$\frac{\nu}{1 - \nu} = \frac{\nu/\nu}{\frac{1}{\nu} - \frac{\nu}{\nu}} = \frac{1}{m - 1} \tag{4.52}$$

Thus, we can write Eq. (4.51) in the form

$$\sigma_h = \frac{\sigma_v}{m - 1} \tag{4.53}$$

Equations (4.51) and (4.53) are commonly used to make a theoretical estimate of the likely horizontal stress at depth in the crust. However, for their realistic application, it must be taken into account that they make the following assumptions:

- The rock body (crustal segment) is confined laterally, that is, it cannot expand laterally. This is the basic assumption and follows from the considerations above, namely that since all the rock cubes tend to expand laterally as a consequence of the vertical stress, then the net expansion is zero. It follows that the lateral strains are zero, that is, $\varepsilon_h = \varepsilon_2 = \varepsilon_3 = 0$. This is commonly approximately true.
- There is no tectonic stress or strain affecting the rock body. Thus, the only loading is the vertical stress as given by Eqs. (4.44)–(4.46) and the overburden pressure is related to gravity. This assumption may sometimes be valid in sedimentary basins and young volcanic rocks, for example. For an active area, however, this assumption is unlikely to hold true for any geologically long period. In a volcanic rift zone, for instance, there are always tectonic stresses operating and these commonly dominate during rifting episodes. Similarly, in a sedimentary basin, there are tectonic stresses associated with episodes of extension and contraction (inversion), although these are normally not continuous in time.
- The stress difference $\sigma_v - \sigma_h$ does not reach the (tensile or shear) strength of the rock. If the stress difference did reach the strength of the rock, then there would be faulting which would reduce the stress difference, perhaps close to zero.
- There is no totally fluid reservoir inside the rock body. In other words, the state of stress indicated by Eqs. (4.51) and (4.53) cannot be reached in the vicinity of a totally fluid body. For example, many magma chambers are, during parts of their lifetimes, totally fluid. The state of stress indicated by Eqs. (4.51) and (4.53) cannot develop in the rock hosting such a magma chamber. This is because as soon as the stress difference reaches the tensile strength of the host rock $T_0$, that is, $\sigma_v - \sigma_h = T_0$, the magma chamber ruptures and forms a magma-driven fracture (a dyke or an inclined sheet). The fluid overpressure in the magma-driven fracture would lessen the stress difference. Since the in-situ tensile strength of rock is quite small, normally 0.5–6 MPa (Appendix E), it follows that at the margin of a fluid-filled reservoir the conditions in Eq. (4.53) cannot be reached at any significant crustal depth.

When these assumptions are satisfied, then for a typical Poisson's ratio of 0.25, that is, a typical Poisson's number of 4.0, Eq. (4.53) yields

$$\sigma_h = \frac{1}{3}\sigma_v \tag{4.54}$$

When the horizontal strains are not assumed zero, a somewhat modified form of Eq. (4.53) can be derived, yielding the approximate formula (Fyfe *et al.*, 1978)

$$\sigma_h = \frac{\sigma_v}{m - 1} \pm \varepsilon_h E \tag{4.55}$$

When using Eq. (4.55), the following should be kept in mind:

- The horizontal strain $\varepsilon_h$ is positive when compressive and negative when tensile.
- The horizontal strain may be regarded as the sum of any tectonic strain and the strain due to the vertical stress.
- The formula has an error of less than about 1%. It is thus not exact. However, this error is very small in comparison with the other assumptions made in calculating rock stresses. For example, in the above formula, the value of Young's modulus may easily vary by a factor of 2–3 within the same rock layer. Thus, we present the formula as an exact equation, although it is strictly not so.

# 4.6 Crustal stress and strain fields

In this section we provide a brief overview of several stress and strain fields assumed to operate in the crust during rock-fracture formation. Many of these fields will be referred to in later chapters when modelling rock-fracture initiation and propagation. Original stresses in surface rocks may be regarded as reference stress states. Such stress states occur, for example, in young sedimentary rocks in large basins. The original stress is then the one that operates shortly after the onset of burial and diagenesis of the sediments in the basin.

Another example of a suitable reference state of stress occurs in young igneous eruptive rocks: in particular, in a young lava pile, close to and at the surface, in a rift zone (Figs. 1.5 and 1.6). In these rocks the condition of Eq. (4.53) may apply, at least for a while and in the absence of tectonic stresses. But when the rocks become buried to greater depths and/or become subject to tectonic stresses of various kinds, the reference or original near-surface stress field becomes modified. This modification results in different types of stress and strain fields. Some of the more common and important fields are discussed in the following sections.

## 4.6.1 Lithostatic stress field

Since lithostatic stress is isotropic, all the principal stresses are equal, and we have

$$\sigma_1 = \sigma_2 = \sigma_3 \tag{4.56}$$

Because the magnitude of the stress at any depth is equal to the overburden pressure, that is, the vertical stress, and since all the stresses are equal, we have

$$\sigma_H = \sigma_h = \sigma_v \tag{4.57}$$

This means that the maximum $\sigma_H$ and the minimum $\sigma_h$ horizontal stresses are equal, and also equal to the vertical stress $\sigma_v$. Isotropic state of stress of this kind is often referred to as hydrostatic because it is the same type of stress or pressure as occurs in a static liquid (Chapter 2). Lithostatic state of stress develops in rocks that have no long-term shear strength, that is, rocks that behave as fluids. All solid rocks have shear strength, so that this stress state is rarely met in detail (except in static molten rocks, magmas, and fluid lavas). However, the yield strength, the conditions for plastic flow, of rocks depends on temperature and can be quite low at comparatively high temperatures. For example, the long-term yield strength of crustal rocks may be as low as about 1 MPa. In experiments, geological strain rates of $10^{-14}\,s^{-1}$ are reached at common mid-crustal temperatures. For example, in the continental crust quartz is very common, and quartz may reach strain rates of $10^{-14}\,s^{-1}$ at temperatures of about 350°C (Kuznir and Park, 1987). This temperature is similar to that at the maximum depth of brittle deformation in a continental crust, that is, the lower boundary or bottom of the seismogenic layer.

We can make the following general comments on the lithostatic stress:

- It corresponds to the state of stress, or pressure, in a static fluid and is thus most likely to be closely approximated where the crust behaves similar to a fluid, that is, it flows.
- This state of stress is generally not reached in solid rocks in presently or recently active parts of the upper crust, but is often approximated in crustal rocks at great depths and high temperatures. The state of stress in some old continental shields where there is not much tectonic activity may also approach lithostatic.
- Rock deformation in general seeks to bring the stress to a lithostatic state. Faulting and fracturing, and at greater depths flow, of rocks all tend to reduce the difference between the principal stresses and thus diminish the shear stress, so as to drive the rock body towards a lithostatic state of stress.

## 4.6.2 Uniaxial strain field

This field is defined by two of the principal strains being zero so that there is only one non-zero principal strain. If, for example, $\varepsilon_1$ is the non-zero principal strain, then we have $\varepsilon_2 = \varepsilon_3 = 0$. If, in addition, $\sigma_1 = \sigma_v$, so that $\varepsilon_1$ is also vertical (for isotropic rocks), and $\sigma_2 = \sigma_3 = \sigma_h$, then, using the derivation in Section 4.5.2, we obtain Eqs. (4.51) and (4.53).

As indicated above, this field may be common in young sedimentary rocks in large basins. It may also occur in young volcanic areas, within lava piles close to the surface. However, over geological time there will be tectonic stresses in active areas, both in sedimentary basins and in volcanic areas, that make the assumption of only one non-zero principal strain untenable.

## 4.6.3 Uniaxial stress field

In this field two of the principal stresses are zero so that there is only one non-zero principal stress. If, for example, $\sigma_1$ is the non-zero principal stress, then we have $\sigma_2 = \sigma_3 = 0$. This means that the rock is free to expand and contract in the plane containing $\sigma_2$ and $\sigma_3$, which is the plane perpendicular to the direction of $\sigma_1$. If, for example, $\sigma_1$ is vertical, as is common in sedimentary basins and rift zones and other areas undergoing extension, then, in a uniaxial stress field, the rock would be free to contract or expand laterally or be laterally unconstrained. This is in contrast to the uniaxial strain field where, by definition, the rock is not free to expand in the plane containing $\varepsilon_2$ and $\varepsilon_3$, which, for isotropic rocks, is also the plane containing $\sigma_2$ and $\sigma_3$.

For a uniaxial stress field, it follows from Eqs. (4.5)–(4.7) that

$$\varepsilon_1 = \frac{\sigma_1}{E} \tag{4.58}$$

and

$$\varepsilon_2 = \varepsilon_3 = -\frac{\nu\sigma_1}{E} \tag{4.59}$$

When $\sigma_1 = \sigma_2 = \rho_r g z$ (cf. Eq. 4.45), we have

$$\varepsilon_1 = \frac{\rho_r g z}{E} \tag{4.60}$$

This field may apply to rocks located between open spaces, such as may form during fluvial erosion (Fig. 4.12). Other examples include rock bodies, such as parts of lava flows, located between open tension fractures or gaping normal faults (Fig. 4.13). Also, it may be an operating stress field in the layers, particularly the upper layers, of isolated mountains, as well as comparatively flat highland areas marked by deep valleys forming their sloping sides.

**Fig. 4.12** Field example of the conditions where a uniaxial stress field might be an appropriate model of the crustal stress. Aerial view north of an 'island' (about 200 m wide) between large channels into a Holocene pahoehoe lava flow in North Iceland. The channels were primarily formed by floods in glacial rivers in early Holocene. Existing gaping normal faults and tension fractures, as are common in the rift zone of Iceland (cf. Fig. 4.13), presumably offered weaknesses that the water (melt water from glaciers) eroded to the present channels. The 'island' is primarily subject to vertical stress due to overburden pressure and is essentially free to expand laterally.

### 4.6.4 Plane stress

This is defined so that one of the principal stresses is zero, whereas the other two are non-zero. If, for example, $\sigma_3 = 0$, then from Eqs. (4.5)–(4.7) we have

$$\varepsilon_1 = \frac{1}{E}(\sigma_1 - \nu\sigma_2) \tag{4.61}$$

$$\varepsilon_1 = \frac{1}{E}(\sigma_2 - \nu\sigma_1) \tag{4.62}$$

$$\varepsilon_1 = -\frac{\nu}{E}(\sigma_1 + \sigma_2) \tag{4.63}$$

This stress field is particularly appropriate for crustal layers that have free surfaces at the top and bottom and have lateral dimensions that greatly exceed the vertical dimension. In other words, because the stresses operate only in the plane of the body, this stress field is very appropriate for fractures in comparatively thin crustal plates (Fig. 4.14). For example, when modelling segments at mid-ocean ridges, or major fractures (Fig. 1.5) or fissure swarms in the rift zone of Iceland (Figs. 3.13 and 3.14), plane-stress models may be very appropriate (Chapter 9).

The main characteristics of plane stress may be summarised as follows:

- The applied stresses are in the plane of the body, usually a comparatively thin plate. Thus, plane-stress formulation implies that one of the dimensions is small in comparison with the other two dimensions (the in-plane dimensions).

**Fig. 4.13** Narrow strip of land inside a major, gaping normal fault. The elliptical strip of land, (about 10 m wide) with the grey moss and a path along its major axis, is free to expand laterally and is mainly subject to overburden pressure due to its own weight. The strip of land is an example where uniaxial stress field may be an appropriate crustal-stress model. Aerial view of a part of the fault Almannagja located in a Holocene pahoehoe lava flow in the rift zone of Southwest Iceland. The entire fault is seen in Fig. 1.5, and a map of the area is given in Fig. 3.13.

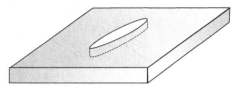

Plane stress is an appropriate model for a fracture in a crustal plate that is thin in comparison with its lateral dimensions.

- The applied stress is assumed constant through the thickness of the plate. For a crustal segment, this means that the applied horizontal stresses do not change with depth, that is, they are constant.
- The upper and lower surfaces of the plate are assumed shear-stress free, that is, principal stress planes (free surfaces). Thus, all the stress components active outward from these principal planes must be zero.
- If $\sigma_3 = 0$ is the normal stress at the upper and lower surfaces, then it follows, on the assumption of plane stress, that $\sigma_3 = 0$ through the plate. Thus, when applied to a crustal segment, the overburden pressure (geostatic stress) is not considered.
- Although the stresses in the direction perpendicular to the plate are zero, the normal strain in that direction is not zero. This is due to Poisson's effect. Thus, even when $\sigma_3 = 0$, then $\varepsilon_3 \neq 0$.
- Plane-stress formulas are not exact. Thus, plane stress is only an approximation to the complete three-dimensional solution. Nevertheless, it is a very useful formulation, and very appropriate, for many rock-fracture problems.

### 4.6.5 Plane strain

This field is defined by one of the principal strains, say $\varepsilon_3$, being zero, whereas the other two are non-zero. Then the principal strains in terms of stresses become

$$\varepsilon_1 = \frac{1 + \nu}{E} [(1 - \nu)\sigma_1 - \nu\sigma_2] \tag{4.64}$$

$$\varepsilon_2 = \frac{1 + \nu}{E} [(1 - \nu)\sigma_2 - \nu\sigma_1] \tag{4.65}$$

$$\varepsilon_3 = 0 \tag{4.66}$$

Plane-strain formulation is most appropriate when modelling structures of interest in the vertical dimension or, more generally, structures with one dimension larger than the other two dimensions. For example, long road tunnels and tunnel cracks (Chapter 9) subject to loading would often be modelled in a plane-strain formulation. Also, if the dip dimension (height) of a fracture is greater than its strike dimension (length), then plane-strain formulation would often be appropriate (Fig. 4.15).

The use of plane-strain and plane-stress formulation for fracture problems is discussed further in Chapter 9. It should be noted, however, that there is no fundamental difference

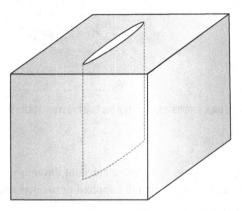

**Fig. 4.15**
Plane strain is an appropriate model for a fracture in a crustal segment that is thick in comparison with its lateral dimensions.

between the plane-strain and plane-stress equations. Plane-stress equations can be transformed into plane-strain equations simply by making the appropriate changes in the elastic constants used in the equations. Thus, solutions obtained in a plane-stress formulation can be rewritten in the form of a plane-strain formulation, and vice versa, by suitable change of expressions of the elastic constants. The appropriate expressions of the elastic constants, to convert from plane-strain to plane-stress solutions, and vice versa, are provided by Slaughter (2002).

## 4.7 Strain energy

One concept that follows directly from Hooke's law is elastic strain energy. This concept is of fundamental importance for understanding rock fractures and fracture propagation (Chapters 7 and 10). The following is a brief explanation of the concept.

When external loads such as forces deform an elastic body they perform work. The work done equals force times displacement in the direction of the force. If the loading is moment, then the work is moment times the rotation angle. The energy absorbed in an elastic body because of work done on it by external forces is called elastic strain energy, or just **strain energy**. Strain energy is the potential energy stored in an elastically deformed body: the energy is equal to the work done (by external forces) in deforming the body.

If we ignore the small amount of energy lost in the form of heat then, when the external loads or forces are removed, the strain energy stored in the rock body can be completely recovered. The strain energy stored in the body during loading can be used to perform work, once the loading is removed, as the body returns to its undeformed or unstrained state. A well-known example of how strain energy is converted into mechanical energy to perform work is a mechanical clock. Winding the clock loads its mainspring, which thereby stores strain energy. As the spring unwinds, its strain energy is converted into the mechanical energy that drives the entire mechanism of the clock.

We make the following general observations about the strain energy of an elastic body:

- Strain energy is stored in an elastic body when it is loaded. The loads may be forces, moments, stresses, strains, or displacements, or combinations of these.
- When the loads are relaxed, that is, removed, a perfectly elastic body returns to its original undeformed and unstrained shape, in which case the strain energy can be completely recovered (apart from a minor energy lost as heat).
- The strain energy stored in the body can, on recovery, be used to perform work. A well-known everyday example is the shooting of an arrow with a bow. When the bow is drawn or stretched, its frame (mainly the limbs) stores elastic strain energy. When the string is released, this energy rapidly goes into driving the arrow. The greater the strain energy that can be stored in an elastic bow, the greater will be the energy that can be transferred into driving the arrow and, other things being equal, the greater the distance travelled by the arrow.
- Strain energy stored in a particular object such as a bow is obtained from the area under a force–extension (displacement) curve (Fig. 1.8), whereas the strain energy per unit volume of a material is equal to the area under the stress–strain curve (Fig. 4.3).
- Strain energy is a measure of the material toughness of the body. That is to say, it indicates how much energy is absorbed by the material (such as a rock) before it fails through fracture propagation (Chapter 10). It is thus a concept of great importance in rock-fracture mechanics.

To establish the fundamental formulas of elastic strain energy, consider the one-dimensional case of a rectangular prism subject to uniaxial tensile stress $\sigma$ that increases slowly from zero to its final value of $\sigma_{xx}$ (here denoted by $\sigma_x$ as is common). The prism has the dimensions $dx$, $dy$, $dz$ and the uniaxial stress is parallel with the $x$-axis of the coordinate system. The body force is assumed to be zero (no gravity effects, for example). Because the stress is increased slowly, the inertial effect may be neglected and the element is assumed, at every instant during the loading, to be in equilibrium. The strain energy stored in the prism during its loading is equal to the work done on it and is given by (Chou and Pagano, 1992; Sadd, 2005)

$$dU = \int_{\sigma=0}^{\sigma=\sigma_x} \sigma \, d\left(u + \frac{\partial u}{\partial x} dx\right) dy \, dx - \int_{\sigma=0}^{\sigma=\sigma_x} \sigma \, du \, dx \, dz = \int_{\sigma=0}^{\sigma=\sigma_x} \sigma \, d\left(\frac{\partial u}{\partial x}\right) dx \, dy \, dz \quad (4.67)$$

where $\partial u/\partial x$ is the displacement gradient. For the integration, the dimensions $dx$, $dy$, $dz$ are constant, whereas $\sigma$, $u$, and $\partial u/\partial x$ are the variables. Because the loading is uniaxial tensile stress, the stresses parallel with the $y$- and $z$-axes ($\sigma_y$ and $\sigma_z$) are zero, so that the displacements in those directions ($v$ and $w$) do not contribute to the elastic strain energy. Using the definition of strain $\partial u/\partial x = \varepsilon_{xx}$ (or $\varepsilon_x$) (Eq. 3.9) and the one-dimensional Hooke's law (Eq. 4.1), namely $\varepsilon = \sigma/E$, we can rewrite Eq. (4.67) in the form

$$dU = \int_{\sigma=0}^{\sigma=\sigma_x} \sigma \frac{d\sigma}{E} dx \, dy \, dz = \frac{\sigma_x^2}{2E} dx \, dy \, dz \quad (4.68)$$

Thus, the strain energy **per unit volume**, which is also referred to as the **strain energy density**, becomes

$$U_0 = \frac{\sigma_x^2}{2E} = \frac{1}{2}\sigma_x \varepsilon_x = \frac{1}{2}E\varepsilon_x^2 \tag{4.69}$$

Equation (4.69) shows that the strain energy per unit volume is equal to the area under the stress–strain curve (Fig. 4.3). The factor $^1\!/_2$ in Eq. (4.69) is because the uniaxial stress $\sigma$ is gradually increased from zero to $\sigma_x$ so that the average loading is not $\sigma_x$ but rather $^1\!/_2\sigma_x$.

Using a similar argument, it can be shown that the strain energy stored in an elastic body as a consequence of shear stress $\tau$, which varies slowly from zero to $\tau_{xy}$ during the loading, is

$$U_0 = \frac{\tau_{xy}^2}{2G} = \frac{1}{2}\tau_{xy}\gamma_{xy} = \frac{1}{2}G\gamma_{xy}^2 \tag{4.70}$$

where $\gamma_{xy}$ is the maximum value of the shear strain generated during the shear-stress loading.

The strain energy stored per unit volume in a material just before failure, either through fracture or plastic flow, is referred to as the modulus of elastic resilience, proof resilience, or just **resilience**. It is the maximum amount of strain energy that can be absorbed in an elastic material and reappear (say, converted into mechanical energy) when the loads are removed. To emphasise that it is always measured per unit volume, we can recast Eq. (4.69) in the form

$$U_0 = \frac{\sigma_x^2}{2E} \times \text{(unit volume)} \tag{4.71}$$

Equation (4.70) could, of course, also be recast in a similar form. Generally, a high resilience depends on either high material strength (tensile, shear, yield), low Young's modulus, or both. The highest resilience occurs in materials with a comparatively high strength and low Young's modulus (Figs. 4.3 and 4.5). Resilience is closely related to the concept of material toughness, which is the basis for understanding the resistance of solid materials in general, and rocks in particular, to fracture initiation and propagation (Chapter 10).

## 4.8 Summary

- When an elastic material is loaded and unloaded, the strain appears and disappears instantaneously. For a loaded isotropic, homogeneous elastic cube with the sides coinciding with the principal stress directions, the three-dimensional Hooke's law gives the principal strains in terms of stresses in Eqs. (4.5)–(4.7) and principal stresses in terms of principal strains in Eqs. (4.15)–(4.17).
- For an isotropic, elastic material there are two independent elastic constants (moduli). When two are known, all the others can be calculated from them. The most commonly

used constants are Young's modulus, Poisson's ratio, shear modulus, bulk modulus, and Lamé's constant. They all have the units of stress apart from Poisson's ratio, which has no units. In addition to these, several other (dependent) constants are commonly used, such as compressibility (the reciprocal of bulk modulus) and Poisson's number (the reciprocal of Poisson's ratio).

- The most commonly used (independent) constants are Young's modulus $E$ and Poisson's ratio $v$. For solid rocks, Poisson's ratio is commonly between 0.10 and 0.35, whereas Young's modulus is commonly between 1 GPa and 100 GPa. Effective, that is, in-situ, elastic constants depend on flaws (such as pores, vesicles, contacts, and fractures). The greater the frequency of fractures in a direction perpendicular to the loading, the lower will be the effective Young's modulus. The procedure of finding in-situ rock properties, using theoretical models and laboratory measurements as a basis, is called upscaling.

- Ideal elastic constants of flawless solids can be obtained through theoretical considerations of the atomic structure of matter. The results show that the elastic constants of metals, ceramics, and other solids depend on atomic bonds. Primary bonds are strong, secondary bonds are weak; primary bonds are of greatest importance in ceramics (including minerals and rocks), which is one reason why they generally are strong and stiff (with high Young's modulus). A theoretical Young's modulus can be derived from the net effect of the repulsive and attractive forces between atoms, which both vary with atomic separation. In particular, the curvature of the potential or bond energy curve at equilibrium separation between atoms is equal to Young's modulus.

- The Earth's surface is free of shear stress, a free surface, and therefore a principal stress plane. It follows that the principal stresses are either parallel or perpendicular to the surface. In areas of horizontal extension, the maximum principal compressive stress is normally vertical at the surface; in areas of horizontal compression, the minimum principal compressive stress is vertical at the surface. The vertical stress is given by Eqs. (4.44) and (4.45).

- Various special stress and strain fields are often appropriate when analysing rock-fracture problems. These include lithostatic stress, plane stress, and plane strain. Lithostatic stress is isotropic and thus cannot, as such, generate fractures. However, it is commonly a suitable reference state of stress, particularly at considerable depths and at the boundaries of many fluid-filled reservoirs, particularly shallow magma chambers. Plane stress is a formulation suitable for analysing fractures that have strike dimensions that are much larger than their dip dimensions. Plane strain is more suitable when the fracture dip dimension is larger than its strike dimension.

- When forces are applied to an elastic material, it stores elastic strain energy. When the forces are removed, the strain energy can be recovered and used to perform work. The stored strain energy per unit volume of an elastic material just before failure is referred to as resilience. Resilience, or the amount of energy stored before fracture, is a measure of material toughness (Chapter 10). The strain energy released when a fracture extends is a major contributor to fracture propagation and a fundamental concept in fracture mechanics.

# 4.9 Main symbols used

| | |
|---|---|
| $A$ | area |
| $E$ | Young's modulus (modulus of elasticity) |
| $F$ | force |
| $G$ | shear modulus (modulus of rigidity) |
| $g$ | acceleration due to gravity |
| $H$ | height of a bar |
| $\Delta H$ | change in height of a bar |
| $K$ | bulk modulus (incompressibility) |
| $k$ | stiffness |
| $L$ | length, length of a bar |
| $\Delta L$ | extension, change in length of a bar |
| $M_r$ | Young's modulus reduction factor |
| $m$ | Poisson's number |
| $m$ | mass, as used in Eq. (4.46) |
| $\bar{s}$ | average spacing of fractures in a set |
| $U$ | potential energy |
| $U_0$ | strain energy density (strain energy per unit volume) |
| $W$ | width (thickness) of a bar |
| $\Delta W$ | change in width (thickness) of a bar |
| $\Delta x$ | contraction (shortening) |
| $z$ | depth below the Earth's surface (crustal depth) |
| $\beta$ | compressibility |
| $\gamma$ | shear strain |
| $\gamma_{ij}$ | shear strain components |
| $\Delta$ | dilation, volume strain |
| $\varepsilon$ | normal strain |
| $\varepsilon_h$ | horizontal strain |
| $\varepsilon_1$ | maximum compressive principal strain |
| $\varepsilon_2$ | intermediate compressive principal strain |
| $\varepsilon_3$ | minimum compressive principal strain |
| $\varepsilon_{ii}$ | components of normal strain |
| $\lambda$ | Lamé's constant |
| $\nu$ | Poisson's ratio |
| $\rho_r$ | rock or crustal density |
| $\sigma$ | normal stress |
| $\sigma_{ii}$ | normal-stress component (example: $\sigma_{xx}$ but sometimes with one index, $\sigma_x$) |
| $\bar{\sigma}$ | mean stress |
| $\sigma_1$ | maximum compressive principal stress |
| $\sigma_2$ | intermediate compressive principal stress |
| $\sigma_3$ | minimum compressive (maximum tensile) principal stress |
| $\sigma_H$ | maximum horizontal compressive stress |
| $\sigma_h$ | minimum horizontal compressive stress |

$\sigma_{\mathrm{v}}$    vertical stress
$\tau$    shear stress
$\tau_{ij}$    components of shear stress

# 4.10 Worked examples

### Example 4.1

Problem

At a certain depth in a rift zone composed primarily of basaltic lava flows (Fig. 2.1), the principal compressive stresses acting on a rock body have magnitudes of 30 MPa, 20 MPa, and 10 MPa. Young's modulus of the rock is 10 GPa, its Poisson's ratio is 0.25, and the average crustal density to this depth is 2500 kg m$^{-3}$. Assuming the rift-zone segment to be approximately homogeneous and isotropic, calculate:

(a) the likely depth of the rock body
(b) the principal strains in the directions of the stresses.

Solution

(a) In a rift zone the maximum principal compressive stress is normally vertical. Thus, we assume that the largest stress, 30 MPa, is vertical. Since the average density is given, we use Eq. (4.45), namely

$$\sigma_{\mathrm{v}} = \rho_{\mathrm{r}} g z$$

and solve it for the depth $z$. Thus

$$z = \frac{\sigma_{\mathrm{v}}}{\rho_{\mathrm{r}} g} = \frac{3 \times 10^7 \, \mathrm{Pa}}{2500 \, \mathrm{kg \, m^{-3}} \times 9.81 \, \mathrm{m \, s^{-2}}} = 1223 \, \mathrm{m}$$

Thus, the rock body with this state of stress would be at a depth of about 1.2 km, or very similar to that seen in the palaeorift zone in East Iceland (Fig. 3.1).

(b) The principal strains in terms of principal stresses and Young's modulus and Poisson's ratio are given by Eqs. (4.5)–(4.7). For the maximum principal compressive strain we have

$$\varepsilon_1 = \frac{1}{E} [\sigma_1 - \nu(\sigma_2 + \sigma_3)]$$

$$= \frac{1}{1 \times 10^{10} \, \mathrm{Pa}} \left[ 3 \times 10^7 \, \mathrm{Pa} - 0.25(2 \times 10^7 \, \mathrm{Pa} + 1 \times 10^7 \, \mathrm{Pa}) \right] = 0.002 \, 25 = 0.225\%$$

For the intermediate principal strain we have

$$\varepsilon_2 = \frac{1}{E} [\sigma_2 - \nu(\sigma_3 + \sigma_1)]$$

$$= \frac{1}{1 \times 10^{10} \, \mathrm{Pa}} \left[ 2 \times 10^7 \, \mathrm{Pa} - 0.25(1 \times 10^7 \, \mathrm{Pa} + 3 \times 10^7 \, \mathrm{Pa}) \right] = 0.001 = 0.1\%$$

And for the minimum principal strain,

$$\varepsilon_3 = \frac{1}{E}[\sigma_3 - \nu(\sigma_1 + \sigma_2)]$$

$$= \frac{1}{1 \times 10^{10} \text{ Pa}}\left[1 \times 10^7 \text{ Pa} - 0.25(3 \times 10^7 \text{ Pa} + 2 \times 10^7 \text{ Pa})\right]$$

$$= -0.000\,25 = -0.025\%$$

This shows that the minimum principal strain is negative, that is, tensile. Tensile strain (and stress) at the depth of about 1.2 km is close to, or slightly in excess of, the maximum depth one would normally expect in a rift zone (Chapters 7 and 8).

---

### Example 4.2

#### Problem

A rod is initially 50 cm long and 1.6 cm in diameter and has a uniform circular cross-sectional area. When subject to an axial load of −12 kN (tensile), the rod increases its length by 0.3 mm and decreases its diameter by 0.0024 mm. If the material in the rod is homogeneous and isotropic

(a) find the Young's modulus of the rod
(b) find the Poisson's ratio of the rod
(c) explain why the rod material is unlikely to be solid rock.

#### Solution

(a) To find Young's modulus, we use the one-dimensional Hooke's law (Eq. 4.1) rewritten as

$$E = \frac{\sigma}{\varepsilon}$$

To use this law, we first have to calculate the stress on a cross-section of the rod and then find the strains. For the stress, we have the following:
Diameter of the rod = 1.6 cm = $1.6 \times 10^{-2}$ m
The radius of the rod, $R$, is thus $8 \times 10^{-3}$ m
The cross-sectional area of the rod is then

$$A = \pi R^2 = \pi(8 \times 10^{-3} \text{ m})^2 = 2.01 \times 10^{-4} \text{ m}^2$$

The force is −12 kN = $-1.2 \times 10^4$ N
Therefore, the stress on the rod is (from Eq. 1.1)

$$\sigma = \frac{F}{A} = \frac{-1.2 \times 10^4 \text{ N}}{2.01 \times 10^{-4} \text{ m}^2} = -5.97 \times 10^7 \text{ Pa} = -59.7 \text{ MPa}$$

For the strains we have, from Eq. (3.4), that the elongation is

$$\varepsilon_{yy} = \frac{\Delta L}{L} = \frac{-3 \times 10^{-4} \text{ m}}{0.5 \text{ m}} = -6 \times 10^{-4} \text{ m}$$

From Hooke's law and the estimated axial strain (elongation) we then get Young's modulus as

$$E = \frac{\sigma}{\varepsilon} = \frac{-5.97 \times 10^7 \, \text{Pa}}{-6 \times 10^{-4}} = 9.95 \times 10^{10} \, \text{Pa} = 99.5 \, \text{GPa}$$

(b) To find Poisson's ratio we need the transverse strain as well as the elongation. The transverse strain, that is, the contraction in the transverse direction, is from Eq. (3.3) or (3.5), the following:

$$\varepsilon_{xx} = \frac{\Delta W}{W} = \frac{2.4 \times 10^{-6} \, \text{m}}{1.6 \times 10^{-2} \, \text{m}} = 1.5 \times 10^{-4}$$

Poisson's ratio is the negative of the lateral contraction over the axial elongation, or

$$\nu = -\frac{\varepsilon_{xx}}{\varepsilon_{yy}} = -\frac{1.5 \times 10^{-4}}{-6 \times 10^{-4}} = 0.25$$

Please notice that the elongation is here parallel with the $y$-axis and the contraction parallel with the $x$-axis, but this is just because this is the orientation of the coordinates in Eqs. (3.3)–(3.5). Commonly, the elongation would be parallel with the $x$-axis, but this is pure convention.

(c) The material has a Poisson's ratio of 0.25 and a Young's modulus of 99.5 GPa. A Poisson's ratio of 0.25 is typical for many solid rocks. A Young's modulus of 99.5 GPa is similar to the laboratory values of some very stiff rocks, such as some gabbros and basalts. However, the rod would be unlikely to be of rock because rocks do not tolerate tensile stresses of 59.7 MPa. In-situ tensile strengths of solid rocks are generally less than 6 MPa, and those of laboratory specimens generally less than about 20 MPa, the highest reported tensile-strength values being around 30 MPa (Jumikis, 1979; Myrvang, 2001; Appendix E.1).

## Example 4.3

### Problem

A specimen of limestone is tested in the laboratory and found to have a Young's modulus of 63 GPa and a Poisson's ratio of 0.27. Find:

(a) the shear modulus
(b) the bulk modulus
(c) Lamé's constant.

### Solution

(a) From Eq. (4.14) we have

$$G = \frac{E}{2(1+\nu)} = \frac{6.3 \times 10^{10} \, \text{Pa}}{2(1+0.27)} \approx 2.5 \times 10^{10} \, \text{Pa} = 25 \, \text{GPa}$$

(b) Rewriting Eq. (4.29) we have

$$K = \frac{E}{3(1 - 2v)} = \frac{6.3 \times 10^{10} \text{ Pa}}{3(1 - 2 \times 0.27)} \approx 4.6 \times 10^{10} \text{Pa} = 46 \text{ GPa}$$

(c) From Eq. (4.31) we have

$$\lambda = \frac{Ev}{(1 + v)(1 - 2v)} = \frac{6.3 \times 10^{10} \text{ Pa} \times 0.27}{(1 + 0.27)(1 - 2 \times 0.27)} \approx 2.9 \times 10^{10} \text{ Pa} = 29 \text{ GPa}$$

In all these examples, we could have used the 63 GPa for Young's modulus directly. This is because the denominator is a pure number. However, normally the best approach, as taken here, is to change all numbers into basic SI units and, therefore, to use pascals rather than megapascals or gigapascals when making the calculations. The final result, however, can be given in megapascals or gigapascals, whichever is appropriate.

---

## Example 4.4

### Problem

The rocks in the uppermost 1 km of a sedimentary basin have an average Young's modulus of 10 GPa, a Poisson's ratio of 0.25, and a density of 2300 kg m$^{-3}$. Find the difference, the differential stress, between the horizontal and vertical stress at the depth of 1 km if:

(a) there is no tectonic strain
(b) there is compressive tectonic strain of 0.1%.

### Solution

From Eq. (4.45) we get the vertical stress as

$$\sigma_v = \rho_r g z = 2300 \text{ kg m}^{-3} \times 9.81 \text{ ms}^{-2} \times 1000 \text{ m} \approx 2.26 \times 10^7 \text{Pa} = 22.6 \text{ MPa}$$

The horizontal stress can be obtained either from Eq. (4.51) or Eq. (4.53). If we use Eq. (4.53) then we have, from Eq. (4.27), Poisson's number $m = 1/v = 1/0.25 = 4.0$. From Eq. (4.53) the horizontal stress is then

$$\sigma_h = \frac{\sigma_v}{m - 1} = \frac{2.26 \times 10^7 \text{ Pa}}{4 - 1} \approx 7.5 \times 10^6 \text{ Pa} = 7.5 \text{ MPa}$$

The stress difference is thus

$$\sigma_v - \sigma_h = 22.6 \text{ MPa} - 7.5 \text{ MPa} = 15.1 \text{ MPa}$$

(b) Here we find again that the vertical stress is about 22.6 MPa, as in (a). But the horizontal compressive stress associated with the horizontal compressive strain in the basin must be added in order to find the stress difference. A strain of 0.1% is equal to 0.001, and Young's modulus is 10 GPa. To find the horizontal stress we use Eq. (4.55), so that

$$\sigma_h = \frac{\sigma_v}{m - 1} \pm \varepsilon_h E = \frac{2.26 \times 10^7 \text{ Pa}}{4 - 1} + 0.001 \times 1 \times 10^{10} \text{ Pa} \approx 1.75 \times 10^7 \text{ Pa} = 17.5 \text{ MPa}$$

It follows that the stress difference is

$$\sigma_v - \sigma_h = 22.6 \text{ MPa} - 17.5 \text{ MPa} = 5.1 \text{ MPa}$$

### Example 4.5

Problem

A hexagonal rock bar, generated by columnar joints in a lava flow (Fig. 4.16), has a hexagonal cross-section with a radius (of an inscribed circle) of 20 cm. The cross-sectional area of a polygon of $n$ sides inscribed in a circle of radius $R$ is given by (Spiegel *et al.*, 2009)

$$A = \frac{1}{2}nR^2 \sin \frac{2\pi}{n}$$

The bar is 5 m tall, has a Young's modulus of 10 GPa, and is subject to a compressive force of 100 kN at its top. Calculate the strain energy in the bar.

Solution

First we must find the stress on the cross-sectional area of the bar, which has the given cross-sectional area

$$A = \frac{1}{2}nR^2 \sin \frac{2\pi}{n}$$

A hexagon has six sides, so that $n = 6$, and the radius $R = 0.2$ m, and $2\pi = 360°$. Thus, the area is

$$A = \frac{1}{2}6 \times 0.2^2 \times \sin \frac{360°}{6} = 0.1 \text{ m}^2$$

**Fig. 4.16** View north, joints (mostly hexagonal) form columns in an interglacial lava flow in South Iceland (the person provides a scale).

From Eq. (1.1) the compressive stress is thus

$$\sigma = \frac{F}{A} = \frac{1 \times 10^5 \, \text{N}}{0.1 \, \text{m}^2} = 1 \times 10^6 \, \text{Pa} = 1 \, \text{MPa}$$

From Eq. (4.69) the strain energy density, or strain energy per unit volume, is

$$U_0 = \frac{\sigma^2}{2E} = \frac{(1 \times 10^6 \, \text{Pa})^2}{2 \times 10^{10} \, \text{Pa}} = 50 \, \text{J m}^{-3}$$

This is per unit volume. The volume of the column is its area times its height, or

$$V = A \times H = 0.1 \, \text{m}^2 \times 5 \, \text{m} = 0.5 \, \text{m}^3$$

The total strain energy stored in the rock column or bar is thus

$$U = 50 \, \text{J m}^{-3} \times 0.5 \, \text{m}^3 = 25 \, \text{J}$$

## 4.11 Exercises

4.1 Define Poisson's ratio, Poisson's number, shear modulus, bulk modulus, compressibility, and Lamé's constant and give their units.

4.2 How do materials with Poisson's ratios of 0.5, 0.0, and with negative values respond to stresses? Give examples of materials with approximately these Poisson's ratios and explain why materials do not have larger ratios than 0.5. Are there known rocks with negative Poisson's ratios?

4.3 What are typical Young's moduli for solid rocks? What is the maximum range of Young's moduli for rocks?

4.4 For a totally fluid magma chamber at a shallow depth in a basaltic lava pile, how would Young's modulus and Poisson's ratio typically change from the host rock to the fluid chamber?

4.5 When modelling crustal stresses in layered rocks, it is often assumed that Poisson's ratio is the same for all the layers (a common value used is 0.25), whereas Young's modulus is assumed to vary between layers. What are the justifications for these assumptions?

4.6 What are primary and secondary bonds? Which are the main bonds in silicates and carbonates?

4.7 Explain the repulsive and attractive forces between atoms in solids. Use a diagram to show how these forces change with separation between atoms.

4.8 Explain Young's modulus in terms of the force separation between atoms in a solid. How does the value (magnitude) of Young's modulus depend on the shape of the bond–energy curve?

4.9 What is a typical difference between a static and a dynamic Young's modulus for rocks? Give examples of rock-fracture processes where modelling using (a) a static modulus and (b) a dynamic modulus would be most appropriate.

4.10 What is an effective elastic constant and how does it relate to the concept of upscaling? How do fractures and other discontinuities normally affect the effective Young's modulus?

4.11 Define lithostatic state of stress. Under what conditions are crustal rocks most likely to approach this state of stress? Why? What is meant by saying that a magma chamber is in lithostatic equilibrium with its host rock? During which periods would we expect the chamber to be in lithostatic equilibrium with the host rock, and when not?

4.12 Define plane stress and plane strain and give examples when in modelling rock fractures it would be appropriate to use plane stress and when appropriate to use plane strain.

4.13 Explain the concept of elastic strain energy and give its units.

4.14 Define strain energy density and resilience.

4.15 In a volcanic rift zone at a certain depth the vertical stress $\sigma_1 = 10$ MPa, and the horizontal principal stresses are $\sigma_2 = 5$ MPa and $\sigma_3 = 0.1$ MPa. If the average crustal density to this depth is 2400 kg m$^{-3}$, Young's modulus is 10 GPa and Poisson's ratio 0.25, calculate (a) the likely depth for this state of stress, and (b) the principal strains in the directions of the principal stresses.

4.16 A rock specimen, of a uniform circular cross-section with an initial diameter of 2 cm and axial length of 10 cm, is subject to an axial compressive stress of 40 MPa. During the test, the cylindrical specimen increases its diameter by 0.003 mm and decreases its length (shortens) by 0.06 mm. Find the Young's modulus and Poisson's ratio of the rock.

4.17 A rock specimen has a bulk modulus of about 29 GPa and a shear modulus of about 15.7 GPa. Find the Poisson's ratio and Young's modulus of the rock.

4.18 The rocks in the uppermost 2 km of a volcanic rift zone have an average Poisson's ratio of 0.25, an average Young's modulus of 20 GPa, and an average density of 2500 kg m$^{-3}$. Assuming that there is no lateral strain, find the theoretical difference between the vertical and the horizontal stress at the depth of 2 km. In case there was a totally fluid magma chamber at the depth of 2 km, could this stress difference be reached? Why?

4.19 A rock specimen with a circular cross-section, 10 cm long and 2 cm in diameter, is subject to an axial compression of 20 MPa. Young's modulus of the specimen is 40 GPa. Calculate (a) the strain energy density and (b) the total strain energy stored in the specimen.

4.20 An axial tensile force of 80 kN is applied to a bar of circular cross-section with a Young's modulus of 80 GPa and Poisson's ratio of 0.25. The initial diameter of the bar is 4 cm and it is 2 m long. Calculate the axial stress, the elongation, and the total strain energy stored in the bar.

# References and suggested reading

Afrouz, A. A., 1992. *Practical Handbook of Rock Mass Classification Systems and Modes of Ground Failure*. London: CRC Press.

Asaro, R. J. and Lubarda, V. A., 2006. *Mechanics of Solids and Materials*. Cambridge: Cambridge University Press.

Barber, J. R., *Elasticity*, 2nd edn. 2002. London: Kluwer.

Bell, F. G., 2000. *Engineering Properties of Soils and Rocks*, 4th edn. Oxford: Blackwell.

Brady, B. H. G. and Brown, E. T., 1993. *Rock Mechanics for Underground Mining*, 2nd edn. London: Kluwer.

Caddell, R. M., 1980. *Deformation and Fracture of Solids*. Upper Saddle River, NJ: Prentice-Hall.

Callister, W. D., 2007. *Materials Science and Engineering*, 7th edn. New York: Wiley.

Carmichael, R. S., 1989. *Practical Handbook of Physical Properties of Rocks and Minerals*. London: CRC Press.

Cheng, C. H. and Johnston, D. H., 1981. Dynamic and static moduli. *Geophysical Research Letters*, **8**, 39–42.

Chilingar, G. V., Serebryakov, V. A., and Robertson, J. O., 2002. *Origin and Prediction of Abnormal Formation Pressures*. Amsterdam: Elsevier.

Chou, P. C. and Pagano, N. J., 1992. *Elasticity. Tensor, Dyadic, and Engineering Approaches*. New York: Dover.

Cottrell, A. H., 1964. *The Mechanical Properties of Matter*. New York: Wiley.

Davis, G. H. and Reynolds, S. J., 1996. *Structural Geology of Rocks and Regions*, 2nd edn. New York: Wiley.

de Podesta, M., 2002. *Understanding the Properties of Matter*, 2nd edn. London: Taylor & Francis.

Fyfe, W. S., Price, N. J., and N. J., Thompson, A. B., 1978. *Fluids in the Earth's Crust*. Amsterdam: Elsevier.

Gilman, J. J., 2003. *Electronic Basis of the Strength of Materials*. Cambridge: Cambridge University Press.

Goodman, R. E., 1989. *Introduction to Rock Mechanics*, 2nd edn. New York: Wiley.

Green, D. J., 1998. *An Introduction to the Mechanical Properties of Ceramics*. Cambridge: Cambridge University Press.

Gudmundsson, A., 2004. Effects of Young's modulus on fault displacement. *C. R. Geoscience*, **336**, 85–92.

Gudmundsson, A., 2006. How local stresses control magma-chamber ruptures, dyke injections, and eruptions in composite volcanoes. *Earth-Science Reviews*, **79**, 1–31.

Jaeger, J. C., Cook, N. G. W., and Zimmerman, R. W., 2007. *Fundamentals of Rock Mechanics*, 4th edn. Oxford: Blackwell.

Jumikis, A. R., 1979. *Rock Mechanics*. Clausthal, Germany: Trans Tech Publications.

Kuznir, N. J. and Park, R. G., 1987. The extensional strength of the continental lithosphere. In: Coward, M. P., Dewey, J. F., and Hancock, P. L. (eds.), *Continental Extension Tectonics*. Geological Society of London Special Publication 28. London: Geological Society of London, pp. 35–52.

Lakes, R., 1987. Foam structures with a negative Poisson's ratio. *Science*, **235**, 1038–1140.

Lautrup, B., 2005. *Physics of Continuous Matter*. Boston: Institute of Physics Publishing.

Link, H., 1968. On the correlation of seismically and statically determined moduli of elasticity of rock masses. *Felsmechanik und Ingenieurgeologie*, **4**, 90–110 (in German).

Mura, T., 1987. *Micromechanics of Defects in Solids*, 2nd edn. London: Kluwer.

Myrvang, A., 2001. *Rock Mechanics*. Trondheim: Norway University of Technology (NTNU) (in Norwegian).

Nemat-Nasser, S. and Hori, M., 1999. *Micromechanics. Overall Properties of Heterogeneous Materials*, 2nd edn. Amsterdam: Elsevier.

Niklas, K. J., 1992. *Plant Biomechanics*. Chicago, IL: The University of Chicago Press.

Philip, M. and Bolton, W., 2002. *Technology of Engineering Materials*. London: Butterworth-Heinemann.

Priest, S. D., 1993. *Discontinuity Analysis for Rock Engineering*. London: Chapman & Hall.

Qu, J. and Cherkaoui, M., 2006. *Fundmentals of Micromechanics of Solids*. New York: Wiley.

Saada, A. S., 2009. *Elasticity Theory and Applications*, 2nd edn. London: Roundhouse.

Sadd, M. H., 2005. *Elasticity: Theory, Applications, and Numerics*. Amsterdam: Elsevier.

Schön, J. H., 2004. *Physical Properties of Rocks: Fundamentals and Principles of Petrophysics*. Amsterdam: Elsevier.

Slaughter, W. S., 2002. *The Linearized Theory of Elasticity*. Berlin: Birkhauser.

Solecki, R. and Conant, R. J., 2003. *Advanced Mechanics of Materials*. Oxford: Oxford University Press.

Spiegel, M. R., Lipschutz, S., and Lin, J., 2009. *Mathematical Handbook of Formulas and Tables*, 3rd edn. New York: McGraw-Hill.

Turcotte, D. L. and Schubert, G., 2002. *Geodynamics*, 2nd edn. Cambridge: Cambridge University Press.

Turton, R., 2000. *The Physics of Solids*. Oxford: Oxford University Press.

Urry, S. A. and Turner, P. J., 1986. *Solving Problems in Solid Mechanics* Volume 2. Harlow, UK: Longman.

van der Pluijm, B. A. and Marshak, S., 1997. *Earth Structure*. New York: McGraw-Hill.

Verhoogen, J., Turner, F. J., Weiss, L. E., Wahrhaftig, C., and Fyfe, W. S., 1970. *The Earth*. New York: Holt, Rinehart and Winston.

# Loading of brittle rocks to failure

## 5.1 Aims

Brittle rocks fail through fracture. Sometimes a single fracture develops during failure, sometimes several or many fractures. How and when rocks fail under loading has been studied extensively in laboratory experiments. Much of the theoretical background derives from studies of small rock specimens subject to various types of loading. The main scientific fields dealing with experimental rock deformation are rock mechanics and rock physics. The principal results are very thoroughly treated in many textbooks and monographs, so that only a brief overview will be given here. The main aims of this chapter are to:

- Describe the general behaviour of rock under loading.
- Define dilatancy.
- Discuss the various types of rock strength.
- Define the secant modulus and the tangent modulus.
- Describe the main stages leading to brittle failure.
- Describe strain hardening, strain softening, and the brittle–ductile transition.
- Define the main factors that affect the depth of the brittle–ductile transition.
- Define and describe the main mechanisms of ductile deformation in rocks.

## 5.2 Behaviour of rock under loading

At the Earth's surface, most solid rocks subject to short-term loading behave elastically. When the load on a rock exceeds a certain critical value, the rock fails through the formation of a fracture. Surface rocks thus generally behave as brittle solids. A well-known type of brittle failure of a solid rock at the surface is the fracturing that occurs when the load of a geological hammer is applied, as an impact, to the rock. Since rocks transmit seismic waves, they clearly behave elastically under dynamic loading in the crust, except where there are fluid-filled reservoirs. When the rate of loading is similar to the velocities of the seismic waves, they are referred to as dynamic; when the rates are much slower, they are referred to as quasi-static or static.

Rocks also behave elastically under static loads. This applies both to surface rocks as well as rocks at considerable crustal depths. During static loads such as associated with magma-chamber deflation and inflation, the crustal rocks from the surface and at least down to depths close to the top of the crustal magma chamber respond approximately elastically. This is well known from numerous geodetic studies of active volcanoes worldwide (Fig. 3.8). Also, crustal rocks close to active fault zones behave, to a first approximation, elastically; this behaviour is the basis of the elastic rebound theory of earthquakes (Scholz, 1990; Turcotte and Schubert, 2002).

Loading to failure in the upper brittle part of the crust results in formation of new fractures or displacement on existing extension or shear fractures. While fracture formation is common, most brittle deformation presumably occurs through opening across existing extension fractures or slip on existing shear fractures or faults. A fault zone subject to appropriate loading generates seismic or aseismic transient fault slip in the brittle part of the crust. Seismogenic faulting means that the slip is accompanied by one or more earthquakes, whereas aseismic slip does not generate a typical earthquake. The aseismic slip may be 'slow', that is, generated through energy release over many hours, weeks, or months, or years, and such slips are sometimes referred to as 'silent earthquakes' (Jordan, 2003; Segall, 2010).

At great depths, sliding through cataclastic flow and, at still greater depths, plastic deformation changes the fault zone into a shear zone. The shell or layer where brittle failure occurs in the crust or lithosphere is referred to as the **seismogenic layer**. Below the seismogenic layer, there is generally ductile deformation in the crust at geological strain rates. Laboratory experiments can be used to produce fractures and fault slips mechanically analogous to those inferred and observed in the seismogenic layer. In particular, the geometry of the stress–strain curve obtained in such experiments gives very important information about the behaviour of the rock under loading.

## 5.3  The experimental stress–strain curve

Consider first a schematic general stress–strain ($\sigma$–$\varepsilon$) curve for the loading to failure under uniaxial compressive loading (Fig. 5.1). This kind of 'complete' stress–strain curve was first obtained in rock-physics laboratory experiments in 1966 (Hudson and Harrison, 1997) and has given a wealth of information on the behaviour of solid rocks under loading.

We will describe the curve in more detail in Section 5.4 but for convenience we summarise the main results first. The main four parts of the curve, discussed below, are numbered 1–4 (Fig. 5.1):

1. The first portion of the $\sigma$–$\varepsilon$ curve is concave upwards and thus non-linear. This portion is non-linear because of 'settling down' of the tested rock specimen or sample. That is to say, the pre-existing microcracks, pores, and other flaws in the rock tend to close (because of the compressive stress). Part of the non-linear behaviour at this stage, however, may be related to the preparation of the rock specimen itself being imperfect in that the ends of the specimen are commonly non-parallel. Except in very porous rocks, this deformation is largely reversible, that is, elastic.

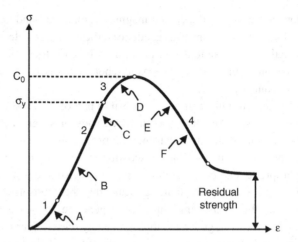

Fig. 5.1 Typical experimental stress–strain curve obtained when a rock specimen is loaded to failure. Portions 1–4 are the main parts of the curve as commonly described. Portions A–F show a more detailed division of the curve into parts. Both sequences, 1–4 and A–F, are described in detail in the text. Residual strength is the strength of the rock specimen after macroscopic failure and corresponds to the kind of strength to be expected in the (highly fractured) crust.

2. The second portion of the $\sigma - \varepsilon$ curve is roughly linear. There is elastic deformation of grains and pores. There may be some sliding on pre-existing cracks resulting in some irreversible deformation, but this is usually a minor effect. It is this linear part of the curve that is used to determine the Young's modulus of the rock. No microcracks are forming at this stage, as is indicated by the absence of any seismicity (acoustic emission) associated with the loading of the specimen.

3. The third portion of the $\sigma - \varepsilon$ curve is again non-linear, partly because of microfracturing and **dilatancy**, that is, inelastic volume increase of the specimen. Dilatancy increases specimen porosity and thus, commonly, its permeability. This portion is characterised by stable microcrack growth, most of the cracks being parallel with the maximum principal compressive stress, $\sigma_1$. The cracks grow incrementally and do not as yet link into a large throughgoing fracture. The departure from linear-elastic behaviour, that is, the non-linear feature of the curve, is reflected in (a) the observed dilatancy, (b) the increased seismicity during the loading of the specimen, and (c) the decreased seismic velocities. The seismic activity (acoustic emission) commonly starts when the stress reaches about 40–50% of the stress at the peak of the curve (the peak or compressive strength, as defined below). From this stress level, the microcracking continues to increase until the complete failure through a macroscopic fracture is reached, as occurs in the fourth portion.

4. The fourth portion of the $\sigma - \varepsilon$ curve is also non-linear and characterised by the development of a macroscopic fracture, that is, a large-scale failure of the entire specimen. At this stage, the microcracks develop into zones that subsequently become shear or extension fractures (depending on the loading conditions). In the compressive regime, the main failure is normally through one or two macroscopic shear fractures, that is, faults. The linking of the microcracks, either extension fractures (parallel with $\sigma_1$) or shear fractures

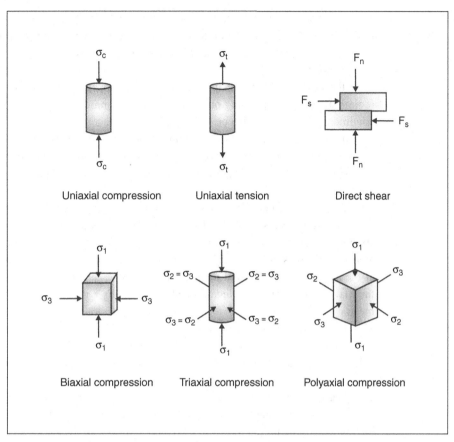

**Fig. 5.2**   Main types of loading of laboratory rock specimens. Among the most common are triaxial compression, where the confining pressure is equal to $\sigma_2 = \sigma_3$, and uniaxial compression. Biaxial compression is suitable for plane-stress experiments and polyaxial compression for the most general state of stress in the Earth's crust. A detailed discussion of these experiments is provided by Hudson and Harrison (1997).

(oblique to $\sigma_1$) or both, into macroscopic fractures, primarily faults, is the main early process at this stage. The main later process is slip along the macroscopic fault that has formed in the specimen. Fault slip occurs only if the frictional resistance along the fault plane is overcome. Thus, the frictional resistance gives rise to strength, referred to as the **residual strength** of the specimen after loading to macroscopic brittle failure.

Here we make the following observations about the $\sigma$–$\varepsilon$ curve:

- Permanent deformation, that is, non-elastic behaviour, begins at the point marked by $\sigma_y$. This is referred to as the **yield stress** or **yield strength** of the specimen.
- The maximum value of the $\sigma$–$\varepsilon$ curve (its highest point) is referred to as the peak strength, ultimate strength, or **uniaxial compressive strength**. It is denoted by $C_0$ in Fig. 5.1. Here 'uniaxial' is used since the experimental results shown in Fig. 5.1 refer to uniaxial compressive tests (cf. Fig. 5.2).

- The development of the macroscopic fracture, a fault, that leads eventually to brittle failure of the specimen is a continuous process from the peak representing the uniaxial compressive strength to the right end of the curve representing the residual strength of the specimen.
- Here the failure is brittle, that is, a fracture develops, and occurs in the compressive regime. However, depending on loading and environmental (pressure and temperature) conditions, failure may also be through plastic flow.
- From the $\sigma-\varepsilon$ curve, the Young's modulus of the specimen can be determined in two ways: either from the slope of the curve at a given point or from the slope of a line connecting two points on the curve. The first method yields a tangent modulus, the second a secant modulus.
- The **tangent modulus** is commonly measured at the point on the linear part of the $\sigma-\varepsilon$ curve where the stress is 50% of the stress corresponding to the uniaxial compressive strength of the specimen. Since the tangent modulus is given by the slope of the curve at a point, it follows that $E = \mathrm{d}\sigma/\mathrm{d}\varepsilon$. The modulus is constant in the linear portion 2, zero at the peak of the curve, and **negative** in portion 4, that is, during the development of the macroscopic fracture. Negative values of Young's modulus have no clear physical meaning, in particular when viewed in terms of behaviour at an atomic scale (Chapter 4). This is because, at an atomic scale, Young's modulus only has a meaning so long as there is no rupture of the atomic structure (Figs. 4.6 and 4.7). Because of these negative values in the fourth portion of the curve, the tangent modulus is less used than the secant modulus.
- The **secant modulus** is a direct measure of the slope of the linear part (portion 2) of the $\sigma-\varepsilon$ curve and thus may be determined anywhere along that part (Fig. 4.5). With repeated loading and unloading of the same specimen, the slope of the linear part of the $\sigma-\varepsilon$ curve gradually decreases, so that the secant modulus decreases, but it neither becomes zero nor negative. The steeper the linear part of the stress–strain curve, the stiffer the rock, as indicated schematically in Fig. 5.3.

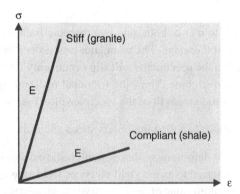

**Fig. 5.3** Schematic version of the linear part of an ideal stress–strain curve, which slope is equal to Young's modulus. When the slope is steep, Young's modulus is high and the rock is referred to as stiff (here granite). When the slope is gentle, Young's modulus is moderate or low, and the rock is referred to as compliant or soft (here shale).

The uniaxial **compressive strength** is often used in rock mechanics, but for rock-fracture mechanics considerations the **tensile strength** is more useful. While the uniaxial compressive strength is easily inferred from the stress–strain curve, there are no simple measures of the uniaxial tensile strength. Uniaxial tests are, in fact, rarely used to determine the tensile strength. It is much more common to use an indirect measure of the tensile strength, such as bending tests. The tensile strength is then inferred from the tensile stress calculated (using elasticity theory) to have been present in the specimen at fracture failure.

More widely used measured tensile strengths are those obtained in hydraulic fracture experiments in drill holes (Chapter 6). Since these tests measure the tensile strength of the rock in response to fluid pressure, the results are suitable for all types of hydrofractures (Chapter 11). In addition, hydraulic-fracture measurements of the in-situ tensile strength yield a surprisingly narrow range of values, mostly between 0.5 and 6 MPa, the maximum being about 9 MPa (Chapter 7).

A detailed description of the methods and limitations for testing of laboratory specimens to determine various rock parameters and constants is beyond the scope of the book; it is provided in several books on rock mechanics and experimental rock deformation (Farmer, 1983; Hudson and Harrison, 1997; Shimada, 2000; Stavrogin and Tarasov, 2001; Peterson and Wong, 2005; Jaeger *et al.*, 2007). Nevertheless, there are some additional points that should be made in order to clarify the use and limitations of specimen testing for understanding rock behaviour, particularly in the field.

The first is that the specimen size affects the magnitudes of the measured parameters. Generally, the larger the size of the specimen, the lower will be its strength (tensile, shear, and compressive). These observations are directly related to the increasing number of flaws such as pores, cracks, and other stress raisers (Chapters 4 and 6) as the size of the specimen is increased. The increase in number of flaws with increasing size or volume of a rock body, and its effect on elastic constants (Chapter 4) and strength (Chapter 7), are well established. The effects of rock-body size on strength may also be related to **Weibull's** statistical weakest-link theory. This theory indicates that one reason why the strength of a larger volume of a solid material is less than that of a small volume of the same material is due to the larger probability of finding a flaw in the large volume than in the small volume. One presentation of Weibull's strength consideration is given by the following equation:

$$T_{\text{L}} = T_0 \left( \frac{V_0}{V_{\text{L}}} \right)^{1/n} \tag{5.1}$$

where, in this case, $T_{\text{L}}$ is the tensile strength of a comparatively large solid (rock) body of volume $V_{\text{L}}$, and $T_0$ and $V_0$ are the tensile strength and volume, respectively, of a smaller body of the same material. The factor $n$ is characteristic of the material. Clearly, as the volume $V_{\text{L}}$ increases in relation to $V_0$, the ratio $V_0/V_{\text{L}}$ gets smaller. With $n$ positive and larger than 1.0 then, as $V_0/V_{\text{L}}$ becomes smaller, the tensile strength $T_{\text{L}}$ also becomes a smaller fraction of the tensile strength $T_0$.

Weibull's theory is entirely statistical in that it does not consider the linking of flaws into macroscopic fractures (Chapter 7) or the fracture propagation through the solid body (Chapters 13 and 14). The theory is suitable for tensile loading, but less so for compressive loading. This follows because, under compressive loading, a stress-raising flaw may give rise to crack initiation. But the propagation and linking of cracks into larger fractures, thus the failure of the entire solid body, is a complex process that needs mechanical considerations (Chapter 7).

The loading conditions in rock-mechanics experiments may also differ and affect the specimen strength. We have already referred to the specimen end effects, and how they may affect measurement results. Also, the type of loading affects the results. There are principally six types of loading conditions (Hudson and Harrison, 1997), namely (Fig. 5.2):

- Uniaxial tension.
- Uniaxial compression.
- Direct shear.
- Biaxial compression.
- Triaxial compression.
- Polyaxial compression.

Uniaxial tension is not used much; for determining the tensile strength, indirect methods are generally preferred. Uniaxial compression is used quite often, and is the basis for the $\sigma-\varepsilon$ curve in Fig. 5.1. Perhaps the most commonly used loading conditions for rock specimens is triaxial compression (Fig. 5.2). Today, this method is normally taken to mean that the maximum principal compressive stress is larger than the intermediate and minimum principal compressive stresses, the latter two being equal. Thus, for the triaxial test we have $\sigma_1 > \sigma_2 = \sigma_3$. The reason why $\sigma_2 = \sigma_3$ is that the test is made in a pressure vessel so that both these stresses are equal to the fluid pressure.

There are also other aspects of the loading conditions that need to be considered. The loading may, for example, be either at a constant stress rate or constant strain rate. Which loading is used reflects on the question of cause and effect. Is the stress causing the strain in laboratory experiments or is the strain causing the stress? If the stress rate is held constant, then the result is normally a runaway fracture following the peak of the $\sigma-\varepsilon$ curve when more stress is applied to the specimen than it can sustain in its fractured state. Then portion 4 of the $\sigma-\varepsilon$ curve (Fig. 5.1) also implies that strain increases as stress decreases, a conclusion that may appear difficult to interpret. One interpretation, however, is that stress concentration around and within the newly formed, low-strength macroscopic fracture is driving the displacement, hence the strain, at a gradually lower specimen loading.

It might be argued that it is easier to interpret the $\sigma-\varepsilon$ curve for strain loading, that is, if the strain is the cause and the stress the effect. Then portion 4 of the curve could be interpreted as showing that the fracturing increases as a consequence of a decrease in its load-bearing capacity (its ability to sustain stress). This interpretation agrees with most stress–strain plots that have strain along the $x$-axis, that is, as an independent variable, and stress along the $y$-axis, that is, as the dependent variable. When the applied strain rate is reduced, Young's modulus as well as the uniaxial compressive strength are both reduced. This means that lowering the strain rate results in a flatter and gentler $\sigma-\varepsilon$ curve, similar to that observed for generally compliant and weak rocks.

While questions of cause and effect are no longer regarded as simple, either in physics or philosophy, it is logical to regard the stress as generally causing strain and fracture formation. This is partly because stress is closely related to force in Newtonian mechanics, whereas strain is not. We normally refer to a mechanical force as giving rise to some effects rather than vice versa – even if, as is well known, the force cannot be defined independently of mass and acceleration. Thus a force per unit area, namely stress, gives rise to elastic strain and, if large enough, to brittle failure of the rock. The way we plot the stresses and strains in a coordinate system is arbitrary and pure convenience. The choice of axes for the variables does not imply cause and effect any more than a plot of fracture lengths and apertures would imply that the length causes the aperture (or vice versa), although they are clearly (linearly) related.

## 5.4  The main stages leading to brittle failure

Many authors divide the $\sigma$–$\varepsilon$ curve into more portions than the four listed above. For the sake of completeness we shall give one example of such a thorough description, which, in addition, throws light on some of the details that were not elaborated in the previous description. The present description divides the curve into six portions (Farmer, 1983), marked A–F (Fig. 5.1). The stages are as follows:

A. As the rock is loaded, the pre-existing flaws, pores, and microcracks initially tend to close. In particular, those cracks that are orientated at high angles to the maximum principal compressive stress. This stage commonly gives rise to a non-linear behaviour, particularly in porous and weak rocks. Thus, the result is a concave upwards curve.

B. Here the rock shows a roughly a linear relation between stress and strain. The linear relations hold for both the axial (along the axis of the specimen) stress and strain as well as for the lateral stress and strain – the latter being related to the Poisson's ratio, which, however, tends to be low, particularly for stiff rocks when subject to uniaxial (unconfined) loading. The strain is essentially recoverable so that the behaviour may be regarded as elastic. There is little if any seismic activity during this stage. The main microcrack propagation only starts at the upper boundary of about 35–50% of the peak stress.

C. This portion is characterised by the beginning of dilation (Chapter 4). There is a near-linear increase in the specimen volume which, however, is offset against the increased compression to which the specimen is subject. In this portion, the near-linear stress–strain curve is, as in portion B, still nearly fully recoverable so that the rock behaviour continues to be approximately elastic. The propagation of microcracks at this stage occurs in a stable manner; most of the cracks propagate for only limited distances and independently of other microcracks. The microcracking is distributed throughout the specimen and the cracks have not as yet started to link into macroscopic fractures. The upper boundary of this portion coincides with the point of maximum specimen compaction and no volume change and is at about 80% of the peak strength.

D. This portion of the stress–strain curve is characterised by a rapid increase in microcracking and linking of the cracks into larger fractures. There is also an increase in the volume of the specimen. The propagation and linking of microcracks into clusters means that the cracking is no longer distributed throughout the specimen but rather con-centrated in high-stress zones that eventually develop into macroscopic fractures and potential faults.

E. In this portion of the stress–strain curve, the specimen has passed its peak strength but is still largely intact. It does, however, contain one or many crack clusters that are develop-ing into macroscopic fractures or faults. This portion is thus regarded as the starting of the development of the macroscopic failure of the specimen through faulting. The failure is characterised by a decreasing load-bearing capacity of the specimen, that is, by **strain softening**. During repeated loading of the specimen, the peak strength decreases with increasing strain and the strain concentrates on the macroscopic fractures and develops them further into faults.

F. In this portion of the stress–strain curve the specimen has essentially parted to form one or more faults that separate the specimen into blocks or **fault** walls. The blocks slide relative to each other and the main deformation mechanism is thus frictional sliding between blocks. During the shearing process, some secondary fractures may also form. The load, that is, the stress (or force), acting on the specimen tends to fall to a constant value, referred to as the residual strength. The residual strength is equal to the frictional resistance of the sliding blocks of the specimen. In active areas of faulting and fractur-ing in the crust, the in-situ shear strength thus corresponds to the residual strength of specimens in laboratory tests.

## 5.5  The brittle–ductile transition

With increasing crustal depth, the rock failure gradually changes from predominantly brittle to predominantly ductile. For rocks and concrete and other similar materials, failure of a specimen in the laboratory, and also along faults in nature, is characterised by strain softening, as indicated above. That is, the load-bearing capacity of the specimen/rock is less after failure than before. For many materials, however, the stress–strain curve continues to rise after the yield stress is reached (Fig. 5.4). This is referred to as **strain hardening** or work hardening because the yield stress increases, the material gets stronger, with further straining or working. This is particularly common in many metals and, in general, materials that undergo plastic deformation. Stiffening of the core of a fault zone (Chapter 14) because of precipitation of secondary minerals and general compaction may result in a mechanical behaviour similar to strain hardening.

The transition from brittle to ductile failure and rock behaviour is generally complex and involves many processes. The ductile processes themselves are mostly outside the scope of this book. However, the brittle–ductile transition is very important in relation with fault-zone behaviour at depth in the crust, and a summary of the main findings is therefore given here.

**Ductility** is the capacity for substantial change of shape in a material such as rock without gross fracturing. The main difference between ductile and brittle deformation is as follows:

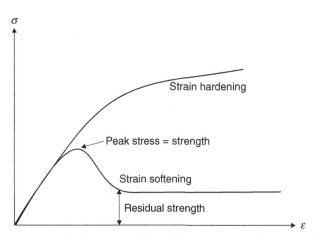

Fig. 5.4 Strain hardening, whereby the slope of the stress–strain curve continues to rise after the yield stress is reached, is common in many materials, particularly in metals. Strain softening, whereby the slope decreases after the peak strength is reached, is common in rocks. The slope of the strain-hardening curve normally becomes gradually less until it approaches zero (the curve becoming horizontal) at failure. The remaining strength in the rock specimen (in the laboratory) or in a fractured crustal segment in general for a strain-softening curve (as is common for rocks) is known as residual strength.

- Ductile deformation is **permanent**, whereas brittle elastic deformation is non-permanent. If the stress on a rock that behaves elastically is removed, the rock returns to its original shape, but not if the rock behaves as ductile.
- Ductile strain is more **uniformly distributed** throughout the rock than is brittle strain, which is always localised around and inside fractures.

The transition from brittle to ductile behaviour is well illustrated by the changes that take place in a major fault zone with increasing depth in the crust (Fig. 5.5; cf. Chapter 7). Close to the surface the deformation is essentially confined to the fault and the surrounding fault rock and is mostly brittle. With increasing depth and temperature, however, the fault zone enters, first, the brittle–ductile transition and, then, the ductile regime. The general change in behaviour is reflected in a combined failure criterion (Chapter 7).

In any particular crustal segment, the depth to the brittle–ductile transition depends on various factors. The main characteristics of the brittle–ductile transition may be summarised as follows:

1. Strain at the brittle–ductile transition may be as large 3–5% before failure. The strain can be so large if the confining pressure is high or if the depth is great. In fact, some take this as a **definition** of the brittle–ductile transition; that is, that the strain to failure is 3–5% and thus nearly an order of magnitude higher than the common strain before failure in the upper, brittle crust.
2. A large brittle **fault zone** changes into a ductile **shear zone** with increasing crustal depth. In this definition, the brittle part is referred to as fault zone, the ductile part as shear zone. The transition zone, however, belongs to both zones. Also, while the deformation in the

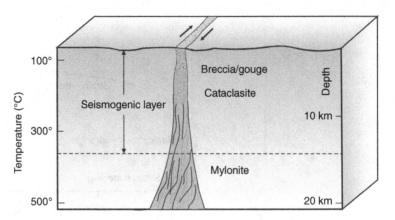

Vertical cross-section of a typical fault zone (here a vertical strike-slip fault) through a continental crustal segment. The upper part where the deformation is primarily brittle is known as fault zone, the lower part where the deformation is primarily ductile is known as shear zone. The maximum depth of seismogenic, brittle deformation marks the bottom of the seismogenic layer. The uppermost part of the fault zone is characterised by breccia and/or gouge, but with increasing depth cataclasite takes over. With increasing depth and pressure and temperature, the deformation becomes gradually more ductile, with mylonite characterising the deeper, ductile parts.

fault zone is predominantly brittle, and deformation in the shear zone predominantly ductile, both zones show some brittle and ductile deformation depending on the scale at which the observation is made. This is discussed further below in connection with cataclastic flow. The ductile part is normally much thicker than the brittle part, and the deformation in the ductile part is more evenly distributed and over larger regions than in the brittle part.

3. The depth to the brittle–ductile transition generally **increases**, that is, it occurs at greater depths if (cf. Paterson and Wong, 2005):
   • The **strain rate increases**, that is, the deformation is fast, such as in earthquakes.
   • The **pore-fluid pressure increases**. This follows because increasing pore-fluid pressure shifts the Mohr's stress circles to the left, towards the brittle regime and brittle failure.
   • If the deformation is in **extension** rather than compression. Generally, much higher confining pressure (and thus a greater crustal depth) is needed to bring about ductile behaviour in rocks during extension than compression. At low temperatures, even sedimentary rocks such as limestone are brittle up to confining pressures of 700 MPa, corresponding to continental crustal depths of more than 20 km.

4. The depth to the brittle–ductile transition generally decreases, that is, it occurs at shallower depths if:
   • The rock **porosity increases**. For the same rock type, a highly porous specimen changes to ductile behaviour at lower confining pressure than one with a low porosity. And porous rocks in general, such as many sedimentary rocks, become ductile at much lower confining pressures or crustal depths than low-porosity rocks, such as many igneous and metamorphic rocks. Essentially non-porous basalt, for example, is

brittle at room temperature to about 1000 MPa, but when its porosity is about 0.08, it may reach the brittle–ductile transition at about 300 MPa, the latter corresponding to an oceanic crustal depth of about 10 km.

- The **grain size increases**. Generally, for a given type of rock the larger the grain size, the more coarse-grained it is, the more ductile is the rock. But this holds only for given rock types: a coarse-grained gabbro is much more brittle than a fine-grained limestone, for example.
- If the rock **temperature increases**. Generally, the higher the temperature the more ductile will be the rock behaviour but only if the confining pressure also increases and/or the strain rate decreases. Thus, some rocks, such as dolerite, are still brittle close to melting temperatures if at atmospheric pressure. Generally, low-porosity igneous rocks such as basalt and granite show limited ductility up to temperatures of 600–900°C in compression experiments at confining pressures of 500 MPa (corresponding to depths of about 17–18 km in the continental crust) for strain rates of $10^{-1}$ to $10^{-3}\text{s}^{-1}$ (Paterson and Wong, 2005). Lower strain rates or higher temperatures are needed to make these rocks behave as ductile.

Generally, most low-porosity igneous rocks at room temperatures are brittle to confining pressures up to 1 GPa or more, which corresponds to continental crustal depths of more than 30 km. Porous lava flows, however, may reach the brittle–ductile transition at as low confining pressures as 30–100 MPa, corresponding to crustal depths of 1.5–4 km. Many porous sedimentary rocks, such as limestone, chalk, rocksalt, sandstone, and siltstone may reach the brittle–ductile transition, at room temperature, at confining pressures of 10–200 MPa, corresponding to crustal depths between about 0.5 km and 8 km.

The main three processes in ductile deformation are **cataclastic flow**, **diffusional flow** (or mass transfer), and **crystal plasticity**. Which mechanism dominates depends on various factors such as temperature, pressure, strain rate, grain size, rock composition, and the abundance of fluids where the deformation takes place, in a manner indicated above. Briefly, the main ductile processes may be described as follows:

**Cataclastic flow:**

- This is permanent strain due to distributed fracturing. The rock is broken into fragments and the deformation is through relative movement (sliding) of these fragments.
- Cataclasis is mesoscopic ductile flow through microscopic brittle fracturing and sliding – analogous to the flow of granular materials such as sand and is sometimes referred to as **granular flow**.
- During flow there is friction between the fragments and dilatancy, that is, volume increase due to porosity increase.
- Resistance to flow increases with overburden pressure, so that it occurs mainly at comparatively shallow depths.

**Diffusional flow:**

- The rock flows and changes shape by diffusion (transport) of material from one part of the body to another part, regardless of the path.

- In ductile deformation of rock, the main three diffusional processes are volume diffusion, grain-boundary diffusion, and pressure solution. Volume and grain-boundary diffusion are solid diffusion (do not depend on fluids being present), whereas pressure solution depends on fluids being in the rock.
- Diffusional flow is temperature dependent; the process commonly occurs at high temperatures (which are not needed for pressure solution).
- In an isotropic stress field, diffusion is non-directional, that is, a random-walk process.

**Crystal plasticity:**
- This is permanent deformation of a crystalline material because of slip and twinning. Both processes are essentially simple shearing and occur at constant volume.
- Is basically due to movement of **dislocations** through the crystal. The dislocations can glide or climb, depending on the temperatures and strain rates.
- A dislocation is a line defect, that is, a linear array of atoms that bounds an area in a crystal that has slipped relative to the rest of the crystal. At any instant, a dislocation is a line that separates slipped from non-slipped region in the crystal. Dislocations are of two main types. In an **edge dislocation**, the dislocation line is perpendicular to the shear direction of the slipping plane. In a **screw dislocation** the dislocation line is parallel with the shear direction of the slipping plane. An edge dislocation is thus similar to a mode II crack, and a screw dislocation similar to a mode III crack (Chapter 9).
- Mechanical twinning occurs when twins form (for example, in calcite) because of stresses; for example, when the shear stress on the potential twin plane reaches a certain limit. Twins can be used as palaeostress indicators.
- Crystal plasticity to a large degree depends on the existence of crystal defects. The conditions for plastic failure or flow, from a purely mathematical or macroscopic point of view, are given by the **Tresca** and **von Mises** criteria (Chapter 7).

# 5.6 Summary

- At and close to the Earth's surface, solid rocks generally behave elastically under geologically short-term loading. When the loading reaches the brittle strength of the rock, however, it fails through the development of one or more new or existing fractures. Depending on the loading conditions, crustal depth, fluid pressure, and other factors the fractures that develop during rock failure may be either extension fractures or shear fractures. The depth to which brittle failure takes place in the crust coincides with the part of the crust where earthquakes occur and is referred to as the seismogenic layer.
- During experimental loading of a rock specimen to failure, the stress–strain curve can be divided into four main portions. The first portion is non-linear and concave upwards and represents the closing of various pores, microcracks, and other flaws in specimen. The second portion is roughly linear and its slope is used to determine the Young's modulus of the specimen, either the tangent modulus or the secant modulus (or both). The third

portion is again non-linear, concave downwards, and is characterised by the growth of microcracks (and associated acoustic emission) and inelastic increase in volume, referred to as dilatancy. The fourth portion is also non-linear and is characterised by the development of a macroscopic fracture, usually a shear fracture (a fault), and represents the large-scale failure of the specimen.

- The remaining strength of the specimen after this stage is known as residual strength. The stress in the specimen at the onset of non-linear deformation in the third portion is known as yield stress or yield strength. The maximum value of the stress–strain curve (its highest point) is known as compressive strength. The tensile strength of a specimen is not inferred directly from the stress–strain curve but rather indirectly using methods such as bending tests. The strength, particularly the tensile strength, depends to a degree on the size of the rock body being tested and is related to Weibull's statistical theory of the weakest link.

- With increasing crustal depth (overburden pressure), rock failure gradually changes from primarily brittle to primarily ductile. The rock strain at the brittle–ductile transition is commonly 3–5%. A fault zone (brittle deformation) in the upper part of the crust normally changes to a shear zone (ductile deformation) at great depths.

- The depth to the brittle–ductile transition increases (is at greater depths) with increasing strain rate, increasing pore-fluid pressure, decreasing grain size (for a given rock type), and when the deformation is in extension rather than compression. The depth to the transition decreases (is at shallower depths) with increasing rock porosity and increasing temperature.

- The main processes in ductile deformation of rocks are cataclastic flow, diffusional flow, and crystal plasticity. Cataclastic flow is granular flow where the rock is broken into fragments and the deformation is due to relative movement of the fragments. Diffusional flow results in shape change of a rock body through material transport from one part of the body to another. Crystal plasticity is primarily related to the movement of dislocations through the crystals.

## 5.7  Main symbols used

$C_0$    compressive strength
$E$    Young's modulus (modulus of elasticity)
$T_L$    tensile strength of a comparatively large rock body in Weibull's weakest link theory
$T_0$    tensile strength
$V_L$    rock volume that is many times larger than $V_0$ in Weibull's weakest link theory
$V_0$    rock volume used to measure $T_0$
$\varepsilon$    normal strain
$\sigma$    normal stress
$\sigma_y$    yield strength (yield stress)

# 5.8 Worked examples

### Example 5.1

Problem

In the laboratory, a rod-like (circular) specimen is subject to an axial tensile force of 8kN (Fig. 5.6). Young's modulus of the rock is 20 GPa, the diameter is 25 mm, and the original axial length is 12 cm. Determine the increase in length of the specimen (Fig. 5.6).

Solution

From Eq. (1.1) the stress is

$$\sigma = \frac{F}{A}$$

We know the force, but not the cross-sectional area $A$ on which the force is operating. The area is given by

$$A = \pi R^2 = \pi \left(\frac{25 \text{ mm}}{2}\right)^2 = 491 \text{ mm}^2 = \frac{491 \text{ mm}^2}{1 \times 10^6} = 4.91 \times 10^{-4} \text{ m}^2$$

Here we first find the area in square millimetres and later change that value into square metres by dividing by a million. We could also have changed the 25 mm immediately into metres and then calculated the area directly in metres. This latter method is the one normally used in the book; the former method is just shown here as an example of a different approach.

The stress is then

$$\sigma = \frac{F}{A} = \frac{-8 \times 10^3 \text{ N}}{4.91 \times 10^{-4} \text{ m}^2} = -1.63 \times 10^7 \text{ Pa} = -16.3 \text{ MPa}$$

From Eq. (4.1) the strain is

$$\varepsilon = \frac{\sigma}{E} = \frac{-1.63 \times 10^7 \text{ Pa}}{2 \times 10^{10} \text{ Pa}} = -0.000\,815$$

The increase in length follows from Eq. (1.3), which can be rewritten in the form

$$\Delta L = \varepsilon \times L = -0.000\,815 \times 0.12 = -9.8 \times 10^5 \text{ m} \approx -0.1 \text{ mm}$$

The tensile stress of about 16 MPa is higher than in-situ rocks would normally tolerate. However, some rock specimens have tensile strengths of about 20 MPa, so that they might tolerate this tensile stress. Since extension is defined as negative, the result has a minus sign.

**Fig. 5.6** Example 5.1. Extension of a specimen or bar of a circular cross-section and subject to an axial tensile force, $F$. The (exaggerated) extension of the bar is indicated by the dark-grey part.

Similarly, since the force is tensile, the stress above has also a minus sign, indicating tension. In some equations and examples, the minus sign is used to indicate tensile stresses, strains, and forces. In many cases, however, the calculations may be performed using absolute values, that is, the magnitudes of the values regardless of their algebraic signs. This follows, because it is normally quite clear if the problem deals with tensile or compressive stresses and strains.

### Example 5.2

Problem

A solid specimen of a circular cross-section is made so that half of the specimen is a cylinder with a diameter of 25 mm, whereas the other half has a diameter of 50 mm (Fig. 5.7). The specimen is subject to an axial tensile force of 50 kN acting on the faces (ends) of the thick and the thin part of the specimen.

(a)  Calculate the stresses in the sections marked A–A′ and B–B′.
(b)  How would the results be relevant to stress variations in the lithosphere?

Solution

For section A–A′, the cross-sectional area $A$ is given by

$$A = \pi R^2 = \pi \left(\frac{50 \text{ mm}}{2}\right)^2 = 1963 \text{ mm}^2 = \frac{1963 \text{ mm}^2}{1 \times 10^6} = 1.96 \times 10^{-3} \text{ m}^2$$

From Eq. (1.1) the stress on area $A$ is

$$\sigma = \frac{F}{A} = \frac{-5 \times 10^4 \text{ N}}{1.96 \times 10^{-3} \text{ m}^2} = -2.55 \times 10^6 \text{ Pa} = -25.5 \text{ MPa}$$

For section B–B′, the cross-sectional area $B$ is given by

$$B = \pi R^2 = \pi \left(\frac{25 \text{ mm}}{2}\right)^2 = 490.8 \text{ mm}^2 = \frac{490.8 \text{ mm}^2}{1 \times 10^6} \approx 4.91 \times 10^{-4} \text{ m}^2$$

From Eq. (1.1) the stress on area $B$ is

$$\sigma = \frac{F}{A} = \frac{-5 \times 10^4 \text{ N}}{4.91 \times 10^{-4} \text{ m}^2} = -1.02 \times 10^8 \text{ Pa} = -102 \text{ MPa}$$

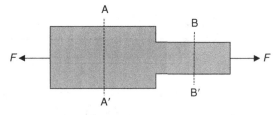

**Fig. 5.7**  Example 5.2. A cylindrical specimen with different cross-sections. The diameter A–A′ is 50 mm, whereas the diameter B–B′ is 25 mm. The axial tensile force $F$ is −50 kN.

A tensile stress of about 25 MPa would very rarely be tolerated by a rock specimen. The maximum measured laboratory tensile strengths of some specimens reach about 30 MPa, but most tensile strengths are below about 20 MPa. No rock specimen would tolerate tensile stress of 102 MPa. Many other materials, however, have much higher tensile strengths and can easily sustain tensile stress of 100 MPa without rupture.

(b) If an elastic rod varies in diameter, or an elastic plate varies in thickness or lateral dimensions, so that the cross-sectional area on which a constant force operates changes from one section through the body to another, then the resulting stress magnitude will also change. This follows directly from the definition of stress in Eq. (1.1). The Earth's lithosphere, for example, shows great variations in thickness. It follows that plate-tectonic forces, such as ridge push or slab pull, give rise to very different stresses depending on which part of the lithosphere they operate. For example, the continental lithosphere is generally much thicker than the oceanic lithosphere, so that abrupt stress variations are expected at contacts between these parts of the lithosphere. Well-known examples of direct contacts between continental and oceanic lithosphere are passive margins. At such margins, the ridge-push forces are likely to decrease rapidly into the continents, as the cross-sectional areas on which the forces operate increase in size. This change shows similarly with the present problem. There are, however, also stress concentrations at continental margins because of difference in the mechanical properties between a typical oceanic lithosphere and a typical continental lithosphere (Chapter 6).

## Example 5.3

### Problem

A roughly homogeneous, isotropic rock plate of width 40 mm, thickness 10 mm, and length 500 mm (Fig. 5.8), with an original, undeformed volume $V_u$ of $0.2 \times 10^6$ mm$^2$ has a Young's modulus of 50 GPa and a Poisson's ratio of 0.25. The plate is subject to a compressive force of 500 N at the faces of its lateral ends (Fig. 5.8). Assume the plate faces (edges) to be principal stress planes (they are free surfaces). Determine:

(a) the changes in the volume of the plate during the loading
(b) whether or not the plate would fail in compression.

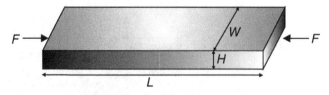

Fig. 5.8    Example 5.3. A homogeneous, isotropic rock plate has a thickness $H = 10$ mm, width $W = 40$ mm, and length $L = 500$ mm (the figure is not to scale). A compressive force $F = 500$ N acts on the edges of the plate and changes its volume.

Solution

(a) The compressive force operates on the faces with areas of 40 mm × 10 mm = 400 mm². This area is equivalent to

$$A = \frac{400 \text{ mm}^2}{1 \times 10^6} = 4 \times 10^{-4} \text{ m}^2$$

From Eq. (1.1) the stress on each area $A$ is

$$\sigma = \frac{F}{A} = \frac{500 \text{ N}}{4 \times 10^{-4} \text{ m}^2} = 1.25 \times 10^6 \text{ Pa} = 1.25 \text{ MPa}$$

Since the force is compressive, the stress is compressive and positive. Because the edges are principal stress planes, this stress must be a principal stress. In fact, since it is the only non-zero stress, it must be $\sigma_1$. For isotropic, homogeneous rocks, the directions of the principal strains coincide with those of the principal stresses. When there is axial compression there will be lateral expansions. To find the change in volume due to the axial compression, we calculate each of the principal strains from Eq. (4.3) thus:

$$\varepsilon_1 = \frac{\sigma_1}{E} = \frac{1.25 \times 10^6 \text{ Pa}}{5 \times 10^{10} \text{ Pa}} = 2.5 \times 10^{-5}$$

The transverse strains due to the axial stress $\sigma_1$ are

$$\varepsilon_2 = \varepsilon_3 = -v\frac{\sigma_1}{E} = -v\varepsilon_1 = -0.25 \times 2.5 \times 10^{-5} = -6.25 \times 10^{-6}$$

The changes in length $\Delta L$, width $\Delta W$, and thickness $\Delta H$ of the plate are as follows (Fig. 5.8):

$\Delta L = 500 \text{ mm} \times 2.5 \times 10^{-5} = 0.0125 \text{ mm}$, which means reduction in length
$\Delta W = 40 \text{ mm} \times (-6.25 \times 10^{-6}) = -0.000\ 25 \text{ mm}$, which means increase in width
$\Delta H = 10 \text{ mm} \times (-6.25 \times 10^{-6}) = -0.000\ 0625 \text{ mm}$, which means increase in height or thickness.

The final or deformed volume $V_d$ of the plate is thus

$$V_d = 499.9875 \times 40.000\ 25 \times 10.000\ 0625 = 199\ 997.4999 \text{ mm}^3$$

Thus the difference between the original and final volume is

$$V_u - V_d = 200\ 000 \text{ mm}^3 - 199\ 997.4999 \text{ mm}^3 = 2.5 \text{ mm}^3$$

Because of the axial compressive stress, the total volume of the plate is thus reduced by about 2.5 mm³ or about $2.5\text{mm}/(1 \times 10^9) = 3 \times 10^{-9} \text{ m}^3$

(b) The applied compressive stress is 1.25 MPa, whereas the laboratory compressive strength is commonly 100 MPa. Thus, the plate would be able to sustain this compressive stress without failure.

## Example 5.4

Problem

A rectangular small rock body of length 8 cm, width 6 cm, and height or thickness 4 cm, is subject to a mean (spherical) compressive stress of 180 MPa. The rock is a part of a comparatively compliant (soft) layer, with a Young's modulus of 20 GPa. Its Poisson's ratio is 0.25. For the given loading, find:

(a) the dilation
(b) the change in volume $\Delta V$ of the rock body
(c) the likely crustal depth of the body if the average crustal density to this depth is 2800 kg m$^{-3}$.

Solution

(a) The initial or undeformed volume of the rock body is $V_{\mathrm{u}} = 8 \times 6 \times 4 = 192$ cm$^3$. From Eq. (4.25) the dilation $\Delta$ is related to the bulk modulus $K$ and the mean stress $\bar{\sigma}$ through the equation

$$K = \frac{\bar{\sigma}}{\Delta}$$

To find the bulk modulus $K$, we rewrite Eq. (4.29) to obtain

$$K = \frac{E}{3(1 - 2v)} = \frac{2 \times 10^{10}\,\mathrm{Pa}}{3(1 - 0.5)} = 1.33 \times 10^{10}\,\mathrm{Pa}$$

Using this value for $K$ and rewriting Eq. (4.25), we get the dilation as

$$\Delta = \frac{\bar{\sigma}}{K} = \frac{1.8 \times 10^8\,\mathrm{Pa}}{1.33 \times 10^{10}\,\mathrm{Pa}} = 1.35 \times 10^{-2}$$

(b) From Eq. (3.14) the dilation is $\Delta = \dfrac{\Delta V}{V_{\mathrm{u}}}$ so that the change in volume is

$$\Delta V = \Delta \times V_{\mathrm{u}} = 1.35 \times 10^{-2} \times 192\ \mathrm{cm}^3 = 2.6\ \mathrm{cm}^3 = 2.6\ \mathrm{cm}^3/(1 \times 10^6)$$
$$= 2.6 \times 10^{-6}\ \mathrm{m}^3$$

(c) For a mean stress of 180 MPa, the depth follows from the formula for vertical stress, namely Eq. (4.45), thus

$$\sigma_{\mathrm{v}} = \rho_{\mathrm{r}} g z \Rightarrow z = \frac{\sigma_{\mathrm{v}}}{\rho_{\mathrm{r}} g} = \frac{1.8 \times 10^8\ \mathrm{Pa}}{2800\ \mathrm{kgm}^{-3} \times 9.81\ \mathrm{m\,s}^{-2}} = 6553\ \mathrm{m} \approx 6.6\ \mathrm{km}$$

# 5.9 Exercises

5.1 Explain the difference between the tangent and the secant Young's modulus. Which one is generally more useful?

5.2 Explain Weibull's weakest-link theory.

5.3 Illustrate the general stress–strain curve, and explain the four main portions that it is commonly divided into.

5.4 Explain the difference between strain (work) hardening and strain (work) softening. Which one is more common during rock failure?

5.5 What is residual strength? In an active area of faulting, which strength on the stress–strain curve is more approximate to the in-situ crustal strength: the uniaxial compressive strength (the peak strength) or the residual strength?

5.6 Define ductility as applied to rock behaviour and describe the main difference between ductile and brittle deformation.

5.7 Draw a vertical cross-section through the crust, showing the main changes in the internal structure of a fault zone as it gradually changes into a shear zone with increasing depth. Indicate the typical temperatures and crustal depths at which the brittle–ductile transition takes place.

5.8 Increasing which of the following brings the brittle–ductile transition to a greater depth, that is, increases the thickness of the seismogenic layer? (a) Strain rate, (b) temperature, (c) pore-fluid pressure, (d) porosity.

5.9 Which are the three main processes responsible for ductile deformation in the crust in general and in shear zones in particular?

5.10 What is a dislocation? What are the two main types of dislocations?

5.11 A cylindrical rock specimen with a radius of 10 mm is subject to an axial compressive stress of 50 MPa. What is the corresponding compressive force?

5.12 If the initial axial length of the specimen in Exercise 5.11 is 10 cm, its Young's modulus is 40 GPa, and its Poisson's ratio 0.27, calculate its axial shortening and strain.

5.13 A rectangular rock specimen has an initial volume of 100 cm$^3$, a Young's modulus of 30 GPa, and a Poisson's ratio of 0.25. If the specimen is loaded by a mean compressive stress (pressure) of 100 MPa, calculate its change in volume.

5.14 Assume that a rock body similar to that in Exercise 5.13 is located at depth in a volcanic rift zone. Make a rough estimate of its likely depth assuming a typical rift-zone crustal density (given in many previous examples).

# References and suggested reading

Bazant, Z. P. and Planas, J., 1998. *Fracture and Size Effect in Concrete and Other Quasibrittle Materials*. New York: CRC Press.

Brady, B. H. G. and Brown, E. T., 1993. *Rock Mechanics for Underground Mining*, 2nd edn. London: Kluwer.

Chen, W. F. and Han, D. J., 2007. *Plasticity for Structural Engineers*. Fort Lauderdale, FL: J. Ross.

Dahlberg, T. and Ekberg, A., 2002. *Failure, Fracture, Fatigue*. Studentlitteratur, Lund, Sweden.

Farmer, I., 1983. *Engineering Behaviour of Rocks*, 2nd edn. London: Chapman & Hall.

Harrison, J. P. and Hudson, J. A., 2000. *Engineering Rock Mechanics*. Part 2: Illustrative Worked Examples. Oxford: Elsevier.

Hudson, J. A. and Harrison, J. P., 1997. *Engineering Rock Mechanics. An Introduction to the Principles*. Oxford: Elsevier.

Jaeger, J. C., Cook, N. G. W., and Zimmerman, R. W., 2007. *Fundamentals of Rock Mechanics*, 4th edn. Oxford: Blackwell.

Jordan, T. H. (chief ed.), 2003. *Living on an Active Earth. Perspectives on Earthquake Science*. Washington, DC: National Academic Press.

Jumikis, A. R., 1979. *Rock Mechanics*. Clausthal, Germany: Trans Tech Publications.

Peterson, M. S. and Wong, T. F., 2005. *Experimental Rock Deformation – The Brittle Field*, 2nd edn. Berlin: Springer-Verlag.

Price, N. J., 1966. *Fault and Joint Development in Brittle and Semi-Brittle Rock*. New York: Pergamon Press.

Price, N. J. and Cosgrove, J. W., 1990. *Analysis of Geological Structures*. Cambridge: Cambridge University Press.

Scholz, C. H., 1990. *The Mechanics of Earthquakes and Faulting*. New York: Cambridge University Press.

Segall, P., 2010. *Earthquake and Volcano Deformation*. Princeton, NJ: Princeton University Press.

Shimada, M., 2000. *Mechanical Behaviour of Rocks Under High Pressure Conditions*. Rotterdam: Balkema.

Stavrogin, A. N. and Tarasov, B. G., 2001. *Experimental Physics and Rock Mechanics. Results of Laboratory Studies*. Abingdon: Balkema.

Turcotte, D. L. and Schubert, G., 2002. *Geodynamics*, 2nd edn. Cambridge: Cambridge University Press.

# 6  Stress concentration

## 6.1  Aims

All brittle deformation is related to stress concentration. The deformation, such as the formation of fractures, occurs because the local stresses in certain parts of the crust or rock bodies are raised above the average (nominal) stress in the surrounding parts. Understanding stress raisers and concentrations is of fundamental importance for assessing likely sites for rock-fracture initiation and propagation; in fact, the Griffith theory of fracture (Chapter 7) is partly based on the idea of stress concentrations around small elliptical flaws in brittle materials. The main aims of this chapter are to:

- Define the concepts of a stress concentration and a stress raiser.
- Provide the basic equations for calculating stress concentrations around elliptical and circular holes and cavities.
- Explain stress concentrations in terms of the atomic structure of solids.
- Provide an analogy between stress concentrations and fluid flow around obstacles.
- Use the equations for stresses around elliptical cavities to explain stress concentrations around fluid-filled reservoirs and magma chambers.
- Use the equations for stresses around circular holes to explain stress measurements using hydraulic fractures injected from drill holes.

## 6.2  Basic definitions

**Stress concentration** occurs when a part of a loaded material has properties that differ from those of the surrounding parts. The easiest way to visualise stress concentration is to consider stresses in a body with a hole or a notch. A **hole** is a two-dimensional cavity. The properties of the hole differ from those of the surrounding part of the body in that the hole is either empty or filled with a fluid such as air, water, or oil. A **notch** is an opening that projects into the solid body from one of its surfaces. Many notches are roughly half-circular or half-elliptical in shape, but they may also be very sharp (crack-like). Notches and holes raise the stresses around them above the normal values and are thus referred to as **stress raisers**.

A three-dimensional version of a hole is a **cavity**. An empty or a fluid-filled cavity is a stress raiser. If the cavity (or the hole) is filled with solid material, it is referred to as

Fig. 6.1  Gabbro pluton in Southeast Iceland, part of an extinct magma chamber. View north, there is a sharp contact between the gabbro and a very intense swarm of inclined sheets, showing that when the gabbro was a shallow crustal magma chamber it acted as a source for the sheets. A magma chamber, while fluid, functions as a cavity that concentrates stresses and acts as a stress raiser. When the chamber has solidified to form a pluton (here gabbro), the pluton acts as an elastic inclusion that modifies the local stresses in its vicinity.

an **elastic inclusion**. Elastic inclusions also modify the stresses in their vicinity because their mechanical properties differ from those of the surrounding material. An example of a typical small inclusion in sedimentary rocks is a fossil. The stiffness of a fossil may differ considerably from that of its host rock such as a limestone layer. The stress concentration around the fossil can then function as the source (the point of origin) of a fracture. An example of a much larger elastic inclusion is a pluton. An old shallow magma chamber, a gabbro body for example, functions as an elastic inclusion (Fig. 6.1; Andrew and Gudmundsson, 2008). A gabbro body normally is stiffer than the host rock, for example a lava pile. Thus, in contrast with the (commonly) soft inclusion of a fossil, a gabbro body would normally act as a stiff inclusion. Later in this chapter some models of volcanoes and magma chambers and plutons as inclusions and cavities will be presented.

To summarise, some of the main reasons for stress concentrations in a rock body are as follows:

- Abrupt changes in its geometry. For example, a part of a rock layer may be much thinner than the rest of the layer. The cross-sectional area of the thin part is then smaller than the normal cross-sectional area of the layer, and stresses concentrate around the thin part. This applies also if the 'notch' on either side is filled with softer (more compliant) material. For example, stress concentration in rift zones (Fig. 6.2) is partly because they

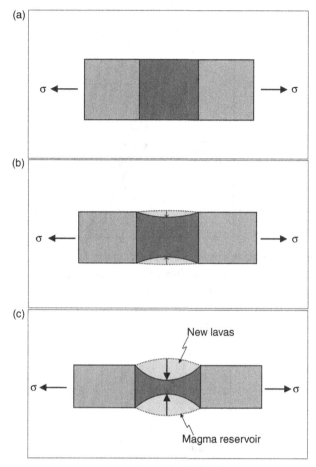

**Fig. 6.2** Rift zone is normally thinner than the surroundings and thus concentrates stresses during tensile loading associated with divergent plate movements. Because of plate pull or tension (a) the zone becomes thin partly through 'plastic necking' (b). The new lava flows at the upper surface and a magma reservoir at the lower surface of the rift zone both function as notches that concentrate stresses (c). The notch filled with eruptive materials functions as a soft, elastic inclusion (Andrew and Gudmundsson, 2008), whereas the notch filled with magma functions as a cavity.

are thinner and also softer (with young eruptive layers forming the uppermost part and a magma reservoir at the lower boundary) than the surrounding crust.

- Differences in mechanical properties between layers. During loading, the stiff (high Young's modulus) layers tend to take on much of the loading and thus become highly stressed in comparison with the more compliant or softer layers. Fault zones, with comparatively soft fault cores and damage zones (Chapters 10 and 14), for example, function as soft inclusions and concentrate stresses (Fig. 6.3).
- The effects of layering are both in tension and compression. When the loading is tensile, the stiff layers take on much of the tensile stress. When the loading is compressive, the stiff layers take on much of the compressive stress (Fig. 6.4).

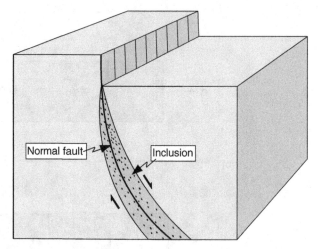

Fig. 6.3 Fault zone is an elastic inclusion. The fault rocks that constitute the core and the damage zone have mechanical properties that differ from those of the host rock. Fault zones may thus develop local stress fields that are very different from the associated regional fields.

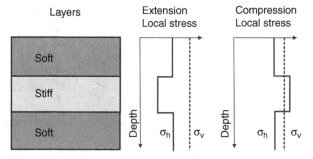

Fig. 6.4 Rocks composed of layers with different mechanical properties commonly show abrupt local stress variations when loaded (cf. Figs. 6.21 and 6.22). Generally, stiff layers take up much of the load during extension and compression, and thus become highly stressed, whereas the compliant or soft layers remain less stressed (in both extension or compression). Here $\sigma_v$ is the vertical stress and $\sigma_h$ the minimum horizontal stress.

- Cavities, holes, and inclusions in the rock normally raise the stress, that is, generate stress concentrations (Fig. 6.5).

## 6.3 Analogies and elliptical holes

Consider first the stress concentration around a circular hole located in a plate or bar of uniform thickness and width (Fig. 6.5). The tensile stress acting on the cross-sections of the bar is $\sigma = F/A$, where $F$ is the force and $A$ is the cross-sectional area of the bar on which the force is acting. Since the bar is of uniform cross-sectional area, it follows that the stress

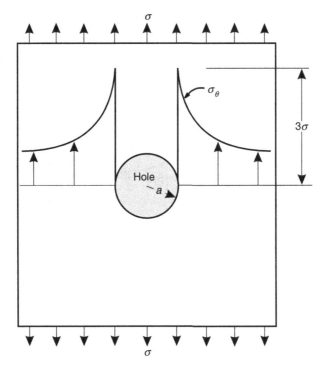

**Fig. 6.5**    Maximum local tensile stress ($\sigma_\theta$) around a circular hole of radius $a$ subject to remote tensile stress is three times the remote (nominal) stress. This local stress decreases rapidly with distance from the hole, and reaches the remote, nominal stress at a distance that is similar to the size (diameter) of the hole.

distribution is the same on any cross-section perpendicular to the long axis of the bar, so long as that cross-sectional area is sufficiently far from the end sections of the bar.

We show the stress at the ends of the plate as uniform, but in geological context they may in fact be somewhat non-uniform. One reason is the common irregularities in the geometry of the end sections. For example, a fault plane that slips and generates horizontal compressive stresses may be curved (Fig. 6.3) or irregular in shape. Also, a dyke intrusion that causes horizontal compressive stresses in the surrounding rock layers may vary in overpressure and thickness along its dip dimension (Fig. 1.6). The stress $\sigma$ is sometimes referred to as the **nominal stress**. It is the average stress in the plate or the bar, for the given loading, if the plate is everywhere of the same geometry, that is, has the same cross-sectional area.

We notice that the stress is higher close to the hole than away from it. This is because the hole is a stress raiser and, as explained below, the maximum tensile stress around a circular hole is three times the nominal applied stress. We also notice that the high stress is only close to the hole: away from the hole the stress is equal to the nominal stress. This is a very important observation and follows partly from Saint-Venant's principle (Chapter 3). It means that stress concentrations are local; at a distance that is large in relation to the size of the stress raiser (the diameter of a hole, for example), the stress field is unchanged.

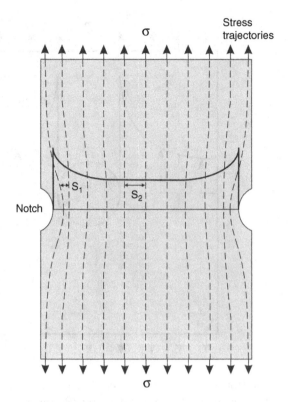

Lines of force, stress trajectories, cluster around the tips of holes and notches and thus raise the stress, that is, give rise to stress concentration. The spacing $S_1 < S_2$.

It is easiest to visualise stress concentrations as changes in the spacing of **stress trajectories**. Stress trajectories are based on an analogy with **lines of forces**. Lines of forces, such as magnetic lines of forces, define a **field**, a fundamental concept in physics. The paths of stress trajectories must lie in the solid material. When the trajectories or lines of forces have to go around a hole or a cavity, they must cluster together. The clustering results in the average spacing between the lines of forces decreasing. Since the stress intensity is force per unit area, it follows that the greater the number of lines of force that pass through a given area, the more concentrated is the stress, that is, the greater the stress concentration (Fig. 6.6).

For understanding stress concentration at an atomic level (Chapter 4), consider a notch or a crack as shown in Fig. 6.7. Here the lines of force coincide with rows of atoms in the material. Since the forces cannot be transmitted through the open crack, they have to be transmitted along the rows of atoms close to the crack tip. Thus, the lines AB and CD become subject to high loads and strains. The loads on the atomic rows, however, decrease rapidly with distance from the crack tip; the load is much less on the row MN than on the ones close to the crack tip. At a distance similar to the strike dimension of the crack, there is little extra loading of the atomic rows, that is, the stress concentration approaches zero

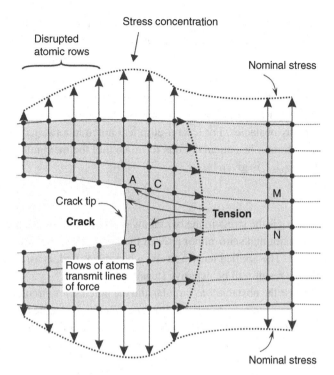

Stress concentration

Disrupted
atomic rows

Nominal stress

Crack tip

**Crack**

A C

B D

M

N

**Tension**

Rows of atoms
transmit lines
of force

Nominal stress

**Fig. 6.7**   Atomic view of stress concentration at the tip of a sharp notch or a crack. The tensile loading $\sigma$ results in stretching
the atomic material lines AB and CD, but also in the lines AC and BD. As a result, there is concentration of tensile stress
but also shear stress at the tip of the crack.

and the stress becomes the nominal stress. This result is analogous to the one obtained for
the circular hole in Fig. 6.5.

The tensile loading is uniaxial, yet it gives rise not only to tensile-stress concentration
but also to shear-stress concentration. This is so because the resulting tensile stress close
to the crack tip will not only be perpendicular, but also parallel, with strike of the crack.
This follows from considering the extension of the atomic rows in Fig. 6.7. In addition to
stretching of the vertical rows AB and CD, there will also be stretching of the horizontal
rows AC, BD, and others nearby. Thus, the tensile stress concentration around the crack is
not only crack-perpendicular but also crack-parallel. This latter is the atomic explanation
for the well-studied **Cook-Gordon debonding** or **delamination** mechanism (Chapter 13).
Because of the distortion of the vertical and horizontal atomic rows in the vicinity of the crack
tip, there will be, in addition to the tensile stress concentration, a significant **shear-stress
concentration**.

There is also a direct analogy between **streamlines** in laminar flow of a fluid and stress
concentration. Laminar flow of a fluid around an elliptical obstacle, such as the flow of
river water around an ellipsoidal stone, shows two stagnant points, that is, points where
there is no fluid flow. These points occur where the streamlines are widely spaced, at the
ends of the minor axis of the ellipse. In addition, there are two points, located at the ends of

the major axis of the ellipse, where the streamlines are clustered together and the velocity reaches a maximum, $V_{max}$, given by (den Hartog, 1987)

$$V_{max} = V_0 \left(1 + \frac{a}{b}\right) \tag{6.1}$$

where $a$ is the semi-major axis of the elliptical obstacle (the 'stone' in the stream), $b$ is the semi-minor axis of the obstacle, and $V_0$ is the velocity of the undisturbed stream flow far from any obstacles. For a semi-elliptical notch in a twisted bar, the equation for the stress concentration has an analogous form, so that the maximum tensile stress at the tip of the notch, $\sigma_{max}$, is given by

$$\sigma_{max} = \sigma_0 \left(1 + \frac{a}{b}\right) \tag{6.2}$$

where $\sigma_0$ is the stress in the bar far away from the semi-elliptical notch, and $a$ and $b$ are the semi-major and semi-minor axes of the notch. Here the minus sign for the tensile stress is omitted. The applied tensile stress, when quantified, is assigned a negative sign, in which case the result will be negative (tensile stress).

In case the obstacle was circular and the notch semi-circular, then $V_{max} = 2V_0$, and $\sigma_{max} = 2\sigma_0$. That is to say, the maximum flow velocity is double the undisturbed velocity, and the maximum tensile stress is double the remote nominal stress. The ratio of the maximum stress to the nominal stress is referred to as the **stress concentration factor** $K_t$, defined as

$$K_t = \frac{\sigma_{max}}{\sigma_0} \tag{6.3}$$

We can now expand these ideas so as to discuss circular and elliptical holes of any kind. Some of these can be used as very crude approximations for rock fractures and pores, as well as for fluid-filled reservoirs and drill holes. Consider first a rock fracture modelled as an **elliptical hole**. The rock is assumed to behave as linear elastic. For an elliptical hole of major axis $2a$ and minor axis $2b$ subject to remote tensile stress $\sigma_0$ in a direction parallel with the minor axis (Fig. 6.8), the stress concentration factor $K_t$ is given by

$$K_t = \frac{\sigma_{max}}{\sigma_0} = \frac{2a}{b} + 1 \tag{6.4}$$

with the maximum tensile stress occurring at the end of the major axis of the hole. Since the **radius of curvature** $\rho_c$ at the end of an elliptical hole, such as a hydrofracture (Fig. 6.9), is given by

$$\rho_c = \frac{b^2}{a} \tag{6.5}$$

we also have

$$\sigma_{max} = \sigma_0 \left(1 + 2\sqrt{a/\rho_c}\right) \tag{6.6}$$

For an elliptical hole or a crack subject to tensile stress $\sigma$ in a direction perpendicular to the major axis, the maximum tensile stress is

$$\sigma_{max} = \sigma_0 \left(\frac{2a}{b} + 1\right) \tag{6.7}$$

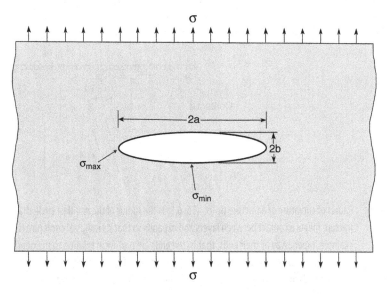

Elliptic hole subject to tensile stress $\sigma$. At the tips (lateral ends or vertices) of the hole the local tensile stress becomes many times the applied or nominal tensile stress $\sigma$. For example, if $a/b = 5$, that is, the hole is five times longer than its opening, then the maximum $\sigma_{max}$ stress at the tips is tensile (and thus strictly with a minus sign) and $11\sigma$, whereas the minimum stress ($\sigma_{min}$) is compressive (with a plus sign) and $1\sigma$.

where, as in the case of a hydrofracture (elliptical hole with an internal fluid pressure overpressure $p_0$ as the only loading), the maximum tensile stress is

$$\sigma_{max} = p_0 \left( \frac{2a}{b} - 1 \right) \tag{6.8}$$

If the applied remote tensile stress is parallel with the major axis of the elliptical hole, the maximum tensile stress occurs at the ends of the minor axis and is

$$\sigma_{max} = \sigma_0 \left( \frac{2b}{a} + 1 \right) \tag{6.9}$$

If an elliptical hole is subjected to uniform shearing stress $\tau$ parallel to the major and minor axes of the hole, then maximum and minimum tensile stress occur at a point just to the side of the end of the elliptical hole. The absolute value of this stress is (Boresi and Sidebottom, 1985)

$$|\sigma|_{max,\,min} = \tau \frac{(a+b)^2}{ab} \tag{6.10}$$

This result has important implications for the development of faults from sets of joints because it means that the maximum tensile stress around an elliptical joint that is subject to shear stress is no longer at the tips of the joint (the ends of the major axis), but a short distance to the side of the tip. From these points of maximum tensile stress, extension fractures may propagate so as to link sets of joints into major shear fractures, that is, faults (Chapter 14).

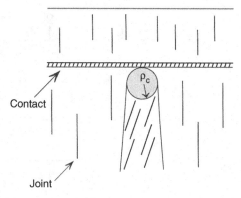

Fig. 6.9 Radius of curvature of a fracture tip $\rho_c$ (Eq. 6.5) is the radius of the smallest circle that can be fitted into the tip. Here a fracture meets a contact between layers and expands so that its radius of curvature is unusually large. Many dykes, for example, taper away at their ends, that is, the ends are narrow in relation to the general thickness of the dyke (Figs. 3.1 and 6.20). However, there are many examples of dykes with comparatively large radius of curvature when the dyke ends in soft rocks such as breccias, sediments, and soil, either inside such layers or at contacts with them (Gudmundsson, 2002).

## 6.4 Circular holes and stress measurements

From Eq. (6.7) above it follows that in the case of a circular hole with $a = b$, the magnitude of maximum tensile stress is three times the applied stress, namely $\sigma_{max} = 3\sigma_0$ (Fig. 6.5). We can now generalise the results for a circular hole and apply them to fluid-filled reservoirs and drill holes.

If the rock body containing the hole is large in comparison with the hole, so that the free surfaces do not affect the hole, then the circumferential tensile stress is given by (Fig. 6.10; Boresi and Sidebottom, 1985)

$$\sigma_\theta = (1 - 2\cos 2\theta)\sigma_0 \qquad (6.11)$$

where the angle $\theta$ is a polar coordinate, measured between the direction of the remote tensile stress $\sigma_0$ and the radius vector $r$, that is, the radial distance from the centre of the hole. Clearly, when $\theta = 90°$ and $270°$, then $\sigma_\theta = 3\sigma_0$ (Figs. 6.5, 6.10, 6.11). Thus, the stress concentration factor of a circular hole under remote tensile stress is $K_t = 3$. The maximum tensile stress occurs at the points on the circle that make an angle of $90°$ to the direction of the remotely applied tensile stress $\sigma_0$. When, however, $\theta = 0°$ and $180°$, then $\sigma_\theta = 1\sigma_0$ (Fig. 6.11). The minimum tensile stress $\theta = 0°$ and $180°$ is actually compressive stress and equal in magnitude (but opposite in sign) to the applied tensile stress $\sigma_0$.

We can use these results to analyse stress around drill holes and tunnels. Most drill holes are vertical or close to vertical, except those that are used for targeting certain petroleum reservoirs. In the latter case, the drill holes, usually at considerable depth, may have any dip or plunge, and often follow subhorizontal target layers for long distances. Many water

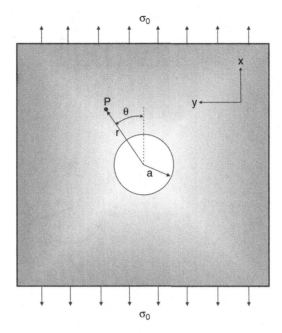

Fig. 6.10 Plate with a circular hole subject to tensile stress $\sigma_0$. Using polar coordinates, the radius vector $r$ and the angle $\theta$, between the radius vector and a fixed axis (here the direction of $\sigma_0$) are indicated for a given point P in the plate.

tunnels are inclined, but road tunnels are generally close to horizontal. All the results are applicable to tunnels with suitable modifications, particularly if they are close to the surface. However, here the focus is on drill holes.

Most drill holes used for stress measurements are vertical, so we assume our hole to be vertical. Stress measurements are made so that a part of the drilled hole is sealed off, using rubber packers, and the fluid (commonly water) pressure in the sealed part is increased until the rock ruptures through the formation of a hydraulic fracture. The term **hydraulic fracture** is normally used for any man-made or artificial hydrofracture. Because hydrofractures are extension fractures, it follows that the stress perpendicular to the hydrofracture is $\sigma_3$, whereas $\sigma_1$ and $\sigma_2$ are in the plane of the hydrofracture. If $\sigma_H$ is the maximum remote horizontal compressive stress on the drill-hole walls, and $\sigma_h$ the minimum horizontal stress (and thus corresponding to the tensile stress $\sigma_0$ in the analysis above), then it follows from Eq. (6.11) and Figs. 6.10 and 6.11 that the maximum circumferential compressive stress $\sigma_\theta^{\max}$ occurs at $\theta = 0°$ and $180°$ and is given by

$$\sigma_\theta^{\max} = 3\sigma_H - \sigma_h \tag{6.12}$$

and that the minimum circumferential compressive (maximum tensile) stress $\sigma_\theta^{\min}$ occurs at $\theta = 90°$ and $270°$ and is given by

$$\sigma_\theta^{\min} = 3\sigma_h - \sigma_H \tag{6.13}$$

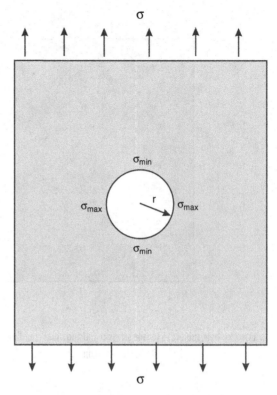

For tension $\sigma$ (or $\sigma_0$) the maximum tensile stress ($\sigma_{max}$) around a circular hole (of radius $r$) is $3\sigma$ and the maximum compressive stress ($\sigma_{min}$) is equal in magnitude to $\sigma$.

There are several results which follow from Eqs. (6.12) and (6.13) that are important for understanding deformation around circular drill holes. These include the following:

- If $\sigma_H > 3\sigma_h$ then, from Eq. (6.13), there will be tensile stress at points $\theta = 90°$ and $270°$. If this tensile stress is larger than the tensile strength of the rock, then **tensile fractures** will be induced at these points, even if there is no fluid pressure as yet in the drill hole. When there is fluid pressure in the drill hole, the resulting fracture is an extension fracture, a hydraulic fracture.
- In points $\theta = 90°$ and $270°$, by contrast, the stress concentration may become so high that the shear strength of the rock is reached, resulting in **breakouts** (see below).
- If the horizontal stress field is isotropic, that is, $\sigma_H = \sigma_h$, then $\sigma_\theta = 2\sigma_H$ along the entire perimeter of the circle. If the drilled rock has low shear strength, then this stress can result in shear fractures all around the drill hole.

If the drill hole is located in a crustal segment of an area that may be regarded as infinite in comparison with the diameter of the drill hole and there is fluid total pressure $p_t$ in the drill hole (between the packers, for example), then this fluid pressure generates a circumferential tensile stress of magnitude $-p_t$ in the hole walls. If we denote the points $\theta = 90°$ and $270°$

by A and A′ and the points $\theta = 0°$ and $180°$ by B and B′, then it follows from Eq. (6.12) that at points B and B′ the maximum circumferential compressive stress around a drill hole subject to remote horizontal maximum and minimum compressive stresses, $\sigma_H$ and $\sigma_h$, respectively, and the fluid total pressure $p_t$ is

$$\sigma_\theta^{max} = 3\sigma_H - \sigma_h - p_t \tag{6.14}$$

Similarly, at points A and A′ the minimum compressive (maximum tensile) circumferential stress is

$$\sigma_\theta^{min} = 3\sigma_h - \sigma_H - p_t \tag{6.15}$$

It follows that at points A and A′ there will be tensile stress when the following condition is satisfied:

$$p_t > 3\sigma_h - \sigma_H \tag{6.16}$$

The condition for formation of a hydrofracture (a hydraulic fracture) at points A and A′ may thus be written as

$$p_t = 3\sigma_h - \sigma_H + T_0 \tag{6.17}$$

where $T_0$ is the in-situ tensile strength of the rock hosting the drill hole. Normally, the hydraulic fracture thus generated is vertical and parallel with $\sigma_H$ and perpendicular to $\sigma_h$. Based on the assumptions used, the hydraulic fracture is thus perpendicular to $\sigma_3$. Away from the drill hole, however, the hydraulic fracture may curve and change its attitude if the regional stresses are different from the local stresses generated around the drill hole (Valko and Economides, 1995). Also, if there is significant pore pressure in the host rock, as is common, then the total horizontal stresses in the equations above become effective stresses (Chapter 7).

The results can also be used for analysing the stresses and conditions for rupture and hydrofracture propagation from two-dimensional fluid-filled reservoirs of circular cross-sections. Examples of such reservoirs include plug-like, cylindrical vertical magma chambers or conduits, which are common in the uppermost parts of many volcanic edifices. Some geothermal, ground water, and petroleum reservoirs may also be approximated by this geometry, in which case the long axis of the cylinder is normally assumed horizontal. To make the problem analytically easier, particularly as regards the coordinate system, the cylinder is regarded as being located inside a crustal segment (or a rock body) that is also cylindrical. We denote the radius of the fluid-filled hole or cylinder, the inner radius, by $R_1$ and the radius of the cylindrical crustal segment, the outer radius, by $R_2$ (Fig. 6.13). Since the radius of a crustal segment, for example a segment holding a composite volcano, is much larger than that of either the drill hole or the cylindrical plug-like conduit in the volcano, it follows that $R_2 \gg R_1$, and $R_2$ may be regarded as being effectively infinite. Again, we use polar coordinates with a radius vector $r$ and polar angle $\theta$ (Fig. 6.10). At its margin at $R_1$, the cylinder is under total fluid pressure $p_t$, whereas the circular margin of the crustal segment (the outer cylinder) is subject to horizontal compressive stress $\sigma_H$ at

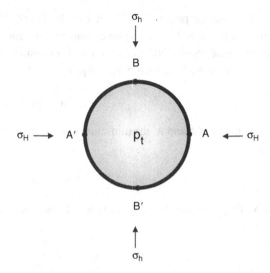

**Fig. 6.12**  Circular drill hole subject to total fluid pressure $p_t$. A–A', and B–B' are the reference points used in the analysis of the stress around the hole.

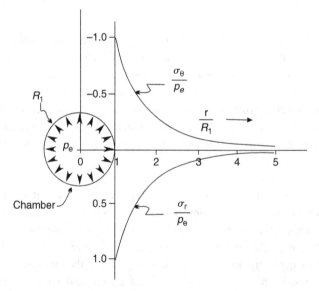

**Fig. 6.13**  Variation (decrease) in radial $\sigma_r$ and circumferential $\sigma_\theta$ stress around a circular hole (a cross-section through a cylinder) of radius $R_1$ subject to excess fluid pressure $p_e$.

$R_2$. For an isotropic state of stress, where $\sigma_H = \sigma_h$, then the radial compressive stress $\sigma_r$ is given by (Saada, 2009)

$$\sigma_r = p_t \left(\frac{R_1}{r}\right)^2 + \sigma_H \left[1 - \left(\frac{R_1}{r}\right)^2\right] \tag{6.18}$$

and the circumferential tensile stress $\sigma_\theta$ by

$$\sigma_\theta = -p_t \left(\frac{R_1}{r}\right)^2 + \sigma_H \left[1 + \left(\frac{R_1}{r}\right)^2\right] \tag{6.19}$$

For fluid-filled reservoirs such as magma chambers, the excess fluid pressure $p_e$, which is the pressure in excess of the compressive stress in surrounding host rock, equals the tensile strength of that rock, $p_e = T_0$, at the time of rupture (Gudmundsson, 2006). For a reservoir that is initially in lithostatic equilibrium with its surroundings, as is common, the relation between the total fluid pressure $p_t$, the lithostatic pressure $p_l$, and the excess pressure is $p_t = p_e + p_l$. When the excess fluid pressure $p_e$ is used rather than the total magmatic pressure $p_t$ then, substituting $p_e$ for $p_t$, the radial compressive stress becomes

$$\sigma_r = p_e \left(\frac{R_1}{r}\right)^2 \tag{6.20}$$

and the circumferential tensile stress becomes

$$\sigma_\theta = -p_e \left(\frac{R_1}{r}\right)^2 \tag{6.21}$$

Equations (6.20) and (6.21) show that the intensity of the stress field due to the excess pressure $p_e$ in a vertical, pressured cylindrical reservoir falls off as the square of the distance $r$ from its margin (Fig. 6.13). These results indicate that while the conditions of hydrofracture initiation and propagation may be met at the margin of, and in the vicinity of, the cylinder, the local tensile stress decreases rapidly away from the cylinder. Thus, at a certain distance the hydrofacture may become arrested.

## 6.5 Cavities

Cavities are three-dimensional holes. Ideally, any rock cavity can be modelled as an ellipsoid with the shape of a sphere, an oblate ellipsoid, or a prolate ellipsoidal (Sadowsky and Sternberg, 1947, 1949; Tsuchida and Nakahara, 1970; Soutas-Little, 1973). Such models can be used, for example, for many fluid-filled reservoirs or, on a much smaller scale, for fluid-filled pores in rocks. Here we mainly use magma chambers as examples in the discussion, but the results are applicable to any type of fluid-filled cavity.

For a spherical cavity comparatively close to the surface, a cavity in an elastic half space with a free top surface, analytical closed-form solutions have been obtained by Keer *et al.* (1998). Similarly, Tsuchida and Nakahara (1970) provide analytical solutions for a cavity in a semi-infinite plate (with free surfaces at its top and bottom) and elastic half spaces. These solutions, however, are too complex to be given here. In the next section, we present solutions to pressured cylinders close to the surface.

For ellipsoidal cavities at depths that are great in relation to their sizes or, in other words, for cavities that are small in relation to their depths, simple analytical solutions for the stress fields are available. Consider first a spherical magma chamber with a radius $R_1$

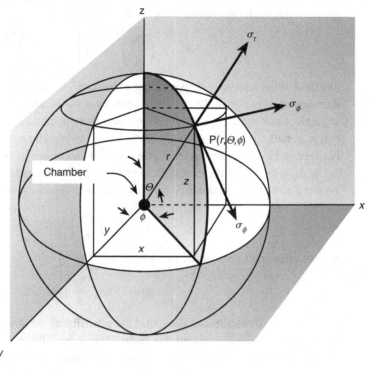

**Fig. 6.14** Spherical coordinates of a given point in the crust $P(r, \theta, \phi)$ as well as the radial compressive stress $\sigma_r$ and the circumferential tensile stresses $\sigma_\theta$ and $\sigma_\varphi$ associated with a magma chamber (or any other reservoir). The chamber is subject to fluid excess pressure $p_e$ as the only loading.

which is much smaller than the distance $d$ to its centre from the free surface of the crustal segment hosting the chamber. Let the spherical chamber be subject to a total fluid pressure $p_t = p_e + p_l$ as the only loading. The chamber radius is $R_1$ and the margin of the elastic crustal segment containing the chamber is also assumed to be a sphere with a radius $R_2$. In comparison with $R_1$, $R_2$ is effectively infinite, so that $R_2 \gg R_1$, and there is lithostatic stress $p_l$ at $R_2$.

We use spherical polar coordinates $(r, \theta, \phi)$, where $r$ is the radius vector (distance), $\theta$ is the angle between the radius vector $r$ and a fixed axis $z$, and $\phi$ is the angle measured around this axis (Fig. 6.14). The radial stress $\sigma_r$ away from the chamber and due to the chamber's total pressure $p_t$ is then

$$\sigma_r = p_t \left( \frac{R_1}{r} \right)^3 + p_l \left[ 1 - \left( \frac{R_1}{r} \right)^3 \right] \tag{6.22}$$

For spherical symmetry the two other principal stresses, $\sigma_\theta$ and $\sigma_\phi$, must be equal and given by

$$\sigma_\theta = \sigma_\phi = -\frac{p_t}{2} \left( \frac{R_1}{r} \right)^3 + \frac{p_l}{2} \left[ \left( \frac{R_1}{r} \right)^3 + 2 \right] \tag{6.23}$$

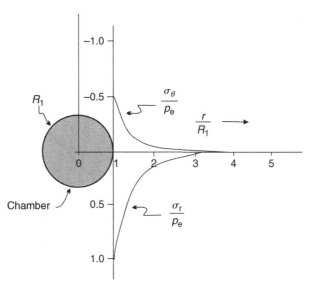

Variation (decrease) in radial $\sigma_r$ and circumferential $\sigma_\theta$ stress around a spherical cavity, here a magma chamber, of radius $R_1$ subject to excess fluid pressure $p_e$. The radial distance as measured from the chamber, $r$, is made dimensionless through dividing by the chamber radius $R_1$. The stresses generated by the excess pressure in the chamber decrease as the cube of the distance from the margin of the chamber.

Again, if the excess magma pressure $p_e$ is used rather than the total pressure $p_t$, Eqs. (6.22) and (6.23) become

$$\sigma_r = p_e \left( \frac{R_1}{r} \right)^3 \tag{6.24}$$

$$\sigma_\theta = \sigma_\phi = -\frac{p_e}{2} \left( \frac{R_1}{r} \right)^3 \tag{6.25}$$

For a very small spherical chamber, where $R_1 \ll d$ and $R_1 \to 0$ but $p_e R_1^3$ is still finite, the intensity of the point excess pressure $S$ of the cavity is given by $S = p_e R_1^3$. The units of the point pressure are Nm, as for work (energy). A point pressure of this type is the basis of the 'Mogi model' (Mogi, 1958), which is widely used in volcanology to model surface deformation (inflation and deflation) above magma chambers during periods of unrest (Fig. 3.8).

Equations (6.24) and (6.25) show that the intensity of the stress field around a spherical chamber subject to excess pressure $p_e$ as the only loading falls of inversely as the cube of the distance (radius vector) $r$ from the surface of the chamber. Using $r = R_1$ in Eqs. (6.24) and (6.25) we see that the compressive stress at the surface of the magma chamber becomes $\sigma_r = p_e$, whereas the tensile stress is $\sigma_\theta = \sigma_\varphi = -0.5 p_e$ (Fig. 6.15).

The analysis above focuses on a cavity such as a magma chamber subject to internal fluid pressure as the only loading. Commonly, however, the loading on the cavity is external.

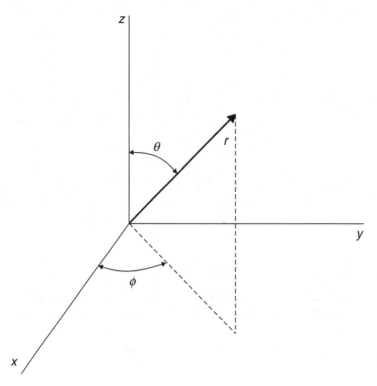

**Fig. 6.16**   Spherical coordinate system $(r, \theta, \phi)$ as used in this chapter (cf. Fig. 6.14).

For example, cavities such as pores or fluid-filled reservoirs that are in lithostatic stress equilibrium with the surrounding rocks and located within sedimentary basins or rift zones may be subject external tensile stress. The resulting stress concentrations are then commonly quite different from those generated by internal fluid excess pressure.

For example, consider a spherical magma chamber located at a considerable depth in a rift zone subject to external tensile stress $-\sigma$ as the only loading. The tensile stress then coincides with the direction of the maximum principal tensile stress, $\sigma_3$, that is, with the spreading vector. We use the coordinate system defined in Fig. 6.16 so that the uniaxial tensile loading $-\sigma$ is parallel with the $z$-axis. Since the tensile loading direction in a rift zone normally coincides with that of the spreading vector, which is horizontal, the applied tensile stress $-\sigma$ should, for such a chamber, be horizontal. However, the results presented here are completely general and apply to any spherical cavity subject to tensile loading. It is thus immaterial with which particular axis of the coordinate system $-\sigma$ coincides. For ease of understanding the particular case of a magma chamber in a rift zone, it may be helpful to imagine that the sphere is rotated by 90° so that the $z$-axis becomes horizontal and parallel with the spreading vector.

If the chamber is initially in a lithostatic equilibrium, as we assume here, then the vertical stress due to overburden pressure is balanced by the magma pressure and may, therefore, be ignored in the analysis. Then the only loading that needs to be considered is the tensile stress $-\sigma$. Using the coordinates defined in Fig. 6.16 the stresses at the surface of the sphere are (Timoshenko and Goodier, 1970; Soutas-Little, 1973)

$$\sigma_\theta = -\frac{\sigma(27-15\nu)}{2(7-5\nu)} + \frac{15\sigma}{(7-5\nu)}\cos^2\theta \tag{6.26}$$

$$\sigma_\phi = -\frac{\sigma(15\nu-3)}{2(7-5\nu)} + \frac{15\nu\sigma}{(7-5\nu)}\cos^2\theta \tag{6.27}$$

When $\theta = 90°$ the second terms in Eqs. (6.26) and (6.27) are zero and the tensile stresses are thus maximum at the equatorial plane of the spherical chamber. Thus, for $\theta = 90°$, the maximum tensile stresses are

$$\sigma_\theta = -\frac{\sigma(27-15\nu)}{2(7-5\nu)} \tag{6.28}$$

$$\sigma_\phi = -\frac{\sigma(15\nu-3)}{2(7-5\nu)} \tag{6.29}$$

At the top and bottom of the spherical chamber, however, the external tensile stress $-\sigma$ generates a compressive stress of magnitude

$$\sigma_\theta = \sigma_\phi = \frac{\sigma(15\nu+3)}{2(7-5\nu)} \tag{6.30}$$

This is thus a three-dimensional version of the effect seen in two dimensions in Fig. 6.11, namely that external tensile loading generates not only tensile stresses around a circular hole and a spherical cavity, but also compressive stresses. When these results are applied to a spherical magma chamber in a rift zone, it follows from the considerations above that the top and bottom of the chamber are along the horizontal, equatorial plane of the real magma chamber. Similarly, the top and bottom of the magma chamber lie along the equatorial plane of the sphere. Consequently, the compressive stress obtained from Eq. (6.30) is generated at the equatorial plane of the rift-zone magma chamber, whereas the tensile stresses obtained from Eqs. (6.28) and (6.29) are generated at the top and bottom of the rift-zone magma chamber.

Simple, closed-form analytical solutions of this type are not available for the stress concentration around a cavity of a general ellipsoidal form (Fig. 6.17). Equations derived by Sadowsky and Sternberg (1947, 1949) for three-dimensional ellipsoidal cavities, however, can be used to estimate stress concentrations around ellipsoidal pores or deep-seated fluid-filled ellipsoidal reservoirs such as magma chambers. If a two-axial, prolate ellipsoidal cavity has horizontal width $2c$, height (dip dimension) $2b$, and length (strike dimension) $2a$, then $c/b$ is referred to as its shape ratio. For a two-axial cavity, the dimension $a$ is equal to either $c$ or $b$. When applied to magma chambers, the focus is on the uppermost part of the chamber where, as before, rupture and dyke or sheet injection is most likely to take place (Figs. 6.1 and 6.17).

The tensile stress concentration at the top of the prolate ellipsoidal magma chamber is calculated for host rocks with Poisson's ratios of 0.25 and 0.30 in Fig. 6.18. For the special

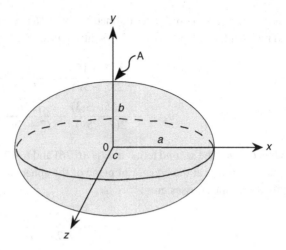

General three-dimensional ellipsoidal cavity where all the axes (2$a$, 2$b$, and 2$c$) have different lengths. The tensile stress concentration at point A is calculated and presented in Fig. 6.18.

case when $c/b = 1$ and $a = 1$, the chamber is spherical and the tensile stress concentration can be calculated from Eqs. (6.28) and (6.29).

For a cavity or chamber with the shape of a triaxial ellipsoid, then $a \geq b \geq c$ (Fig. 6.17), and the stress concentration factor $k$ at point $A$ on the ellipsoidal magma cavity can be calculated. In the case of a triaxial magma chamber that is elongate, such as a lava tube or a tunnel, or a magma chamber of circular vertical cross-section and elongate parallel with the axis of a rift zone, two-dimensional (hole) models may be used to calculate the stress concentration. For example, if the length of the magma chamber parallel with the axis of the rift zone, that is, the chamber strike dimension, is much greater than its dip dimension (so that $2a \gg 2b$), the tensile stress at point $A$ at the top of the cavity can be calculated approximately using a modified version of Eq. (6.7), namely

$$\sigma_3 = -\sigma_0 \left[ \frac{2b}{c} + 1 \right] \tag{6.31}$$

Here, $\sigma_3$ is the maximum principal tensile stress, which occurs at point A, and $-\sigma_0$ is the remote tensile stress related to plate pull. For example, when $c/b = 0.1$ and a very elongate magma chamber (so that $b/a \approx 0.0$), Eq. (6.31) gives $\sigma_3 = -21\sigma_0$, in agreement with Fig. 6.18.

## 6.6  Holes close to a free surface

Consider a two-dimensional cylindrical hole with a circular vertical cross-section located comparatively close to the Earth's surface. Here, again, the focus is on magma chambers in rift zones, but any type of fluid-filled reservoir could be analysed in a similar way.

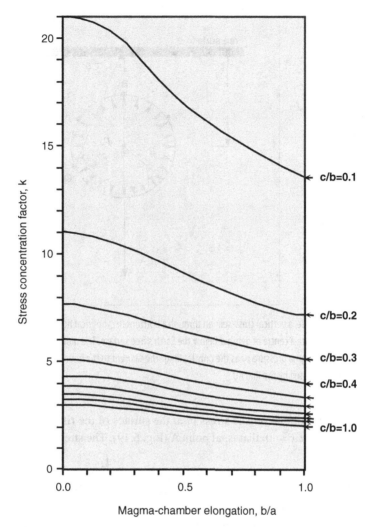

**Fig. 6.18** Concentration of tensile stress at point A in Fig. 6.17. The tensile stress presented by the stress-concentration factor $k$ (Eq. 6.3) is given for various shape ratios ($c/b$), aspect ratios or elongation ($b/a$) or the ellipsoidal cavity, here a magma chamber. The data are from Sadowsky and Sternberg (1947, 1949) and modified and applied to magma chambers by Gudmundsson (2006).

We consider a cylinder, or rather its vertical circular cross-section, with a radius $R_1$ and depth to centre $d$ (Fig. 6.19). Inside the cylindrical chamber is magma with an excess pressure $p_e$. This is the only loading. The stress $\sigma_x$ at the surface of the hosting rift zone is then (Savin, 1961)

$$\sigma_x = 4 p_e \left[ \frac{R_1^2 (x^2 - d^2 + R_1^2)}{z(x^2 + d^2 - R_1^2)^2} \right] \qquad (6.32)$$

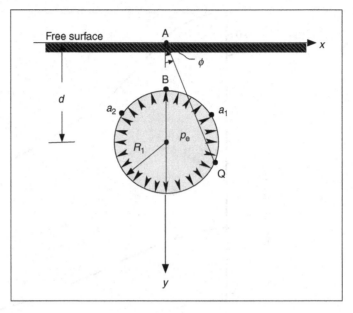

**Fig. 6.19** Circular hole, a vertical cross-section through a horizontal, cylindrical fluid-filled reservoir or a magma chamber, located with a centre at depth $d$ below the Earth's free surface. The radius of the chamber is $R_1$, which is subject to excess magma pressure $p_e$ as the only loading. The tangent to the boundary of the magma chamber at points $a_1$ and $a_2$ is indicated by the line AQ.

The maximum tensile stress $\sigma_t$ at the surface of the rift zone hosting the magma chamber occurs at $x = 0$, that is, at point A (Fig. 6.19). The stress magnitude is

$$\sigma_t = -\frac{4 p_e R_1^2}{d^2 - R_1^2} \tag{6.33}$$

During increases in the chamber excess pressure or magma-chamber inflation, the stress $\sigma_t$ may generate tension fractures (Fig. 1.1) or widen existing tension fractures and normal faults (Figs. 1.5, 3.9, 3.13, and 3.14) at the surface of the rift zone. At the surface of the rift zone at points $x = |(d^2 - R_1^2)^{1/2}|$ $\sigma_t$ changes from tensile to compressive. The largest compressive stress, $\sigma_c$, occurs at points $x = |3(d^2 - R_1^2)^{1/2}|$ and its magnitude is

$$\sigma_c = \frac{p_e R_1^2}{2(d^2 - R_1^2)} \tag{6.34}$$

and its absolute value $1/8\sigma_t$.

The rupture and dyke injection normally occur where, at the boundary of the chamber, the tensile stress concentration during a particular period of unrest reaches a maximum. By analogy with Eq. (6.11), the circumferential stress at the boundary of the chamber $\sigma_\theta$ is given by

$$\sigma_\theta = -p_e(1 + 2\tan^2 \phi) \tag{6.35}$$

the angle $\phi$ being defined in Fig. 6.19. The upper part of the magma chamber is normally more likely to rupture and inject dykes than the lower part, so that we focus on the upper part. From Eq. (6.35) we see that the peak value of $\sigma_\theta$, denoted by $\sigma_b$, occurs at those points where the angle $\phi$ is maximum, namely at $a_1$ and $a_2$ where the line AQ is tangent to the boundary of the magma chamber (Fig. 6.19). At points $a_1$ and $a_2$ the maximum tensile stress reaches the value

$$\sigma_b = -\frac{p_e(d^2 + R_1^2)}{d^2 - R_1^2} \tag{6.36}$$

Equations (6.35, 6.36) indicate as follows:

- When $d > 1.73R_1$, the maximum chamber-induced tensile stress occurs at its boundary (perimeter), at points $a_1$ and $a_2$ (Fig. 6.19), and is given by Eq. (6.36). Since the maximum tensile stress is at the chamber boundary, rather than at the surface of the rift zone, this stress field favours dyke or sheet injection (Fig. 6.1).
- When $d < 1.73R_1$, the maximum chamber-induced tensile stress occurs not at its boundary but rather at the surface of the rift zone at point A (Fig. 6.19), and is given by Eq. (6.33). This stress field favours the formation of tension fractures (Fig. 1.1) or widening of existing tension fractures and normal faults (Figs. 1.5, 3.9, 3.13, and 3.14) at the surface of the rift zone and is unlikely to trigger dyke or sheet injection.
- When $d = 1.73R_1$, the maximum chamber-induced tensile stress at the surface of the rift zone is equal to that at the boundary of the chamber, both having the magnitude $\sigma_b = \sigma_t = 2p_e$.

At point $B$ (Fig. 6.19) the chamber-induced tensile stress is equal to $p_e$, whereas at points A, $a_1$, and $a_2$ the tensile stress invariably exceeds $p_e$. At points A, $a_1$ and $a_2$ the stress depends on the difference between the depth to the centre of the magma chamber $d$ and the chamber radius $R_1$. Equations (6.33) and (6.36) both have $d^2 - R_1^2$ in the denominator, so that when $R_1 \to d$, that is, when the depth to the top (to point $B$) of the magma chamber decreases, then $\sigma_b$ and $\sigma_t$ may, theoretically, become many times greater than $p_e$. In nature, however, the in-situ tensile strength of the host rock forming the boundary of the chamber is normally 0.5–6 MPa (Chapter 7), and the actual tensile stresses generated are limited to these values. When the condition of rupture is reached at points $a_1$ and $a_2$, there will be injection of inclined sheets or dykes. One or more such injections result in compressive stress being generated in the roof of the chamber. This follows because the excess pressure $p_e$ produces compressive stress next to the sheets and dykes. This magma-induced compressive stress immediately relaxes the tensile stress at the boundary of the chamber. Since most magma chambers are at considerable depths compared with their radii, $d$ would rarely be very similar to $R_1$ (Fig. 6.19). Also, it should be noted that tensile stress concentration around real magma chambers, that is, three-dimensional cavities (Fig. 6.18; Eqs. 6.28 and 6.30) is normally less than that indicated by these two-dimensional (circular hole) results.

# 6.7 Holes in anisotropic rocks

The analytical models presented so far in this chapter assume the crust holding the holes and cavities to be homogeneous and isotropic. Numerical models of magma chambers, and fluid-filled reservoirs in general, based on such assumptions generally fit very well with the analytical results (Gudmundsson, 2006; Gudmundsson and Philipp, 2006). No crustal segments or rock bodies, however, are really homogeneous and isotropic. Analytical and numerical models where the crustal segment hosting a cavity or a magma chamber is assumed to be isotropic give certain ideas as to stress concentrations and local stress fields, but are too general to be suitable for the detailed analysis of, for example, the exact paths of propagating hydrofractures such as dykes, inclined sheets, or mineral veins. In fact, if the crust were homogeneous and isotropic, then most buoyant hydrofractures should reach the Earth's surface. But field observations show that most dykes, mineral veins, and other hydrofractures become arrested at various crustal depths and thus never reach the surface (Chapter 13). In fact, most rock fractures of any kind become arrested after short-distance propagation. Commonly, fracture arrest occurs at contacts or interfaces between dissimilar rock layers (Chapters 11 and 13). The anisotropy of a typical crustal segment is thus of fundamental importance for understanding fracture propagation, arrest, and fluid transport.

To analyse the effects of crustal anisotropy on the propagation and arrest of fractures we must normally use numerical models. This is primarily for two reasons:

- First, the mechanical properties of the rock layers that host the magma chambers are normally highly variable (Carmichael, 1989; Bell, 2000; Gudmundsson, 2002, 2003; Schön, 2004). For example, soft pyroclastic rocks subject to loading behave very differently from stiff basaltic lava flows or sills. Both rock types are common host rocks of magma chambers in rift zones and individual volcanoes. Closed-form analytical stress solutions are not available for such complex materials, and even if they were the numerical methods would normally be preferred since they are more flexible and yield results comparatively quickly.
- The local stress fields in anisotropic and heterogeneous rocks, particularly those hosting fluid-filled reservoirs, are commonly very complex. This follows directly from the variation in the elastic properties with location and direction within the rocks. Discontinuities, such as contacts and fractures, can, for example, produce abrupt changes in the local stress fields. Again, it is not feasible to tackle abrupt variations in local stresses using analytical methods, so that numerical methods are preferred.

In this section, the focus is on dyke injections from magma chambers, particularly dyke propagation and arrest, in anisotropic crustal segments. The results are very general, however, and applicable to any fluid-driven fractures.

If a dyke is to supply magma to an eruption, it has to propagate through all the layers between its source magma chamber and the Earth's surface. The same applies to any hydrofracture: to reach the surface it must propagate through all the various layers and discontinuities between its source and the surface. These layers and discontinuities develop

**Fig. 6.20**   Arrested dyke at a contact between compliant or soft pyroclastic layer and a stiff lava flow in Tenerife, Canary Islands. View east, the dyke is 0.26 m at the bottom of its 3.7-m exposure but much thinner at the tip.

local stresses, some of which encourage dyke propagation whereas others encourage dyke arrest. Abrupt changes in the material properties of the layers and their contacts or interfaces, and the associated local stresses, are the primary reasons for so many dykes becoming arrested on their paths to the surface (Figs. 3.1 and 6.20).

In order to explore the effects of local stresses on dyke emplacement, we consider several numerical models of stresses around a magma chamber hosted by a layered crustal segment, such as a composite volcano or a part of a rift zone. Here the focus is on a few comparatively simple results to illustrate the effects of layering on dyke paths and arrest. More complex and detailed models are available in recent papers (Gudmundsson, 2006, 2007; Gudmundsson and Philipp, 2006).

The Young's moduli used in the models are based on the following observations. Laboratory measurements indicate that layers of basalt, such as lava flows, and gabbro, such as large intrusions, may have Young's moduli as high as 110–130 GPa. By contrast, some volcanic tuffs and breccias have Young's moduli as low as 0.05–0.1 GPa (Appendix D.1; Bell, 2000). More extreme values are known, however. For example, some rocks reach stiffnesses of up to 150–200 GPa (Myrvang, 2001). Unconsolidated geomaterials such as sand and gravel may have static stiffnesses as small as 0.08 GPa, and some, such as clay, as small as 0.003 GPa (Appendix D.1). Poisson's ratios of the rocks that commonly constitute composite volcanoes, however, have a much narrower range. For example, Poisson's ratio is about 0.25 for so dissimilar rocks as lava flows and many volcanic tuffs (Bell, 2000).

Most active composite volcanoes and rift zones contain many layers of breccias, ignimbrites, and tuffs (Fig. 6.20). For example, in the rift zone and in the composite volcanoes of Iceland there are layers of basaltic breccias, and scoria and soil are common between the lava flows. In addition, there are some ignimbrite layers in the lava pile. In the rift zone of Iceland, for instance, the most common soft or compliant rocks are hyaloclastites (basaltic breccias) formed during submarine or subglacial eruptions. Static laboratory Young's moduli of young hyaloclastites are mostly between 0.5 GPa and 8 GPa (Gudmundsson, 2006). In the active rift zone, sedimentary rocks, mainly tillites of late Pleistocene age and common outside and inside the composite volcanoes, have stiffnesses similar to those of the hyaloclastites, whereas unconsolidated sediments, common in some composite volcanoes and rift zones, are much more compliant (Schön, 2004). Near-surface Holocene and Pleistocene basaltic lava flows have static Young's moduli mostly between 10 GPa and 35 GPa. The Young's moduli of these rocks generally increase with age and depth of burial in the lava pile (Chapter 4).

The layered models in Figs. 6.20 and 6.21 show the effects of magma-chamber geometry on the local stresses in a composite volcano or a rift zone composed of layers of contrasting stiffnesses. The magma chamber is located in a rock unit of a stiffness typical for the crust at the depth of a few kilometres, 40 GPa. By contrast, the layers between the chamber and the surface form a part of the composite volcano or rift zone and alternate in Young's modulus between 1 GPa, which is very compliant, and 100 GPa, which is very stiff. This large variation in stiffness is well within the limits obtained in laboratory measurements. The compliant layers may represent tuffs, sediments, scoria, hyaloclastites, and soils between lava flows, whereas the very stiff layers may represent basaltic and intermediate lava flows, welded pyroclastics, and intrusions.

The model in Fig. 6.21 consists of a magma chamber of a circular cross-section subject to internal excess magmatic pressure of 5 MPa as the only loading. This excess pressure is chosen so as to be similar to the in-situ tensile strength of a typical crystalline host rock, 0.5–6 MPa (Chapter 7). The chamber is located in a rift-zone segment composed of seven layers of contrasting Young's moduli with 1 GPa for the thin layers, 40 GPa for the layer hosting the chamber, and 100 GPa for the thick layers. The same layer stiffnesses and magmatic pressures are used in the model in Fig. 6.22; the only difference is the geometry of the chamber. The local stress fields are represented by the trajectories of the maximum principal compressive stress $\sigma_1$.

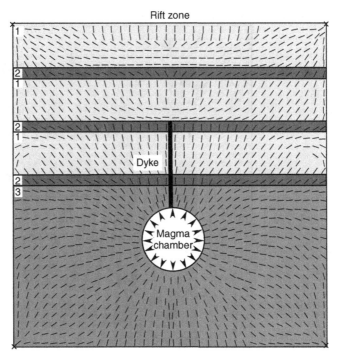

**Fig. 6.21** In a layered (anisotropic) crustal segment, potential dyke-propagation paths may come to an end, that is, the dyke may become arrested, when the local stress field, here indicated by the trajectories of $\sigma_1$, changes abruptly between layers. The dyke path follows the trajectories of $\sigma_1$ so that when it meets the top of the thin layer (marked by 2), where the trajectories of $\sigma_1$ become horizontal, the dyke may change into a sill or, more likely and as indicated here, become arrested. In this numerical model, the thick layers (marked by the number 1) are stiff (100 GPa), the thin layers (marked by the number 2) are soft (1 GPa), and the layer hosting the magma chamber (marked by the number 3) is moderately stiff (40 GPa). The only loading is the internal magma excess pressure $p_e$.

Clearly, there is a strong effect of the magma-chamber geometry on the local stress fields in these models. Dykes generally propagate parallel with the $\sigma_1$-trajectories. It follows that those injected from the circular chamber in Fig. 6.21 would follow a steep or vertical path to the upper contact between the central thin soft layer and the second thick layer where the $\sigma_1$-trajectories change to horizontal. The dyke path would therefore either end at this contact or become horizontal, in which case the dyke might change into a sill. The dyke has to propagate at this contact as a sill because, as an extension fracture, its orientation must be parallel with the $\sigma_1$-trajectories. Alternatively, the dyke could become arrested and the path end at the contact. In the marginal upper parts of the model the $\sigma_1$-trajectories are inclined in all the layers, making it possible for some inclined sheets to reach the uppermost layer and, perhaps, the surface. As the margins and corners of the model are approached, however, the results become less reliable and, generally, only the central part of such a numerical model, far from sides and corners, is used to forecast the likely dyke paths.

The stresses around the sill-like chamber (Fig. 6.22) also show abrupt changes from vertical to horizontal $\sigma_1$-trajectories. These occur already in the first compliant layer above

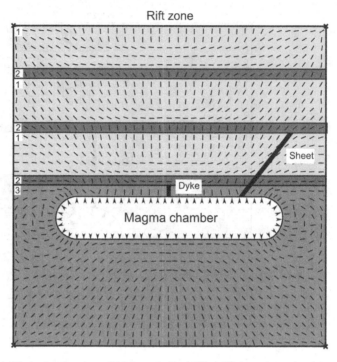

**Fig. 6.22** Numerical model with the same layering and loading as in Fig. 6.22 but different magma-chamber geometry. A dyke from the central upper margin of the chamber propagates for a very short distance before it becomes arrested (in the first thin layer, marked by the number 2, above the magma chamber). An inclined sheet may propagate further into the crust, following the inclined trajectories of $\sigma_1$. How far it extends, however, is uncertain because close to the margin of the model the trends of the trajectories become unreliable.

the magma chamber. Thus, since layering and loading conditions are the same, it is the difference in geometry between the chambers in Figs. 6.21 and 6.22 that gives rise to the very different local stresses in the layers between the chambers and the surface. During periods of unrest, these different local stresses would result in very different dyke-propagation paths. In particular, in the model in Fig. 6.22 most dykes injected from the upper central part of the sill-like chamber would become arrested after a short propagation, just above the margin of the chamber. Some inclined sheets, injected from the lateral ends of the chamber, might propagate to shallower depths in the crust, possibly reaching the surface to feed eruptions.

## 6.8 Summary

- Stresses concentrate, so that their magnitude increases, around cavities, holes, and inclusions, all of which have mechanical properties that differ from those of the host material (the matrix). The different properties are primarily different stiffnesses (Young's modulus). Many holes and cavities in the crust, for example, contain fluids such as air, water, or magma. 'Lines of force' must lie in the solid part and thus cluster around an empty or a fluid-filled hole or cavity. Since stress is force per unit area, clustering of the lines of force, that is, increasing the number of lines of force per unit area, raises the stress and gives rise to stress concentration.

- For a circular hole (two dimensional) subject to tensile stress $\sigma_0$, the maximum tensile stress $3\sigma_0$ occurs at points located on the perimeter at right angles to the direction of the stress $\sigma_0$, whereas the minimum stresses are compressive and occur at points parallel with $\sigma_0$. This information is used when making stress measurements in drill holes. Then a liquid (usually water) is pumped into a part of the drill hole (sealed off with rubber packers) until the hole wall ruptures because of the fluid pressure and a hydrofracture forms. Man-made hydrofracture of this kind is referred to as a hydraulic fracture. The orientation of hydraulic fracture is perpendicular to the local $\sigma_3$ and parallel with $\sigma_1$ and $\sigma_2$. The tensile strength, $T_0$, of the rock containing the drill hole is obtained from the difference in the water pressure needed to initiate the hydraulic fracture (form a new fracture) and subsequently reopen it, and is normally between 0.5 and 6 MPa.
- Local stresses around cavities can be used to model fluid-filled pores but also much larger reservoirs such as confined ground-water aquifers, petroleum reservoirs, and magma chambers. For internal fluid excess pressure, the stresses around a spherical cavity are as given by Eqs. (6.24) and (6.25), and for external tension the stresses are given by Eqs. (6.28)–(6.30).
- When a hole, a magma chamber for example, is close to the Earth's surface in relation to its diameter, that is, the diameter is similar to or larger than the depth to the hole, the free surface has effects on the local stresses around the hole. For a circular hole, a cylindrical magma chamber with the long axis parallel with the axis of a rift zone, subject to excess magma pressure as the only loading, the stresses at the surface of the rift zone are given by Eqs. (6.32)–(6.34) and the stresses at the margin of the chamber by Eqs. (6.35) and (6.36). The chamber can either give rise to tension-fracture formation in the rift zone or, alternatively, the injection of sheets from the chamber margin.
- The local stresses around holes and cavities in heterogeneous and anisotropic crustal segments are best analysed through numerical models. Two simple (two-dimensional) models of magma chambers are used as examples in the chapter. In one (Fig. 6.21) the magma chamber has a circular cross-sectional area, in the other (Fig. 6.22) it has a close-to-elliptical or sill-like cross-sectional area. The local stresses show that a dyke injected from the top of either chamber would soon enter layers where the trajectories of $\sigma_1$ are horizontal and thus unfavourable for vertical dyke propagation. On meeting such a layer, the vertically propagating dyke may either change into a sill or, more commonly, stop its propagation, that is, become arrested.

## 6.9 Main symbols used

$a$    semi-major axis of an elliptical solid or notch (a hole)

$b$    semi-minor axis of an elliptical solid or notch (a hole)

| | |
|---|---|
| $a, b, c$ | semi-axes of a triaxial ellipsoid (a reservoir) |
| $d$ | depth (to the centre of a reservoir) below the Earth's surface |
| $K_t$ | stress concentration factor |
| $p_e$ | fluid excess pressure |
| $p_l$ | lithostatic pressure (geostatic or overburden pressure) |
| $p_t$ | total fluid pressure |
| $p_0$ | fluid overpressure (in a hydrofracture) |
| $R_1$ | radius of a fluid-filled cylindrical or spherical cavity (a reservoir) |
| $R_2$ | radius of a solid cylindrical or spherical crustal segment (hosting a reservoir) |
| $r$ | radius vector, a spherical polar coordinate |
| $T_0$ | tensile strength |
| $V_{max}$ | maximum velocity in a laminar fluid flow |
| $V_0$ | velocity of an undisturbed laminar flow |
| $\nu$ | Poisson's ratio |
| $\rho_c$ | radius of curvature |
| $\sigma_{max}$ | maximum tensile stress at the tip of a semi-elliptical notch |
| $\sigma_r$ | radial stress, a principal stress away from a spherical cavity (a reservoir) |
| $\sigma_0$ | average stress in a bar or plate far from notches or other stress raisers |
| $\sigma_\theta$ | circumferential stress, a principal stress around a spherical cavity (a reservoir) |
| $\sigma_\varphi$ | tangential stress, a principal stress around a spherical cavity (a reservoir) |
| $\sigma_H$ | maximum horizontal compressive stress |
| $\sigma_h$ | minimum horizontal compressive stress |
| $\tau$ | shear stress |
| $\theta$ | angle, a spherical polar coordinate |
| $\varphi$ | angle, a spherical polar coordinate |

# 6.10 Worked examples

## Example 6.1

### Problem

A rift-zone segment has two volcanoes, each with a shallow magma chamber (Fig. 6.23). The horizontal cross-sectional shapes of the shallow magma chambers are indicated by the geometries of the collapse calderas A and B at the surface. A is circular with a diameter of 3 km, whereas B is elliptical with a major axis of 5 km and a minor axis of 2 km. The dimensions of B are similar to those of the extinct magma chamber/pluton of Slaufru-dalur in Southeast Iceland (Fig. 8.6). The only loading is plate-pull (tension) related to the spreading vector, which is perpendicular to the rift zone. Use stress concentration around holes to:

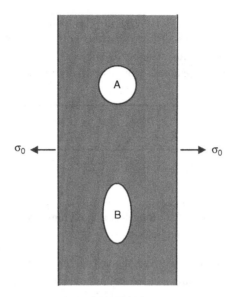

Fig. 6.23 Example 6.1. Two volcanoes A and B, each a collapse caldera with a shallow magma chamber, are located in a rift zone. Volcano A has a magma chamber and caldera with a circular cross-sectional area, whereas volcano B has a magma chamber and caldera with an elliptical cross-sectional area (similar to Slaufrudalur in Fig. 8.6).

(a) calculate the maximum tensile stress around the chambers if the spreading-related tension was equal to 0.5 MPa and 6 MPa
(b) assess if magma-chamber rupture is likely for these tensile stresses.

Solution

(a) From Eq. (6.9), the maximum tensile stress around an elliptical hole is

$$\sigma_{max} = \sigma_0 \left( \frac{2b}{a} + 1 \right)$$

For chamber A, we have $a = b$ (the radius is constant), and for a tension of 0.5 MPa the maximum tensile stress is

$$\sigma_{max} = \sigma_0 \left( \frac{2b}{a} + 1 \right) = -0.5 \text{ MPa} \left( \frac{3 \text{ km}}{1.5 \text{ km}} + 1 \right) = -1.5 \text{ MPa}$$

For a tension of 6 MPa, the maximum tensile stress would be 12 times greater (since 6 MPa is 12 times 0.5 MPa), so the tensile stress would be $-1.5 \times 12 = -18$ MPa. For chamber B, we have

$$\sigma_{max} = \sigma_0 \left( \frac{2b}{a} + 1 \right) = -0.5 \text{ MPa} \left( \frac{5 \text{ km}}{1 \text{ km}} + 1 \right) = -3.0 \text{ MPa}$$

For a tension of 6 MPa, the maximum tensile stress would be 12 times greater, so the tensile stress would be $-3.0 \times 12 = -36$ MPa.

(b) The in-situ tensile strength of solid rocks is 0.5–6 MPa, with the most common values being 2–3 MPa. Thus, rupture is unlikely for chamber A subject to 0.5 MPa tension but possible for chamber B subject to 0.5 MPa tension. In the case of 6 MPa tension, both chambers would be likely to rupture and inject dykes or inclined sheets (Chapter 11).

## Example 6.2

### Problem

An elliptical joint 2 m long has a maximum aperture (opening) of 10 cm. Use the stress concentration around holes to:

(a) calculate the maximum theoretical tensile stress at its lateral ends (tips) of the joint if the only loading is internal fluid overpressure of 1 MPa
(b) calculate the same as in (a) if the only loading is tensile stress of 1 MPa perpendicular to its long axis of the joint.
(c) decide if these maximum theoretical tensile stresses can be reached in crustal rocks? Why? What happens?

### Solution

(a) From Eq. (6.8) the maximum tensile stress for the fracture subject to internal fluid overpressure $p_0$ is

$$\sigma_{max} = p_0 \left( \frac{2a}{b} - 1 \right) = 1 \times 10^6 \text{Pa} \left( \frac{2 \times 1 \text{ m}}{0.05 \text{ m}} - 1 \right) = 3.9 \times 10^7 \text{Pa} = 39 \text{ MPa}$$

(b) From Eq. (6.7) the maximum tensile stress for the fracture under external tension $\sigma_0$ is

$$\sigma_{max} = \sigma_0 \left( \frac{2a}{b} + 1 \right) = 1 \times 10^6 \text{Pa} \left( \frac{2 \times 1 \text{ m}}{0.05 \text{ m}} + 1 \right) = 4.1 \times 10^7 \text{Pa} = 41 \text{ MPa}$$

In both (a) and (b) we are dealing with tensile stresses, here calculated and shown as absolute values (magnitudes). If some compressive stresses were operating on the fractures as well, the tensile stresses should be given negative algebraic signs, that is, minus signs, since we define tensile stress as negative.

(c) Since the in-situ tensile strength of solid rocks is generally between 0.5 MPa and 6 MPa, these tensile stresses are not likely to be reached in nature. Thus, long before the tensile stresses could reach about 40 MPa, the joint tips would propagate. When a fluid-driven fracture propagates, its volume suddenly increases (because of increased length or strike dimension and aperture), so that the fluid pressure drops to zero. When the tensile-stress driven fracture propagates, the local tensile stress around it decreases (relaxes) and approaches or reaches zero (Section 7.6).

---

## Example 6.3

Problem

A vertical section (between packers) of a vertical drill hole is filled with water under pressure at a depth of 3 km. At this depth the minimum horizontal compressive stress is 70 MPa and the maximum horizontal stress 90 MPa. If at one stage the total water pressure in this part of the drill hole is 50 MPa, what would be the stresses at points A and A' and B and B' in Fig. 6.12?

Solution

From Eq. (6.14) the maximum compressive stress is at points B and B' and is given by

$$\sigma_\theta^{\max} = 3\sigma_H - \sigma_h - p_t = 3 \times 90\ \text{MPa} - 70\ \text{MPa} - 50\ \text{MPa} = 150\ \text{MPa}$$

From Eq. (6.15) the minimum compressive stress occurs at points A and A' and is given by

$$\sigma_\theta^{\min} = 3\sigma_h - \sigma_H - p_t = 3 \times 70\ \text{MPa} - 90\ \text{MPa} - 50\ \text{MPa} = 70\ \text{MPa}$$

Both stresses are compressive, that is, for these loading conditions, no tensile stress would be generated around the drill hole. Thus, no hydraulic fracture would form. If there was significant pore-fluid pressure $p_f$ in the rock hosting the drill hole (Chapter 7), then the horizontal stresses would become effective stresses, namely $\sigma_H^e = \sigma_H - p_f$ and $\sigma_h^e = \sigma_h - p_f$.

---

## Example 6.4

Problem

Consider a road tunnel at 500 m depth through plutonic rock with an average rock density to this depth of 2600 kg m$^{-3}$. The horizontal stress in the rock at this depth is 20 MPa.

(a) Calculate the stresses at points 0° and 180° (corresponding to A–A' in Fig. 6.12) with respect to $\sigma_H$ and at 0° and 180° (corresponding to B–B' in Fig. 6.12) with respect to $\sigma_h$.
(b) Make the same calculations if the road tunnel was at a depth of only 20 m.
(c) Determine whether breakouts or extension fractures would be likely to form in (a) and (b).

Solution

(a) The horizontal stress is given but we have to find the vertical stress due to the overburden at the depth of the tunnel. From Eq. (4.45) we get the vertical stress as

$$\sigma_v = \rho_r g z = 2600\ \text{kg m}^{-3} \times 9.81\ \text{m s}^{-2} \times 500\ \text{m} = 1.28 \times 10^7\ \text{Pa} = 12.8\ \text{MPa}$$

This is then the minimum stress since 12.7 MPa is less than 20 MPa. Using Eq. (6.12) and substituting $\sigma_v$ for $\sigma_h$, so as to have more appropriate symbols, we get the stresses at points $0°$ and $180°$ with respect to $\sigma_H$ (B–B′) thus

$$\sigma_\theta^{max} = 3\sigma_H - \sigma_v = 3 \times 20 \text{ MPa} - 12.8 \text{ MPa} = 47.2 \text{ MPa}$$

Here we have made the calculations using megapascals, as is all right because these are the only units used in the calculations. However, if we were also using metres or other basic SI units in the calculations, we would have to change megapascals to pascals in order to get the correct result. For the minimum circumferential compressive stress $\sigma_\theta^{min}$ at $0°$ and $180°$ (corresponding to B–B′ in Fig. 6.12) with respect to $\sigma_h$, we use Eq. (6.13), substituting $\sigma_v$ for $\sigma_h$, to get

$$\sigma_\theta^{min} = 3\sigma_v - \sigma_H = 3 \times 12.8 \text{ MPa} - 20 \text{ MPa} = 18.4 \text{ MPa}$$

(b) For the same tunnel at 20 m depth, the density may be somewhat different. This is because the near-surface part of the pluton may contain numerous fractures that would lower its density in comparison with its deeper parts. However, no information is given on this in the example, so we assume that the density of the uppermost 20 m is the same as the average density of the pluton. Proceeding as in (a) we find the vertical stress as

$$\sigma_v = \rho_r g z = 2600 \text{ kg m}^{-3} \times 9.81 \text{ m s}^{-2} \times 20 \text{ m} = 5.1 \times 10^5 \text{Pa} = 0.51 \text{ MPa}$$

Following the same procedure as above, we get the maximum stress as

$$\sigma_\theta^{max} = 3\sigma_H - \sigma_v = 3 \times 20 \text{ MPa} - 0.51 \text{ MPa} = 59.5 \text{ MPa}$$

And for the minimum compressive stress (maximum tensile stress) we get

$$\sigma_\theta^{min} = 3\sigma_v - \sigma_H = 3 \times 0.51 \text{ MPa} - 20 \text{ MPa} = -18.5 \text{ MPa}$$

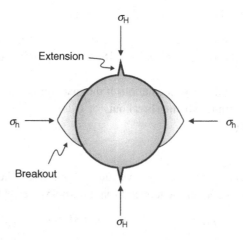

**Fig. 6.24**   Example 6.4. A horizontal road tunnel subject to a biaxial stress field, showing potential regions for the formation of breakouts and extension-fracture formation.

(c) The results for the road tunnel at 500 m depth yield only compressive stresses. So there
would be no extension fractures at points A–A'. However, it is possible that failure might
occur when the compressive stress at points B–B' reaches 47 MPa. The results for the
tunnel at 20 m depth yield tensile stresses (negative values) of 18.5 MPa at points A–A'
and compressive stresses of 59.5 MPa at points A–A'. Horizontal extension fractures
would thus be expected at points A–A'. Since there is no fluid pressure in the tunnel,
these would be tension fractures (Fig. 6.24). Shear failure would also be likely at points
B–B'. In vertical drill holes such a failure is referred to as breakouts (Fig. 6.24), but in
road tunnels as spalling. Spalling as a mechanism is analogous to exfoliation, namely
fracture formation or reopening close to the Earth's surface due to high surface-parallel
compressive stress.

# 6.11 Exercises

6.1 Explain stress concentrations and stress raisers. What are the main factors that give rise
to stress concentrations? Give geological examples of structures that may concentrate
and raise the stresses and initiate fractures.

6.2 What are lines of force? How can they be used to explain stress concentrations? What
is the analogy between stress concentrations and streamlines in a fluid?

6.3 What is an elastic inclusion? Give some geological examples of elastic inclusions and
how they affect fracture formation.

6.4 Define the stress concentration factor. What is the stress concentration factor for an
elliptical hole? Give examples of the use of the elliptical-hole model in rock-fracture
studies.

6.5 Explain breakouts and induced extension fractures in terms of stress concentrations
around circular drill holes. How can we infer the state of stress in the rocks surrounding
the drill hole from these types of brittle deformation?

6.6 What are cavities? Give examples of cavities in rocks and how they affect the local
stress fields.

6.7 What is the Mogi model in volcanology?

6.8 What main effects does the Earth's free surface have on the local stress field around
a shallow magma chamber of a circular, vertical cross-sectional area? That is, how
would the local stress around a chamber at a crustal depth that is large in comparison
with the diameter of the cross-sectional area differ from the local stress around a
chamber of the same geometry and subject to the same loading conditions but located
at a shallow depth in comparison with the diameter?

6.9 A magma chamber with a circular horizontal cross-sectional area is located at a depth
of 2 km in a rift zone (Fig. 6.23). The chamber is filled with magma which is initially
in lithostatic equilibrium with the host rock. The diameter of the chamber is 8 km,
equal to that of an associated caldera. If the only loading is external tensile stress due
to plate pull, how high would this tensile stress have to be so that the chamber would

rupture and inject one or more dykes? Where along the chamber boundary would the dyke injection be most likely to occur?

6.10 An elliptical magma chamber is at a depth of 2 km in a rift zone and subject to plate-pull tension as the only loading (Fig. 6.23). The chamber is filled with magma, initially in lithostatic equilibrium with the host rock, and has a major axis of 4 km, coinciding with the axis of the rift zone, and a minor axis of 3 km. What tensile stress would be needed to rupture the chamber and inject dykes? Where along the boundary of the chamber would dyke injection be most likely to occur?

6.11 A joint has a strike dimension of 0.5 m and a maximum opening of 2 cm. Use stress concentration around elliptical holes to calculate the theoretical tensile stresses at its lateral ends (tips) if the only loading is (a) an external tensile stress of 0.1 MPa, or (b) an internal fluid pressure of 0.1 MPa. Would the joint be likely to propagate for the given loading conditions? Why?

6.12 A horizontal road tunnel is drilled at a depth of 200 m through a mountain whose rock density above the tunnel is $2500 \ \text{kg m}^{-3}$. The horizontal stress perpendicular to the tunnel axis at this depth is 3 MPa. Calculate the stresses at the points B–B' and A–A' (Fig. 6.12) at the boundary of the tunnel. Would spalling (breakouts) or extension fractures be likely to develop at these localities? Why?

# References and suggested reading

Afrouz, A. A., 1992. *Practical Handbook of Rock Mass Classification Systems and Modes of Ground Failure*. London: CRC Press.

Andrew, R. E. B. and Gudmundsson, A., 2008. Volcanoes as elastic inclusions: their effects on the propagation of dykes, volcanic fissures, and volcanic zones in Iceland. *Journal of Volcanology and Geothermal Research*, **177**, 1045–1054.

Bell, F. G., 2000. *Engineering Properties of Soils and Rocks*, 4th edn. Oxford: Blackwell.

Boresi, A.P. and Sidebottom, O. M., 1985. *Advanced Mechanics of Materials*, 4th edn. New York: Wiley.

Carmichael, R. S., 1989. *Practical Handbook of Physical Properties of Rocks and Minerals*. London: CRC Press.

Dahlberg, T. and Ekberg, A., 2002. *Failure, Fracture, Fatigue*. Studentlitteratur, Lund (Sweden).

den Hartog, J. P., 1987. *Advanced Strength of Materials*. New York: Dover.

Economides, M. J. and Nolte, K. G. (eds.), 2000. *Reservoir Stimulation*, 3rd edn. New York: Wiley.

Gudmundsson, A., 2002. Emplacement and arrest of sheets and dykes in central volcanoes. *Journal of Volcanology and Geothermal Research*, **116**, 279–298.

Gudmundsson, A., 2003. Surface stresses associated with arrested dykes in rift zones. *Bulletin of Volcanology*, **65**, 606–619.

Gudmundsson, A., 2006. How local stresses control magma-chamber ruptures, dyke injections, and eruptions in composite volcanoes. *Earth-Science Reviews*, **79**, 1–31.

Gudmundsson, A., 2007. Conceptual and numerical models of ring-fault formation. *Journal of Volcanology and Geothermal Research*, **164**, 142–160.

Gudmundsson, A. and Philipp, S. L., 2006. How local stresses prevent volcanic eruptions. *Journal of Volcanology and Geothermal Research*, **158**, 257–268.

Keer, L. M., Xu, Y., and Luk, V. K., 1998. Boundary effects in penetration or perforation. *Journal of Applied Mechanics*, **65**, 489–496.

Mogi, K., 1958. Relations between eruptions of various volcanoes and the deformations of the ground surfaces around them. *Bulletin of the Earthquake Research Institute*, **36**, 99–134.

Myrvang, A., 2001. *Rock Mechanics*. Trondheim: Norway University of Technology (NTNU) (in Norwegian).

Saada, A.S., 2009. *Elasticity Theory and Applications*. London: Roundhouse.

Sadowsky, M. A. and Sternberg, E., 1947. Stress concentration around an ellipsoidal cavity in an infinite body under arbitrary plane stress perpendicular to the axis of revolution of cavity. *Journal of Applied Mechanics*, **14**, A191–A201.

Sadowsky, M. A. and Sternberg, E., 1949. Stress concentration around a triaxial ellipsoidal cavity. *Journal of Applied Mechanics*, **16**, 149–157.

Savin, G. N., 1961. *Stress Concentration Around Holes*. New York: Pergamon.

Schön, J. H., 2004. *Physical Properties of Rocks: Fundamentals and Principles of Petrophysics*. Amsterdam: Elsevier.

Soutas-Little, R. W., 1973. *Elasticity*. New York: Dover.

Timoshenko, S. and Goodier, J. N., 1970. *Theory of Elasticity*, 3rd edn. New York: McGraw-Hill.

Tsuchida, E. and Nakahara, I., 1970. Three-dimensional stress concentration around a spherical cavity in a semi-infinite elastic body. *Japan Society of Mechnical Engineers Bulletin*, **13**, 499–508.

Valko, P. and Economides, M. J., 1995. *Hydraulic Fracture Mechanics*. New York: Wiley.

# Theories of brittle failure of rocks

## 7.1 Aims

When rock is subject to higher stresses than it can sustain it fails. There are many modes of failure of rocks and other solids, but all belong primarily to one of two broad classes, namely ductile failure and brittle failure. However, some criteria apply both to strictly brittle rocks as well as to materials, such as granular media, that would normally be regarded as ductile. In fact, there is a gradation from strictly brittle (brittle fracture) to strictly ductile (plastic flow) in rocks and many other solids. In this chapter we discuss some of the main types of failure, focusing on the principal criteria for brittle failure, that is, failure through extension fractures and shear fractures. Brittle failure results in fractures which originate from stress concentrations (Chapter 6). The main aims of this chapter are to:

- Define the concepts of strength and failure.
- Explain the concept of a Coulomb material.
- Define and explain the Coulomb failure criterion.
- Explain and discuss some empirical failure criteria.
- Discuss the basic elements of the Griffith theory.
- Present the Griffith criterion for failure of rocks.

## 7.2 Failure and strength

**Failure** may be defined as the stress condition at which the solid either starts to flow or break. This stress condition thus defines the maximum stress or stress difference that the solid can sustain, that is, its load-bearing capacity. When the failure of the solid is through fracture formation, the failure is **brittle**. When the failure of the solid is through plastic flow, the failure is **ductile**. In a fault zone there is a gradual change from predominantly brittle failure, through brittle–ductile transition, to predominantly ductile failure.

Before we introduce some of the main criteria for rock failure, we should clarify the concept of rock strength. There are several types of rock strength; these refer to different types of loadings. The basic definitions may be summarised as follows:

1. **Rock strength** is the maximum load that the rock can withstand before failure. The load can be specified in terms of stress, strain, or rates of these. Here we shall mostly define

strengths in terms of applied stresses. There are three principal concepts of strength that we refer to in this book: tensile strength, shear strength, and compressive strength. These are commonly measured in the laboratory, but the **in-situ (site or field)** strengths, that is, the strengths as measured or calculated for a rock body or a crustal segment in nature, are more important for understanding rock fractures.

2. **Tensile strength** $T_0$ is the tensile stress in a rock when it ruptures and forms an extension fracture. More specifically, it is the tensile stress at the location of the fracture. The tensile stress resulting in fracture may be generated in many ways, for example by the bending of a laboratory specimen. The most reliable and useful results at depth in the crust are obtained from hydraulic fracture measurements (Chapter 6), made in a section of a drill hole, the section being sealed off from those above and below by rubber packers. A fluid is injected into a section of a drill hole and its pressure increased until the wall-rock ruptures. Following rupture, the tensile strength across the hydrofracture formed in the wall of the drill hole is zero. To find the tensile strength, the fluid pressure is again increased until water starts to flow into the just-formed fracture. The difference in fluid pressure in the two experiments is the tensile strength of the rock (Fig. 7.1). The tensile strength of rocks obtained from hydraulic fracture experiments in solid rocks is commonly from 0.5 to 6 MPa, with a maximum value of about 9 MPa. The maximum value was obtained at a depth of about 9 km in the KTB drill hole in Germany (Amadei and Stephansson, 1997).

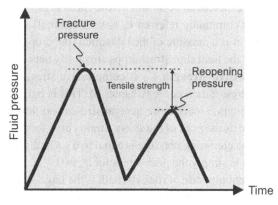

**Fig. 7.1**　Schematic illustration of how tensile strength is inferred from pressure changes in a drill hole. The **fracture pressure** is the total fluid pressure in the drill hole needed to rupture the walls and initiate a hydrofracture. The walls may contain some joints and other flaws, so the hydrofracture basically forms by linking small-scale flaws into an outcrop-scale hydraulic fracture. The pressure then falls as the fluid (normally water) flows into the newly formed fracture (and partly into the host rock). When water is injected again at the same location in the drill hole (between the packers), the total pressure needed for the water to restart to flow out of the hole through reopening the newly formed fracture is known as the **reopening pressure**. Since a large-scale hydrofracture, an extension fracture, already exists this second fluid pressure is less than the first fracture pressure by a magnitude equal to the in-situ tensile strength of the rock. Thus, the difference between the two pressure peaks (the fracture pressure and the reopening pressure) is the **in-situ tensile strength** of the rock at the location of the hydrofracture. This schematic illustration omits various complications in the pressure curves (cf. Valko and Economides, 1995; Amadei and Stephansson, 1997).

The average in-situ tensile strength in most solid rocks is normally around 3 MPa. Thus, in comparison with many other rock-physics 'constants', such as Young's modulus, the tensile strength has a narrow range and is close to being a real constant. Since it is measured through the injection of fluids, the in-situ tensile strength is very suitable for modelling fluid-driven fractures, the only type of large extension fractures that forms at depths in the crust (Chapter 8). The range of measured in-situ tensile strengths is very close to being half the range of typical shear strengths of active faults, as inferred from stress drops, as is predicted by the Modified Griffith criterion (Section 7.4).

3. **Inherent shear strength** or **cohesion** $\tau_0$ is the shear stress in a rock when it ruptures and forms or reactivates a shear fracture, that is, a fault. More specifically, it is the shear stress at the particular rupture location of the fault when the effective normal stress is zero. Shear stress can be measured either in the laboratory or in the field. Experiments on rock specimens loaded until they fail in shear give the shear strength of intact rocks. This is commonly about twice the tensile strength of intact rocks (Section 7.4). In the field, a common measure of the shear strength is the **driving shear stress**, that is, the shear stress available for generating fault slip. When a tension fracture forms, the tensile stress generating the fracture is normally relaxed through the fracture opening displacement (Section 7.6, Chapter 9). Under static or quasi-static conditions, the tensile stress is equal to the tensile strength of the rock. Similarly, when a shear fracture slips, the driving shear stress is normally relaxed through the slip. Thus, the driving shear stress for fault slip may be expected to be similar to the shear strength of the fault zone. The driving shear stress is commonly referred to as the (nominal) **stress drop** in earthquake mechanics because it is a measure of the relaxation (the drop) in shear stress on the fault plane as a result of the fault slip. Stress drops are mostly between about 1 and 12 MPa (Section 7.4).

4. **Compressive strength** $C_0$ is compressive stress in a rock when it reaches the peak of the stress–strain curve (Chapter 6). This is particularly so for the uniaxial compressive strength. Under more general stress conditions, the compressive strength may be regarded as the peak of the stress–strain curve for any given value of the confining pressure. The confining pressure is equal to $\sigma_3$. On this definition, the uniaxial compressive strength is simply the peak stress for $\sigma_3 = 0$.

5. A different measure of rock strength is the large-scale **crustal strength** as inferred from the **stress difference** $\sigma_1 - \sigma_3$. In this view, rock strength is the maximum stress difference that a rock can withstand before it fails. It is common to plot the stress difference as a function of crustal depth (Fig. 7.2), which normally shows that the maximum stress difference at failure increases with depth. This is in accordance with empirical theories of slip or shear failure, such as the Coulomb and the Mohr theories, discussed below, both of which show an increase in stress difference (diameter of the Mohr's circles) at greater normal stress, that is, at greater depth. Generally, the stress difference increases to the lower margin or bottom of the brittle crust. The brittle part of the crust coincides essentially with the **seismogenic layer**, which is that part of the lithosphere where most tectonic earthquakes are generated. Below the lower margin or bottom of the seismogenic layer, the crustal/lithospheric deformation is increasingly ductile, and the yield strength, or the condition for plastic flow, are reached at gradually lower stress differences. Consequently, the strength of the crust/lithosphere, as measured by the maximum stress difference before failure, decreases with increasing crustal depth below the lower

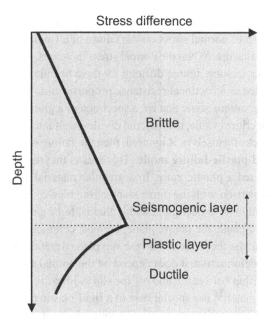

Fig. 7.2 Schematic presentation of crustal strength, as measured by the maximum stress difference $\sigma_1 - \sigma_3$ that the rock can sustain before failure, as a function of depth. The seismogenic layer is that part of the crust (or the lithosphere) where the crust responds to stresses primarily through fracture formation and slip, many of which produce earthquakes. The seismogenic layer is thus a measure of the thickness of the brittle part of the crust. The plastic layer is where the crust responds to stresses primarily through flow, and so is the ductile part of the crust. In many active continental areas the seismogenic layer (sometimes called the seismogenic zone) has a thickness of about 20 km. The thickness, however, depends on temperature (geothermal gradient), rock properties, and strain rate and varies much between regions. For example, earthquakes are rare at temperatures above about 350–400°C. Worldwide, the thickness of the seismogenic layer varies from a few kilometres (fast-spreading ocean ridges) to many tens of kilometres (in comparatively stable continental areas).

margin of the seismogenic layer. Because of variation in the mechanical properties of the crust, and in geothermal gradients, there may be several relative strength peaks in the crust before the maximum peak is reached. Such variations in crustal strength with depth are particularly common in continental areas.

Now that we have defined rock strength, we turn to criteria for failure. Most of these are empirical. The best known, which is equally applicable to granular materials such as sands and soils, is the Coulomb criterion. Before we discuss the criterion itself, however, it is necessary to discuss briefly the concept of a Coulomb material, whose behaviour connects directly with the criterion.

## 7.3  The Coulomb material

Coulomb material is a **granular material** whose failure depends on the pressure or normal stress. This is in contrast with many other yield criteria, such as von Mises and Tresca, where isotropic pressure or normal stress does not affect the conditions of yield. The reason for the effect of normal stress on the failure of a Coulomb material is primarily because of its granular nature. When the normal stress increases, the grains become more densely packed together, making it more difficult for them to slide over one another. Thus, the sliding may be treated as a frictional resistance proportional to the normal stress on the sliding plane.

Observations show that for a specimen of a granular material subject to shear stress of a certain critical value, the material divides itself into two blocks. If the elastic deformation in the blocks themselves is ignored, then the failure of the granular material may be regarded as **rigid-plastic failure mode**. This means that the slip plane between the rigid blocks is considered a **plastic zone**. In a granular material, the plastic zone is normally very thin in comparison with the dimensions of the blocks. Thus, the plastic zone itself is normally assumed to have negligible width, that is, to be a surface or plane. This plane is known as the yield plane, the failure plane, or, most commonly, the **slip plane**.

While the shear stress $\tau$ on the slip plane depends neither on the size of the plane nor the rate of deformation, it does depend on the normal stress, $\sigma_n$. A granular material where the relationship between $\tau$ and $\sigma_n$ for slip is linear is known as an **ideal Coulomb material**. Such a material is a special case of a rigid-plastic material. The Coulomb two-dimensional yield criterion (assuming $\sigma_2 = 0$) for such a material is

$$\tau = \tau_0 + \mu\sigma_n \tag{7.1}$$

where $\tau$ is the shear stress on the slip plane, $\tau_0$ is the cohesion, cohesive strength, or inherent **shear strength** of the material, $\mu$ is the **coefficient of internal friction**, and $\sigma_n$ is the normal stress on the slip plane. The (inherent) shear strength of the material is its shear strength when **no normal stress is applied**.

For many coarse granular materials $\tau_0$ is so small that they are known as **cohesionless materials**. Using $\tau_0 = 0$ and multiplying the shear and normal stresses in Eq. (7.1) by unit areas to get the forces (since force = stress × area) we obtain

$$F = \mu N \tag{7.2}$$

where $F$ is the frictional force and $N$ the normal force on the slip plane. This is the **law of friction** between large bodies in contact.

For materials with cohesion, the shear strength $\tau_0$ is non-zero. The reason why there is cohesion or shear strength is not fully understood, but the cohesion is commonly attributed to cohesive forces between the grains. The cohesive forces may include van der Waals bonding, that is, the attractive forces between particles or atoms, and in particular the surface tension between particles. This latter applies particularly to partially wet granular materials where liquid bridges form between nearby adjacent grains and give rise to tensile forces that contribute to the cohesion or strength of the material.

Equation (7.1) is also commonly written in a form where the tangent of the angle of internal friction, $\phi$, is substituted for $\mu$, that is, $\tan \phi = \mu$, so that we get

$$\tau = \tau_0 + \sigma_n \tan \phi \tag{7.3}$$

For cohesionless materials, the angle of internal friction $\phi$ equals its **angle of repose**, defined as the steepest angle a hill of such a material (such as dry sand) can sustain without slipping.

For rocks, the angle of repose is referred to as the **angle of internal friction**. The angle of repose depends on the nature of the cohesionless material: it is about 20° for granular material composed of smooth spherical grains but as much as 50° for material composed of angular grains. For dry sand, $\phi$ is typically about 30°.

There is a connection between the Coulomb yield criterion and the other yield criteria. For example, for a material such as a soft, wet clay there is significant cohesion, but the angle of internal friction is zero, that is, $\phi = 0$. Then Eq. (7.3) reduces to $\tau = \tau_0$, which is the **Tresca criterion** for plastic yield. In fact, a plot of the Coulomb criterion indicates that yielding occurs when the stress states lie on a certain yield surface, which in the case of the Coulomb and the Tresca criteria are hexagonal in shape. The Tresca criterion may be regarded as a special case of the Coulomb criterion. In other words, the Coulomb criterion may be regarded as a generalised Tresca criterion where the lithostatic stress or pressure, presented in terms of the normal stress $\sigma_n$, is taken into account (cf. Chen and Han, 2007).

## 7.4 The Coulomb criterion for rocks

This is also referred to as the Navier–Coulomb criterion or as the Mohr–Columb criterion. It is essentially a yield criterion for flow or slip and thus in many ways analogous to the criterion for yield of plastic materials. The two-dimensional Coulomb criterion has been widely applied to explain shear fracture or faulting in soils and rocks. Then Eq. (7.1) is sometimes rewritten so that the cohesion or shear strength is substituted with twice the tensile strength, that is, $\tau_0 = 2T_0$. This modification follows directly from the Griffith theory and is in good agreement with observations which indicate that the in-situ tensile strength of rocks is commonly about double their shear strength. Thus, the **Modified Griffith criterion** becomes

$$\tau = 2T_0 + \mu\sigma_n \tag{7.4}$$

Since this is the two-dimensional version of the Coulomb criterion, the intermediate principal stress is zero, that is, $\sigma_2 = 0$. The criterion is shown graphically, using Mohr's diagram or circles (Fig. 7.3). We make the following observations about this criterion:

1. The line that marks the Coulomb criterion for failure is referred to as the **Coulomb line or envelope of failure**. Here, $\tau$ is the shear stress needed for slip along a plane (the fault plane) with a normal stress $\sigma_n$ and tensile strength $T_0$ (or cohesion or shear strength $\tau_0$). When $\tau_0 = 2T_0 = 0$, the material is cohesionless and the Coulomb line passes through the origin. The angle of internal friction, $\phi$, is the angle between the Coulomb line of failure and the horizontal axis, that is, the $\sigma_n$-axis. $\phi$ is also the minimum angle of a plane in the rock for downward sliding of the hanging wall block (of the same or similar rock type) along that plane – which thereby becomes a fault plane. The coefficient of internal friction, $\mu$, is the slope of the Coulomb line of failure.

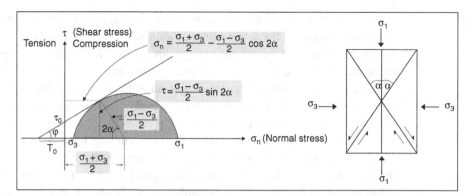

**Fig. 7.3** Coulomb failure line is a reasonable criterion in the compressive regime, but less suitable for the tensile regime. In particular, the Coulomb failure line overestimates the tensile strength of rocks, $T_0$. The normal and shear stresses on a slip plane are indicated, as well as (on the figure to the right) the angle $\alpha$ between (here, conjugate) slip planes and $\sigma_1$.

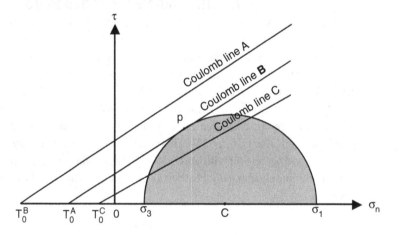

**Fig. 7.4** Rock stability in relation to the Coulomb line. For line A the rock is stable and for line B the rock is at failure. Under static conditions, line C cannot exist because failure and slip would occur as soon as one point touched the failure line and, thereby, reduce the stress difference and thus the diameter of the Mohr's circle. $T_0$ with superscripts A, B, and C are different theoretical tensile strengths.

2. The centre of a Mohr's circle is at $(\sigma_1 + \sigma_3)/2$, which is equal to the two-dimensional mean stress (Chapter 2). The radius of the circle is $(\sigma_1 - \sigma_3)/2$, which is also the maximum shear stress, and its diameter thus $\sigma_1 - \sigma_3$.

3. The Mohr's circle, representing a particular state of two-dimensional stress in the crust, may lie entirely below the straight line A that marks the Coulomb **line of failure** (Fig. 7.4). Then from Eq. (7.4) we have the condition

$$\tau < 2T_0 + \mu\sigma_n \tag{7.5}$$

and no slip will occur on any plane, that is, no faulting will take place. The rock body subject to this stress condition is thus in a state of **stable static equilibrium**.

4. The Mohr circle may touch the Coulomb line of failure B at one point, in which case there is one plane on which the conditions of Eq. (7.4) is satisfied, namely

$$\tau = 2T_0 + \mu\sigma_n$$

This plane is marked by point p on the Mohr's circle in Fig. 7.4 and the rock is said to be in the state of **incipient shear failure** or fault slip. Thus, if fault slip occurs at this stage, it will occur along the plane marked by point p.

5. If the Mohr's circle is so large that the Coulomb line C cuts it, there should be many planes, namely all those belonging to the points along the arc above line C, where the following condition is satisfied:

$$\tau > 2T_0 + \mu\sigma_n \tag{7.6}$$

However, according to the Coulomb criterion this cannot happen. That is to say, no material that satisfies the Coulomb criterion and thus behaves as a Coulomb material can have planes that satisfy the inequality (7.6). This follows because as soon as one point satisfies the condition of Eq. (7.4), there will be slip on the associated plane. And this slip will relax the shear stress, which, thereby, can never exceed the limit given by Eq. (7.4).

Thus, from this discussion it follows that:

(a) circles defined by $\sigma_n - \tau$ coordinates below the Coulomb line represent stable stress conditions;

(b) a circle defined by $\sigma_n - \tau$ coordinates so that one point of the circle touches the Coulomb line represents incipient shear failure or limiting equilibrium stress conditions; and

(c) a circle defined by $\sigma_n - \tau$ coordinates that lie above the Coulomb line represents stress conditions that cannot be reached under static loading.

6. The Coulomb criterion thus specifies the maximum magnitude that geomaterials such as rock can tolerate before slip. The criterion, however, does not specify the direction of slip. In fact, experiments on rocks and granular materials show that there are always two nascent slip planes. When these are fully developed fault planes, they are referred to as **conjugate faults** (Fig. 7.5). These faults are inclined at $90° - \phi$ to each other, $45° - \phi$ to the maximum principal compressive stress, $\sigma_1$. To take the conjugate planes into account, the Coulomb criterion is sometimes written using the absolute value of the shear stress, that is, $|\tau|$, in the form

$$|\tau| = 2T_0 + \mu\sigma_n \tag{7.7}$$

Conjugate faults are quite common in nature. The boundary faults of grabens, for example, are conjugate (Fig. 2.1). Similarly, in many strike-slip fault zones, conjugate dextral and sinistral faults are common (Fig. 7.6). But rocks are generally heterogeneous and anisotropic, so that in many regions of faulting, only one of the two potential conjugate planes develops into faults. If both develop, one becomes the dominant set of planes (Fig. 7.7).

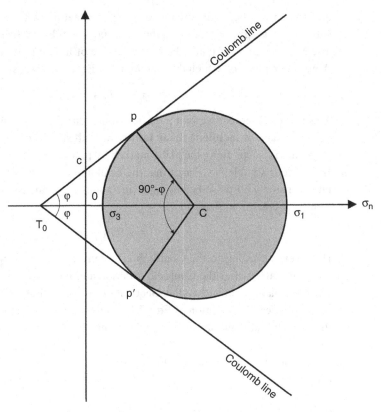

Fig. 7.5 In the Coulomb criterion, there are always two possible slip planes, here presented by points $p$ and $p'$. In a homogeneous, isotropic rock neither is preferred and a common result is the formation of conjugate faults (Figs. 7.6 and 7.7).

7. The Coulomb criterion is most suitable for materials where the tensile strength is only a small fraction of the compressive strength. In fact, for cohesionless materials, the tensile (and inherent shear) strength is zero. Generally, the Coulomb criterion is thus suitable for rocks since they, in contrast to metals for example where the tensile and compressive strengths are generally similar, have tensile strengths that are commonly only about 10% of their compressive strengths. Nevertheless, the linear Coulomb criterion generally overestimates the tensile strengths of rocks.

8. For rocks, the Coulomb criterion is thus normally presented with a **tensile cut-off** (Fig. 7.8). This is done so as to fit the failure criterion in the tensile regime with the actual measured tensile strength of rocks. The Coulomb criterion is therefore not very suitable for the tensile regime, but rather good for the compressive regime, in particular at comparatively high confining stress. High confining stress means greater overburden pressure, that is, greater crustal depth. Close to the surface, and in the tensile regime, faulting occurs through the linking up of small, en echelon extension fractures (Chapter 14), so that the Coulomb criterion of frictional sliding is there less suitable.

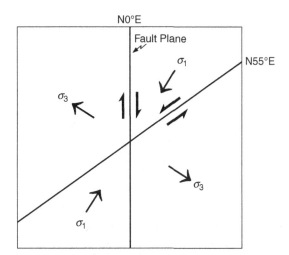

**Fig. 7.6**  Conjugate faults. This schematic illustration is based on actual observations of conjugate strike-slip faults in a seismic zone in South Iceland (Fig. 7.7). The north-striking dextral faults are better developed, however, than the northeast-striking sinistral faults.

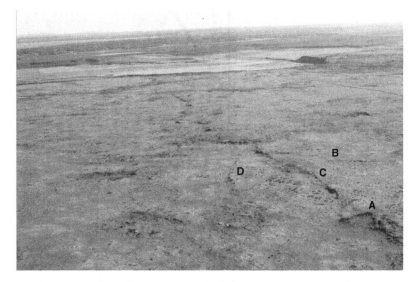

**Fig. 7.7**  Aerial view south-southwest of part of a Holocene strike-slip fault in South Iceland. The main fault (AC) is a dextral fault, whereas (D) is a sinistral conjugate fault to the main fault. Segment B is a fracture of unknown type. Fault AC trends north, fault D northeast, as indicated in Fig. 7.6.

The Coulomb criterion may be written in different ways. For example, instead of showing the shear stress as a function of the normal stress on the failure plane (the fault), the magnitude of the maximum principal compressive stress $\sigma_1$ at failure can be given in terms of the minimum principal compressive stress $\sigma_3$. Thus, we may write (Jaeger

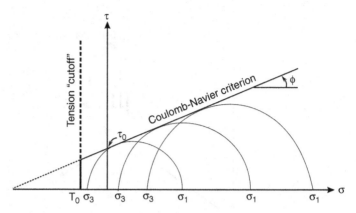

**Fig. 7.8**    Coulomb failure line and Mohr's circles. Because the Coulomb criterion overestimates the rock tensile strength $T_0$, there is commonly an added 'tensile (or tension) cut-off' to show the real tensile strength.

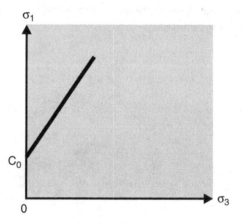

**Fig. 7.9**    Coulomb criterion can be presented in various ways. One is to show it as a function of the two principal stresses and the uniaxial compressive strength, $C_0$.

*et al.*, 2007)

$$\sigma_1 = C_0 + q\sigma_3 \tag{7.8}$$

where $C_0$ is the uniaxial compressive strength of the rock, which can be presented in terms of the shear strength and the coefficient of internal friction in the form

$$C_0 = 2\tau_0 \left[ \left( \mu^2 + 1 \right)^{\frac{1}{2}} + \mu \right] \tag{7.9}$$

$C_0$ is also the intercept of the failure line with the $\sigma_1$-axis (Fig. 7.9). The constant $q$ is the equal to the tangent of the angle $\theta$, which is the angle between $\sigma_1$ and a normal to the slip or fault plane (Chapter 2). The constant $q$ can also be presented in terms of the coefficient

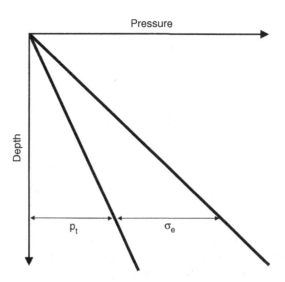

Pressure

Depth

$p_t$

$\sigma_e$

**Fig. 7.10** Effective stress $\sigma_e$ is the total normal stress, here the lithostatic stress or overburden pressure, minus the total fluid pressure, here $p_t$. When the total fluid pressure equals the lithostatic stress or overburden pressure, as is common in the roofs of magma chambers and possibly at the top of parts of the asthenosphere and some gas-filled petroleum reservoirs, the effective stress is zero.

of internal friction, namely so that

$$q = \left[ \left( \mu^2 + 1 \right)^{\frac{1}{2}} + \mu \right]^2 = \tan^2 \theta \tag{7.10}$$

and

$$\theta = 45^\circ + \frac{1}{2} \varphi \tag{7.11}$$

where, as before, $\varphi$ is the angle of internal friction.

Fluid pressure in the pores of a rock or other geomaterials can have a large effect on the conditions of shear failure as presented by the Coulomb criterion. Pore-fluid pressure affects the normal stress but not the shear stress. Thus, in a porous, fluid-filled rock, the **effective normal stresses** (Fig. 7.10) become

$$\begin{aligned} \sigma_1 - p_t \\ \sigma_2 - p_t \\ \sigma_3 - p_t \end{aligned} \tag{7.12}$$

where $p_t$ is the **total fluid pressure** in the pores. Because we use mainly fluid excess pressure $p_e$ for reservoirs (its maximum being equal to the tensile strength) and fluid overpressure $p_o$ for hydrofractures (equal to the excess pressure plus the buoyancy effects, Chapter 8), it is important to notice that when we discuss the effects of fluid pressure on shear failure using the Coulomb criterion, it is the total fluid pressure that enters the discussion.

The total fluid pressure operates equally in all directions and reduces the effective normal stresses in the rocks. As a consequence, the Mohr's circles are **shifted to the left** (Fig. 7.11).

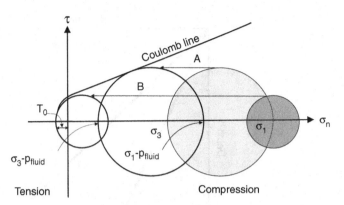

Fig. 7.11 Increasing the pore-fluid pressure in a rock body or crustal segment shifts the Mohr's circles to the left, that is, towards the Coulomb failure line. This is the principal reason why landslides and earthquakes are common when the fluid pressure has increased, such as in the rocks surrounding dams, in man-made geothermal reservoirs, and during and following heavy rainfalls ($p_{\text{fluid}}$ is the total fluid pressure $p_t$).

Thus, if the original state of stress was so that the Mohr's circle was well below the Coulomb failure line, and the rock thus in a stable equilibrium, increasing the pore-fluid pressure may shift the Mohr's circle so much to the left on the diagram that one point on the Mohr's circle touches the Coulomb line, resulting in fault slip (shift $A$, Fig. 7.11). And if the original stress difference is small, the Mohr's circle may be shifted all the way into the tensile regime and generate either a pure extension fracture (hydrofracture) or a hybrid (mixed-mode, Chapter 9) fracture, that is, a fracture formed partly in extension and partly in shear (shift $B$, Fig. 7.11). Such fractures are very common in rift zones, as seen in the open normal faults (Fig. 1.5).

Shifting of the Mohr's circles to the left is exactly what usually happens when the fluid pressure in the rock body or a crustal segment is increased, such as during the filling of a dam with water. Some of the water in the dam migrates into the surrounding rock and increases its pore-fluid pressure. The pore-fluid pressure also increases during the pumping of fluids into porous layers or reservoirs. The latter is common in the petroleum industry as well as during the development of man-made geothermal reservoirs. We may summarise some of main points related to pore-fluid pressure in the crust as follows:

• Pore-fluid pressure changes are common in the crust. They happen in all subsurface fluid-filled reservoirs (for ground water, geothermal water, gas, oil, and magma) and in surface waters in lakes, dams, rivers, and in open ground-water aquifers. Pore-fluid pressure changes are also common following heavy rainfalls, which is one reason why landslides and other near-surface failures are common during very rainy periods. Fluid-pressure changes affect the probability of fault slip and associated earthquakes. In fact, all tectonic earthquakes are thought to be related to zones of high fluid pressure.

• Increasing pore-fluid pressure reduces the normal stresses in the host rock but not the associated shear stresses (Fig. 7.11). Consequently, the Mohr's circle for a given state of stress in the crust is shifted to the left and towards the Coulomb line of failure. This means that increasing the pore-fluid pressure increases the chances of fault slip and thus commonly an earthquake (aseismic slips are also common in the crust; not all fault slips give rise to earthquakes).

- If the initial stress difference between $\sigma_1$ and $\sigma_3$ is very small, increasing the pore-fluid pressure may shift the Mohr circle all the way into the tensile part of the diagram, that is, partly or wholly to the left of the $\tau$-axis (Fig. 7.11). Then the formation of an extension fracture, a hydrofracture, becomes possible when $-\sigma_3 = T_0$, that is, when maximum principal tensile stress reaches the tensile strength of the rock.

- In a totally fluid-filled reservoir, such as many magma chambers, the fluid pressure balances the overburden pressure. This means that the crustal segment above the reservoir is effectively 'floating' or resting on the fluid in the reservoir. It follows that the stress difference in the roof and walls of the reservoir next to the fluid is always small. When the fluid in the reservoir is in lithostatic equilibrium with the host rock, the effective stress is zero. Thus, the stress state is marked by a point at the origin on the Mohr's diagram. If the fluid pressure in the reservoir increases, generating an excess pressure, there will be tensile stress in the roof and walls of the reservoir, thereby shifting $\sigma_3$ into the tensile regime. As soon as $-\sigma_3 = T_0$, a hydrofracture (for a magma chamber, a dyke or an inclined sheet) will form and reduce the stress difference to zero again.

- In the case the reservoir develops underpressure (fluid pressure in the reservoir that is less than the overburden pressure, or lithostatic stress), then $\sigma_1$ becomes shifted to the right, that is, into the compressive regime of the diagram. Underpressure in this sense is quite common in petroleum and ground-water reservoirs (closed aquifers) where all the fluid is contained in the pores and the weight of the overburden is partly supported by the elastic rock matrix (the rock containing the pores). In totally fluid reservoirs, such as the top parts of many magma chambers and, presumably, the gas caps of some petroleum reservoirs, underpressure is unlikely because the full weight of the crust above the reservoir (the overburden) is resting on the fluid.

## 7.5  Some empirical criteria

There are many empirical criteria in addition to the Coulomb criterion that have been used to forecast and explain rock failure. Many such criteria are discussed in detail by Sheorey (1997). Among the best known are the Mohr's criterion and the Hoek–Brown criterion, which we shall now discuss briefly.

The **Mohr's criterion** states that shear failure takes place along a plane on which the shear stress $\tau$ and the normal stress $\sigma_n$ are related by the following function:

$$\tau = f(\sigma_n) \tag{7.13}$$

The function is regarded as characteristic of the tested rock type. The curve represented by Eq. (7.13) is found experimentally as the envelope of Mohr's circles corresponding to failure under a variety of conditions (Fig. 7.12). Normally, **Mohr's envelope** is non-linear,

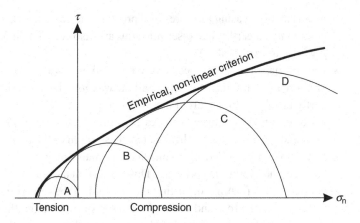

**Fig. 7.12**   Empirical, non-linear Mohr's failure criterion.

concave downward, and often parabolic. A few points may be mentioned regarding this criterion:

- The Coulomb criterion may be regarded as a special case (a straight line) of the more general Mohr's envelope.
- The intermediate stress, $\sigma_2$, does not affect failure and is regarded as being parallel with the fracture plane.
- The envelope is strictly based on empirical data, that is, rock-mechanics laboratory tests where failure occurs at each point where a Mohr's circle touches the Mohr's envelope.
- The possible stress difference $\sigma_1 - \sigma_3$ before failure increases with increasing confining pressure, that is, with increasing $\sigma_3$ or, in a geological context, crustal depth. This means that the curve opens to the right. This is also the primary reason for the large-scale crustal strength increasing with depth down to the lower boundary of the seismogenic layer (Fig. 7.2).

Another well-known empirical criterion for rock failure is the **Hoek–Brown criterion**. This criterion may be expressed as (Hudson and Harrison, 1997; Sheorey, 1997)

$$\sigma_1 = \sigma_3 + (mC_0\sigma_3 + sC_0^2)^{1/2} \tag{7.14}$$

or, alternatively, as

$$(\sigma_1 - \sigma_3)^2 = sC_0^2 + mC_0\sigma_3 \tag{7.15}$$

where $C_0$ is the uniaxial compressive strength of the intact rock, $m$ and $s$ are constants for the particular rock type that arise from the curve-fitting procedure, and $\sigma_1$ and $\sigma_3$ are the maximum and minimum principal stresses. The constants also have a physical meaning. Thus, $s$ is related to the cohesion of the rock and is thus a measure of how fractured the rock is. For uniaxial compression, we have $\sigma_3 = 0$, in which case Eq. (7.15) gives $\sigma_1 = C_0 s^{1/2}$. From this it follows that, since the uniaxial compressive strength must be equal to the uniaxial compressive stress at failure, $s = 1.0$ for intact rocks. For highly fractured rocks $s \to 0$, that is, $s$ approaches zero. Similarly, the constant $m$ relates to how well the rock particles are interlocked, which, for intact rocks, is also high, but becomes gradually lower for more fractured rocks. This constant also depends on the rock type and general mechanical properties and has no very specific bounds.

The Hoek–Brown criterion predicts a relationship between the tensile strength $T_0$ and the compressive strength $C_0$ of the rock. In this criterion, the tensile strength of the rock is obtained from the equation

$$T_0 = \frac{C_0}{2} \left[ (m^2 - 4s)^{1/2} - m \right] \tag{7.16}$$

Equation (7.16) predicts a high ratio of $C_0/T_0$. For example, for a typical intact granite, with $m = 20$ and $s = 1$, we obtain $C_0 \approx 20T_0$. This is higher than commonly measured and predicted by other criteria, where commonly $C_0 \approx 10T_0$.

None of the criteria above is entirely appropriate for the tensile regime. For example, the Coulomb criterion significantly overestimates the tensile strength. Also, they are all empirical criteria with no clear physical basis that can be derived from first principles. The only criterion that has a clear physical basis is the Griffith criterion, derived from the Griffith theory, which was initially proposed to explain why the observed tensile strength of solids such as glass is many orders of magnitude less than their theoretical strength of about 10% of Young's modulus. Here we first discuss the basis of the Griffith theory and subsequently explain his criterion as applied to the failure of rock.

## 7.6 The theory of Griffith

When the repulsive and attractive forces between atoms are considered (Chapter 4), it follows that the theoretical tensile strength of solids is of the order of $0.1E$, that is, 10% of Young's modulus (e.g. Caddell, 1980; Lawn, 1993). For minerals and rocks this would imply a typical theoretical tensile strength of about 1–10 GPa whereas the actual laboratory strength is normally 10–20 MPa (0.001–0.002 GPa) or less than one percent of the theoretical value. Similarly, for other solids, the tensile (and shear) strengths are typically less than 1% of the theoretical strengths.

Griffith (1920, 1924) proposed the first theory to explain this difference between observed and theoretical strengths of solids. In the **theory of Griffith** it is postulated that every solid contains elliptical flaws or very fine cracks, **Griffith cracks** (Fig. 7.13). Because of stress concentrations at the tips of these cracks (Chapter 6), the theoretical tensile strength is reached even though the average or nominal tensile stress in the specimen is quite low. According to the Griffith theory, the measured tensile strengths are small simply because they refer to the nominal tensile stresses in the specimens, whereas the specimen failure is due to local stresses at the tips of the Griffith cracks. These local stresses are, because of stress concentrations around the tips of the (elliptical) cracks, equal to the theoretical tensile strengths. This theory, which elegantly explains the apparent discrepancy between the observed and predicted tensile strength, was initially proposed for glass, but has since been supported by observations of various solids and is now firmly established. Since

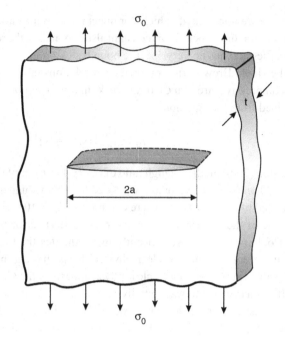

**Fig. 7.13** Griffith crack of strike dimension 2$a$. This crack goes through the thickness of the elastic body, from one stress surface to another, and is referred to as a through crack (Chapter 9).

Griffith's is the founding theory of **fracture mechanics**, we shall here discuss it in some detail before presenting the derived Griffith criterion for rock failure. The Griffith theory has subsequently been derived and developed in various ways. A more detailed version of the theory, presented using a somewhat different approach, is presented in Chapter 10.

More specifically, Griffith suggested that fracture initiation in a brittle material in a biaxial stress field occurs from points of highest tensile stress concentration on the surface of already-formed flaws in the material. He proposed that the material contained a population of elliptical, randomly oriented flaws (the Griffith cracks) and assumed that a suitable oriented crack would start to propagate when the tensile stress at any point around the crack reached the tensile strength of the material at that point. Griffith initially applied his theory to the failure of glass, which behaves as a very brittle material under normal daily pressure and temperature conditions. But his theory of failure has been found to be applicable to a variety of other brittle and quasi-brittle materials, such as rocks and ceramics in general.

### 7.6.1 Stress concentrations

We know from Eq. (6.4) that an elliptical hole raises the stress around it according to the equation

$$K_t = \frac{\sigma_{max}}{\sigma_0} = \frac{2a}{b} + 1 \tag{7.17}$$

where $K_t$ is the stress concentration factor and the elliptical hole has a major axis $2a$ and a minor axis $2b$ and is subject to remote tensile stress $\sigma_0$ in a direction parallel with the minor axis (Fig. 6.8). The maximum tensile stress occurs at the end of the major axis $2a$ of the hole. Using the radius of curvature $\rho_c$ at the end of the elliptical hole, given by (Eq. 6.5),

$$\rho_c = \frac{b^2}{a} \tag{7.18}$$

the maximum tensile stress at the lateral ends of the major axis of the hole may also be written as

$$\sigma_{max} = \sigma_0 \left(1 + 2\sqrt{a/\rho_c}\right) \tag{7.19}$$

For an elliptical crack subject to tensile stress $\sigma_0$ in a direction perpendicular to the major axis, the maximum tensile stress is thus

$$\sigma_{max} = \sigma_0 \left(\frac{2a}{b} + 1\right) \approx 2\sigma_0 \left(\frac{a}{\rho_c}\right)^{1/2} \tag{7.20}$$

These results were used by Griffith (1920, 1924) in his analysis of the conditions for propagation of cracks from flaws or Griffith cracks. Griffith considered the global balance of energy in a solid body with a crack or flaw, taking into account the following:

- The energy stored as elastic strain energy in the solid body hosting the crack.
- The energy needed to generate a new crack surface.
- The work performed during the crack growth by the loads (stress or displacement) on the body.

## 7.6.2 Strain energy

If a plate without a crack is subject to tensile loading $\sigma$, the strain energy density (energy per unit volume) (from Eqs. 4.69 and 4.71) is

$$U_0 = \frac{\sigma^2}{2E} \tag{7.21}$$

If a crack of strike dimension $2a$ is introduced into the elastic plate (Fig. 7.14), then a constant tensile stress will open the crack into a flat ellipse (Chapter 9). There will thus be relaxation of the tensile stress around the crack, a concept sometimes referred to in geology as a **stress shadow**. This relaxation results in some of the elastic strain energy in the plate being released. The stress-relaxation or stress-shadow zone around the crack has approximately the geometry of a triangle of height $ka$, that is, the height is equal to the semi-major axis of the (elliptical) crack multiplied by a certain constant $k$ (Fig. 7.14). The area of a triangle is equal to half its base, here $a$, times its height, here $ka$. So if the plate has a thickness $t$ (Fig. 7.13) then the volume of stress-relaxed region above the crack is $ka^2t$. The strain energy density released, that is, the strain energy $U_0$ released per unit thickness ($t = 1$) is then

$$U_0 = \frac{\sigma^2}{2E}ka^2 \tag{7.22}$$

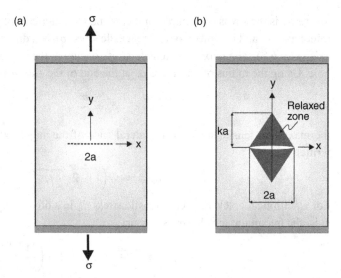

**Fig. 7.14** Elastic plate subject to constant tensile stress $\sigma$ at its edges. In (a) there is no crack, but in (b) a crack of the type in Fig. 7.13 has been introduced. Opening of the crack results in a relaxation of the tensile stress in an approximately triangular area or zone around the crack. The height of this area is $ka$.

This is in good agreement with Griffith's more accurate analysis which gave, for plane stress,

$$U_0 = \frac{\sigma^2}{2E}\pi a^2 \tag{7.23}$$

and for plain strain (Chapter 4)

$$U_0 = \frac{\sigma^2(1 - \nu^2)}{2E}\pi a^2 \tag{7.24}$$

The variation of $U_0$ with crack length is given in Fig. 7.15. Because $U_0$ is energy release, in the plot in Fig. 7.15 it is shown as negative.

### 7.6.3 Energy needed to generate a new crack surface

While strain energy $U_0$ is released (and thus an output) during relaxation associated with crack formation (and propagation), energy is needed as input into the solid body to create the new crack surfaces. The reason why energy is needed to generate fracture surfaces can be explained as follows (Chapter 4). To separate two atomic planes from each other so as to form a fracture, the planes must be moved away from each other to a distance where there are no interacting forces between the planes. This large separation requires work, that is, force times distance, which is also a measure of energy (Figs. 4.7 and 4.8). The required energy is denoted schematically in Fig. 7.16, which is effectively a replot of Fig. 4.8. The

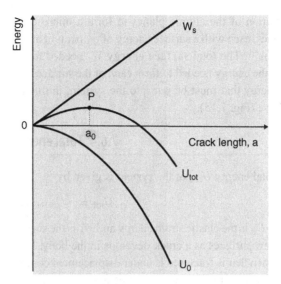

**Fig. 7.15** Variation in energy as a crack extends. $W_s$ is the energy input needed to generate two new crack surfaces, $U_0$ is the strain energy released, and $U_{tot}$ is the change in the total energy of the body as the crack extends. When the crack length is $a_0$ the system is in equilibrium, indicated by point p on the $U_{tot}$ curve. To the left of this point, energy must be put into the system (to form crack surfaces) for the crack to propagate; to the right of this point, more energy is released as the crack propagates than is needed to generate its surfaces.

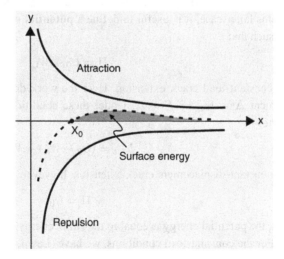

**Fig. 7.16** Schematic presentation of the interatomic attraction and repulsion, here as forces per unit area, that is, stress (on the $y$-axis). When the stresses are in balance, the distance between the atomic planes is $x_0$, which corresponds to $a_0$ on the force–distance diagram (Fig. 4.8). In this presentation, the shaded area corresponds to the surface energy $W_s$ in Fig. 7.15.

separation of the atomic planes to form a microscopic fracture clearly creates two new surfaces, each with a surface energy of $w_s$ per unit area $A$, and thus with the units of N m/m$^2$ or N m$^{-1}$. The total **surface energy** $W_s$ needed to form the microscopic fracture is thus twice the energy needed to form each of the surfaces, or $W_s = 2Aw_s$. Because $W_s$ represents the energy that must be put into the system, in this case the solid body, it is regarded as positive (Fig. 7.15).

## 7.6.4 Total energy

The total energy $U_{tot}$ of the system is given by

$$U_{\text{tot}} = -U_0 + W_{\text{s}} \tag{7.25}$$

where $U_0$ is the elastic strain energy and $W_s$ is the surface energy or work needed to generate two new surfaces as a crack develops in the body. In this formulation, we assume that the specimen that is fractured is under displacement control, that is, in a fixed-grip or **constant displacement** type of test. This means that the specimen is subject to a prescribed constant displacement, so that the external loads do not perform any work. If, however, the specimen is in a load control or **constant load** test, so that the load is prescribed, then in addition to the strain energy $U_0$ there is work $W_L$ done by external displacements of the loads as the crack grows. This follows because if the outer boundary of the body containing the crack is free to move, the boundary would be expected to undergo some displacement as the crack extends. The points on the boundary where the loads are applied will thus move, and the loads will do work which equals the load times the displacement.

In this latter case, it is useful to define a **potential energy** $\Pi$ of the specimen or solid body such that

$$\Pi = U_0 - W_{\text{L}} \tag{7.26}$$

for a constant-load crack extension. Here the work done by the load $F$ during the displacement $\Delta$ is $W_{\text{L}} = F \times \Delta$. Under these conditions, the total energy in Eq. (7.25) becomes

$$U_{\text{tot}} = (U_0 - W_{\text{L}}) + W_{\text{s}} \tag{7.27}$$

For a constant-displacement crack extension, however, we have

$$\Pi = U_0 \tag{7.28}$$

that is, the potential energy is equal to the strain energy, since the work done by the load is zero. For the constant-load conditions, we have (Lawn, 1993)

$$W_{\text{L}} = 2U_0 \tag{7.29}$$

So the mechanical energy is $-W_{\text{L}} + U_0 = -U_0$.

With reference to the curves for strain energy, surface energy, and the total energy of the solid body within which the crack is developing (Fig. 7.15), it follows that for a crack to extend or grow within the region $0 \le a \le a_0$, energy must be transferred to or put into the hosting elastic body. By contrast, for the region $a > a_0$ the crack growth results in energy release. Consequently, once the crack has reached the critical length $a_0$, further growth releases energy so that the crack will automatically propagate.

### 7.6.5 The energy release rate $G$

The strain energy release rate largely determines whether a fracture is able to grow in a solid body. The value of $\partial U_0 / \partial a$ is a measure of the strain-energy release rate for an incremental extension of the crack. The strain-energy release rate during crack growth is given the symbol $G$ and defined as

$$G = \frac{\partial U_0}{\partial a} \tag{7.30}$$

From the plane-stress formula for the strain energy density (Eq. 7.23) we get

$$G = \frac{\partial}{\partial a} \left( \frac{\sigma^2}{2E} \pi a^2 \right) = \frac{\sigma^2 \pi a}{E} \tag{7.31}$$

and, similarly, from the plane-strain formula for strain energy density (Eq. 7.24) we get

$$G = \frac{\partial}{\partial a} \left( \frac{\sigma^2}{2E} \pi a^2 (1 - v^2) \right) = \frac{\sigma^2 (1 - v^2) \pi a}{E} \tag{7.32}$$

The energy needed to form the crack surfaces as the crack grows is absorbed in the solid body and given the symbol $R$, which is defined as

$$R = \frac{\partial W_s}{\partial a} \tag{7.33}$$

and referred to as the **crack resistance**. The threshold **condition for unstable crack propagation** is that the strain energy release $G$ is equal to the crack resistance $R$, that is,

$$G = R \tag{7.34}$$

For a facture to propagate, the energy release rate of the rock must reach a certain minimum critical value. This critical value $G_c$ is, like $R$, a material constant and depends on the energy needed to rupture the atomic bonds within the solid material. $G_c$ is known as the **material toughness** of the solid and is the critical energy release rate for a crack to propagate. Using Eqs. (7.31) and (7.32) and solving for the tensile stress $\sigma$, we obtain the minimum tensile stress needed to initiate fracture propagation as

$$\sigma = \left( \frac{EG_c}{\pi a} \right)^{1/2} \tag{7.35}$$

for plane stress. The corresponding formula for plane strain is

$$\sigma = \left( \frac{EG_c}{\pi (1 - v^2) a} \right)^{1/2} \tag{7.36}$$

These equations show that tensile stress for failure varies inversely as the square root of the initial crack length, $a$. It thus follows that the tensile strength is inversely proportional to the square root of the initial crack length, that is,

$$T_0 \propto a^{-1/2} \tag{7.37}$$

which may be regarded as a formal, theoretical explanation for tensile strength of a rock body decreasing with its size up to a certain limit (Chapter 5). As the body size increases, the larger will be the initial size $a$ of Griffith cracks or other weaknesses that it can contain, and thus the smaller the tensile strength. Rewriting Eq. (7.35), we obtain

$$\sigma\sqrt{\pi a} = \sqrt{EG_c} \tag{7.38}$$

where the term $\sigma\sqrt{\pi a}$ is known as the **stress intensity factor** and denoted by $K$. For a fracture to propagate, the stress intensity factor must reach a certain minimum critical value. The critical stress intensity factor for a fracture to propagate is known as the **fracture toughness**, denoted by $K_c$. Thus, from Eq. (7.38) the fracture toughness is

$$K_c = \sigma\sqrt{\pi a} \tag{7.39}$$

These results are elaborated in Chapter 10.

## 7.7 The Griffith criterion for rocks

The theory presented above refers to tensile loading and thus, when applied to rock fractures, to the tensile regime. However, Griffith (1924) was able to extend his theory so that it also included failure in the compressive regime. Consider first the Griffith criteria for rock failure in the **tensile regime**. There the two-dimensional Griffith criterion of fracture initiation is as follows:

$$\text{if } \sigma_1 < -3\sigma_3 \quad \text{then} \quad \sigma_3 = -T_0 \tag{7.40}$$

where $\sigma_1$ is the maximum principal compressive stress, $\sigma_3$ is the minimum principal compressive stress, and $T_0$ is the tensile strength of the host rock. Substituting $-T_0$ for $\sigma_3$ in the inequality above, we obtain

$$\sigma_1 - \sigma_3 < 4T_0 \tag{7.41}$$

which shows that the stress difference, for the failure criterion in the tensile regime to apply, cannot exceed $4T_0$. Using

$$\sigma_1 = \rho_r g d \tag{7.42}$$

where $\rho_r$ is the host-rock density, $g$ is the acceleration due to gravity, and $d$ is the fracture depth, the maximum depth $d_{max}$ that a large-scale tension fracture can reach (in the absence of internal fluid overpressure) is

$$d_{max} = \frac{3T_0}{\rho_r g} \tag{7.43}$$

For the **compressive regime**, the two-dimensional Griffith criterion for facture initiation is as follows:

$$\text{if } \sigma_1 > -3\sigma_3 \quad \text{then} \quad (\sigma_1 - \sigma_3)^2 = 8T_0(\sigma_1 + \sigma_3) \tag{7.44}$$

If $\alpha$ is the angle between the major axis of an elliptical crack and the direction of $\sigma_1$, it follows from the formula above that the orientation of the crack when it starts to propagate from the point of maximum tension stress on its boundary is given by

$$\cos 2\alpha = \frac{\sigma_1 - \sigma_3}{2(\sigma_1 + \sigma_3)} \tag{7.45}$$

In the compressive regime, the crack that is most likely to start propagating is thus oblique to the direction of $\sigma_1$ and a potential shear crack.

The criteria for tensile and compressive regimes, as presented in Eqs. (7.41) and (7.45), can also be expressed in terms of the normal stress $\sigma_n$ and the shear stress $\tau$ acting on the plane that coincides with the major axis of the crack, thus

$$\tau = 4T_0(\sigma_n + T_0) \tag{7.46}$$

The Griffith criterion for failure or peak strength in the compressive regime has not been very useful for rocks. However, the Griffith criterion, as expressed in terms of energy considerations, is the basis for fracture mechanics and its application to rock fractures.

## 7.8  A combined rock-failure criterion

There is no single theory, empirical or theoretical, that provides a satisfactory framework for explaining and predicting failure of rocks. Even in the brittle field, we normally use a combination of the theories of Griffith (primarily for the tensile regime) and Coloumb (primarily for the compressive regime). In a major fault zone that extends from the surface through the brittle crust (the seismogenic layer) and down into the more ductile part of the crust or lithosphere (Fig. 7.17), there are different theories by which we could normally explain failure in the different parts of the fault zone.

In the uppermost part of the fault zone the Griffith theory normally offers a suitable criterion, particularly if this part is in the tensile regime. The fault zone in Fig. 7.17 is, based on its vertical dip, most likely to be a strike-slip fault. Many such faults, for example in Iceland (Fig. 7.7), enter a tensile regime close to and at the surface. For the near-surface parts, where tension fractures are commonly associated with the strike-slip fault (Fig. 7.7; Chapter 12), the Griffith criterion would thus be most suitable.

At greater crustal depths and temperatures, with breccias and other cataclastic rocks, failure is best described using the Coulomb criterion. This criterion is particularly suitable for granular media as well as rocks in the compressive brittle regime, and thus appropriate for the fault slip at depths that depend on the tensile strength of the rock but are generally deeper than the uppermost several hundred metres to a few kilometres of the crust.

At still greater depths, there is a change in the fault zone, as well as in host-rock behaviour, from primarily brittle to primarily ductile behaviour. This change is gradual and depends on temperature and strain rates (Appendix E.2 provides data on the brittle–ductile transition in some crustal rocks). Either the Coulomb criterion or a criterion for plastic flow, such as the Tresca or the von Mises criteria, may be used at this depth. The most suitable failure

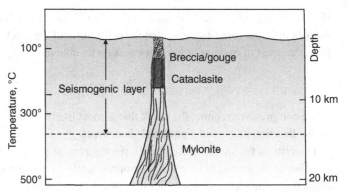

Fig. 7.17 Schematic vertical section through a crustal segment containing a continental fault zone gradating into a ductile shear zone. The fault zone is confined to the brittle, that is, seismogenic layer, where the main deformation is through fracture and fault slip. The shear zone is confined to the ductile layer where the main deformation is through plastic flow. The boundaries between these zones, however, are not sharp and depend on temperature, strain rate, pore-fluid pressure, and other factors.

criterion for the part of the fault zone that is well into the ductile field, however, is the von Mises criterion, which is generally regarded as more accurate than the Tresca criterion. In the ductile field, the fault zone is no longer referred to as such but rather as a shear zone. Thus, we distinguish between a predominantly **brittle fault zone** and a predominantly **ductile shear zone**. The boundaries between these different regimes within a fault zone (Fig. 7.17), and thus between the various criteria, are gradual and vary with time as the fault rocks evolve and change their properties. To give an idea of the most suitable rock-failure criteria as a function of normal stress or depth in the crust, a combined rock-failure criterion in presented in Fig. 7.18.

# 7.9  Tresca and von Mises criteria

We have several times mentioned the yield criteria of Tresca and von Mises. Here we shall explain them briefly.

## 7.9.1  Tresca criterion

This criterion is also known as the maximum-shear-stress criterion, but we shall refer to it as the Tresca criterion, since that is the name most commonly used. The **Tresca criterion** states that plastic deformation or yield or flow starts when the maximum shear stress in a rock (or any other solid material) reaches the critical value known as the **yield stress** (or yield strength) $\sigma_y$ (cf. data in Appendix E.2). The yield stress is a constant for perfectly

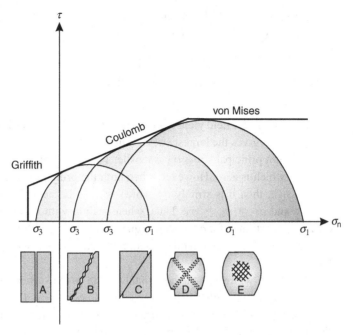

Fig. 7.18 Combined rock-failure criterion as a function of normal stress, that is, crustal depth (and thus overburden pressure). As the depth/pressure (and, normally, also the temperature) increase to the right on the diagram, the most useful failure criterion changes from that of Griffith, through Coulomb, to von Mises. The typical types of associated deformation are indicated very schematically by a tension fracture (A), a tension-and-shear or hybrid fracture (B), brittle shear fracture (C), conjugate brittle–ductile shear fractures (D), and plastic deformation (E).

plastic materials. Analytically, the criterion may be expressed in the form

$$\tau_{\max} = \frac{\sigma_1 - \sigma_3}{2} = C_y \tag{7.47}$$

where $C_y$ is a yield constant. The yield constant can be related to the **yield stress** $\sigma_y$ in uniaxial tension or compression (Fig. 5.1). Then yield occurs when the absolute value of either $\sigma_1$ or $\sigma_3$ is equal to $\sigma_y$, that is,

$$|\sigma_1| = \sigma_y \tag{7.48}$$

or

$$|\sigma_3| = \sigma_y \tag{7.49}$$

The maximum shear stress is then $\sigma_y/2$, so that, from Eq. (7.47), we have

$$\tau_{\max} = \frac{\sigma_1 - \sigma_3}{2} = \frac{\sigma_y}{2} \tag{7.50}$$

Rearranging, we get

$$|\sigma_1 - \sigma_3| = \sigma_y \tag{7.51}$$

Combining Eqs. (7.48), (7.49), and (7.51), the Tresca criterion can be formulated so that yield occurs when the following condition is satisfied:

$$\left( \begin{array}{c} |\sigma_1 - \sigma_3| \\ |\sigma_1| \\ |\sigma_3| \end{array} \right) = \sigma_y \tag{7.52}$$

In a biaxial stress field yield begins where one of these equations is first satisfied, that is, the one which gives the largest stress. Equations (7.49)–(7.52) are the correct forms when the minimum principal stress $\sigma_3$ is tensile. Then, for a biaxial stress, it is the intermediate stress $\sigma_2$ which is zero. However, when both the principal stresses of the biaxial field are compressive, then it is strictly $\sigma_3$ which is zero, because, by definition, compression is positive and $\sigma_1 \geq \sigma_2 \geq \sigma_3$. Thus, when dealing with a **biaxial compressive field**, $\sigma_2$ should be substituted for $\sigma_3$ in these equations. Strictly, therefore, this yield criterion is not very suitable, although sometimes used, for biaxial compression (cf. Harrison and Hudson, 2000).

In a diagram with $\sigma_1$ as the horizontal and $\sigma_3$ as the vertical coordinate axes, Eq. (7.48) is presented by two (rather than one) vertical lines and Eq. (7.49) by two horizontal lines. This is because of the absolute values used. Similarly, Eq. (7.51) is presented by two lines dipping at 45° (Fig. 7.19). From this we obtain the hexagonal in Fig. 7.20 that, according to the Tresca critierion, can be interpreted as follows:

- The hexagonal lines at 45° to the principal stresses represent the planes of maximum shear stress.
- Yield occurs when a point in the $\sigma_1$–$\sigma_3$ plot falls anywhere on the hexagonal.
- More specifically, yield occurs when the maximum shear stress at a point in the rock is equal to the maximum shear stress in the rock when it yields under uniaxial tension (or compression). From Eq. (7.50), this shear stress is $\sigma_y/2$.
- For stresses at all points inside the hexagonal, the material behaves elastically.
- For a material whose behaviour is elastic up to yield, and then plastic, that is, an elastic–perfectly plastic material, stress states represented by points outside the hexagonal cannot be reached. No stresses beyond those that generate yield according to Eq. (7.52) are possible.

## 7.9.2  Von Mises criterion

This criterion is also known as the von Mises–Henky criterion, the maximum-distortion-energy criterion, and as the maximum shear strain energy criterion. We shall refer to it as the von Mises criterion since that is the name most commonly used (Fig. 7.21). The criterion can be stated in several ways. One statement is as follows: yield begins in a material when the sum of the squares of the principal components of the deviatoric stress reaches a specified critical value. Deviatoric stress is discussed in Chapter 2. The corresponding equation is

$$(\sigma_1 - \sigma_2)^2 + (\sigma_2 - \sigma_3)^2 + (\sigma_3 - \sigma_1)^2 = C_y \tag{7.53}$$

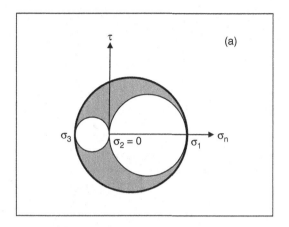

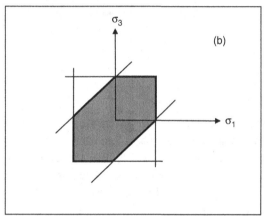

**Fig. 7.19**   Construction of the Tresca hexagonal. (a) For this biaxial state of stress, the intermediate principal stress, $\sigma_2$, is zero. (b) The diagonal (inclined at 45°) lines in the illustrations represent the top term in Eq. (7.52). The vertical and horizontal lines represent the middle and the bottom terms (equations) in Eq. (7.52). The Tresca criterion states that yield or plastic flow starts as soon as a point in the principal-stress space touches or falls on the indicated hexagon.

To relate the yield constant $C_y$ to the yield stress $\sigma_y$, we again use uniaxial tension (or compression). Then $\sigma_1 = \sigma_y$ and $\sigma_2 = \sigma_3 = 0$. Then we have, from Eq. (7.53) that

$$2\sigma_1^2 = 2\sigma_y^2 \tag{7.54}$$

Using Eq. (7.54) and $\sigma_2 = \sigma_3 = 0$ in Eq. (7.53) we get

$$(\sigma_1 - \sigma_2)^2 + (\sigma_2 - \sigma_3)^2 + (\sigma_3 - \sigma_1)^2 = 2\sigma_y^2 \tag{7.55}$$

For pure shear $\sigma_1 = -\sigma_3 = k$, a constant, and $\sigma_2 = 0$. From Eq. (7.54) we then get

$$\sigma_1^2 + \sigma_1^2 + 4\sigma_1^2 = 6\sigma_1^2 = 6k^2 \tag{7.56}$$

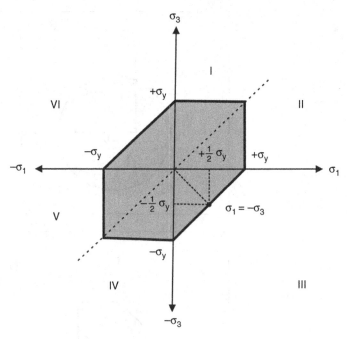

Fig. 7.20 The details of the Tresca hexagonal. For any state of stress that is represented by a point in the principal-stress space that lies within the shaded hexagon, there will be no yielding. But yield occurs as soon as any such point touches the hexagon. The various regions (I–VI) are marked by the following lines (making up the hexagon and indicating yield): (I) the horizontal line is $\sigma_3 = \sigma_y$; (II) the vertical line is $\sigma_1 = \sigma_y$; (III) the diagonal or 45° line is $\sigma_1 = -\sigma_3 = \frac{1}{2}\sigma_y$; (IV) the horizontal line is $\sigma_3 = -\sigma_y$; (V) the vertical line is $\sigma_1 = -\sigma_y$; (VI) the diagonal or 45° line is $\sigma_1 = -\sigma_3 = \frac{1}{2}\sigma_y$.

From Eqs. (7.55) and (7.56) we can thus write one version of the von Mises criterion as

$$(\sigma_1 - \sigma_2)^2 + (\sigma_2 - \sigma_3)^2 + (\sigma_3 - \sigma_1)^2 = 2\sigma_y^2 = 6k^2 \qquad (7.57)$$

From Eq. (7.57) it follows that the uniaxial tensile (or compressive) yield $\sigma_y$ and the pure shear yield $k$ are related thus:

$$\sigma_y = k\sqrt{3} \qquad (7.58)$$

Equation (7.55) can be simplified for the plane-stress (biaxial stress) condition, taking one of the principal stresses as zero, say $\sigma_2 = 0$. Substituting $\sigma_2 = 0$ in Eq. (7.55), we get

$$2\sigma_1^2 + 2\sigma_3^2 + 2\sigma_1\sigma_3 = 2\sigma_y^2 \qquad (7.59)$$

We can rewrite Eq. (7.59) in the form (cf. Solecki and Conant, 2003)

$$\sigma_y^2 = \frac{(\sigma_1 + \sigma_3)^2}{4} + \frac{(\sigma_1 - \sigma_3)^2}{4/3} \qquad (7.60)$$

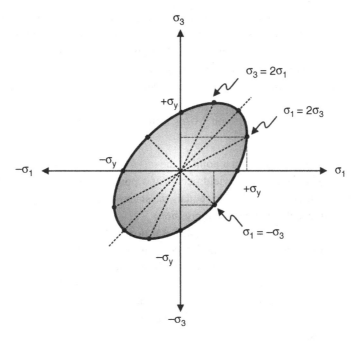

**Fig. 7.21** The von Mises yield ellipse. For any state of stress that is represented by a point in the principal-stress space that lies within the shaded ellipse, there will be no yielding. But yield occurs as soon as any such point touches the ellipse. Several stress conditions at the onset of plastic yield (points falling on the ellipse) are indicated. For example, when $\sigma_1 = -\sigma_3$, there is pure shear.

or as

$$\frac{\left[\frac{\sqrt{2}}{2}(\sigma_1 + \sigma_3)\right]^2}{(\sigma_y\sqrt{2})^2} + \frac{\left[\frac{\sqrt{2}}{2}(\sigma_1 - \sigma_3)\right]^2}{\left[(\sqrt{3})^{-1}\sigma_y\sqrt{2}\right]^2} = 1 \qquad (7.61)$$

Equation (7.62) is clearly in the form of an ellipse, namely

$$\frac{x^2}{a^2} + \frac{y^2}{b^2} = 1 \qquad (7.62)$$

with the semi-major and semi-minor axes given by the denominators in Eq. (7.60). Comparing Eqs. (7.60) and (7.62), we see that $x = 0$ is the line $\sigma_1 = -\sigma_3$ and that $y = 0$ is the line $\sigma_1 = \sigma_3$. This means that the ellipse in Eq. (7.60) is rotated 45° counterclockwise from the coordinate system given by the axes $\sigma_1$ and $\sigma_3$ (Fig. 7.21).

Another statement of the von Mises criterion refers to the **octahedral shear stress** which, with reference to Eq. (7.57), is defined as (cf. Jaeger, 1969; Soleki and Conant, 2003; Chen and Han, 2007)

$$\tau_{\text{oct}} = \frac{1}{3}[(\sigma_1 - \sigma_2)^2 + (\sigma_2 - \sigma_3)^2 + (\sigma_3 - \sigma_1)^2]^{1/2} \qquad (7.63)$$

The von Mises criterion then says that yield occurs at a point in the rock where the octahedral shear stress is equal to $\sqrt{2}/3$ or about 0.47 times the yield stress under uniaxial tension (or compression). An octahedron is a polyhedron with eight faces which, for a regular octahedron, are equilateral triangles (equilateral means that all the sides of the triangle have the same length). Octahedral shear stress thus refers to the stress on the face of an octahedron.

According to the von Mises criterion the ellipse in Fig. 7.21 can be interpreted as follows:

- Yield occurs when a point in the $\sigma_1$–$\sigma_3$ plot falls anywhere on the ellipse itself.
- More specifically, yield occurs at a point in the rock when the shear-strain energy per unit volume in the rock equals the shear-strain energy at yielding of the same rock under uniaxial tension (or compression).
- Alternatively, the rock yields at a point when the octahedral shear stress is about 0.47 times the yield stress at that point under uniaxial tension or compression.
- For stresses at all points inside the ellipse, the material behaviour is elastic.
- For a material whose behaviour is elastic up to yield, and then plastic, that is, an elastic–perfectly plastic material, stress states represented by points outside the ellipse cannot be reached. In terms of this criterion, no stresses beyond those that generate yield are possible.

An illustration with both criteria is shown in Fig. 7.22. Since the ellipse lies somewhat outside the hexagonal, except in the six points where they touch, the von Mises criterion normally indicates higher stress for yield to occur than the Tresca criterion. That is, the Tresca criterion is more conservative in that it predicts yield at generally lower stresses.

Experiments show that the von Mises criterion is generally more accurate than the Tresca criterion (Solecki and Conant, 2003). Both theories assume that plastic yielding is **independent of** hydrostatic pressure or **mean stress**, that is, yield is only a function of shear stress or deviatoric stress. Thus, plastic yielding is associated with **distortion or shape change** rather than dilation or volume change. These criteria are thus different from the Coulomb criterion for granular materials, such as sand, soil, and for rocks that **depends on** the **mean stress** or hydrostatic pressure or, as expressed in the failure criterion itself, normal stress. However, as indicated above, the Tresca criterion may be regarded as a special case of the Coulomb criterion, namely the one with $\phi = 0$ in Eq. (7.3).

Geometrically, the Tresca criterion is represented in three dimensions as an infinitely long, regular hexagonal of a constant cross-sectional area. By contrast, the Coulomb criterion is represented in three dimensions by an irregular hexagonal of a cross-sectional area that increases in size with increasing mean stress or crustal depth (Davis and Selvadurai, 2002; Chen and Han, 2007). Thus, the Tresca criterion (and the von Mises criterion as well) is independent of mean stress and thus independent of crustal depth. By contrast, the Coulomb criterion depends on crustal depth; a larger stress difference is needed to cause shear failure at greater depths than at shallower depths. In the two-dimensional version of the Coulomb criterion, this follows from the gradual increase in the diameter of the Mohr's circle needed to reach the failure line or failure envelope (in Mohr's version of the criterion) as the depth, that is, $\sigma_n$, increases (Figs. 7.5, 7.8, and 7.12).

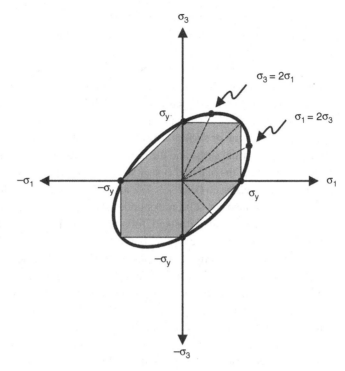

**Fig. 7.22**   Comparison of the Tresca and the von Mises yield criteria, as represented graphically. Generally, except in the points where the hexagon touches the ellipse, the von Mises criterion predicts a somewhat higher stress for the onset of yielding than the Tresca criterion. Generally, the von Mises criterion fits better with measurements, primarily of the yield of metals, and is considered to be more accurate.

## 7.10  Summary

- Rock fails when subject to stresses that reach the rock strength. Failure is the stress condition by which a solid such as rock begins to flow or break. Brittle failure is through breaking, that is, fracture formation; ductile failure is through plastic flow. In a major fault zone, the upper part fails through fracturing and is a proper fault zone. The deeper part fails through plastic flow and is a shear zone.
- Rock strength may be defined in various ways. When the tensile strength $T_0$ is reached, the rock ruptures and forms an extension fracture. When the shear strength $\tau_0$ is reached, the rock ruptures and forms a shear fracture, that is, a fault. A common measure of the shear strength is the driving shear stress, or stress drop, associated with seismogenic slip on faults. The compressive strength $C_0$ normally coincides with the peak of the stress–strain curve in a laboratory experiment. Crustal strength is often interpreted as the difference between the principal stresses $\sigma_1 - \sigma_3$ and is the maximum stress difference that the rock can withstand before it fails. The crustal strength generally increases with

crustal depth and is maximum at the bottom of the seismogenic layer, below which the crustal deformation gradually becomes more ductile.

- A Coulomb material is a granular material whose failure, when subject to stress, depends on the normal stress (and thus the mean stress). This dependence is because of its granular nature. As the normal stress or mean stress increases the grains become more densely packed, making it more difficult for them to slide over one another. Coulomb-material failure may be regarded as similar to rigid-plastic behaviour, the failure plane being referred to as the slip plane or surface. An ideal Coulomb material is one where the relationship between the normal stress and the failure shear stress is linear. If the cohesion or inherent shear strength of the material is very small or zero, it is known as a cohesionless material.

- The Coulomb criterion (Eqs. 7.1, 7.3, and 7.4) is widely used for soils and other granular materials, but also for solid rocks under shear failure. The criterion specifies the shear stress that the rock can withstand, for a given normal stress and shear strength (or cohesion), before shear failure or slip, but it does not specify the direction of slip. In fact, at failure there are theoretically two potential slip planes, which, if both develop into faults, generate conjugate faults. Because of rock anisotropy and heterogeneity, as well as variations in loading conditions, it is commonly only one of the slip planes that develops into a macroscopic fault. The dependence of the Coulomb criterion on the normal stress (and the mean stress) distinguishes it from other yield criteria, such as Tresca and von Mises, which are independent of mean stress and normal stress.

- Additional empirical criteria include the Mohr's criterion, which states that rock failure occurs on a plane where the shear stress and the normal stress are related by a certain function, and the Hoek–Brown criterion, which is given by Eq. (7.15).

- The only non-empirical physical theory of brittle failure of solids, including rocks, is the Griffith criterion, which is based on the Griffith theory. This theory was set forth to explain the difference between the high theoretical and the comparatively low measured tensile strength of solids. Griffith postulated that each solid contains minute elliptical flaws or cracks, known as Griffith cracks, that concentrate stresses and, for certain conditions, propagate and link into macroscopic fractures. The existence of Griffith cracks has been confirmed in numerous observations. Generally, a Griffith crack will begin to propagate and continue to extend so long as the strain energy released in doing so is greater than the surface energy (the input energy) needed to generate the new crack surfaces. That is, the strain energy release rate $G$ must be larger than the crack resistance $R$, the latter being the energy needed to form crack surfaces as the crack grows and thus the energy absorbed by the solid. The critical energy release rate for a crack to propagate is known as the material toughness and denoted by $G_c$.

- When applied to rocks, the Griffith criterion for the tensile regime is given by Eq. (7.41) and for the compressive regime by Eq. (7.45). The Coulomb and Griffith criteria are best applied to different tectonic conditions. Thus, for a fault zone, the Griffith criterion works best in its uppermost part, particularly in a tensile regime. At greater depths and temperatures, the Coulomb criterion may be more suitable, in particular for slip within the granular breccia and gouge in the fault core. At still greater depths, the deformation is no longer primarily brittle but rather ductile (plastic), whereby the fault zone changes into a ductile shear zone. In these deep parts of the crust, the plastic yield criteria of Tresca and von Mises may be the most appropriate failure criteria. Griffith–Coulomb–von Mises together form a combined rock-failure criterion.

- There are many yield criteria (criteria for plastic deformation or flow), but the best known and most used are known by the names of Tresca and von Mises. The Tresca criterion states that plastic yield occurs when the maximum shear stress in the rock (or other solid) equals the yield stress (or yield strength) of the rock. It is known as the maximum shear stress criterion (Eq. 7.48). The von Mises criterion is given by Eq. (7.57) and states that plastic yield occurs when the octahedral shear stress or the strain energy of distortion reaches a critical value. Both criteria give satisfactory predictions but the von Mises is generally regarded as more accurate.

## 7.11 Main symbols used

| | |
|---|---|
| $a$ | semi-major axis of an elliptical hole |
| $a$ | half length (strike dimension) of a crack |
| $b$ | semi-minor axis of an elliptical hole |
| $C_0$ | compressive strength |
| $d$ | depth below the Earth's surface |
| $E$ | Young's modulus (modulus of elasticity) |
| $F$ | force, load |
| $G$ | strain energy release rate during crack extension (propagation) |
| $G_c$ | critical strain energy release rate, material toughness |
| $g$ | acceleration due to gravity |
| $K$ | stress intensity factor |
| $K_c$ | critical stress intensity factor, fracture toughness |
| $K_t$ | stress concentration factor |
| $k$ | constant in Eq. (7.22) |
| $m$ | constant in Eq. (7.14) |
| $N$ | normal force |
| $p_t$ | total fluid pressure, also $p_{fluid}$ and $p_f$ |
| $q$ | parameter defined in Eq. (7.10) |
| $R$ | crack resistance |
| $s$ | constant in (Eq. 7.14) |
| $T_0$ | tensile strength |
| $U_{tot}$ | total energy |
| $U_0$ | strain energy density (strain energy per unit volume) |
| $W_L$ | work done by a load on a body with a crack |
| $W_s$ | surface energy (work) needed to generate two new crack surfaces |
| $\alpha$ | angle between the major axis of an elliptical crack and $\sigma_1$ |
| $\mu$ | coefficient of internal friction |
| $\nu$ | Poisson's ratio |
| $\Pi$ | potential energy of a body with cracks |
| $\pi$ | $3.141\,59\ldots \approx 3.1416$ |

$\rho_c$     radius of curvature
$\rho_f$     fluid density
$\rho_r$     rock or crustal density
$\sigma$     tensile loading, tensile stress
$\sigma_n$   normal stress (mostly used in connection with a slip along fault planes)
$\sigma_1$   maximum compressive principal stress
$\sigma_2$   intermediate compressive principal stress
$\sigma_3$   minimum compressive (maximum tensile) principal stress
$\sigma_0$   remote tensile stress
$\tau$       shear stress
$\tau_0$     inherent shear strength (or cohesion or cohesive strength)
$\phi$       angle of internal friction

# 7.12 Worked examples

## Example 7.1

### Problem

The fault in Fig. 2.4 dips 73° where it dissects the upper basaltic lava flow. Given the level of erosion in this area, the estimated normal stress on this part of the fault plane is estimated at 0.8 MPa (Example 2.3). For basalt, a typical coefficient of internal friction is 1.11 and a typical in-situ tensile strength is about 3 MPa. Calculate the driving shear stress $\tau$ needed to cause slip on this part of the fault if:

(a) the fault plane is dry
(b) the total fluid pressure on the fault plane equals the normal stress on the plane.

### Solution

(a) From Eq. (7.4) we get the driving shear stress for slip as

$$\tau = 2T_0 + \mu\sigma_n = 2 \times 3 \times 10^6 \text{ Pa} + 1.11 \times 8 \times 10^5 \text{ Pa} = 6.89 \times 10^6 \text{ Pa} \approx 7 \text{ MPa}$$

(b) When the total fluid pressure $p_f$ equals the normal stress then we have

$$(\sigma_n - p_f) = 0 \Rightarrow \mu(\sigma_n - p_f) = 0$$

so that the second term on the right-hand side of Eq. (7.4) is zero and only the first term on the right-hand side is non-zero, namely $2T_0 = 6$ MPa. It is likely that in many fault zones the total fluid pressure will approach or reach (and occasionally exceed) the normal stress prior to slip, which may partly explain why the driving shear stress $\tau$ for fault slip, as measured by the stress drop, does not depend much on crustal depth (and thus the normal stress) and has a remarkably narrow range, with typical values between about 1 MPa and 12 MPa.

---

**Example 7.2**

**Problem**

A normal fault plane in a flat rift zone that has a cohesion or shear strength of 1 MPa, dips 70° and thus makes an angle of 20° to the direction of $\sigma_1$ – similar to the normal faults in Fig. 2.1. If the minimum principal compressive stress is 5 MPa, find the maximum principal compressive stress needed to generate fault slip on the plane. At roughly what depth would that stress be reached?

**Solution**

From Eq. (7.8), the maximum principal compressive stress $\sigma_1$ for slip to occur is given by

$$\sigma_1 = C_0 + q\sigma_3$$

Neither $C_0$, the uniaxial compressive strength of the fault rock, nor the parameter $q$ are known, so that we need to find them using information given in the example. We know that $\alpha$ is 20°, and we know from Example 2.7 that $\alpha + \theta = 90°$, so that $\theta$ is 70°, where $\theta$ is the angle between the normal to the fault plane and $\sigma_1$ (Chapter 2). From Eq. (7.10) we know that the parameter $q$ is a function of $\theta$ through the relation

$$q = \left[ \left( \mu^2 + 1 \right)^{\frac{1}{2}} + \mu \right]^2 = \tan^2 \theta$$

From Eqs. (7.9) and (7.10), we obtain an expression for the uniaxial compressive strength thus

$$C_0 = 2\tau_0 \left[ \left( \mu^2 + 1 \right)^{\frac{1}{2}} + \mu \right] = 2\tau_0 \tan \theta \tag{7.64}$$

We give this equation a new number since it is different from Eq. (7.9). We can now either use Eq. (7.8) directly or use Eqs. (7.10) and (7.64) to rewrite it in a form that is perhaps easier to use, namely

$$\sigma_1 = 2\tau_0 \tan \theta + \sigma_3 \tan^2 \theta \tag{7.65}$$

Since we know all the values on the right-hand side of this equation, we can now use Eq. (7.65) to find the value of $\sigma_1$ for fault slip thus:

$$\sigma_1 = 2\tau_0 \tan \theta + \sigma_3 \tan^2 \theta = 2 \times 1 \text{ MPa} \times 2.7475 + 5 \text{ MPa} \times 7.5486 \approx 43 \text{ MPa}$$

For an average crustal density of 2500 kg m$^{-3}$, $\sigma_1$, here the vertical stress, would reach this value at a depth of about (Eq. 4.45)

$$z = \frac{\sigma_1}{\rho_r g} = \frac{4.3 \times 10^7 \text{ Pa}}{2500 \text{ kg m}^{-3} \times 9.81 \text{ m s}^{-2}} \approx 1750 \text{ m}$$

This level is about 300 m deeper than that of the outcrops of the normal faults in Fig. 2.1.

## Example 7.3

### Problem

A certain crustal rock has an angle of internal friction of 30°, a cohesion or shear strength of 8 MPa, and is located at a depth where the minimum compressive principal stress is 25 MPa. Find the value of the maximum principal compressive stress at fault slip.

### Solution

To use Eqs. (7.8) or (7.65), we need to know the angle $\theta$ that the normal to the fault plane makes with the maximum principal compressive stress $\sigma_1$. From Eq. (7.11) we have

$$\theta = 45° + \frac{1}{2}\phi = 45° + \frac{30°}{2} = 60°$$

Using this value and $\sigma_3$ as 25 MPa, from Eq. (7.65) we get

$$\sigma_1 = 2\tau_0 \tan\theta + \sigma_3 \tan^2\theta = 2 \times 8 \text{ MPa} \times \tan 60° + 25 \text{ MPa} \times \tan^2 60°$$
$$= 102.7 \text{ MPa}$$

Notice that here we can use megapascals in our calculations because basic SI units are not used in the calculations. In this example, we cannot use Eq. (4.44) to determine the depth at which this value of $\sigma_1$ is reached because no information is provided on the type of fault that we are dealing with. We cannot, therefore, assume that $\sigma_1$ is the vertical stress, which is usually the case only for normal faults (Chapter 8).

## Example 7.4

### Problem

A 1 m long tension fracture in the rift zone of North Iceland (cf. Fig. 1.1) is subject to tensile loading parallel with the spreading vector. Since the Holocene pahoehoe lava flow hosting the fracture has numerous, open contacts, we assume that the fracture is a through crack (Chapter 9) and that a plane-strain formulation is appropriate. The static Young's modulus of the lava flow is estimated at 5 GPa and Poisson's ratio at 0.25. Calculate the spreading-vector related tensile stress needed to propagate this fracture further using:

(a) a typical laboratory material toughness value of 200 J m$^{-2}$, based on millimetre-scale fractures
(b) an estimated in-situ material toughness for large fractures of 2 MJ m$^{-2}$ (Atkinson, 1987; Gudmundsson, 2009) based on kilometre-scale fractures.

### Solution

(a) From Eq. (7.36) we get the rift-zone tensile stress for fracture propagation as

$$\sigma = \left(\frac{EG_c}{\pi(1-\nu^2)a}\right)^{1/2} = \left(\frac{5 \times 10^9 \times 200 \text{ J m}^{-2}}{3.1415(1-0.25^2) \times 0.5}\right)^{1/2} = 8.24 \times 10^5 \text{ Pa} \approx 0.8 \text{ MPa}$$

(b) Again from Eq. (7.36) we get the tensile stress as

$$\sigma = \left(\frac{EG_c}{\pi(1-\nu^2)a}\right)^{1/2} = \left(\frac{5\times10^9\times2\times10^6\,\mathrm{J\,m^{-2}}}{3.1415(1-0.25^2)\times0.5}\right)^{1/2} = 8.24\times10^8\ \mathrm{Pa} \approx 80\ \mathrm{MPa}$$

Here, again, we use the absolute values, so that the minus sign is omitted. However, in calculations where compressive stresses are involved, the tensile stresses, such as here, must be given with their minus signs.

Since the tensile stresses during rifting episodes are normally of the order of several megapascals, these results indicate that neither material toughness estimate is entirely suitable for this size of a fracture. This is not surprising since the laboratory estimates are based on millimetre-scale fractures in close-to-homogeneous, isotropic specimens, whereas the in-situ estimates are based on kilometre-scale faults and dykes. Thus, the material toughness suitable for metre-scale fractures, like this one, is somewhere in between the estimates based on the millimetre-scale and the kilometre-scale fractures.

## Example 7.5

### Problem

At a certain depth in the crust the stress conditions are so that the maximum principal compressive stress is twice the minimum principal compressive stress. If the in-situ tensile strength of the crustal rocks is 3 MPa, their average density 2700 kg m$^{-3}$, and the fracture or fault initiation takes place in the compressive regime, find:

(a) the principal stresses $\sigma_1$ and $\sigma_3$ for fracture initiation
(b) the crustal depth of fracture initiation if the faulting occurs in a rift zone.

### Solution

(a) From Eq. (7.45) we see that since $\sigma_3 = \sigma_1/2$, and both are therefore compressive. It follows that $\sigma_1 > -3\sigma_3$, so the following equation should be used:

$$(\sigma_1 - \sigma_3)^2 = 8T_0(\sigma_1 + \sigma_3)$$

This equation can be rewritten in the form

$$(\sigma_1 - \sigma_3)^2 - 8T_0(\sigma_1 + \sigma_3) = 0$$

Using the relations $(a-b)^2 = a^2 - 2ab + b^2$, $T_0 = 3$ MPa, $\sigma_3 = \sigma_1/2$, and solving for the principal stresses, we get

$$\left(\sigma_1 - \frac{\sigma_1}{2}\right)^2 - 8T_0\left(\sigma_1 + \frac{\sigma_1}{2}\right) = 0$$

$$\sigma_1^2 - 2\sigma_1\frac{\sigma_1}{2} + \frac{\sigma_1^2}{4} - 24\sigma_1 - 12\sigma_1 = 0$$

So that

$$\frac{\sigma_1^2}{4} - 36\sigma_1 = 0$$

Dividing by $\sigma_1$ and then solving we get

$$\frac{\sigma_1}{4} = 36$$

Thus, $\sigma_1 = 36 \times 4 = 144$ MPa and, since $\sigma_3 = \sigma_1/2$, we get $\sigma_3 = 72$ MPa.

$\sigma_1/\sigma_3$ ratios of 2 or more are common in the uppermost 1–2 km, particularly in continental areas (Amadei and Stephansson, 1997), and also at deeper crustal levels in various environments (Zoback, 2007). Similar ratios occur in the palaeorift zone in East Iceland, where the maximum principal stress is about double the minimum principal stress at a depth of about 0.4 km (Haimson and Rummel, 1982). Remarkably, the maximum and minimum principal compressive stresses at about 9 km in the KTB drill hole in Germany are 285 MPa and 147 MPa, yielding a ratio of 1.94 (Amadei and Stephansson, 1997). These are horizontal stresses. This is the deepest in-situ stress measurement worldwide obtained so far.

(b) Rewriting Eq. (4.45) and using the crustal density of 2700 kg m$^{-3}$, we get

$$z = \frac{\sigma_1}{\rho_r g} = \frac{1.44 \times 10^8 \text{ Pa}}{2700 \text{ kg m}^{-3} \times 9.81 \text{ m s}^{-2}} = 5437 \text{ m} \approx 5.4 \text{ km}$$

Thus, for normal faulting, this state of stress at fault initiation could be found at a depth of about 5.4 km in a rift zone. The crustal density used here is appropriate for the rift zone in Iceland, but is not much different in other active rift zones. In (b) we had to change megapascals to pascals because we used basic SI units for the acceleration and the crustal density.

---

### Example 7.6

#### Problem

The yield stress in uniaxial compression of a limestone at room temperature is 60 MPa (Chapter 5). A thin-plate specimen of this limestone is loaded in a biaxial stress field. One principal stress is compressive with a magnitude of 50 MPa. Since the stress field is biaxial, the vertical principal stress is assumed to be zero, so that $\sigma_2 = 0$.

(a) Use the Tresca criterion to find the magnitude of the principal compressive stress at right angles to the 50 MPa stress for yielding to occur.

(b) Use the von Mises criterion to find the other principal stress, as in (a).

#### Solution

(a) One expression of the Tresca criterion is Eq. (7.51), from which we can obtain the magnitude of the compressive stress needed to cause yielding as follows:

$$|\sigma_1 - \sigma_3| = \sigma_y = 60 \text{ MPa} = |50 \text{ MPa} - \sigma_3| = |50 \text{ MPa} - (-10 \text{ MPa})|$$

So $\sigma_3 = -10$ MPa, which is similar to a common laboratory tensile strength.

(b) From Eq. (7.55) we have

$$(\sigma_1 - \sigma_2)^2 + (\sigma_2 - \sigma_3)^2 + (\sigma_3 - \sigma_1)^2 = 2\sigma_y^2$$

So that we have (here omitting the MPa units)

$$(50 - 0)^2 + (0 - \sigma_3)^2 + (\sigma_3 - 50)^2 = 2(60)^2 = 7200$$

Or

$$2500 + \sigma_3^2 + \sigma_3^2 - 100\sigma_3 + 2500 = 7200$$

Rearranging, we get

$$2\sigma_3^2 - 100\sigma_3 - 2200 = 0$$

This is a quadratic equation of the form $ax^2 + bx + c = 0$, which has the solution

$$x = \frac{-b \pm \left(b^2 - 4ac\right)^{1/2}}{2a}$$

In the present case, $x = \sigma_3$, $a = 2$, $b = -100$, and $c = -2200$.
The solution for $\sigma_3$ is thus

$$\sigma_3 = \frac{100 \pm \left((-100)^2 - 4 \times 2 \times (-2200)\right)^{1/2}}{2 \times 2}$$

Or, showing the MPa units,

$$\sigma_3 = \frac{100 + 166}{4} = 66.5 \text{ MPa}$$

Or

$$\sigma_3 = \frac{100 - 166}{4} = -16.5 \text{ MPa}$$

Thus a tensile stress of about 16.5 MPa is required to cause yielding of the limestone. The value 66.5 MPa indicates the compressive stress needed, if applied, to cause yielding in the same specimen. The results thus indicate that the magnitude of the tensile stress for yielding given by the von Mises criterion is higher (16.5 MPa) than that given by the Tresca criterion (10 MPa). This is always so, since the Tresca criterion is more conservative, that is, it predicts yielding to occur at a lower stress difference than the von Mises criterion. Today, the von Mises criterion is much more widely used, particularly in materials science.

## 7.13 Exercises

7.1 What is failure of a solid? Explain the difference between brittle failure and ductile failure.

7.2 Define rock strength. What are the main types or concepts of rock strength? What is the general difference between laboratory strength and in-situ strength? Which is normally greater and why?

7.3 Define rock tensile strength. What symbol is used for tensile strength, how is it measured in situ (in nature), and what are typical in-situ tensile-strength values of rocks?

7.4 Define rock shear strength. What are its typical in-situ values and how does it relate to tensile strength and stress drop in earthquakes?

7.5 Define rock compressive strength.

7.6 How does maximum stress difference that the crustal rocks can sustain before failure change with depth in the crust?

7.7 What is a Coulomb material? What are an ideal Coulomb material and a cohesionless material?

7.8 Define (a) inherent shear strength, (b) coefficient of internal friction, (c) slip plane, (d) the law of friction, (e) angle of repose, and (f) angle of internal friction.

7.9 Give the equation for the Modified Griffith criterion and explain all the symbols used. Explain what is meant by the Coulomb line (or envelope) of failure.

7.10 What are conjugate faults? How does their formation relate to the Coulomb criterion? Give geological examples of conjugate faults.

7.11 What is a tensile cut-off and why is it used?

7.12 What is effective normal stress? How does pore-fluid pressure affect the probability of rock failure? Give geological examples of how changing the pore-fluid pressure may affect brittle deformation.

7.13 Describe the Mohr's and the Hoek–Brown failure criteria.

7.14 Briefly describe the Griffith theory of failure of brittle solids. What were the principal observations that Griffith wanted to explain? What are Griffith cracks? How does the Griffith theory relate to stress concentrations? How does the Griffith theory relate to elastic strain energy?

7.15 Explain the strain energy release generated by a propagating crack and the energy needed to generate a new crack surface. How do these energies add up to the total energy associated with crack propagation in a solid?

7.16 Write the plane-stress equation for strain energy release rate $G$ and explain all the symbols used and their units.

7.17 Define crack resistance.

7.18 Explain the combined rock-failure criterion and how it relates to the changes from brittle deformation to ductile deformation as a fracture zone becomes a shear zone with increasing depth in the crust.

7.19 What is the Tresca criterion for yield or plastic flow?

7.20 What is the von Mises criterion for yield or plastic flow? What is the main difference between the Tresca and the von Mises criteria as regards their graphic presentation and which one is thought to be generally more accurate?

7.21 Calculate the shear stress needed to generate slip on a normal fault if the normal stress on the fault plane is 1 MPa, the tensile strength of the fault rock is 2 MPa, and the coefficient of internal friction is 0.8. How does this stress compare with common stress drops associated with earthquakes?

7.22 Consider a normal fault dipping 65° at a depth of 4 km in a rift zone where the average crustal density to that depth is 2700 kg m$^{-3}$. The tensile strength of the fault rock is 5 MPa and the coefficient of internal friction if 1.0. If $\sigma_3$ at that depth is 70 MPa and $\sigma_1$ is equal to the overburden pressure, would the shear stress on a dry fault be high enough for it to slip assuming that the shear-stress condition for slip is given by Eq. (7.4)? If not, what total pore-fluid pressure in the fault rock would be needed to cause slip?

7.23 A normal fault makes an angle of 15° to $\sigma_1$, which is vertical. If the in-situ tensile strength of the fault rock is 1 MPa and $\sigma_3$ is 0.5 MPa at the depth of nucleation of the potential fault slip, what would $\sigma_1$ have to be to cause fault slip? If the average rock density to the depth of potential fault slip is 2400 kg m$^{-3}$, what would that depth be?

7.24 A thin-plate specimen of gypsum with a yield strength of 40 MPa is subject to biaxial stress field (plane stress) where $\sigma_1$ is 35 MPa and $\sigma_2$ (the vertical stress) zero. Use the Tresca criterion to find the magnitude of $\sigma_3$ for yielding to occur.

# References and further reading

Amadei, B. and Stephansson, O., 1997. *Rock Stress and its Measurement.* London: Chapman & Hall.

Anderson, E. M., 1951. *The Dynamics of Faulting and Dyke Formation with Applications to Britain*, 2nd edn. Edinburgh: Oliver and Boyd.

Asaro, R. J. and Lubarda, V. A., 2006. *Mechanics of Solids and Materials.* Cambridge: Cambridge University Press.

Atkinson, B. K. (ed.), 1987. *Fracture Mechanics of Rock.* London: Academic Press.

Bell, F. G., 2000. *Engineering Properties of Soils and Rocks*, 4th edn. Oxford: Blackwell.

Brady, B. H. G. and Brown, E. T., 1993. *Rock Mechanics for Underground Mining*, 2nd edn. London: Kluwer.

Caddell, R. M., 1980. *Deformation and Fracture of Solids.* Upper Saddle River, NJ: Prentice-Hall.

Carmichael, R. S., 1989. *Practical Handbook of Physical Properties of Rocks and Minerals.* CRC Press, London.

Chen, W. F., 2008. *Limit Analysis and Soil Plasticity.* Fort Lauderdale, FL: J. Ross.

Chen, W. F. and Han, D. J., 2007. *Plasticity for Structural Engineers.* Fort Lauderdale, FL: J. Ross.

Cottrell, A. H., 1964. *The Mechanical Properties of Matter.* New York: Wiley.

Dahlberg, T. and Ekberg, A., 2002. *Failure, Fracture, Fatigue.* Studentlitteratur, Lund (Sweden).

Davis, R. O. and Selvadurai, A. P. S., 2002. *Plasticity and Geomechanics.* Cambridge: Cambridge University Press.

Green, D. J., 1998. *An Introduction to the Mechanical Properties of Ceramics.* Cambridge: Cambridge University Press.

Griffith, A. A., 1920. The phenomena of rupture and flow in solids. *Philosophical Transactions of the Royal Society of London*, **A221**, 163–198.

Griffith, A. A., 1924. Theory of rupture: In: Biezeno C. B. and Burgers, J. M. (eds.), *Proceedings of the First International Congress on Applied Mechanics*. Delft: Waltman, pp. 55–63.

Gudmundsson, A., 2009. Toughness and failure of volcanic edifices. *Tectonophysics*, **471**, 27–35.

Haimson, B. and Rummel, F., 1982. Hydrofracturing stress measurements in the Iceland research drilling project drill hole at Reydarfjordur, Iceland. *Journal of Geophysical Research*, **87**, 6631–6649.

Harrison, J. P. and Hudson, J. A., 2000. *Engineering Rock Mechanics. Part 2: Illustrative Worked Examples*. Oxford: Elsevier.

Hosford, W. F., 2005. *Mechanical Behaviour of Materials*. Cambridge: Cambridge University Press.

Hudson, J. A. and Harrison, J. P., 1997. *Engineering Rock Mechanics. An Introduction to the Principles*. Oxford: Elsevier.

Jaeger, J. C., 1969. *Elasticity, Fracture, and Flow*, 3rd edn. London: Chapman & Hall.

Jaeger, J. C., Cook, N. G. W., and Zimmerman, R. W., 2007. *Fundamentals of Rock Mechanics*, 4th edn. Oxford: Blackwell.

Janssen, M., Zuidema, J., and Wanhill, R., 2004. *Fracture Mechanics*, 2nd edn. Abingdon, UK: Spon.

Jianqiao, Y., 2008. *Structural and Stress Analysis*. London: Taylor & Francis.

Jumikis, A. R., 1979. *Rock Mechanics*. Clausthal, Germany: Trans Tech Publications.

Kanamori, H. and Anderson, D. L., 1975. Theoretical basis of some empirical relations in seismology. *Bulletin of the Seismological Society of America*, **65**, 1073–1095.

Lawn, B., 1993. *Fracture of Brittle Solids*, 2nd edn. Cambridge: Cambridge University Press.

Mase, G. E., 1970. *Continuum Mechanics*. New York: McGraw-Hill.

Meyers, M. A. and Chawla, K. K., 2009. *Mechanical Behaviour of Materials*. Cambridge: Cambridge University Press.

Myrvang, A., 2001. *Rock Mechanics*. Trondheim: Norway University of Technology (NTNU) (in Norwegian).

Nedderman, R. M., 2005. *Statics and Kinematics of Granular Materials*. Cambridge: Cambridge University Press.

Nemat-Nasser, S. and Hori, M., 1999. *Micromechanics. Overall Properties of Heterogeneous Materials*, 2nd edn. Amsterdam: Elsevier.

Parker, A. P., 1981. *The Mechanics of Fracture and Fatigue*. London: Spon.

Priest, S. D., 1993. *Discontinuity Analysis for Rock Engineering*. London: Chapman & Hall.

Schön, J. H., 2004. *Physical Properties of Rocks: Fundamentals and Principles of Petrophysics*. Amsterdam: Elsevier.

Sheorey, P. R., 1997. *Empirical Rock Failure Criteria*. Rotterdam: Balkema.

Sibson, R. H., 1984. Roughness at the base of the seismogenic zone: contributing factors. *Journal of Geophysical Research*, **89**, 5791–5799.

Solecki, R. and Conant R. J., 2003. *Advanced Mechanics of Materials*. Oxford: Oxford University Press.

Valko, P. and Economides, M. J., 1995. *Hydraulic Fracture Mechanics*. New York: Wiley.

Yeats, R. S., Sieh, K., and Allen, C. R., 1997. *The Geology of Earthquakes*. Oxford: Oxford University Press.

Zoback, M. D., 2007. *Reservoir Geomechanics*. Cambridge: Cambridge University Press.

# Extension fractures and shear fractures

## 8.1 Aims

Depending on the relative displacement across the fracture plane, all tectonic fractures are of two main mechanical types: extension fractures and shear fractures. In an extension fracture, the sense of displacement is perpendicular to, and away from, the fracture plane. In a shear fracture the sense of displacement is parallel with the fracture plane. To be able to distinguish between these two main types theoretically and in the field is of fundamental importance for understanding a variety of fracture-related problems and processes. These two types will be referred to many times in subsequent chapters. This brief chapter focuses on learning to recognise them in the field with reference to the basic theories for their formation discussed in earlier chapters. The main aims of this chapter are to:

- Explain the mechanical difference between extension fractures and shear fractures.
- Discuss the two main types of extension fractures, namely: (a) tension fractures, which are formed by absolute tension and usually close to the Earth's surface, and (b) fluid-driven fractures or hydrofractures, which are generated by fluid pressure and can form at any crustal depth.
- Show that shear fractures comprise all faults as well as some fractures that are normally classified as joints in the field.
- Provide field examples of the main mechanical types of fractures.

## 8.2 Basic types of rock fractures

When rock is subject to a gradually increasing loading, it eventually fails. The failure can be either brittle or ductile (Chapter 7). Brittle failure takes place through the development or slip on one or more extension fractures, shear fractures, or both (Fig. 8.1). The stress that a rock can tolerate before it fails in a brittle manner depends on the loading conditions and on whether the stress is compressive, tensile, or shear. The stress needed to form a shear fracture is greater than that needed to form a tension fracture, for example, and the stress needed to generate a fracture, to break the rock, depends on the type of fracture that is formed.

We would like to know the conditions under which fractures such as those in Fig. 8.1 formed. One of the fractures, the dyke, was clearly filled with magma. Since the dyke was

**Fig. 8.1**  Dyke and a graben (two normal faults) in the Tertiary lava pile, seen in a sea cliff, in North Iceland. The dyke is about 6 m thick; the fault displacement is about 3 m, and the sea cliff is about 120 m high (cf. Fig. 2.1).

formed at great depth – here the dyke is seen at a depth of about 1.5 km below the original top of the lava pile – the magma cannot simply have flowed from above into an existing tension fracture. This follows because tension fractures form only at shallow depths, usually at depths much less than 1 km (Section 8.3). Thus, we infer that the magma had a role in the dyke-fracture formation. The dyke dip is vertical and thus different from that of a typical fault, such as the normal faults of the nearby graben. Normal faults, except at and close to the surface, where they may be vertical (Fig. 1.5; Chapter 14), are usually inclined, that is, with a dip less than 90° (Figs. 2.4 and 8.1). We know that rift zones, such as the one where these normal faults and the dyke formed about 12 Ma ago, tend to have flat surfaces with the maximum principal compressive stress $\sigma_1$ vertical and the minimum compressive stress $\sigma_3$ horizontal (Chapter 4). Thus, the path of the vertical dyke is very likely to have been parallel with $\sigma_1$ and perpendicular to $\sigma_3$. This is also indicated by numerical models (Fig. 6.21). The dyke thus formed in a principal stress plane which, by definition, does not have any shear stress acting upon it, and is a fluid-driven (magma-driven) fracture. That is to say, the dyke-fracture was almost entirely formed by the overpressure of the magma, and the result is a hydrofracture and, therefore, an extension fracture. A fluid-driven fracture, of course, uses existing weaknesses in the rocks for its development; in this case mainly the columnar joints in the lava flows. But the opening of the fracture is almost entirely the result of the fluid overpressure and not the result of tensile stresses. In contrast with the dyke, the two normal faults forming the graben in Fig. 8.1 were oblique to the principal stresses at their time of formation (assuming that the dyke and graben formed in a similar stress field). They thus had shear stresses acting upon them, making them shear fractures.

**Fig. 8.2**  Ground-water filled tension fracture, with an opening that is mostly about 10 m wide, in the rift zone of Southwest Iceland (located in Figs. 3.13 and 3.14). For an extension fracture such as this tension fracture, the fracture plane is parallel with the maximum ($\sigma_1$) and intermediate ($\sigma_2$), and perpendicular to the minimum ($\sigma_3$), principal compressive stresses. During the tension-fracture formation, and during rifting episodes in general, $\sigma_3$ is tensile, that is, negative. Photo: Valerio Acocella.

Thus, in this outcrop (Fig. 8.1) we have clear examples of the two main mechanical types of rock fractures: an extension fracture, a dyke, and a shear fracture, namely the two normal faults that form the graben. Let us now discuss their main mechanical characteristics.

## 8.2.1  Extension fractures

An **extension fracture** forms in a plane that is perpendicular to the minimum principal compressive stress, $\sigma_3$. Consequently, since the principal stresses are mutually perpendicular, the plane of the extension fracture must include the other principal stresses, namely the maximum ($\sigma_1$) and the intermediate ($\sigma_2$) principal stresses (Fig. 8.2). It follows that an extension fracture has no shear stress acting upon it, so that there cannot be any movement parallel with the fracture. The only movement possible, when an extension fracture forms, is opening or movement perpendicular to the fracture walls (parallel with the direction of $\sigma_3$).

Extension fractures are generated by normal stresses and are of two main types: **fluid-driven fractures** and **tension fractures**. Fluid-driven fractures are opened by fluid pressure and can form at any depth in the crust and mantle. Here we refer to them mainly as **hydrofractures**. Their condition of formation is that the fluid pressure in a part of the crust, such as in the host rock of a fluid-filled reservoir, exceeds the minimum principal compressive stress plus the tensile strength in that part of the crust. This condition is commonly satisfied in the crust, for example during dyke emplacement (Fig. 8.1). Most dykes, inclined sheets, and sills in volcanic areas, as well as many mineral veins in geothermal fields and many joints, are hydrofractures.

Tension fractures are opened by an absolute tension. That is to say, for a tension fracture to form, the minimum principal compressive stress, $\sigma_3$, must be negative. At an outcrop scale, $\sigma_3$ can be negative only in the uppermost 1 km of the crust, and under most conditions to a depth of only a few hundred metres or less. Outcrop-scale tension fractures therefore form only at shallow depths and at the surface, particularly in areas of active rifting such as divergent plate boundaries. Many joints at shallow depths are tension fractures in areas of uplift and extension. Tension fractures are also very common at and close to the surface of rift zones (Fig. 8.2). However, when a tension fracture propagates to depths exceeding this critical depth (of less than 1 km), the tension fracture must change into a normal fault, that is, into a shear fracture.

## 8.2.2 Shear fractures

Shear fractures are generated by shear stresses. The most common and best known shear fractures are the main types of faults, found everywhere in the crust (Fig. 8.1). These are normal faults, reverse faults, and strike-slip faults (Fig. 8.3). Many fractures classified in the field as joints are also shear fractures.

Shear fractures can form at any depth within the crust. In the mantle, however, shear fractures that generate earthquakes, that is, seismogenic faults, are mainly associated with down-going slabs at convergent plate boundaries. A shear fracture makes an angle with the principal stress directions, so that a shear-fracture cannot be in (coincide with) a principal stress plane. Since many areas of active tectonics, particularly divergent plate boundaries, have comparatively flat surfaces, one of the principal stresses is commonly vertical, the others being horizontal (Fig. 8.3). As a consequence, many shear fractures, particularly normal and reverse faults, are seen as steeply to shallowly dipping in vertical sections (Figs. 8.1 and 8.3). Strike-slip faults, however, tend to be close to vertical in such sections, as is understandable given their relationship with the principal stresses (Fig. 8.3).

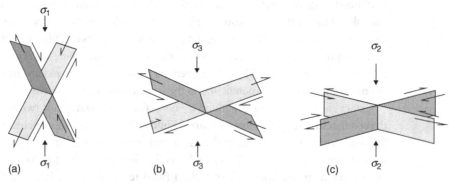

**Fig. 8.3**  Main fault types and the associated principal stress orientations according to Anderson's (1951) theory of faulting: (a) normal, (b) reverse (or thrust), and (c) strike-slip.

Although shear fractures are formed by shear stress, many, and perhaps most, are related to zones of high fluid pressure. Not only is shear failure encouraged by increasing pore-fluid pressure (Chapter 7), but the shear-fracture plane itself is commonly subject to fluid transport and high pressure. This is presumably one reason why the driving shear stress for slip on faults, as inferred from the stress drop during earthquakes, is small and within a comparatively small range. The total range is only about 30 MPa, and most driving-stress values as inferred from stress drops of recorded earthquakes are between about 1 and 12 MPa (Kanamori and Anderson, 1975; Kasahara, 1981; Scholz, 1990).

## 8.3 Tension fractures

A tension fracture forms when the tensile stress in a direction perpendicular to the potential fracture plane reaches the tensile strength of the rock (Fig. 8.2). When the conditions of Eq. (7.41) are satisfied, the Griffith criterion for the tensile regime implies that a tension fracture forms when the tensile stress $\sigma_3$ reaches the tensile strength $T_0$ of the rock, namely when

$$-\sigma_3 = T_0 \qquad (8.1)$$

Laboratory measurements of small specimens yield rock tensile strengths of as much as 30 MPa, the common values being between about 5 and 25 MPa (Appendix E). However, small specimens cannot contain any of the joints, contacts, or other outcrop-scale weaknesses that in situ rocks almost always do. So the in-situ tensile strength is much lower than the laboratory strength. Measurements worldwide, mostly using hydraulic fracturing, yield in-situ tensile strengths of rocks that suggest maximum in-situ tensile strengths of about 9 MPa, the most common values being in the range 0.5–6 MPa (Chapter 7).

These values are essentially constant, indicating an average tensile strength of about 3 MPa. In addition, this limited range of tensile strength is also important in view of the relationship between tensile strength and shear strength. The Modified Griffith criterion (Chapter 7) indicates that the inherent shear strength or cohesion, $\tau_0$, is twice the tensile strength, in good agreement with the stress drops and inferred shear strengths $\tau_0$, which are mostly between 1 and 12 MPa. Thus, the common range of in-situ tensile strengths, 0.5–6 MPa, is in good agreement with observations of shear stresses associated with seismogenic faulting and the predictions based on the Modified Griffith criterion.

From the Griffith crack theory (Eq. 7.44) it can be shown that the maximum depth that a tension fracture can reach, $d_{\max}$, before it changes into a normal fault is

$$d_{\max} = \frac{3T_0}{\rho_r g} \qquad (8.2)$$

where $\rho_r$ is the average density of the host rock to the depth of $d_{\max}$ and $g$ is the acceleration due to gravity. The average tensile strength of solid rocks is around 3 MPa, and the uppermost part of the crust has a density of 2100–2500 kg m$^{-3}$ (Appendix E). With $g = 9.81$ m s$^{-2}$, it follows from Eq. (8.2) that for typical solid rocks $d_{\max}$ is about 400 m (Example 8.1).

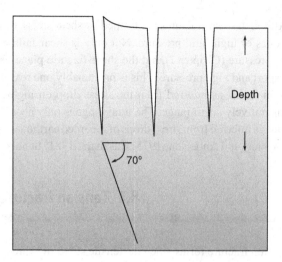

Fig. 8.4  As soon as a surface tension fracture reaches a critical depth it must change into a normal fault, in which case some vertical displacement starts to develop along the fracture. A common dip of normal faults in basaltic rift zones is about 70°.

To get the maximum possible depth, we can use the maximum tensile strength for the general range, 6 MPa, and the minimum density, $2100 \text{ kg m}^{-3}$, in which case $d_{max}$ would be close to 900 m (Example 8.1). Thus, in a typical solid rock with typical tensile strength and density, a tension fracture that starts propagating from the surface (of a rift zone, for example) and into the crust must change into a normal fault at a depth of several hundred metres (Fig. 8.4). If we used the maximum measured in-situ tensile strength of 9 MPa, we would have to use the comparatively high density of the rocks within which this tensile strength was measured (at a depth of 9 km in the KTB drill hole in Germany; Chapters 7 and 11) and the maximum depth would be about the same (less than 1 km). A tension fracture is thus very unlikely to propagate deeper than about one kilometre (900 m in our example) before it changes into a normal fault.

These simple considerations have implications for processes of fluid transport to the surface, such as through dykes and mineral veins. The results show that all outcrop-scale extension fractures exposed at depths greater than about 1 km must be formed by fluid pressure, that is, they must be hydrofractures. There is no way that extension fractures formed under static or quasi-static tensile loading can be formed by absolute tension at depths of several kilometres. Dynamic loadings, such as are associated with seismogenic slip on earthquake faults, may help generate extension fractures, such as wing cracks or fractures (Chapter 16). But most mineral veins and dykes are formed under quasi-static stress conditions, rather than dynamic ones. Veins and dykes exposed to depths of many kilometres must therefore be primarily the result of fluid overpressure rather than absolute tectonic tensile stresses. The idea sometimes raised, that volcanic fissures, or feeder dykes, may form as a result of absolute tensile stresses generating fractures that reach magma chambers at depths of many kilometres and tap off the magma, is therefore neither in agreement with theoretical considerations, nor, in fact, with observations of dykes (Chapter 11).

Typical rift zones contain many tension fractures. Some are large and reach apertures (openings) exceeding ten metres (Fig. 8.2). But even tension fractures with apertures of ten metres or more are unlikely to reach depths that exceed several hundred metres. This conclusion is, in fact, in good agreement with the strike dimensions (lengths) of tension fractures in rift zones. The average length of large tension fractures is several hundred metres in many rift-zone segments, such as in Iceland (Chapter 14), whereas in the same rift-zone segments the average length of normal faults is about two kilometres (Gudmundsson, 1995). This implies that the strike dimensions of many tension fractures are similar to their maximum depths, so that their cross-sectional geometry is that of a half circle or half ellipse, in agreement with theoretical studies (Chapters 9 and 11). Extension fractures at greater depths must, therefore, be formed by fluid pressure, that is, they must be hydrofractures.

## 8.4 Hydrofractures

Most extension fractures in the Earth's crust are hydrofractures. They can form at any crustal depth, as well as in the mantle, provided the fluid pressure is high enough for the rock to rupture in extension. Most joints and many mineral veins are probably hydrofractures as well as all dykes, inclined sheets, and sills in volcanic areas. All these fractures are primarily formed by internal fluid overpressure that is so high as to rupture the rock and drive the fracture propagation.

Consider first some field examples of hydrofractures. We have concluded that the dyke in Fig. 8.1 is a hydrofracture. A dyke-fracture of this kind is initiated when the excess pressure in the associated magma chamber reaches the tensile strength of the host rock. Subsequently, the dyke-fracture propagation through the crustal layers is driven by the magma overpressure (defined below). Another clear indication that dykes form as hydrofractures is given by the many dykes injected through the roof of an extinct felsic magma chamber, a pluton, in Southeast Iceland (Fig. 8.5). This exceptionally well-exposed chamber, a granophyre pluton formed about 8 Ma ago, shows many felsic dykes cutting through its roof (Fig. 8.6). These are excellent examples of hydrofractures. The field observations show very clearly that the roof of the fluid-filled magma chamber ruptured many times, each time resulting in the formation of a hydrofracture, a dyke, that, because of its driving pressure, propagated up into the basaltic lava pile (the roof) above the chamber.

Here it is important to distinguish between three terms related to fluid pressure in reservoirs and hydrofractures. These are excess pressure, overpressure, and total pressure (cf. Chapter 6). These may be defined as follows (Fig. 8.7):

- **Excess pressure** $p_e$ is the fluid pressure in the reservoir in excess of the overburden pressure or lithostatic pressure. At rupture it is normally equal to the tensile strength of the host rock of the reservoir and is thus generally in the range 0.5–6 MPa, and most commonly about 3 MPa.
- **Overpressure** $p_o$ (also referred to as **driving pressure** and **net pressure**) is the pressure that drives the hydrofracture propagation. Overpressure is the combined effect of the initial excess pressure in the fluid source, such as a magma chamber, and the magma

**Slaufrudalur**

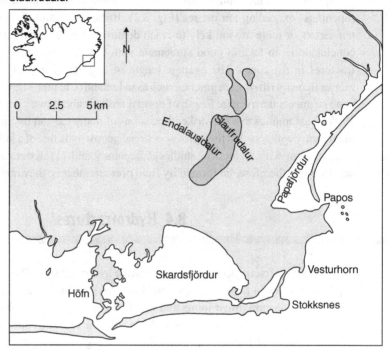

**Fig. 8.5**   Location of the elongate, felsic (granophyre) Slaufrudalur pluton, an extinct shallow magma chamber, in Southeast Iceland (Gudmundsson, 2010).

buoyancy, which is due to the density difference between the fluid in the fracture and the rock through which the fracture propagates. It is referred to the normal stress on the fracture, which, for an extension fracture, is the minimum principal compressive stress, $\sigma_3$. Although the excess pressure normally decreases along the flow direction in the fracture (for example, up the dip dimension of the fracture), the buoyancy term increases so long as the average density of the rock through which the fracture propagates is greater than the density of the fluid. The overpressure may thus reach several tens of megapascals even if the excess pressure at the fluid source is equal to the rock tensile strength and only a few megapascals.

- **Total pressure** $p_t$ in a reservoir is the excess pressure plus the lithostatic stress or pressure (overburden pressure). When a fluid-filled reservoir is in a lithostatic equilibrium with its host rock, no extension fracturing or faulting can normally take place. It is only when there is some excess pressure in the reservoir that stresses build up in the surrounding rocks, which, eventually, may result in rock failure (reservoir rupture) and fracture formation. It is therefore the excess pressure in the reservoir, and the overpressure in a hydrofracture, that are of importance for brittle deformation. In models and discussions of hydrofractures and reservoirs, the focus is on the excess pressure and the overpressure, whereas the total pressure is of little importance.

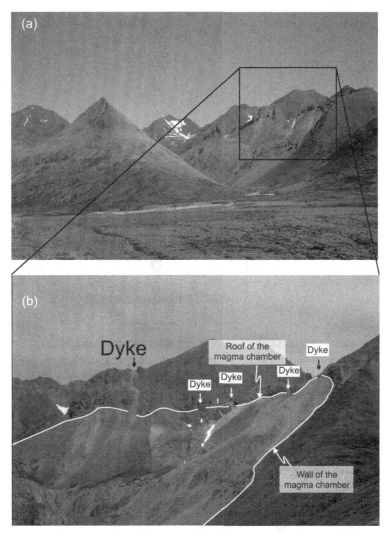

**Fig. 8.6** (a) Part of the Slaufrudalur pluton/magma chamber (Fig. 8.5). (b) The walls and the roof are exposed. Many felsic dykes (indicated by arrows) dissect the roof and propagate up into the host rock above. Some of these may have reached the surface to supply magma to fissure eruptions (Gudmundsson, 2010).

In the simplest terms, a reservoir ruptures and a hydrofracture initiates when the following condition is satisfied:

$$p_1 + p_e = \sigma_3 + T_0 \tag{8.3}$$

Here, $p_1$ is the lithostatic stress or overburden pressure at the rupture site, $p_e = p_t - p_1$ is the difference between the total fluid pressure $p_t$ in the reservoir and the lithostatic stress at the time of reservoir rupture, $\sigma_3$ is the minimum compressive or maximum tensile principal stress, and $T_0$ the local in-situ tensile strength at the rupture site. If a hydrofracture becomes

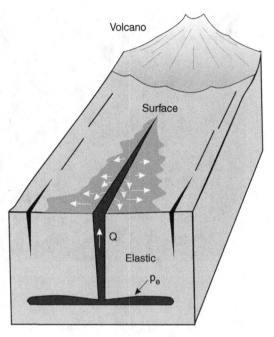

**Fig. 8.7** Hydrofracture in an elastic crust. The fracture initiates when the excess pressure $p_e$ in the source (here a sill-like reservoir) reaches the tensile strength of the roof. When the hydrofracture propagates vertically up into the host rock above, its fluid overpressure may increase significantly because of buoyancy effects. Q stands for the volumetric flow rate through the fracture (cf. Chapter 17).

injected into the roof of the reservoir and begins to propagate up into the crustal layers above, its overpressure $p_o$ becomes

$$p_o = p_e + (\rho_r - \rho_f)gh + \sigma_d \tag{8.4}$$

where $p_e$ is the fluid excess pressure in the source, $\rho_r$ is the host-rock density, $\rho_f$ is the fluid density, $g$ is acceleration due to gravity, $h$ is the dip dimension or height of that part of the hydrofracture, above the point of rupture and hydrofracture initiation, which is being considered, and $\sigma_d$ is the differential stress at the level where the hydrofracture is examined (Fig. 8.7).

Equation (8.4) can be used to estimate the likely overpressure of a hydrofracture that has reached a certain dip dimension or height above its source reservoir. For example, a vertically propagating hydrofracture may become arrested at a certain depth below the surface (Figs. 3.1 and 6.20). Then the overpressure in the hydrofracture, at any depth down its dip dimension, can be estimated crudely using Eq. (8.4), keeping the following factors in mind:

- At the time of hydrofracture initiation from a large reservoir, the excess pressure $p_e = p_t - p_l$ is, as a rule, positive. This follows because if the excess pressure was negative, the implication would be that the fluid pressure in the reservoir at the time of rupture

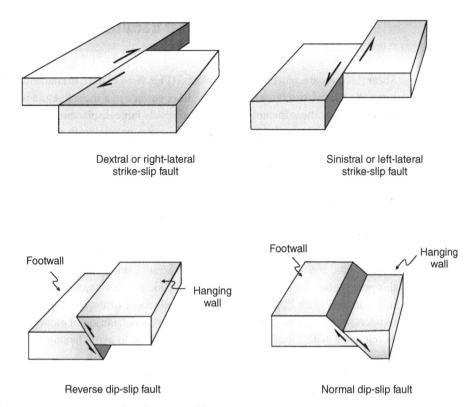

**Fig. 8.8** Main types of faults illustrated: dextral and sinistral strike-slip, and normal and reverse dip-slip.

was less than $\sigma_3$, in which case there would be no obvious mechanical reason for the hydrofracture to initiate in the first place. However, $p_e$ may be zero, in which case the reservoir is in lithostatic equilibrium with the host rock. This is presumably a common pressure condition for many totally fluid reservoirs during much of their lifetimes. The effective stress in the rocks resting on the reservoir is then zero, so that hydrofracture initiation is unlikely.

- A positive excess fluid pressure in a reservoir is normally less than the tensile strength of the rock, that is, $p_e < T_0$. When the excess pressure reaches the tensile strength, so that $p_e = T_0$, the reservoir ruptures in tension and a hydrofracture forms.
- The differential stress is defined as $\sigma_d = \sigma_1 - \sigma_3$. It is either zero or positive. It cannot be negative because, by definition, $\sigma_1$ cannot be less than $\sigma_3$. When $\sigma_d$ is zero, $\sigma_1$ and $\sigma_3$ are equal and the stress is isotropic (in two dimensions at least).
- The density difference $\rho_r - \rho_f$ can be negative, when the fluid is denser than rock; zero, when the density of the fluid is equal to that of the rock; or positive, when the rock is denser than the fluid.
- The density of water $\rho_w$ is generally in the range $900 \le \rho_w \le 1100 \ \mathrm{kg \ m^{-3}}$, whereas the density of most crustal rocks $\rho_r$ is in the range $2000 \le \rho_r \le 3000 \ \mathrm{kg \ m^{-3}}$. Thus, when the fluid is water $\rho_r - \rho_f$ is always positive.

# 8.5 Shear fractures

Shear fractures are generated by shear stress (Figs. 8.1 and 8.3). In contrast to extension fractures, the relative displacement of the walls or surface of a shear fracture is parallel with the fracture plane. Shear fractures with comparatively large displacements are known as faults. Joints normally do not have a visible displacement parallel with the fracture plane. Many, perhaps most, joints are likely to be extension fractures and, in particular, hydrofractures (Secor, 1965). Some joints, however, may be shear fractures. The displacement parallel with the fracture plane is then simply not visible in the field. This is so either because the displacement is too small to be detected by the eye, or because the rock hosting the joint is comparatively homogeneous and isotropic and lacks marker layers or structures to identify the joint-parallel displacement.

The main types of faults are well known (Fig. 8.8). They are normal faults, reverse or thrust faults, and strike-slip faults. The associated stress fields are shown in Fig. 8.3. In this presentation the faults are assumed to be located in an area with a flat surface. This is in accordance with E. M. Anderson's original model of the mechanics of fault formation (Anderson, 1951), where flat surfaces, that is, areas outside major mountains, were assumed as reference states. Anderson made this assumption to avoid having to deal with inclined principal stresses, as are common in Alpine landscapes.

One way to remember and understand the orientation of the fault planes in relation to the stresses in crustal segments within flat surfaces is to consider the trends of the principal stresses $\sigma_3$ and $\sigma_1$ ($\sigma_2$ is parallel with the fault plane), namely:

- Normal faults occur in areas of horizontal extension and (normally) crustal thinning. It follows that $\sigma_3$ is horizontal and parallel with the direction of extension, whereas $\sigma_1$ is vertical and parallel with the direction of crustal thinning. Examples include continental rift zones and grabens and divergent plate boundaries at mid-ocean ridges, in which case $\sigma_3$ is normally parallel with the local spreading vector.
- Reverse and thrust faults occur in areas of horizontal compression, so that $\sigma_3$ is vertical and parallel with the direction of uplift or crustal thickening, whereas $\sigma_1$ is horizontal and parallel with the direction of crustal shortening. Examples include convergent plate boundaries such as continental-continental collision zones and associated mountain belts.
- Strike-slip faults occur in areas of horizontal extension in one direction and horizontal compression in the perpendicular direction. Thus, both $\sigma_3$ and $\sigma_1$ are horizontal. Examples include oceanic transform faults.
- Thus, for normal faults, $\sigma_1$ is vertical; for reverse faults, $\sigma_3$ is vertical; and for strike-slip faults, $\sigma_2$ is vertical.

The shear stress driving shear-fracture formation, and slip on an existing fault, is given by (Chapter 7)

$$\tau = \tau_0 + \mu\sigma_n \tag{8.5}$$

where $\tau$ is the shear stress on the potential or actual fault plane, $\tau_0$ is the shear or cohesive strength of the rock at that plane, $\mu$ is the coefficient of internal friction, and $\sigma_n$ is the normal stress on the plane. When there is total fluid pressure $p_t$ in the rock, on and in the vicinity of the fault plane, at the time of fault slip Eq. (8.5) becomes

$$\tau = \tau_0 + \mu(\sigma_n - p_t) \tag{8.6}$$

When the total fluid pressure $p_t$ approaches or equals the normal stress $\sigma_n$, the term $\mu(\sigma_n - p_t)$ approaches or equals zero and the driving shear stress for slip becomes $\tau_0$, which is equal to $2T_0$ (Chapter 7). Since the in-situ tensile strength of rocks is commonly in the range 0.5–6 MPa, it follows that, for high-fluid-pressure fault zones, the driving shear stress for slip should be 1–12 MPa. As indicated, these values are in agreement with common stress drops. Although the low Young's modulus in many damage zones and cores (Chapter 14) yields comparatively low shear stresses in many active fault zones, they tend to slip because of the existing weak fault plane (or planes), the high fluid pressure (and thus the low friction), and the low effective normal stress on the fault plane.

## 8.6  Summary

- Mechanically there are two types of fractures: extension fractures and shear fractures. Extension fractures are of two main subtypes: tension fractures and hydrofractures. Shear fractures are faults. Their main subtypes are normal faults, reverse (or thrust) faults, and strike-slip faults (dextral and sinistral).
- Outcrop-scale tension fractures can only form close to the surface and normally in areas of extension, such as divergent plate boundaries, sedimentary basins, and continental rift zones. If a tension fracture tries to propagate from the surface of an area undergoing extension to a crustal depth of more than about 0.5–1 km, it must change into a normal fault. Hydrofractures, by contrast, can form at any depth if the excess fluid pressure in a reservoir or a pore reaches the local tensile strength of the host rock. Hydrofractures include all fluid-formed fractures such as dykes, many mineral veins, and many joints.
- Shear fractures, faults, form by shear stress. The driving shear stress is thought to be similar to the measured stress drops in seismogenic faulting, commonly 1–12 MPa. Shear fractures can form at any depth in the brittle crust. However, normal faults are mainly confined to areas of extension, reverse (thrust) faults to areas of compression, and strike-slip faults to areas of horizontal extension in one direction and a horizontal compression in the orthogonal direction.

## 8.7  Main symbols used

| | |
|---|---|
| $g$ | acceleration due to gravity |
| $h$ | height or dip dimension of a hydrofracture (above its fluid source) |
| $p_e$ | excess fluid pressure |
| $p_l$ | lithostatic pressure (or stress) |

$p_t$    total fluid pressure
$T_0$    tensile strength
$\mu$    coefficient of internal friction
$\rho_f$    fluid density
$\rho_r$    rock or crustal density
$\sigma$    tensile loading, tensile stress
$\sigma_d$    differential stress ($\sigma_1 - \sigma_3$)
$\sigma_n$    normal stress (mostly used in connection with slip or fault planes)
$\sigma_1$    maximum compressive principal stress
$\sigma_2$    intermediate compressive principal stress
$\sigma_3$    minimum compressive (maximum tensile) principal stress
$\sigma_0$    remote tensile stress
$\tau$    shear stress
$\tau_0$    inherent shear strength (or cohesion or cohesive strength)
$\varphi$    angle of internal friction

# 8.8 Worked examples

## Example 8.1

### Problem

Fissure swarms at divergent plate boundaries contain many open tension fractures (Fig. 8.2). When lava erupts inside a fissure swarm, some of it may flow from the surface and into existing, open tension fractures and normal faults, as was observed during the Krafla Fires in North Iceland (Figs. 3.4–3.6, and 8.9). The average density of the uppermost 0.5 km of the crust in the rift zone of Iceland is 2300 kg m$^{-3}$, and for the next 1 km below, the density is 2600 kg m$^{-3}$. If lava with a density of 2500 kg m$^{-3}$ fills a 200-m-deep tension fracture:

(a) calculate the fluid pressure at the bottom of the lava-filled fracture
(b) find the theoretical depths at which the tension fracture should change into a normal fault based on likely tensile strengths and host-rock densities
(c) explain why this tension fracture may not change into a normal fault at the depths in (b).

### Solution

(a) From Eq. (4.45) we have

$$\sigma_v = \rho_r g z$$

In this case the vertical stress or total fluid pressure, $p_t$, in the tension fracture is due to the weight of the new lava that fills it. Thus, we have

$$\sigma_v = p_t = \rho_r g z = 2500 \text{ kg m}^{-3} \times 9.81 \text{ m s}^{-2} \times 200 \text{ m} = 4.9 \times 10^6 \text{ Pa} \approx 5 \text{ MPa}$$

**Fig. 8.9**  View north, basaltic lava flow (black) formed in the Krafla Fires, 1975–1984 (Figs. 3.4–3.6), in the rift zone of North Iceland flows into a tension fracture and nearly fills it to form a pseudodyke. The opening (aperture) of the tension fracture, located in a (light-grey) Holocene pahoehoe lava flow, is about 1.5 m. The person provides a scale.

(b)  From Eq. (8.2) and with a density range of 2300–2500 kg m$^{-3}$, the estimated depth for the change of a tension fracture into a normal fault is 400–900 m. If we assume first that the change from a tension fracture to a normal fault is likely to take place in the uppermost 0.5 km and use its density of 2300 kg m$^{-3}$, then from Eq. (8.2) we get

$$d_{max} = \frac{3T_0}{\rho_r g} = \frac{3 \times 3 \times 10^6 \text{ Pa}}{2300 \text{ kg m}^{-3} \times 9.81 \text{ m s}^{-2}} = 399 \text{ m} \approx 400 \text{ m}$$

If, however, we use the maximum tensile strength, 6 MPa, then clearly the above result would be nearly double. However, then density of the uppermost 700–800 m is close to 2400 kg m$^{-3}$, so that the maximum depth for a tension fracture to change into a normal fault, based on Eq. (4.45), is

$$d_{max} = \frac{3T_0}{\rho_r g} = \frac{3 \times 6 \times 10^6 \text{ Pa}}{2400 \text{ kg m}^{-3} \times 9.81 \text{ m s}^{-2}} = 765 \text{ m}$$

the depth being somewhat less for a density of 2500 kg m$^{-3}$. The absolute maximum theoretical depth where this change might happen is obtained if we use the minimum possible density, say 2100 kg m$^{-3}$ (such as might occur in the top part of sedimentary rocks or in some hyaloclastite breccias, as are common in Iceland), and yet assume a maximum tensile strength of 6 MPa. In this case we would obtain a theoretical depth of

$$d_{max} = \frac{3T_0}{\rho_r g} = \frac{3 \times 6 \times 10^6 \text{ Pa}}{2100 \text{ kg m}^{-3} \times 9.81 \text{ m s}^{-2}} = 874 \text{ m}$$

This theoretical depth, close to 900 m, is a maximum and generally rather unlikely to be reached.

(c) Because of the fluid pressure generated by the lava, the lava-filled tension fracture becomes a hydrofracture. So long as the lava in the tension fracture is fluid and has a density greater than that of the host rock (the uppermost part of the rift zone), the tension fracture may continue to propagate as a dyke-like fracture to greater crustal depths. The average rift-zone density in Iceland reaches the assumed density of the lava, $2500 \text{ kg m}^{-3}$, at crustal depths of about 1.5 km. Theoretically, therefore, the lava-filled fracture could propagate as a hydrofracture, and thus not change into a normal fault, to at least this depth. On solidification, the lava-filled fracture becomes a **pseudodyke**, a false dyke (Fig. 8.9). These have been observed in Hawaii, but are, apparently, not very common. They can be distinguished from normal dykes in that the pseudodykes are very vesicular (due to degassing of the lava during its transport towards and into the surface fracture) and normally join their source lava flow without the formation of a crater cone. The junction between the source lava and the pseudodyke is thus more similar to that between a sill and its feeder dyke than that between a typical feeder and its lava flow (cf. Figs. 1.3 and 1.6).

## Example 8.2

### Problem

Consider the two dykes in Figs. 8.10 and 8.11. Both have exposed vertical tips at about the same depth in the lava pile in Iceland. One dyke (Fig. 8.10) is exposed in Southwest Iceland, the other dyke (Fig. 8.11) in North Iceland at a similar depth of erosion (about 1200 m) below the original surface of the rift zone within which they were emplaced. Both tips can be measured. The tip of dyke (i) in Fig. 8.10 has a radius of curvature of 1 cm, whereas the tip of dyke (ii) in Fig. 8.11 has a radius of curvature of about 50 cm. The dykes are basaltic and with a magma density of about $2650 \text{ kg m}^{-3}$. The dykes were injected either from shallow magma chambers, at about 1 km below the present dyke exposures, or from deep-seated magma reservoirs (Fig. 8.12) at depths of about 8–20 km below the present exposures. Assume that the excess pressure in the chamber/reservoir before rupture and dyke injection is 3 MPa.

(a) Calculate the magma overpressure at the level of exposure assuming that the dykes were injected from a magma reservoir at a depth of 10 km below the present exposure. Use the average density of $2800 \text{ kg m}^{-3}$ for the density of the layers between the roof of the magma reservoir and the present exposures.

(b) Calculate the magma overpressure in the dykes at the level of exposure if they were injected from a shallow chamber at a depth of about 1 km below the present exposure. Use the crustal density of $2650 \text{ kg m}^{-3}$ for the crustal layer between the present exposure and the roof of the magma chamber.

(c) Use stress-concentration formulas to calculate the theoretical tensile stress at each dyke tip on the assumption that they were injected from the shallow magma chambers in (a).

Dyke tip

**Fig. 8.10**  Illustration for Example 8.2. View southwest, a basaltic dyke thins towards its vertical tip at about 1200 m below the original top of the lava pile in Southwest Iceland. The hammer provides a scale.

Solution

(a) The magma overpressure is obtained from Eq. (8.4). To simplify the problem, we assume that the stress difference $\sigma_d$ is equal to the tensile strength, 3 MPa, an assumption that has a stronger basis in case (b). From Eq. (8.4) we get

$$p_0 = p_e + (\rho_r - \rho_f)gh + \sigma_d$$
$$= 3 \times 10^6 \, \text{Pa} + (2800 \, \text{kg m}^{-3} - 2650 \, \text{kg m}^{-3}) \times 9.81 \, \text{m s}^{-2}$$
$$\times 9000 \, \text{m} + 3 \times 10^6 \, \text{Pa}$$
$$= 1.92 \times 10^7 \, \text{Pa} = 19.2 \, \text{MPa}$$

(b) Again the magma overpressure is obtained from Eq. (8.4) as follows:

$$p_0 = p_e + (\rho_r - \rho_f)gh + \sigma_d$$
$$= 3 \times 10^6 \, \text{Pa} + (2650 \, \text{kg m}^{-3} - 2650 \, \text{kg m}^{-3}) \times 9.81 \, \text{m s}^{-2}$$
$$\times 1000 \, \text{m} + 3 \times 10^6 \, \text{Pa}$$
$$= 6 \times 10^6 \, \text{Pa} = 6 \, \text{MPa}$$

**Fig. 8.11**   Illustration for Example 8.2. View north, a basaltic dyke is offset across a red soil/scoria layer between basaltic lava flows at about 1300 m below the original top of the lava pile in North Iceland. The lava flows are stiff, the soil/scoria layer is soft, and the dyke tip is rounded where it meets with the soft layer. The backpack at the bottom of the lower segment provides a scale.

In this case, since the average crustal density of the layers between the top of the chamber and the dyke exposures is the same as the magma density, an assumption that is very commonly true for such shallow chambers, buoyancy is zero and the second term in Eq. (8.4) becomes zero. It follows that the overpressure is solely due to the excess pressure in the chamber at the time of dyke injection, 3 MPa, and the stress difference, which is also taken as equal to the tensile strength and is thus 3 MPa. It may be argued that,

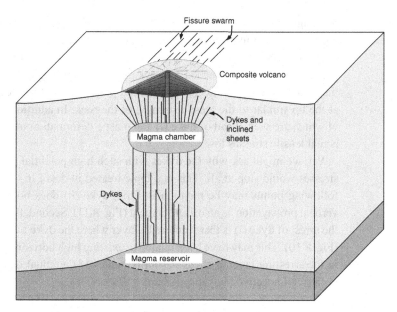

**Fig. 8.12**   Illustration for Example 8.2. Many dykes in Iceland originate from deep-seated magma reservoirs, at depths of 10–20 km, whereas others are injected from shallow magma chambers at depths of 1.5–3 km below the surface. The shallow chambers, in turn, are supplied with magma from the deep-seated reservoirs through dykes (Gudmundsson, 2006).

for such a shallow chamber, the stress difference is unlikely to be much greater than the tensile strength of the host rock, because as soon as $\sigma_3$ in the roof of the chamber (initially equal to the vertical stress, $\sigma_1$) is reduced by a magnitude equal to the tensile strength, a magma-chamber rupture and dyke injection will take place.

(c)  From Eq. (6.6) we have

$$\sigma_{\max} = \sigma_0 \left(1 + 2\sqrt{a/\rho_c}\right)$$

For dyke (i), in Southwest Iceland, the radius of curvature is 0.01 m, the half dip dimension (the semi-major axis of the elliptical dyke) is 500 m (the depth to the source chamber is 1000 m), and the magma overpressure is $6 \times 10^6$ Pa. Thus, the theoretical tensile stress at the dyke tip, from Eq. (6.6), is

$$\sigma_{\max} = 6 \times 10^6 \text{ Pa} \times \left(1 + 2\sqrt{\frac{500 \text{ m}}{0.01 \text{ m}}}\right) = 2.7 \times 10^9 \text{ Pa} = 2.7 \text{ GPa}$$

Similarly, for dyke (ii), in North Iceland, the radius of curvature is 0.5 m but the other values are the same as for dyke (i). Thus, from Eq. (6.6) the theoretical tensile stress at the tip is

$$\sigma_{\max} = 6 \times 10^6 \text{ Pa} \times \left(1 + 2\sqrt{\frac{500 \text{ m}}{0.5 \text{ m}}}\right) = 3.8 \times 10^8 \text{ Pa} = 0.38 \text{ GPa}$$

Although this second value is nearly an order of magnitude lower than that for dyke (i), both are two to three orders of magnitude greater than the in-situ tensile strength of the host rock. So these theoretical tensile stresses are never reached in nature. Long before they could be reached, the dyke tip would propagate, increase the volume of the dyke-fracture and thus lower the fluid pressure, and thus keep the tensile stresses at the tip similar to the tensile strength of the rock. In addition, microcracking (some of which are seen ahead of dyke (i) and plastic deformation of the host rock keeps the actual tensile stresses low.

Yet, we might ask why the dykes with such high potential or theoretical tensile tip stresses would stop at all. This is a topic treated in detail in Chapter 13, but here the following points may be mentioned. First, dyke (ii) does not stop – it continues its vertical propagation as an offset segment (Fig. 8.11). Second, the most likely reason for the arrest of dyke (i) is that it entered a layer where the dyke overpressure became zero (Fig. 8.10). This may have been related to existing high horizontal compressive stresses in the arresting layer, a stress barrier (Chapter 13), gradual decrease in overpressure because of negative buoyancy (Chapter 17), or both.

# 8.9 Exercises

8.1 What are the two basic mechanical types of rock fractures? What types of stresses generate these fractures? Give geological examples of the main rock types.

8.2 What field criteria can be used to determine if a fracture is an extension fracture?

8.3 What field criteria can be used to determine if a fracture is a shear fracture?

8.4 To what crustal depths can a large-scale tension fracture at the surface of a rift zone normally propagate before it changes into a normal fault?

8.5 Define a hydrofracture (a fluid-driven fracture) and give examples.

8.6 What is excess (fluid) pressure? What are the typical values of an excess pressure in a reservoir before rupture and hydrofracture formation?

8.7 What is (fluid) overpressure (driving pressure, net pressure) in a hydrofracture? How does it relate to excess pressure in the source reservoir or chamber?

8.8 What is total (fluid) pressure in a reservoir? Why is it of less importance as regards fracture formation than overpressure?

8.9 Define and illustrate the main types of faults in relation to the principal stresses.

8.10 In what tectonic environments would normal faults be common?

8.11 In what tectonic environment would reverse and thrust faults be common?

8.12 In what tectonic environment would strike-slip faults be common?

8.13 Consider tension fractures at the surface of a young sedimentary basin. If the tensile strength of the sedimentary layers is 0.5 MPa and the density of the uppermost layers is 2100 kg m$^3$, calculate the depth at which the tension fractures would be likely to change into normal faults.

8.14 Two basaltic dykes are exposed side by side in a lava pile, the estimated depth of exposure below the original surface of the volcanic zone being 1.5 km. Both have an estimated magma density at their time of emplacement of 2650 kg m$^{-3}$. One is 1 m thick and thought to be injected from a shallow magma chamber at a depth 2 km below the present exposure, whereas the other dyke is 5 m thick and thought to have been injected (either before the formation of the shallow chamber or after it became extinct and solidified as a pluton) as a regional dyke from a magma reservoir at a depth 10 km below the present exposure (Gudmundsson, 2002, 2006). The average crustal density between the exposure and the top of the shallow chamber is 2650 kg m$^{-3}$, whereas that between the exposure and the top of the deep-seated reservoir is 2850 kg m$^{-3}$. If the stress difference $\sigma_d$ was 1 MPa at the time of dyke injection (at the level of exposure), calculate the overpressure in each dyke. If these dykes were feeders, how would the overpressure (assumed similar at the surface) be reflected in the eruption intensity?

# References and suggested reading

Anderson, E. M., 1951. *The Dynamics of Faulting and Dyke Formation with Applications to Britain*, 2nd edn. Edinburgh: Oliver and Boyd.

Andrew, R. E. B. and Gudmundsson, A., 2008. Volcanoes as elastic inclusions: Their effects on the propagation of dykes, volcanic fissures, and volcanic zones in Iceland. *Journal of Volcanology and Geothermal Research*, **177**, 1045–1054.

Bell, F. G., 2000. *Engineering Properties of Soils and Rocks*, 4th edn. Oxford: Blackwell.

Boresi, A. P. and Sidebottom, O. M., 1985. *Advanced Mechanics of Materials*, 4th edn. New York: Wiley.

Carmichael, R. S., 1989. *Practical Handbook of Physical Properties of Rocks and Minerals*. London: CRC Press.

Economides, M. J. and Nolte, K. G. (eds.), 2000. *Reservoir Stimulation*, 3rd edn. New York: Wiley.

Geshi, N., Kusumoto, S., and Gudmundsson, A., 2010. The geometric difference between non-feeders and feeder dikes. *Geology*, **38**, 195–198.

Gudmundsson, A., 1995. Infrastructure and mechanics of volcanic systems in Iceland. *Journal of Volcanology and Geothermal Resarch*, **64**, 1–22.

Gudmundsson, A., 2002. Emplacement and arrest of sheets and dykes in central volcanoes. *Journal of Volcanology and Geothermal Research*, **116**, 279–298.

Gudmundsson, A., 2003. Surface stresses associated with arrested dykes in rift zones. *Bull. Volcanol.* **65**, 606–619.

Gudmundsson, A., 2006. How local stresses control magma-chamber ruptures, dyke injections, and eruptions in composite volcanoes. *Earth-Science Reviews*, **79**, 1–31.

Gudmundsson, A., 2010. Deflection of dykes into sills at discontinuities and magma-chamber formation. *Tectonophysics*, doi:10.1016/j.tecto.2009.10.015.

Gudmundsson, A. and Philipp, S. L., 2006. How local stresses prevent volcanic eruptions. *Journal of Volcanology and Geothermal Research*, **158**, 257–268.

Jumikis, A. R., 1979. *Rock Mechanics*. Clausthal, Germany: Trans Tech Publications.

Kanamori, H. and Anderson, D. L., 1975. Theoretical basis of some empirical relations in seismology. *Bulletin of the Seismological Society of America*, **65**, 1073–1095.

Kasahara, K., 1981. *Earthquake Mechanics*. New York: Cambridge University Press.

Price, N. J., 1966. *Fault and Joint Development in Brittle and Semi-Brittle Rock*. Oxford: Pergamon.

Savin, G. N., 1961. *Stress Concentration Around Holes*. New York: Pergamon.

Scholz, C. H., 1990. *The Mechanics of Earthquakes and Faulting*. New York: Cambridge University Press.

Schön, J. H., 2004. *Physical Properties of Rocks: Fundamentals and Principles of Petrophysics*. Amsterdam: Elsevier.

Secor, D. T., 1965. The role of fluid pressure in jointing. *American Journal of Science*, **263**, 633–646.

Valko, P. and Economides, M. J., 1995. *Hydraulic Fracture Mechanics*. New York: Wiley.

# Displacements and driving stresses of fractures

## 9.1 Aims

How do we model fractures observed in the field? For example how do we determine the tensile stress that opened the fracture in Fig. 8.2 or the shear stress responsible for the vertical displacement on the normal fault in Fig. 8.1? Similarly, how can we estimate the magma overpressure that generated the dyke in Fig. 8.1? To answer these questions we need mathematical models of rock fractures. Such models are normally referred to as crack models, even though the words 'fracture' and 'crack' are often used as synonyms. The conceptual and analytical models provided in this chapter are fundamental for the remainder of the book. The main aims of this chapter are to:

- Explain the three main geometric crack models used for rock fractures, namely through-the-thickness cracks, part-through cracks, and interior cracks.
- Clarify further the concepts of strike dimension, dip dimension, and controlling dimension and how these are used for rock fractures.
- Explain the three main modes of displacement of the fracture surfaces or walls, namely mode I, mode II, and mode III, and for which types of rock fractures each mode is most appropriate.
- Explain what is meant by mixed-mode (hybrid) cracks and for which types of rock fractures these are applicable.
- Provide basic formulas and numerical examples on the use of the geometric models and crack-displacement modes to estimate the driving stresses and pressures associated with rock-fracture formation and development.

## 9.2 Crack geometries

There are three basic ideal geometric crack models, namely a through-the-thickness crack, a part-through crack, and an interior crack (Fig. 9.1):

- A through-the-thickness crack goes right through the elastic body which contains it. The model is also referred to as a through-thickness crack or a **through crack**. The crack extends from one free surface (a surface in contact with a fluid, for example air or magma) to another free surface. This is an appropriate model, for instance, for a feeder

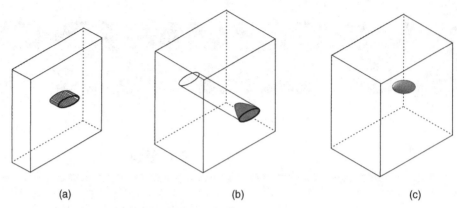

Fig. 9.1 Ideal crack geometries for modelling fractures: (a) a through-the-thickness (or through) crack, (b) a part-through crack (thumbnail crack), and (c) a penny-shaped (interior) crack.

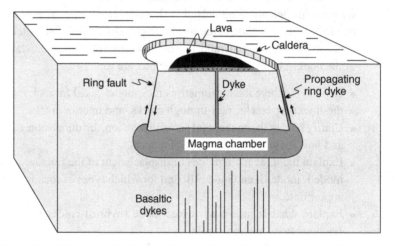

Fig. 9.2 Feeder dyke seen here in the centre of a collapse caldera and the ring dyke/ring fault are best modelled as through cracks. All ring faults are dip-slip faults, either reverse or normal faults. This one is a reverse fault (Gudmundsson, 2008).

dyke (Figs. 1.3, 1.6, and 9.2) and a ring fault (Figs. 9.2 and 9.3). The dyke extends from one fluid–solid free surface, namely the magma–host rock contact at the top of the magma reservoir to another fluid–solid free surface, namely the surface of the rift zone or volcano (the atmosphere–rock contact) to which the dyke supplies magma. The same applies to the ring fault (Figs. 9.2 and 9.3). Here it is assumed that the top part of the magma reservoir, where the dyke and the ring fault (and ring dyke) originate, is totally molten.

- A **part-through crack** extends only partly into the elastic body from its surface. This model is also referred to as a thumbnail crack. The crack extends from one fluid–solid free surface into the host body. This is an appropriate model, for example, for many non-feeder dykes. The dyke extends from the fluid–solid free surface at its source magma

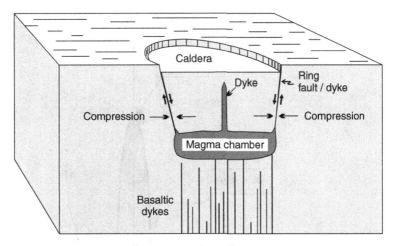

Fig. 9.3 Non-feeder dyke, here in the centre of a collapse caldera, is best modelled as a part-through crack. The dyke originates at a free surface, namely at the top of the shallow, fluid magma chamber, and then extends partly up into the host rock. Here the ring fault of the collapse caldera, a dip-slip fault, is shown as a normal fault (Gudmundsson, 2008).

reservoir or chamber, and ends at some depth in the crust (Figs. 3.1, 6.20, and 9.3). The model is also suitable for hydraulic fractures injected from drill holes (Chapters 6 and 7; Fig. 13.28).

- An **interior crack** is located in the interior of the elastic body that contains it. It does not terminate at a free surface, and the elastic body hosting it is regarded as infinite. The general shape of an interior crack is elliptic. The circular **penny-shaped crack** is a special case of the general elliptical interior crack. Many, and perhaps most, fractures in the crust are appropriately modelled as interior cracks because they do not meet any fluid–solid free surfaces. An example would be a small joint located within, and far from the surface of, a large pluton.

When modelling fractures, it is important to know which dimension of the fracture is the one that primarily controls its displacement. A fracture, like any other crustal structure, has three dimensions: length, height, and opening (Fig. 9.4). Fracture length and height have different meanings depending on context. For example, fluid flow along the length of a dyke might refer to flow along its vertical dimension, whereas for a dyke exposed at the surface, length would commonly mean its lateral dimension, that is, along the dyke strike. Similarly, the height of an inclined sheet may either refer to its vertical dimension measured from its source magma chamber or to its dimension measured along dip. To avoid this confusion, here and elsewhere in the book, the terms **strike and dip dimensions** are used for the maximum size of the fracture parallel with its strike and dip, respectively.

When one of these dimensions is significantly larger than the other, then the smaller of the two is referred to as the **controlling dimension**. What this means is that during loading of an elliptical crack, that is, one with unequal strike and dip dimensions (Fig. 9.5), the smaller of the two dimensions has much larger effects, that is, it largely controls the crack

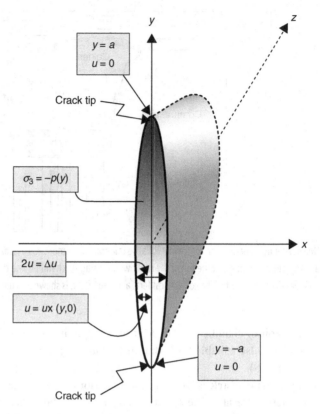

**Fig. 9.4** Crack model of a rock fracture, showing half the crack. Here $x$, $y$, and $z$ are the coordinate axes, so that the fracture is in the $yz$-plane and opens up parallel with the direction of the $x$-axis. The total opening is $2u = \Delta u$, and the fracture is opened by a fluid pressure $p$ (in rock, fluid overpressure $p_0$), which varies as a function of $y$, that is, $\sigma_3 = -p(y)$. In other examples, the crack may be in the $xz$-plane or in the $xy$-plane.

displacement. For example, if the strike dimension of a large strike-slip fault is much larger than its dip dimension, then the dip dimension is the one that enters the equation used to calculate the displacement of the fault during an earthquake. The strike dimension is then effectively taken as infinite, in which case the other dimension (the dip dimension in this case) must control the slip (Example 9.4).

A difference between the strike and dip dimensions of a fracture is common. For example, many studies of normal faults worldwide indicate that they tend to be somewhat elongated along strike. More specifically, the ratio of the strike dimension to the dip dimension commonly varies from about 0.3 to 3.0, with most values between 1 and 3 (Nicol *et al.*, 1996). This means that although some normal faults have larger dip dimensions (are deeper) than strike dimensions, most normal faults range from being with equal dimensions (the ratio 1.0 indicates a circular crack) to being elliptical with the strike dimension up to three times the dip dimension.

A fracture located in a homogeneous, isotropic rock far from any other discontinuities or free surfaces and subject to uniform loading is ideally circular. An appropriate model

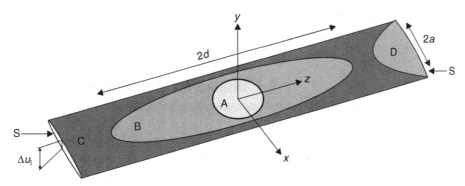

Ideal crack geometries for modelling rock fractures (cf. Fig. 9.1). A is a penny-shaped (circular) interior crack; B is an elliptical interior crack; C is a tunnel-crack, that is, a through-crack extending between the free surfaces marked by S; and D is a part-through crack, extending from one free surface, S, and partly into the host rock. The major axis of the elliptical crack B is $2d$ and the minor axis $2a$. The tunnel crack C is a mode I crack with an aperture (opening) of $\Delta u_I$ (cf. Gudmundsson, 2000).

for such a fracture is the penny-shaped interior crack (Figs. 9.1 and 9.5). As indicated for normal faults (Nicol *et al.*, 1996), many, and perhaps most, fractures are not circular but rather elliptical or, if through cracks, with a **tunnel shape** (Figs. 9.5 and 13.18). There are many reasons for the deviation of a fracture from an ideal circular shape, including the following:

- No rocks are strictly homogeneous and isotropic. At the scale of a fracture, however, some rocks may behave approximately as homogeneous and isotropic.
- Many rocks are layered. And if the layers have different mechanical properties, the fractures often become arrested at contacts between layers. Such fractures may grow into elliptical cracks (Figs. 9.4, 9.5, 13.18, and 13.28). If the fractures are restricted to a single layer, they can only grow, if at all, along strike and, therefore, may become very elongated (Fig. 13.28).
- The effects of layering on fracture geometry apply at various scales. For example, in sedimentary rocks joints and mineral veins are often elongated in layers that have thicknesses in the order of fraction of a metre. But large seismogenic faults are often highly elongated because the seismogenic layer is comparatively thin (commonly a few tens of kilometres). Thus, once the fault has propagated through the entire seismogenic layer, it can only grow along strike and its length may become hundreds of kilometres along strike (the strike dimension) while its height is restricted to a seismogenic layer (Fig. 7.2) with a thickness of a few tens of kilometres or less (the dip dimension).
- Many fractures are subject to loading (stress, pressure) that is not uniform, but rather varies along the fracture. For example, the fluid overpressure that drives the propagation of many hydrofractures varies along the fracture (Chapter 13). For some loading conditions the variation is smooth, for example linear from a maximum in the centre of the fracture to

zero at the fracture tips. In other hydrofractures, the fluid-overpressure variation may be irregular, partly because of layering of the host rock (Chapter 11).

## 9.3  Displacement modes of cracks

There are three ideal displacement modes of cracks: mode I, mode II, and mode III (Fig. 9.6). In **mode I** displacement, referred to as opening or tensile mode, the crack surfaces or walls move directly apart. In **mode II** displacement, referred to as sliding, in-plane or forward shear mode, the fracture surfaces or walls slide over one another in a direction perpendicular to the leading edge (tip) of the crack. In **mode III** displacement, referred to as tearing, anti-plane or transverse shear mode, the crack surfaces or walls move relative to one another in a direction that is parallel with the leading edge (tip) of the crack.

For the simple modelling of an extension fracture, for example a pure tension fracture, mineral vein or dyke, the appropriate model is a mode I crack. For a large strike-slip fault, the appropriate model is normally a mode III crack. For a dip-slip fault, however, the appropriate model may be either a mode II crack or a mode III crack, depending on the controlling dimension of the fault. Furthermore, the controlling dimension depends on the shape of the fracture.

These ideal displacement models apply to pure extension and pure shear fractures. Many fractures, however, are best modelled as mixed-mode (hybrid) cracks. These fractures are the result of displacement in more than one mode, for example modes I and II, or modes I and III. An illustration of all the possible mixed-modes is given by Hudson and Harrison (1997). Even for mixed-mode fractures, however, one mode may be dominant (Fig. 9.7), in which case that mode can, to a first approximation, be used to model the fracture displacement.

Consider a few examples of fractures with appropriate crack models. The dyke in Fig. 9.8 is an extension fracture and can be modelled as a mode I crack. If the dyke is a feeder (Fig. 9.2), it would be best modelled as a through crack. By contrast, if the dyke is a non-feeder (Fig. 9.3), that is, became arrested at a certain depth below the free surface at the time of its emplacement, then the appropriate geometric model would be a part-through crack.

Similarly, the graben in Figs. 2.1 and 8.1 is composed of two normal faults that can be modelled as mode II cracks. The normal faults of this graben clearly end vertically at about

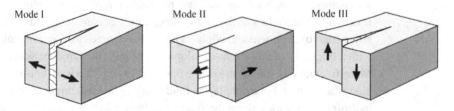

**Fig. 9.6**    Ideal wall or surface displacements of a fracture are denoted by mode I, mode II, and mode III.

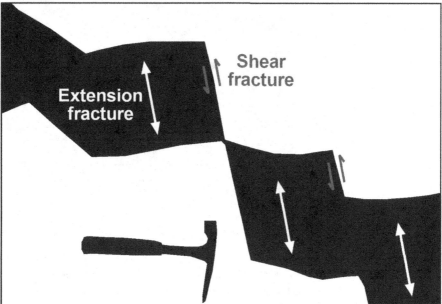

**Fig. 9.7**   During a rifting event an extension (here a tension) fracture forms at the surface of the rift zone in Iceland. Because the fracture uses the existing weaknesses in the pahoehoe lava flow, namely the hexagonal columnar joints, the original shape of the fracture is irregular and parts of it give rise to shear displacement. This fracture is thus mixed-mode, partly extension (mode I) and partly shear (mode II), although the dominant mode is I. The similarities with the opening of an ocean-ridge, with the shear part corresponding to an oceanic transform fault and the extension fracture corresponding to a ridge segment, are clear. As opening of the fracture increases in further rifting events, the irregularities become proportionally smaller and the fracture changes into an essentially pure tension fracture (cf. Figs. 1.1 and 8.2).

Fig. 9.8 Most dykes, including this one, are extension fractures and thus best modelled as mode I cracks. If they reach the surface as feeders, they should be modelled as mode I through-cracks, but if they become arrested and end as non-feeders, they should be modelled as mode I part-through cracks. Photo: a basaltic dyke in Tenerife, Canary Islands.

sea level, so that they did not reach to a free surface (for example, a fluid magma chamber) at their lower tips. Since they have vertical displacements of several metres, and are located in the present exposure at only about 1.5 km below the original surface of the rift zone within which they formed, these normal faults are likely to have reached the surface. It follows that a part-through crack is an appropriate model for each normal fault.

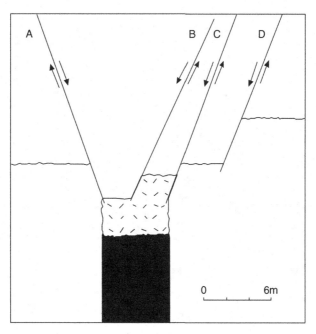

**Fig. 9.9**   Four normal faults and a dyke in the Tertiary lava pile in Northwest Iceland (cf. Gudmundsson, 2003).

The normal faults in Fig. 9.9, however, might be modelled in several ways. Faults A, B, and C meet the uppermost part of the dyke, whereas fault D ends within a lava flow. If the dyke existed and was fluid at the time of fault formation, then faults A–C could be modelled as mode III through cracks. The faults are at less than one kilometre below the initial top of the rift zone, so that given their displacements, they presumably reached the surface at their time of formation. Thus, if there was fluid magma at their lower ends, then the faults had free surfaces at their upper (air) and lower (magma) ends and were thus through cracks, and their mode of displacement would be mode III. By contrast, fault D should be modelled as a part-through mode II crack, since it extends (from the Earth's surface) partly into the elastic body (the lava pile) hosting it and does not reach a free surface (a fluid) at its bottom. Field data and mechanical considerations, however, indicate that the faults formed before the dyke emplacement, and that the dyke was captured by the graben (Gudmundsson, 2003), in which case all the normal faults should be modelled as fault D, namely, as part-through mode II cracks.

Many rock fractures are mixed-mode (Fig. 9.7). One fine large example is the normal fault Almannagja in Southwest Iceland (Figs. 9.10 and 9.11). This fault, which forms one of the boundary faults of the Thingvellir Graben (Figs. 3.13 and 3.14), is located in a Holocene lava flow and has a vertical surface displacement of as much as 40 m. However, the fault has also a very large aperture (opening), in places as great as 60 m. While part of this aperture may be apparent, that is, due to collapse of parts of the vertical walls, the aperture, as measured, is mostly genuine. That is, the observed aperture is largely, or entirely, due to the separation (or opening) of the fracture walls. Because Almannagja is a fault that is likely

Part of the Almannagja normal fault, forming the western boundary of the Thingvellir Graben in the Holocene part of
the rift zone of Southwest Iceland (cf. Figs. 3.13 and 3.14). View southwest, the fault has an opening of up to 60 m
and a vertical displacement of up to 40 m. The fault is thus a mixed-mode fracture; partly mode I, a tension fracture,
and partly mode III, that is, a dip-slip fault that is also a through-crack that extends from the surface down to a
magma reservoir – from one free surface to another.

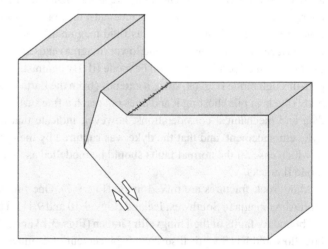

Schematic model of an open (gaping) normal fault such as Almannagja (Fig. 9.10). The tilting of the hanging wall is
presumably partly related to friction along the fault plane where it changes from a tension fracture to a shear fracture,
a normal fault.

to extend to a magma reservoir in the lower crust or upper mantle, it is best modelled as a through crack. The upper free surface is the fluid–solid contact between the atmosphere and the Holocene lava flow, whereas the lower free surface is the contact between the rocks of the lower crust and the magma in the reservoir. As indicated above, for a dip-slip fault modelled as a through crack, the displacement is modelled as mode III. This follows because the dip dimension is then effectively infinite, a tunnel crack, so that the strike dimension is the controlling dimension, and the displacement is in a direction parallel with the leading edges of the crack (Gudmundsson, 2000). For a mode III crack model of a dip-slip fault, the leading edges are the lateral ends or tip line of the fault (Fig. 9.6).

Almannagja is thus part mode III (the vertical displacement) and part mode I (the opening). To get a crude estimate of the driving shear stress necessary to generate the vertical displacement of 40 m, a through-crack mode III model can be used. Similarly, to model the driving tensile stress necessary to generate an aperture of 60 m, a through-crack mode I model can be used (Fig. 9.6). In crude calculations for mixed-mode fractures, the different displacements can be calculated independently of each other. However, the mixed-mode displacement can also be obtained from formulas that include both opening and vertical displacements. When making estimates of this kind of the driving stresses associated with Almannagja, as for other rock fractures, it is assumed that the displacements are reached in single slips. It is known that some normal faults in the rift zone of Iceland have reached vertical and horizontal displacements of as much as 10 m in single rifting event, but displacements of 40–60 m in single events are highly unlikely. In fact, it is known that Almannagja reached its present vertical displacement in at least several episodes, the last one being in AD 1789. Similarly, rifting episodes in the rift zone of North Iceland yielded vertical displacements of only 1–2 m in individual rifting events. It is thus likely that large faults in general form in many slip events, in which case only the individual events can be used to estimate the driving shear and tensile stresses, as is supported by palaeoseismic studies worldwide (Yeats *et al.*, 1997; McCalpin, 2009).

To estimate the driving stresses associated with fractures of various kinds, we need the appropriate formulas. In the following, the main formulas for assessing fluid overpressure and tensile and shear stresses associated with measured displacements are given and discussed. These formulas can also be used, in a rewritten form, to calculate the likely displacement for a fracture of given dimensions, when the driving stresses or fluid overpressures are known. The discussion begins with extension fractures, that is, tension fractures and hydrofractures, and then moves on to the various types of faults.

## 9.4  Tension fractures

Since many fractures are elliptical, we consider an elliptical crack as the most general shape (Fig. 9.5). As will become apparent, the analysis presented here can be used for factures of other shapes as well. Although the equations derived below are for tension fractures, the results form the basis for analysing all the other fracture displacements, including hydrofractures and shear fractures. The solution to the problem of an elliptical crack in an infinite elastic solid subject to tensile loading, that is, mode I loading, was provided by Green and Sneddon (1950) and for arbitrary (mixed-mode) loading by Kassir and Sih (1966).

The periphery or tip-line of an elliptical fracture is described by the formula for an ellipse with the origin at the centre of the coordinate system (Fig. 9.5). The major axis of the ellipse is $2d$ and coincides with the $z$-axis; the minor axis of the ellipse is $2a$ and coincides with the $x$-axis. Here, $2a$ is the axis of the plane of the elliptical crack, that is, the dip dimension of the crack (Fig. 9.5) and not the aperture variation of an elliptical opening of a crack. Using the notation in Fig. 9.5, it follows that the geometry of the elliptical crack is defined by

$$1 - \frac{z^2}{d^2} - \frac{x^2}{a^2} = 0 \tag{9.1}$$

The semi-axes are defined as $d \geq a > 0$ and the eccentricity of the ellipse, $\varepsilon$, is given by

$$\varepsilon = \left(1 - \frac{a^2}{d^2}\right)^{1/2} \tag{9.2}$$

where $0 < \varepsilon < 1$. The normal opening displacement $u$ of the elliptical crack, as a function of location along the crack surface, $u = u(x, z, 0)$, is

$$u = -\frac{\sigma(1 - \nu)a}{GE(\varepsilon)}\left(1 - \frac{z^2}{d^2} - \frac{x^2}{a^2}\right)^{1/2} \tag{9.3}$$

where $-\sigma$ is the tensile stress (denoted as negative) opening the crack, $\nu$ is Poisson's ratio, and $G$ is the shear modulus of the rock within which the crack is located. Here $E(\varepsilon)$ denotes the complete elliptic integral of the second kind, defined as

$$E(\varepsilon) = \int_0^{\pi/2} \sqrt{1 - \varepsilon^2 \sin^2 \theta}\, d\theta \tag{9.4}$$

This integral cannot be evaluated by elementary techniques for general values of $\varepsilon$. However, standard mathematical tables (Beyer, 1976) give $E(\varepsilon)$ for various values of $\varepsilon$.

Consider first a circular interior crack, a penny-shaped crack (Fig. 9.5). For a circle the semi-axes are equal, that is, $a = d$. It follows from Eq. (9.2) that $\varepsilon = 0$, in which case $E(0) = \pi/2 \cong 1.57$ (Beyer, 1976). Denote the radial coordinate of a circle with a centre at the origin by $r$ and use the relations $x^2 + z^2 = r^2$ and $G = E/[2(1 + \nu)]$, where $E$ is Young's modulus. Then, from Eq. (9.3) the opening displacement $u = u(r)$ is

$$u = -\frac{4\sigma(1 - \nu^2)}{\pi E}(a^2 - r^2)^{1/2} \tag{9.5}$$

This is the formula for the opening displacement (half the aperture) of a circular (penny-shaped) crack opened under constant tensile stress, $-\sigma$.

Penny-shaped tension fractures occur in many rocks at shallow depths. They are likely to be particularly common in rocks that are close to being homogeneous and isotropic, but less frequent in layered rocks. This follows because an initial penny-shaped fracture would tend to become elliptical or tunnel-shaped (Fig. 9.5) in layered rocks. Many models on fluid flow in fractured reservoir assume the ideal cracks to be penny-shaped (Chapter 15).

Let us now modify the results presented in Eq. (9.5) so that they become useful for tension fractures that differ in geometry from penny-shaped cracks. For elliptical cracks, one dimension is larger, and commonly much larger, than the other dimension (Fig. 9.5). For example, if the dip dimension is much larger than the strike dimension, $d \gg a$, then, in relation to $a$, the dip dimension may often be assumed infinite, that is, $d \rightarrow \infty$. Using this in Eq. (9.2), the eccentricity approaches unity, that is, $\varepsilon \rightarrow 1$. For the complete integral of the second kind, $E(1) = 1$. Using this, as well as the above relation between Young's modulus and shear modulus, from Eq. (9.3) we get

$$u = -\frac{2\sigma(1 - v^2)}{E}(a^2 - x^2)^{1/2} \tag{9.6}$$

This is the plane-strain formula (Fig. 4.15) for a two-dimensional elliptical through crack, a tunnel crack, subject to a constant opening-mode or tensile loading, $-\sigma$. If the true dip dimension is short (but still assumed infinite), a slightly different plane-stress formula (Fig. 4.14) with $(1 - v^2) = 1$ is commonly used (Paris and Sih, 1965; Tada *et al.*, 2000). In Eq. (9.6), $u$ depends only on the smaller dimension $a$, the strike dimension, indicating that the smaller of the strike and dip dimensions controls the crack displacement.

For an elliptical crack, the maximum displacement $u_{max}$ occurs at its centre. In modelling rock fractures, it is commonly known, or assumed, that the displacement measured in the field is the maximum displacement of the fracture. This is normally justified. The present model (Eq. 9.7) and field measurements show that the displacements of many fractures vary roughly as in a flat ellipse (Fig. 9.12). There may be considerable variation in the displacement if the host rock is heterogeneous. In particular, for a hydrofracture dissecting layered rocks, the aperture in each layer depends on its Young's modulus, so that the aperture variation may be irregular. Despite irregularities of this kind, the overall variation in displacement in general, and aperture in particular, is so that the largest displacement is normally not much different from the average displacement, thereby supporting the assumption.

Substituting $x = z = 0$ in Eq. (9.3), and using the above relation between Young's modulus $E$ and shear modulus $G$, we obtain

$$u_{max} = -\frac{2\sigma(1 - v^2)a}{EE(\varepsilon)} \tag{9.7}$$

showing, again, that the maximum displacement depends primarily on the smaller dimension, here the strike dimension $a$. The only effect of the larger (here the dip) dimension $d$ is through the eccentricity $\varepsilon$ and the value of $E(\varepsilon)$. For instance, if $d = 2a$ then $E(\varepsilon) \cong 1.21$, if $d = 3a$ then $E(\varepsilon) \cong 1.11$, and if $d = 4a$ then $E(\varepsilon) \cong 1.06$, and less for greater $d/a$ ratios. For the aspect ratios of interest, $E(\varepsilon)$ is therefore always close to unity and the resulting displacement is very similar to that obtained in the centre of a through crack (Eq. 9.7) with $x = 0$. We can therefore conclude that, for elliptical cracks, the smaller of the strike and dip dimensions has the greatest effect on the displacement, particularly when the $d/a$ ratio is large. The smaller dimension may therefore be regarded as the dimension controlling fracture displacements.

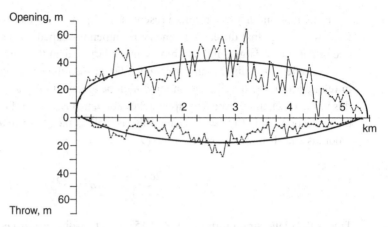

Fig. 9.12 Measured variation in opening or aperture and throw or vertical displacement along a part of Almannagja (Figs. 9.10 and 9.11). Also indicated, schematically, is the general flat-ellipse variation in the opening and the throw (Gudmundsson, 1987).

When studying tension fractures (and hydrofractures) the variation in stress or overpressure at the time of fracture formation is normally not known. While the stress can sometimes be inferred based on the details of the fracture geometry, a common practical approach is to assume that the driving tensile stress (fluid overpressure, for hydrofractures) was constant at the time of fracture formation. As a rough first approximation, this may be helpful; however, the fluid overpressures are known to vary, particularly in layered rocks (Chapter 13). As we shall see below, similar assumptions are commonly made as regards slip on faults during earthquakes, or the total displacement on a seismogenic fault. It may often be reasonable to assume that the driving shear stress was constant during the fault development and, in particular, individual slip events. If the driving stress and overpressure stress are assumed to be constant, then very simple mathematical formulas are obtained. These formulas can then be applied directly to field measurements of fractures and provide simple relations between the driving stress/overpressure and the fracture displacements.

In all the formulas in this and the following sections, the symbol $E$ denotes Young's modulus and $v$ denotes Poisson's ratio. Also, $L$ denotes the strike dimension $2a$ of the fracture and $R$ its dip dimension $2d$ (Fig. 9.5). If the strike dimension is much larger than the dip dimension, a plane-stress formulation is used (Chapter 4; Paris and Sih, 1965; Tada et al., 2000); if the dip dimension is the larger a plane-strain formulation, with the term $(1 - v^2)$, is used. The relative (total) displacement $\Delta u$ is twice the displacement $u$, that is, $\Delta u = 2u$. Thus, for example, if $u$ is the normal displacement of a tension fracture, then $\Delta u_I = 2u$ is the aperture of the fracture. The same applies to the displacements of mode II and mode III cracks: $\Delta u = 2u$.

A through-the-thickness mode I crack (Fig. 9.5) is an appropriate model for many tension fractures, such as the many fractures at the surface of the fissure swarms in Holocene pahoehoe lava flows in Iceland (Figs. 1.1 and 8.2). Many of these fractures extend from the surface to a horizontal, open contact between the flow units (Fig. 9.13). If a shallow flow

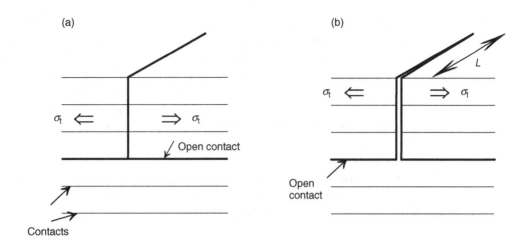

(a)

σ_t ⇐       ⇒ σ_t

Open contact

Contacts

(b)

L

σ_t ⇐       ⇒ σ_t

Open contact

Fig. 9.13 A fracture, such as a tension fracture or a joint, that extends from the surface to an open contact may be modelled as a through-crack. Here the fracture is subject to tensile stress $\sigma_t$. (a) Open or very weak contacts are common at shallow depths, such as in pahoehoe lava flows (Fig. 13.14), and may function essentially as free the fracture surfaces. (b) The opening increases because of the tensile stress $\sigma_t$, but the fracture does not propagate through the open contact at depth. The fracture becomes a through crack with a controlling (strike) dimension $L$.

unit, or a group of units, is largely decoupled from the underlying units by an open contact, a discontinuity, the unit may behave similar to a solid body with free surfaces at its top and bottom. The tension fracture can then be modelled as a mode I through crack. From Eq. (9.7) it follows that when $d > 2\text{–}3a$, Eq. (9.6) may be used to calculate the displacement in the centre of the crack. Thus if the strike dimension $L = 2a$ is much smaller than the dip dimension $R = 2d$, the maximum (central) opening displacement of the crack $\Delta u_I$, that is, the maximum aperture of the fracture, depends entirely on its strike dimension. With $x = 0$, Eq. (9.6) yields

$$\Delta u_I = \frac{-2\sigma(1 - v^2)L}{E} \tag{9.8}$$

A different formula is obtained when an extension fracture goes only partly into the hosting rock body. If the strike dimension of the extension fracture is much greater than its dip dimension, so that, for example, $L > 2\text{–}3R$, then the controlling dimension is the dip dimension $R$. A plane-stress mode I crack model then gives the following equation (Tada et al., 2000):

$$\Delta u_I = -\frac{4\sigma R V}{E} \tag{9.9}$$

The function $V = V(R/T)$, where $T$ is the total thickness of the rock body hosting the fracture, is defined (using radians) as

$$V\left(\frac{R}{T}\right) = \frac{1.46 + 3.42\left[1 - \cos\left(\frac{\pi R}{2T}\right)\right]}{\left[\cos\left(\frac{\pi R}{2T}\right)\right]^2} \tag{9.10}$$

Equations (9.8)–(9.11) are very useful for crude estimates of the aperture of a tension fracture, modelled either as a through crack or as a part-through crack, when the associated tensile stress is known. In general, the tensile stress likely to operate during tension-fracture formation ranges from about 0.5 to 6 MPa, and is commonly about 2 to 3 MPa for surface rocks. This follows because the field or in-situ tensile strength of solid rocks is generally within this range, namely between 0.5 and 6 MPa (Appendix E). Thus, in the absence of other information, it may be fairly accurately assumed that the tensile stress associated with the formation of a tension fracture, at or close to the surface, was between 0.5 and 6 MPa.

In field studies, it is normally the fracture aperture that is measured. So if we want to use the measured fracture apertures to estimate the driving tensile stresses associated with their formation, Eqs. (9.8) and (9.9) can be rewritten in the forms

$$-\sigma = \frac{\Delta u_I E}{2L\left(1 - \nu^2\right)} \tag{9.11}$$

and

$$-\sigma = \frac{\Delta u_I E}{4VR} \tag{9.12}$$

where all the symbols are as defined above, and the function $V$ is given by Eq. (9.10). These equations, together with Eqs. (9.8)–(9.11) allow us to deal with many topics related to tension fractures as measured in the field. And simply by changing the driving tensile stress to fluid overpressure, we can obtain similar equations for hydrofractures.

## 9.5 Hydrofractures

When studying palaeofluid-transporting fractures such as dykes and mineral veins in the field, it is often important to estimate the fluid overpressure that formed the fractures. For example, when we study a dyke in the field, we would like to know the magmatic overpressure associated with its emplacement. During volcanic eruptions, the overpressure of the feeder dyke is an important constraint on the mechanism of the eruption, since it has strong effects on the volumetric flow rate through the associated volcanic fissure.

Similarly, although some veins are originally shear fractures or tension fractures, many and perhaps most of those exposed at considerable depth in eroded sedimentary basins and lava piles, are hydrofractures. In such cases, it is important to know the fluid overpressure associated with the fracture formation because it controls the fracture aperture and thereby has a large effect on the volumetric flow rate through the fracture.

Consider a hydrofracture (mode I) with a strike dimension $L$ smaller than its dip dimension $R$ (Fig. 9.14). If the fracture is a through crack, it follows by analogy with Eq. (9.8) that its aperture or relative opening displacement $\Delta u_I$ is related to the overpressure $p_o$ of

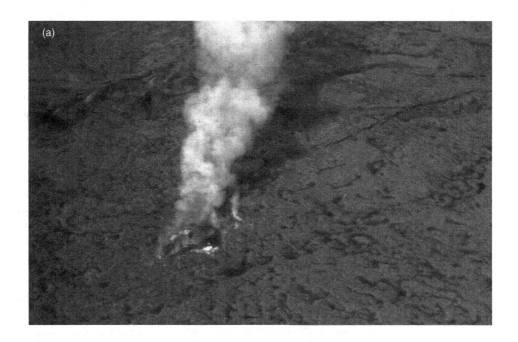

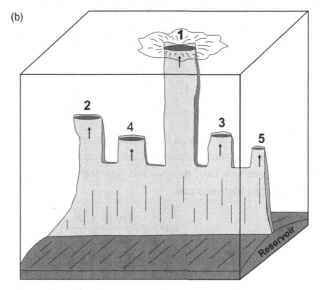

**Fig. 9.14** (a) Beginning of the volcanic-fissure development during the July 1980 eruption in the Krafla Volcano, North Iceland. The volcanic fissure seen here is the first 'finger' of a fissure that subsequently became many kilometres long and consisted of offset segments (photo: Aevar Johannsson). (b) One interpretation of the feeder dyke in (a). The part of the fissure seen in (a) is the first 'finger' that has reached the surface (number 1). The feeder dyke, as seen here, is a mode I, through-crack with a strike dimension (length at the surface) that is much smaller than the dip dimension (depth to the magma reservoir). Consequently, the strike dimension is the controlling dimension.

the fluid through the formula

$$\Delta u_{\mathrm{I}} = \frac{2p_0(1 - v^2)L}{E}$$ (9.13)

If, however, the hydrofracture is a part-through crack and its strike dimension $L$ is greater than its dip dimension $R$, the controlling dimension is $R$ and, by analogy with Eq. (9.9), the fracture aperture is given by

$$\Delta u_{\mathrm{I}} = \frac{4p_0 V R}{E}$$ (9.14)

Equations (9.13) and (9.14) show that fluid overpressure can have a great effect on the aperture of a hydrofracture. These equations are, in a modified form, also used for artificial hydraulic fractures, such as are commonly used for stress measurements (Amadei and Stephansson, 1997), for increasing the permeability of petroleum reservoirs (Valko and Economides, 1995), and for the development of enhanced geothermal systems (Tester *et al.*, 2007).

Often we want to find the fluid overpressure from field measurements of hydrofractures such as dykes and sills and mineral veins. Then we can use the measured fracture apertures for a crude estimate of the driving fluid pressure associated with their formation. Equations (9.13) and (9.14) are then rewritten in the forms

$$p_0 = \frac{\Delta u_{\mathrm{I}} E}{2L \left(1 - v^2\right)}$$ (9.15)

and

$$p_0 = \frac{\Delta u_{\mathrm{I}} E}{4V R}$$ (9.16)

where all the symbols are as defined above, and the function $V$ is given by Eq. (9.11). The special case of a penny-shaped hydrofracture is also of great interest. This follows, partly, because some sills have roughly this geometry. By analogy with Eq. (9.5), we have for a penny-shaped hydrofracture

$$u(r) = \frac{4p_0 \left(1 - v^2\right)}{\pi E} \left(a^2 - r^2\right)^{1/2}$$ (9.17)

in which case the maximum aperture, in the centre of the hydrofracture, becomes

$$\Delta u_{\mathrm{I}} = \frac{8p_0 \left(1 - v^2\right) a}{\pi E}$$ (9.18)

The volume $V_{\mathrm{f}}$ of the penny-shaped hydrofracture can be estimated from the formula

$$V_{\mathrm{f}} = \frac{16(1 - v^2)p_0 a^3}{3E}$$ (9.19)

Equations (9.13)–(9.19) allow us to quantify the likely fluid overpressure, shape, and volume of hydrofractures measured in the field. The results have a variety of applications, as we shall see later in this and subsequent chapters.

# 9.6 Dip-slip faults

For faults and other shear fractures, the driving shear stress is the difference between the remotely applied shear stress and the residual frictional strength on the fault after sliding. Then residual frictional strength is considered equal to the coefficient of sliding friction multiplied by the normal stress, that is, equal to the term $\mu\sigma_n$ in the Modified Griffith criterion (Chapter 7; Nur, 1974). Thus, we have

$$\tau = 2T_0 + \mu\sigma_n \tag{9.20}$$

Fault slip in the crust, however, normally occurs on planes of high fluid pressure. It is well known that most tectonic earthquakes are related to zones of high fluid pressure, so that, from Eq. (9.20), the shear stress for fault slip, $\tau$, becomes

$$\tau = 2T_0 + \mu(\sigma_n - p_t) \tag{9.21}$$

where $T_0$ is the tensile strength of the rock, $\mu$ is the coefficient of internal friction, $\sigma_n$ is the normal stress on the fault plane, and $p_t$ is the total fluid pressure on the fault plane at the time of slip. When the fluid pressure approaches or equals the normal stress, the term $\mu(\sigma_n - p_t)$ approaches or equals zero. In fact, the total fluid pressure may become so high as to exceed the normal stress on the fault plane, generating an overpressure with respect to $\sigma_n$. Then the term $\mu(\sigma_n - p_t)$ becomes negative. This is possible, and may explain some unusually low driving stresses (stress drops), as indicated below. However, for most fault slips, the term $\mu(\sigma_n - p_t)$ is presumably close to zero. Then the driving shear stress for slip becomes double the tensile strength, that is, equal to $2T_0$.

Since the in-situ tensile strength of rocks is commonly in the range 0.5–6 MPa it follows from these considerations that, for most high-fluid-pressure fault zones, the driving shear stress for slip should be 1–12 MPa. As said, this range is in agreement with common static stress drops in earthquakes, which are mostly in the range 1–12 MPa (Kanamori and Anderson, 1975; Kasahara, 1981; Scholz, 1990).

While the general stress drops are thus between 1 and 12 MPa, and fit very well with the driving shear stress being equal to $2T_0$, some are as high as 30 MPa, others as low as 0.3 MPa. These may be related to difference in friction or, rather, different residual shear strength on the fault plane after slip. When the fluid pressure on the fault plane before slip is very high, so that the term $\mu(\sigma_n - p_t)$ becomes negative, then it follows from Eq. (9.21) that less shear stress is needed to generate the slip. Thus, the static stress drop, or driving shear stress, may become lower. By contrast, the comparatively high stress drops, of up to 30 MPa, are likely to be related to slips on faults where the residual shear strength is comparatively high or, which effectively means the same, the fluid pressure is relatively low. While asperities may also contribute to the residual shear strength, in major fault zones these are, for the most part, eroded away and thus not likely to be of as great significance as fluid pressure (Chapters 14 and 16).

In applying the models of mode II and mode III cracks to dip-slip and strike-slip faults, and other shear fractures, we will assume that the shear stress in Eqs. (9.20) and (9.21) is equal to the driving shear stress. To make this clear, we denote the driving shear stress in the formulas by $\tau_d$. If the residual frictional or shear strength, $\tau_f$, on a fault plane after slip is known, then of course the driving shear stress should be reduced to

$$\tau_d = \tau - \tau_f \tag{9.22}$$

However, the residual frictional strength is normally not known. Also, as argued above, the residual frictional strength in many fault zones, commonly interpreted as $\mu(\sigma_n - p_t)$ (Nur, 1974), is close to zero and thus negligible.

For a dip-slip fault where the strike dimension $L$ controls the displacement, a mode III crack is normally the appropriate model. Then the displacement $\Delta u_{III}$ and the strike dimension $L$ are related through the formula

$$\Delta u_{III} = \frac{2\tau_d(1 + \nu)L}{E} \tag{9.23}$$

For a dip-slip fault where the dip dimension $R$ controls the displacement, a mode II crack is normally the appropriate model. Then the displacement $\Delta u_{II}$ is related to the dip dimension $R$ of the fault through the formula

$$\Delta u_{II} = \frac{4\tau_d R V}{E} \tag{9.24}$$

## 9.7 Strike-slip faults

A strike-slip fault that reaches the surface is normally modelled as a mode III crack (Fig. 9.15). The displacement $\Delta u_{III}$ is then related to the dip dimension $R$ of the fault through the formula

$$\Delta u_{III} = \frac{4\tau_d(1 + \nu)R}{E} \tag{9.25}$$

where all the symbols are as defined above. The difference by a factor of 2 between Eqs. (9.23) and (9.25) is because the strike dimension (trace length) $L$ is equal to the diameter of the crack, whereas the dip dimension $R$ is equal to only half the diameter of the crack. This crack model applies equally well to sinistral and dextral strike-slip faults. Equation (9.25) is particularly appropriate for strike-slip faults with the upper tip at the surface (reaching the surface), and the lower tip at a certain depth in the crust. These are thus part-through faults. For very large strike-slip faults of this type, the lower tip is at the bottom of the seismogenic layer, that is, the brittle layer that generates earthquakes (Kasahara, 1981; Scholz, 1990).

Some strike-slip faults are better modelled as through cracks. For example, a strike-slip fault that extends from the surface to a fluid-filled reservoir is better modelled as a through crack, since the fluid–solid contact of the reservoir makes for a free surface. Strike-slip faults of this kind include some that exist within large, active volcanoes. For example, strike-slip faults located within large, active collapse calderas would commonly fit this category. Also, many transform faults at mid-ocean ridges, particularly fast-spreading ridges with magma beneath the transform fault, might be modelled as through cracks.

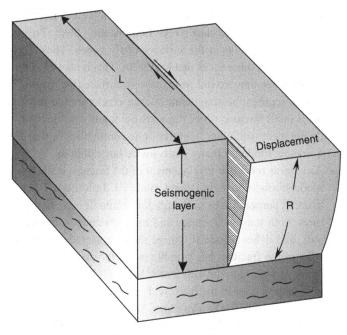

**Fig. 9.15**  Model of a strike-slip fault. Most strike-slip faults, here a dextral fault, can be modelled as mode III cracks. *R* denotes the dip dimension of the fault, here equal to the thickness of the seismogenic layer. The strike dimension of the fault, *L*, can be, and commonly is, many times the dip dimension. The total displacement at the surface is denoted by $\Delta u_{\mathrm{III}}$ (cf. Fig. 9.18).

For strike-slip faults extending from one free surface to another, the appropriate model is a mode II crack. For a mode II crack, the maximum plane-strain displacement $\Delta u_{\mathrm{II}}$ is given by

$$\Delta u_{\mathrm{II}} = \frac{2\tau_{\mathrm{d}}(1-\nu^2)L}{E} \tag{9.26}$$

Here $L$ is the strike dimension of the fault, and the other symbols are as defined above.

## 9.8 Summary

- The three basic geometrical models used for modelling rock fractures are: (1) through cracks, (2) part-through cracks, and (3) interior cracks. Through cracks are suitable for modelling fractures that extend from one free surface to another. For example, they are very appropriate models for feeder dykes, which extend from a fluid magma chamber (a free surface) to the surface (air). They can also be used when modelling fractures that are confined to a single layer or unit where the upper and lower part of the layer are in contact with very ductile layers, or open contacts, at the time of fracture formation. This applies to many mineral veins and joints and some tension fractures in rift zones.

- Part-through cracks are suitable models for most fractures that extend from one free surface and into the elastic rock body hosting them. This is a suitable model for many fractures and faults at the surfaces of rift zones. These fractures extend partly into the lava flows or other rocks that host them. This is also an appropriate model for non-feeder dykes, since they extend from one free surface (the magma chamber) and partly into the crustal segment hosting them. Interior cracks are the best models for many and perhaps most small fractures since these are commonly inside the rock body that hosts them and do not meet with other fractures or discontinuities or free surfaces. This is, for example, a suitable mode for many joints inside large plutons.
- As regards modes of displacement of the fracture walls or surfaces, there are three ideal models: mode I, mode II, and mode III. In a mode I model, the displacement is pure opening perpendicular to the fracture plane. It is a suitable model for all extension fractures, both tension fractures and hydrofractures – for example, for all inclined sheets, sills, and dykes. Mode II, where the displacement is perpendicular to the propagating edge of the fracture, is a suitable mode for many part-through fractures – for example, many part-through dip-slip faults. Mode III, where the displacement is parallel with the propagating edge of the fracture, is a suitable model for through-crack dip-slip faults, as well as for most strike-slip faults. On rare occasions, when a strike-slip fault has a free surface at its top (air) and bottom (a magma chamber), a through-crack (plane-stress) mode II model is more suitable.
- Many rock fractures are mixed mode. When one of the displacement modes is dominant, then that mode may be used to make an approximate model of the fracture.
- Using the formulas in this chapter makes it possible to model the displacements, or estimate the driving stresses or pressures, associated with fracture formation or fault slip. Before the appropriate formula is obtained, we must first assess the most suitable model as regards fracture geometry and the mode of surface displacement. Such an assessment requires geological understanding of the type of fracture we are dealing with as well as the loading conditions during the fracture formation or slip.

## 9.9 Main symbols used

| | |
|---|---|
| $a$ | radius of a circular (penny-shaped) interior crack |
| $a$ | minor semi-axis of an elliptical crack, half its dip dimension |
| $d$ | major semi-axis of an elliptical crack, half its strike dimension |
| $E$ | Young's modulus (modulus of elasticity) |
| $E(\varepsilon)$ | complete elliptic integral of the second kind |
| $G$ | shear modulus (modulus of rigidity) |
| $L$ | the strike dimension of a crack, $2a$ |
| $p_t$ | total fluid pressure |
| $p_0$ | fluid overpressure |
| $R$ | the dip dimension of a crack, $2d$ |
| $r$ | radial coordinate of a circular (penny-shaped) crack |

| | |
|---|---|
| $T$ | total thickness of a crustal segment or a rock body hosting a part-through crack |
| $T_0$ | tensile strength |
| $u$ | normal opening displacement of a crack, half the aperture $\Delta u$ |
| $\Delta u_{\mathrm{I}}$ | maximum aperture (total opening displacement) of a mode I crack |
| $\Delta u_{\mathrm{II}}$ | maximum total displacement of a mode II crack |
| $\Delta u_{\mathrm{III}}$ | maximum total displacement of a mode III crack |
| $V$ | function for a part-through crack |
| $V_{\mathrm{f}}$ | volume of a circular (penny-shaped) crack |
| $\varepsilon$ | eccentricity of an ellipse in Eq. (9.2) |
| $\mu$ | coefficient of internal friction |
| $\nu$ | Poisson's ratio |
| $\pi$ | 3.1416... |
| $\sigma$ | normal tensile stress |
| $\sigma_n$ | normal stress (usually compressive) on a fault plane |
| $\tau$ | shear stress for faulting |
| $\tau_d$ | driving shear stress associated with a given fault slip or displacement |
| $\tau_f$ | residual frictional or shear strength on a fault plane after slip |

## 9.10 Worked examples

### Example 9.1

Problem

The mineral vein in Fig. 9.16 has a strike dimension (outcrop length) of 80 cm and a maximum thickness of 2 mm. Suppose, at the time of vein formation, that the Young's modulus of the basaltic lava flow hosting the vein was 15 GPa and its Poisson's ratio was 0.25. Assume that the vein is a through-crack, that is, that it extends from one free surface (presumably a horizontal cooling joint) to another free surface. If the entire opening of the vein, as measured by its present maximum thickness, is due to the geothermal-water overpressure at that time, how large was that overpressure?

Solution

If we assume that the strike dimension (outcrop length) is that controlling dimension, that is, that the strike dimension is smaller than the dip dimension (height), then from Eq. (9.15) we calculate the fluid overpressure, $p_0$, as follows:

$$p_{\mathrm{o}} = \frac{\Delta u_{\mathrm{I}} E}{2L\left(1 - v^2\right)} = \frac{0.002 \text{ m} \times 1.5 \times 10^{10} \text{Pa}}{2 \times 0.8 \text{ m} \times (1 - 0.25^2)} = 2 \times 10^7 \text{Pa} = 20 \text{ MPa}$$

This result is very similar to other overpressure estimates from mineral veins. The assumption that the strike dimension is smaller than the dip dimensions follows from comparing

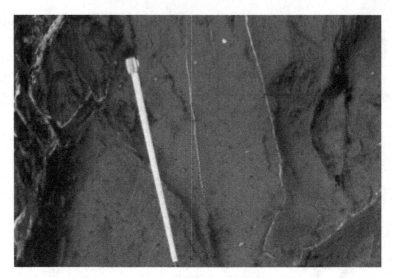

Fig. 9.16 Example 9.1. A mineral vein exposed at a depth of 1.5 km below the original surface in the Husavik-Flatey Fault in North Iceland. View vertical, the measuring tape is 1 m long.

the outcrop length of the vein with its likely height. If the vein goes from one horizontal joint to another, then the spacing of these joints is an indication of the dip dimension of the vein. A typical spacing of horizontal joints is 0.5–2 m, so that, on average, the dip dimension might be expected to be somewhat larger than the strike dimension.

## Example 9.2

### Problem

During the drilling of a road tunnel, a tension fracture propagates from the Earth's surface down into the roof of the tunnel (Fig. 9.17). The tunnel is horizontal and a depth of 200 m below the surface. The strike dimension or outcrop length of the fracture at the surface, and in the roof of the tunnel, is 100 m, and its maximum opening (aperture) is 10 cm. If the host rock has a Young's modulus of 10 GPa and a Poisson's ratio of 0.25, find the tensile stress that generated the fracture.

### Solution

The tension fracture is a through-crack; it extends from the free surface of the Earth to the free surface of the (air-filled) tunnel. Since the dip dimension is 200 m and the strike dimension 100 m, the strike dimension is the controlling dimension. From Eq. (9.11) the tensile stress forming the fracture is calculated as follows:

$$-\sigma = \frac{\Delta u_{\mathrm{I}} E}{2L\left(1 - \nu^2\right)} = -\frac{0.1 \text{ m} \times 1 \times 10^{10}\text{Pa}}{2 \times 100 \times (1 - 0.25^2)} = 5.3 \times 10^6\text{Pa} = 5.3 \text{ MPa}$$

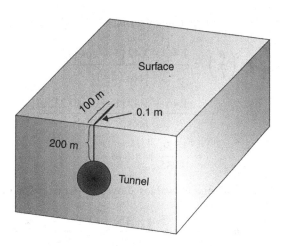

Fig. 9.17   Example 9.2. A tension fracture extends from a road tunnel to the surface.

This result is very reasonable. It indicates that the estimated tensile stress at the time of fracture formation was within the range of commonly estimated in-situ tensile strengths of solid rocks (0.5–6 MPa).

### Example 9.3

Problem

A tension fracture 50 m long in a pahoehoe lava flow opens during a rifting episode (cf. Fig. 1.1). The pahoehoe lava flow is 200 m thick, as in Fig. 13.14. The lava flow has a Young's modulus of 5 GPa. The dip dimension or depth of the fracture is 30 m, that is, it extends only partly into the lava flow from the surface of the rift zone. Calculate the likely opening or aperture of the fracture.

Solution

Since the fracture extends from one free surface, that is, the surface of the rift zone and partly into the elastic body hosting it, namely into the pahoehoe lava flow, it is a part-through crack. From Eq. (9.9) we have

$$\Delta u_{\mathrm{I}} = \frac{-4\sigma\,RV}{E}$$

We take the tensile stress $\sigma$ as being equal to the average tensile strength, that is, about 3 MPa. We could of course assume it to be either higher (up to at least 6 MPa) or lower (down to 0.5–1 MPa), but it is reasonable to use the average value. The depth or dip dimension $R$ is known as 30 m and Young's modulus of the lava flow as $5 \times 10^9$ Pa.

   To solve the problem, we need to determine the function $V = V\left(R/T\right)$. The parameter $T$ is the total thickness of the rock body hosting the fracture, which in this case is the pahoehoe lava flow, so that $T$ is 200 m (and $2T$ in the equation below, thus 400 m). Using radians ($360° = 2\pi$ radians), we find the function $V$ from Eq. (9.11) as follows:

$$V\left(\frac{R}{T}\right) = \frac{1.46 + 3.42\left[1 - \cos\left(\frac{\pi R}{2T}\right)\right]}{\left[\cos\left(\frac{\pi R}{2T}\right)\right]^2}$$

$$= \frac{1.46 + 3.42\left[1 - \cos(0.2356)\right]}{\left[\cos(0.2356)\right]^2} = \frac{1.46 + 3.42(1 - 0.972)}{0.945}$$

$$= \frac{1.556}{0.945} = 1.646$$

Using this result and the values above, we get from Eq. (9.9) the fracture opening as

$$\Delta u_\mathrm{I} = \frac{4\sigma R V}{E} = \frac{4 \times 3 \times 10^6 \mathrm{Pa} \times 30\ \mathrm{m} \times 1.646}{5 \times 10^9 \mathrm{Pa}} = 0.12\ \mathrm{m} = 12\ \mathrm{cm}$$

Thus, for the given tensile stress, the maximum opening displacement or aperture of this fracture would be 12 cm. If instead we had used the maximum tensile stress, 6 MPa, then the calculated maximum opening would have been twice this value, or 24 cm. These calculated openings are similar to those observed on small tension fractures formed during rifting episodes such as in the fissure swarms of Iceland.

---

### Example 9.4

#### Problem

Figure 9.18 shows part of a 25-km-long dextral strike-slip fault in the South Iceland Seismic Zone. At the location of this fault, the seismogenic layer is about 15 km thick and was ruptured when this fault slipped. If the average dynamic Young's modulus of the seismogenic layer is 100 GPa, its Poisson's ratio 0.25, and the stress drop (driving stress) associated with the fault slip is 3 MPa (a typical value), what would be the surface slip (displacement)?

#### Solution

A strike-slip fault that reaches the surface and extends through the entire seismogenic layer is normally modelled as a mode III crack (Fig. 9.19). The dip dimension of the fault or, which is here the same, the thickness of the seismogenic layer, denoted R, is the controlling dimension of the slip. Using Eq. (9.25) we calculate the slip as follows:

$$\Delta u_\mathrm{III} = \frac{4\tau_\mathrm{d}(1 + v)R}{E} = \frac{4 \times 3 \times 10^6 \mathrm{Pa} \times (1 + 0.25) \times 15\ 000\ \mathrm{m}}{1 \times 10^{11}\mathrm{Pa}} = 2.25\ \mathrm{m}$$

This fits very well with actual estimates of the fault displacement based on geometric considerations of the push ups and pull-aparts along the fault (Fig. 9.18). These estimates give the slip on the fault, which was presumably formed in a single earthquake, as about 2.67 m (Bergerat and Angelier, 2003).

**Fig. 9.18**   Example 9.4. Dextral strike-slip fault in the Holocene lava flows of the South Iceland Seismic Zone. View northwest.

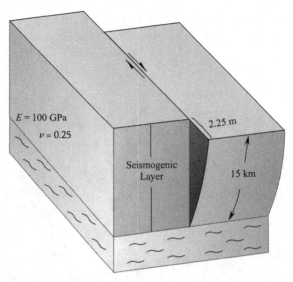

$E = 100$ GPa
$\nu = 0.25$

Seismogenic
Layer

2.25 m

15 km

**Fig. 9.19**　Example 9.4 (cf. Fig. 9.15).

There are several points that should be considered when making the above calculation of slip:

1. In this model, the maximum slip is supposed to occur at the surface (Fig. 9.19). This is a reasonable assumption because the Earth's surface is a free surface, so that, for constant loading, any fracture that reaches it tends to increase its displacement. This is simply because, at the free surface of an elastic half space, the elastic constraints on the displacement as they would be for a crack in a full space (say, an interior crack) are relaxed.
2. The elastic constants used are average values. If the layering is considered, the results could be different. For example, in the crust of the South Iceland Seismic Zone, the dynamic Young's modulus gradually increases with depth. Thus, in the uppermost 3–4 km, the dynamic modulus increases from about 10 GPa to about 60 GPa, but at greater depths it exceeds 100 GPa (Gudmundsson, 1988). Thus, the displacement might vary with depth and if the changes in the modulus are abrupt at contacts between crustal units or 'layers' (where a crustal 'layer' here has a thickness of the order of kilometres), then the slip may also vary abruptly. The layering is also likely to affect the surface displacement. The present example is thus an exercise showing how to use a simple analytical model to calculate slip for a homogeneous, isotropic crust, and does not take the crustal layering into account.
3. As a general rule, large strike-slip faults reaching the surface are modelled as mode III cracks, as is done here. The only exception would be if there was a free surface not only at the top of the crustal segment hosting the fault, but also at the bottom of the segment. Such conditions are met, for example, where a strike-slip fault is located within a large active caldera with a fluid magma chamber.

## Example 9.5

### Problem

A normal fault has a strike dimension of 5.8 km at the surface of a rift zone, similar to the one seen in Fig. 1.5 (cf. Fig. 9.10). Following an earthquake on this fault, aftershocks, that is, smaller earthquakes occurring after the main one and located close to or at the same fault plane, indicate that the fault is 3 km deep, that is, its dip dimension is 3 km. General seismic studies in the rift-zone segment hosting the fault show that the thickness of its seismogenic layer is 10 km.

(a) If the average Young's modulus of the upper 3 km of the crust is 25 GPa and Poisson's ratio is 0.25, what is the slip on the fault for a stress drop of 3 MPa?

(b) What is the slip on the fault if its dip dimension is 10 km and all other parameters and constants are the same as in (a)?

### Solution

(a) The dip dimension is smaller than the strike dimension, so that the dip dimension is the controlling dimension. Since the fault extends from the free surface of the rift zone and to a depth of 3 km in a seismogenic layer that is 10 km thick, the fault is modelled as a part-through crack. It follows that the model for the fault displacement is a mode II crack. From Eq. (9.24) we have

$$\Delta u_{\mathrm{II}} = \frac{4\tau_{\mathrm{d}} R V}{E}$$

To solve the problem, we must find the value of the function $V = V(R/T)$. The parameter $T$ is the total thickness of the rock body hosting the fracture; in this case the seismogenic layer, so that $T$ is 10 000 m and the dip dimension $R$ of the fault is 3000 m. Using radians ($360° = 2\pi$ radians), we find the value of the function $V$ from Eq. (9.11) thus:

$$V\left(\frac{R}{T}\right) = \frac{1.46 + 3.42\left[1 - \cos\left(\frac{\pi R}{2T}\right)\right]}{\left[\cos\left(\frac{\pi R}{2T}\right)\right]^2}$$

$$= \frac{1.46 + 3.42\left[1 - \cos\left(\frac{\pi \times 3000 \text{ m}}{2 \times 10\,000 \text{ m}}\right)\right]}{\left[\cos\left(\frac{\pi \times 3000 \text{ m}}{2 \times 10\,000 \text{ m}}\right)\right]^2} = \frac{1.8328}{0.7939} = 2.31$$

Using this result and the values given above for the other parameters, we obtain the slip or total displacement from Eq. (9.24) as

$$\Delta u_{\mathrm{II}} = \frac{4\tau_{\mathrm{d}} R V}{E} = \frac{4 \times 3 \times 10^6 \text{Pa} \times 3000 \text{ m} \times 2.31}{2.5 \times 10^{10} \text{Pa}} = 3.3 \text{ m}$$

This is a very reasonable value and similar to a common vertical displacement on normal faults during rifting events.

(b) If the dip dimension is 10 km and equal to the thickness of the seismogenic layer while the strike dimension is still 5.8 km, then the strike dimension is smaller and thus the controlling dimension. A through-crack mode III model would then be most appropriate. From Eq. (9.23) the slip is calculated as follows:

$$\Delta u_{\text{III}} = \frac{2\tau_d(1+v)L}{E} = \frac{2 \times 3 \times 10^6 \text{Pa} \times (1+0.25) \times 5800 \text{ m}}{2.5 \times 10^{10} \text{Pa}} = 1.7 \text{ m}$$

Thus, the displacement would be considerably smaller for a through-crack mode III model than for the part-through crack model.

# 9.11 Exercises

9.1 Describe, using illustrations, the three main ideal geometric crack models. Provide geological examples of each model.

9.2 What is a controlling dimension?

9.3 Consider a feeder dyke injected from a magma chamber at a depth of 3 km below the surface of a volcanic zone. The feeder dyke has a strike dimension of 1.5 km, that is, the associated volcanic fissure is 1.5 km long, and an estimated thickness (volcanic-fissure opening or aperture) of 1 m. What is the controlling dimension of this dyke?

9.4 Describe, using illustrations, the three ideal displacement modes of cracks. Provide geological examples for each mode.

9.5 Many normal faults in rift zones are open at the surface. For a large normal fault of this kind, what mode or modes are most appropriate for modelling its displacements?

9.6 A small tension fracture, a joint, at the surface of a pahoehoe lava flow is 1.4 m long and extends to an open contact between flow units at a depth of 2 m. The flow unit has a Young's modulus of 5 GPa and a Poisson's ratio of 0.25. During a rifting episode the joint is subject to a tensile stress of 3 MPa, in a direction perpendicular to the joint strike. Calculate the resulting maximum opening of the joint.

9.7 Calculate the maximum aperture (opening) of the joint in Exercise 9.6 if it is subject to a fluid overpressure of 20 MPa.

9.8 A tension fracture at the surface of a rift zone is hosted by a 100-m-thick hyaloclastite (basaltic breccia) layer. The tension fracture is 20 m long and 10 m deep and thus goes only partly into the hosting layer. Young's modulus of the hyaloclastite layer is 3 GPa. During a rifting episode the fracture is subject to a tensile stress of 3 MPa in a direction perpendicular to its strike. Calculate the resulting maximum opening of the tension fracture.

9.9 A sinistral strike-slip fault is 50 km long at the surface and hosted by a seismogenic layer that is 20 km thick. The average dynamic Young's modulus of the layer is 90 GPa and its Poisson's ratio is 0.27. During seismogenic faulting with an estimated stress drop of 6 MPa, the entire seismogenic layer is ruptured. Calculate the associated surface slip of the fault.

9.10 A normal fault dissects a collapse caldera associated with a shallow magma chamber with a top at a depth of 3 km. The strike dimension of the normal fault is 6 km and it extends to the depth of the magma chamber. If the host rock of the fault has an average dynamic Young's modulus of 25 GPa and a Poisson's ratio of 0.25, calculate the slip on the fault for a driving shear stress of 5 MPa.

# References and suggested reading

Amadei, B. and Stephansson, O., 1997. *Rock Stress and its Measurements*. New York: Chapman & Hall.

Bergerat, F. and Angelier, J., 2003. Mechanical behaviour of the Arnes and Hestfjall Faults of the June 2000 earthquakes in Southern Iceland: inferences from surface traces and tectonic model. *Journal of Structural Geology*, **25**, 1507–1523.

Beyer, W. H. (ed.), 1976. *Standard Mathematical Tables*, 24th edn. Cleveland, OH: CRC Press.

Bonafede, M. and Rivalta, E., 1999. On tensile cracks close to and across the interface between two welded elastic half-spaces. *Geophysical Journal International*, **138**, 410–434.

Broberg, K. B., 1999. *Cracks and Fracture*. London: Academic Press.

Cartwright, J. A., Trudgill, B. D., and Mansfield, C. S., 1995. Fault growth by segment linkage: an explanation for scatter in maximum displacement and trace length data from the Canyonlands Grabens of SE Utah. *Journal of Structural Geology*, **17**, 1319–1326.

Clark, R. M. and Cox, S. J. D., 1996. A modern regression approach to determining fault displacement-length scaling relationships. *Journal of Structural Geology*, **18**, 147–152.

Cowie, P. A. and Scholz, C. H., 1992. Displacement-length scaling relationships for faults: data synthesis and discussion. *Journal of Structural Geology*, **14**, 1149–1156.

Dawers, N. H., Anders, M. H., and Scholz, C. H., 1993. Growth of normal faults: displacement-length scaling. *Geology*, **21**, 1107–1110.

Gray, T. G. F., 1992. *Handbook of Crack Opening Data*. Cambridge: Abingdon.

Green, A. E. and Sneddon, I. N., 1950. The distribution of stress in the neighbourhood of a flat elliptical crack in an elastic body. *Proceedings of the Cambridge Philosophical Society*, **46**, 159–163.

Gudmundsson, A., 1987. Tectonics of the Thingvellir Fissure Swarm, SW Iceland. *Journal of Structural Geology*, **9**, 61–69.

Gudmundsson, A., 1988. Effect of tensile stress concentration around magma chambers on intrusion and extrusion frequencies. *Journal of Volcanology and Geothermal Research*, **35**, 179–194.

Gudmundsson, A., 2000. Fracture dimensions, displacements and fluid transport. *Journal of Structural Geology*, **22**, 1221–1231.

Gudmundsson, A., 2003. Surface stresses associated with arrested dykes in rift zones. *Bulletin of Volcanology*, **65**, 606–619.

Gudmundsson, A., 2008. Magma-chamber geometry, fluid transport, local stresses, and rock behaviour during collapse–caldera formation. In: Gottsman, J. and Marti, J. (eds.), *Caldera Volcanism: Analysis, Modelling and Response*. Developments in Volcanology 10. Amsterdam: Elsevier, pp. 313–349.

Haimson, B. C. and Rummel, F., 1982. Hydrofracturing stress measurements in the Iceland research drilling project drill hole at Reydarfjordur, Iceland. *Journal of Geophysics Research*, **87**, 6631–6649.

Hudson, J. A. and Harrison, J. P., 1997. *Engineering Rock Mechanics. An Introduction to the Principles*. Oxford: Elsevier.

Kanamori, H. and Anderson, D. L., 1975. Theoretical basis of some empirical relations in seismology. *Bulletin of the Seismological Society of America*, **65**, 1073–1095.

Kasahara, K., 1981. *Earthquake Mechanics*. New York: Cambridge University Press.

Kassir, M. and Sih, G. C., 1966. Three-dimensional stress distribution around an elliptical crack under arbitrary loadings. *Journal of Applied Mechanics*, **33**, 601–611.

Kassir, M. and Sih, G. C., 1975. *Three-dimensional Crack Problems*. Leyden, The Netherlands: Noordhoff.

McCalpin, J. P. (ed.), 2009. *Paleoseismology*, 2nd edn. New York: Academic Press.

Nicol, A., Watterson, J., Walsh, J. J., and Childs, C., 1996. The shapes, major axis orientations and displacement pattern of fault surfaces. *Journal of Structural Geology*, **18**, 235–248.

Nur, A., 1974. Tectonophysics: the study of relations between deformation and forces in the earth. In: *Advances in Rock Mechanics, Vol. 1, Part A*. Washington, DC: US National Committee for Rock Mechanics, National Academy of Sciences, pp. 243–317.

Odling, N. E., 1997. Scaling and connectivity of joint systems in sandstones from western Norway. *Journal of Structural Geology*, **19**, 1257–1271.

Paris, P. C. and Sih, G. C., 1965. Stress analysis of cracks. In: *Fracture Toughness Testing and its Application*. Philadelphia, PA: American Society for Testing of Materials, pp. 30–81.

Pollard, D. D., Segall, P., and Delaney, P.T., 1982. Formation and interpretation of dilatant echelon cracks. *Geological Society of America Bulletin*, **93**, 1291–1303.

Pollard, D. D. and Segall, P., 1987. Theoretical displacements and stresses near fractures in rock: with applications to faults, joints, veins, dikes and solution surfaces. In: Atkinson, B. (ed.), *Fracture Mechanics of Rock*. London: Academic Press, pp. 277–349.

Rice, J. R., 1980. The mechanics of earthquake rupture. In: Dziewonski, A. M. and Boschi, E. (eds.), *Physics of the Earth's Interior*. Amsterdam: North Holland, pp. 555–649.

Rippon, J. H., 1985. Contoured patterns of the throw and hade of normal faults in the Coal Measures (Westphalian) of north-east Derbyshire. *Proceedings of the Yorkshire Geological Society*, **45**, 147–161.

Schlische, R. W., Young, S. S., Ackermann, R. V., and Gupta, A., 1996. Geometry and scaling relations of a population of very small rift-related normal faults. *Geology*, **24**, 683–686.

Scholz, C. H., 1990. *The Mechanics of Earthquakes and Faulting*. New York: Cambridge University Press.

Schultz, R. A., 1995. Limits on strength and deformation properties of jointed basaltic rock masses. *Rock Mechancis and Rock Engineering*, **28**, 1–15.

Schultz, R. A., 1997. Displacement-length scaling for terrestrial and Martian faults: implications for Valles Marineris and shallow planetary grabens. *Journal of Geophysical Research*, **102**, 12 009–12 015.

Sibson, R.H., 1996. Structural permeability of fluid-driven fault-fracture meshes. *Journal of Structural Geology*, **18**, 1031–1042.

Sih, G. C. and Liebowitz, H., 1968. Mathematical theories of brittle fracture. In: Liebowitz, H. (ed.), *Fracture: An Advanced Treatise, Vol. 2*. New York: Academic Press, pp. 67–190.

Sneddon, I. N. and Lowengrub, M., 1969. *Crack Problems in the Classical Theory of Elasticity*. New York: Wiley.

Tada, H., Paris, P. C., and Irwin, G. R., 2000. *The Stress Analysis of Cracks Handbook*. Hellertown, PA: Del Research Corporation.

Tester, J. W. *et al.*, 2007. Impact of enhanced geothermal systems on the US energy supply in the twenty-first century. *Philosophical Transactions of the Royal Society*, **A365**, 1057–1094.

Tsang, C. F. and Neretnieks, I., 1998. Flow channeling in heterogeneous fractured rocks. *Reviews of Geophysics*, **36**, 275–298.

Valko, P. and Economides, M. J., 1995. *Hydraulic Fracture Mechanics*. New York: Wiley.

Vermilye, J. M. and Scholz, C. H., 1995. Relation between vein length and aperture. *Journal of Structural Geology*, **17**, 423–434.

Willemse, E. J. M., Pollard, D. D., and Aydin, A., 1996. Three-dimensional analysis of slip distributions on normal fault arrays with consequences for fault scaling. *Journal of Structural Geology*, **18**, 295–309.

Yeats, R. S., Sieh, K., and Allen, C. R., 1997. *The Geology of Earthquakes*. Oxford: Oxford University Press.

# Toughness and fracture mechanics

## 10.1 Aims

Rock strengths, such as tensile and shear strength, are one measure of its resistance to brittle failure. Another measure of rock resistance to brittle failure is the energy absorbed by the rock during fracture propagation. The critical energy absorbed is referred to as material toughness. A third measure of resistance to fracture is the stress intensity at the fracture tip that must be reached if the fracture is to propagate. The critical stress intensity is referred to as fracture toughness. Toughness is a basic concept in fracture mechanics that provides a physical framework for understanding many processes associated with rock fractures. The main aims of this chapter are to:

- Explain the concept of material toughness.
- Explain the concept of fracture toughness.
- Show how material and fracture toughness are related.
- Relate the concept of toughness to the initiation of a fracture.
- Introduce the concepts of process zone, fault core, and damage zone.
- Show how to calculate the toughness associated with extension fractures.
- Show how to calculate the toughness associated with faults.

## 10.2 Toughness

The toughness of a rock is a measure of its resistance to fracture. In other words, for a tough rock, large amounts of energy are needed to cause failure; much energy is absorbed by a tough rock as it fractures. This concept and the related concept of fracture toughness derive from materials science (Hull and Clyne, 1996; Chawla, 1998; Kobayashi, 2004) and fracture mechanics (Atkins and Mai, 1985; Broberg, 1999; Anderson, 2005).

The main characteristics of toughness may be summarised as follows:

- **Material toughness** is a measure of the energy absorbed in a material per unit area of crack in that material. It is also referred to as the critical strain energy release rate and has units of $J\ m^{-2}$, that is, the same as of energy (work) per unit area. The units $J\ m^{-2}$ can also be expressed as $N\ m^{-1}$, that is, as force per unit length of crack. Material toughness is thus sometimes referred to as **crack extension force**. One measure of toughness is the area under the stress–strain curve, that is, the elastic strain energy (Fig. 10.1). (Material-toughness data on rocks are given in Appendix E.2.)

- The **energy release rate** of a material is usually denoted by $G$ and its critical value, the material toughness, by $G_c$. In this book, energy release rate $G$ and material toughness $G_c$ will be used interchangeably, although the measured or estimated material toughness is strictly the critical energy release rate needed for a crack to propagate.
- **Fracture toughness**, usually denoted by the symbol $K_c$, is the critical **stress intensity factor** for a fracture to propagate. It has units of N m$^{-3/2}$ or Pa m$^{1/2}$. Since pascal (Pa) is a very small measure of stress, fracture toughness is normally given in MPa m$^{1/2}$. (Fracture-toughness data on rocks are given in Appendix E.2.)
- Fracture toughness varies positively with increasing volume of material at the fracture tip that deforms plastically or though microcracking (Fig. 10.2). The energy release rate and stress intensity, and therefore the material toughness and fracture toughness, are related. They do, however, provide different measures of the fracture resistance of a material and have different units.
- The energy release rate $G$ measures the energy per unit area of crack extension; the stress intensity factor $K$ measures the magnitude (intensity) of the stress field close to the crack tip.

A tough material is not necessary ductile. Many highly stiff materials are tough. Many composite materials, although stiff and light, are made tough, that is, resistant to fracture, through a special arrangement of the layers that constitute them. The high toughness of many composites is closely linked to their interfacial effects, such as debonding (delamination), whereby a crack propagates along the interface between two different materials and/or becomes arrested (Hull and Clyne, 1996; Blaint and Hutchinson, 2001; Tvergaard and Hutchinson, 2008).

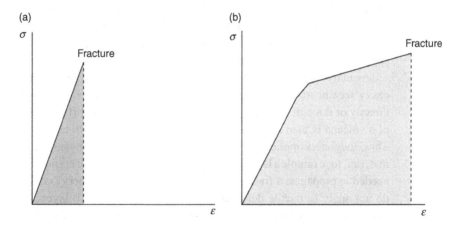

**Fig. 10.1** (a) Brittle material has a small area under the stress–strain curve before failure. That is to say, there is comparatively little elastic strain energy stored in the material before a fracture forms. (b) A tough material has a comparatively large area under the stress–strain curve before failure. That is to say, a comparatively large elastic strain energy is stored in the material before a fracture forms.

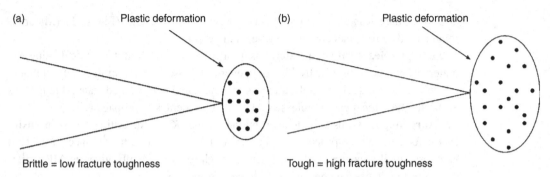

(a) Plastic deformation

(b) Plastic deformation

Brittle = low fracture toughness                    Tough = high fracture toughness

**Fig. 10.2** Fracture toughness varies positively with increasing volume of material at the fracture tip that deforms plastically or though microcracking. (a) If the volume is small, the toughness is low and the material brittle. (b) If the volume is large, the toughness is high and the material tough.

For the same reason, layered rocks, particularly those composed of layers with contrasting mechanical properties, may be tough (Fig. 10.3). Fractures commonly propagate along contacts between layers; for example, forming offset dykes (Fig. 10.4) or dyke–sill contacts (Fig. 10.5), or become arrested on meeting the contact (Fig. 8.10; Chapter 13). Propagation of a fracture along an interface gives rise to energy release rate in much the same way as for a crack in a homogeneous material. This energy, however, is not as readily available as for cracks in homogeneous materials because (i) the toughness of the interface is sensitive to the way it is formed, and (ii) interfacial cracks often propagate under mixed mode loading conditions (Balint and Hutchinson, 2001; Tvergaard and Hutchinson, 2008). This is in contrast with cracks in homogeneous materials which tend to propagate as mode I cracks. The energy release rate of an interfacial crack is greater when there is a shear stress component (mode II) added to the opening (mode I) component (Fig. 10.6). For a homogeneous, isotropic material with an interface, the interface tends to open up if its strength is less than 20% of the strength of the rest of the material (Hull and Clyne, 1996).

It is important to distinguish between toughness and ductile yield of layers. Here the focus is on the energy needed for a fracture to propagate through many layers. The results are applicable to fracture propagation in any layered rocks, not only volcanic (Fig. 10.3) but also sedimentary (Fig. 15.11) and metamorphic rocks (Fig. 10.6). The effects of layering are very easily seen in stratovolcanoes, where many of the injected dykes become arrested, either directly or through changing into sills (Figs. 10.4, 10.5, and 10.7). The material toughness of a volcano is then a measure of the energy associated with brittle failure of the edifice. Thus, toughness applies primarily to brittle or quasi-brittle behaviour; the more brittle the material, for example a layered rock mass or an entire volcanic edifice, the smaller the energy needed to propagate a fracture through the edifice. These energy considerations, however, do not apply to ductile deformation along weak (low-yield) layers, a widely considered mechanism for volcano spreading (Wooller *et al.*, 2004; Oehler *et al.*, 2005).

Fracture and material toughness have been analysed and estimated for various rock types (Fourney, 1983; Atkinson, 1987; Paterson and Wong, 2005). Most values for fracture toughness are 0.5–3 MPa $m^{1/2}$ and for material toughness 20–400 J $m^{-2}$ (Appendix E.2; Fourney, 1983; Atkinson and Meredith, 1987; Li, 1987; Balme *et al.*, 2004). They refer to

**Fig. 10.3**  Volcanic edifices, particularly stratovolcanoes, are commonly composed of layers with widely different mechanical properties, particularly different stiffnesses. Abrupt changes in stiffness at layer contacts encourages fracture deflection and arrest (Figs. 10.4–10.7). This may be one reason why fracture propagation to the surface, resulting in dyke-fed eruptions, earthquakes, or landslides, is more rare in stratovolcanoes, which are composed of layers with different mechanical properties, than in basaltic edifices, which are composed of layers with similar mechanical properties (Gudmundsson, 2009, 2010). Photo: Ines Galindo.

mode I cracks in small, essentially homogeneous laboratory rock samples. To obtain values appropriate for large parts of, or entire, volcanic edifices, large natural fractures such as faults and dykes must be used.

## 10.3  Fracture mechanics

### 10.3.1  Introduction

Toughness plays an important rock in the science of fracture mechanics, initiated by the work of Griffith (1920) and developed, first by Irwin and co-workers and later by numerous

**Fig. 10.4**    Dyke, about 60 cm thick, becomes deflected along the contact between pyroclastic layers in Tenerife. At the contact the dyke forms a sill that is as thin as 2 cm.

**Fig. 10.5**    Dyke–sill contact. The dyke (15 cm thick) becomes deflected into a sill (40 cm thick) at the contact between layers in the Tertiary lava pile in Southwest Iceland.

Fig. 10.6 Fracture deflection at contacts. For fracture deflection in sedimentary rocks, see Fig. 15.11. Generally, long and geometrically irregular fracture paths indicate that much energy was needed to propagate the fractures, that is, that the rocks had comparatively large material toughnesses at the time of fracture formation.

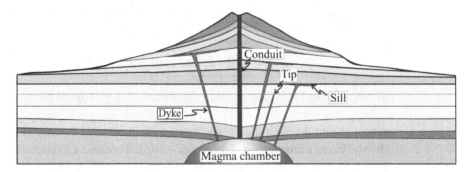

Fig. 10.7 In a stratovolcano the rock layers are commonly with widely different mechanical properties (Fig. 10.3). It follows that the material toughness of a stratovolcano is generally larger than that of a basaltic edifice, that is, it requires comparatively great energy to propagate any type of fracture through the stratovolcano. This is presumably one reason why large landslides and dyke-fed eruptions are more infrequent in typical stratovolcanoes than in basaltic edifices (Gudmundsson, 2009, 2010).

workers worldwide, during the past 60 years (Rossmanith, 1997). Today, fracture mechanics is a major field of science and engineering. Its application to rock fractures, however, has been limited (Rossmanith, 1983; Atkinson, 1987). One reason is that fracture mechanics was developed for, and is still mainly used in connection with, essentially homogeneous, isotropic materials. By contrast, rocks are generally heterogeneous and anisotropic, particularly at outcrop scale, which is the scale of interest for rock fractures in most geological processes (Figs. 10.3, 10.6, and 10.7).

Another reason for the limited use of fracture mechanics for analysing rock fractures is that fracture mechanics has very much focused on mode I fractures, that is, extension fractures. While these are very important in geological processes, faults are at least equally important, and for these fracture mechanics has not been applied much as yet.

The third reason for its limited use for rock fractures is that some of the basic concepts of fracture mechanics, such as material toughness and fracture toughness, have not been very easy to apply to rocks at an outcrop scale. These concepts are, of course, well adapted to laboratory rock samples, and many estimates of toughness have been made in the laboratory (Rossmanith, 1983; Atkinson, 1987). But it is not so easy to scale the laboratory results up to outcrops for analysing the associated fractures. This follows because rocks are normally heterogeneous and, in particular, layered. And the layering may have great effects on the toughness of rock masses as seen in outcrops through delamination (debonding), stress rotation, and other factors that affect fracture propagation, deflection, and arrest.

There is, however, a great similarity between the behaviour of layered rocks and some artificial materials, in particular composite materials. Many theories on toughness and failure of composite materials can therefore, with suitable modifications, be applied to crustal rocks and rock fractures. In this section we discuss some of these approaches and how they can be used to advance our understanding of the development of rock fractures. The discussion focuses primarily on mode I cracks, that is, extension fractures, but the principles apply to all fractures. For example, the opening or changes in aperture as a mode I fracture propagates, and associated work, can be also be formulated in terms of work related to shear displacement on mode II and mode III cracks.

## 10.3.2 The energy approach

Griffith's (1920) basic idea was that, for a fracture to propagate, the free energy (energy available to do mechanical work) of the solid body hosting the fracture (glass in his initial studies) should decrease during the fracture extension. This energy is then available to drive the fracture propagation. The energy is needed because a fracture can only propagate through generating new surfaces. And to form new surfaces, energy is needed, namely surface energy.

There are two main energy sources for generating new surfaces and thus for fracture propagation. As regards the rock body hosting a potential fracture, these sources are as follows:

- Internal **elastic strain energy**. This is the energy stored in the body when subject to elastic deformation. It is equal to the work done to produce the deformation, and thus also equal to the area under the elastic portion of the stress–strain curve (Fig. 10.1; Chapter 4). As is shown below, part of the elastic strain energy is released when a fracture forms and initiates in the body and can be used to generate new fracture surfaces.

- **Work** done on the body. The work done on the body is normally through loads applied to the body, such as forces, moments, stresses, displacements, or pressure, before fracture. When a mode I fracture propagates, it increases not only its length but also its aperture (Chapter 9), so that the points of attachment of the loads become displaced (move). Since work is defined as force times displacement, it follows that the loads perform work during their displacement. Part of this work can be used to generate new fracture surfaces.

In what follows, we will derive the general expressions of the material toughness and fracture toughness from energy principles. The derivation is somewhat simplified so as to make it easily understandable and, as indicated above, the focus is on mode I cracks. This simplification follows that commonly used in elementary textbooks in fracture mechanics and materials science.

### 10.3.3 Material toughness and fracture toughness

Consider an ideal mode I crack of length $a$ in a rock specimen of thickness $t$ (Fig. 10.8). We assume that the crack is propagating so that its tip is advancing and the crack length increasing. During any instant, the tip may advance so that the length of the crack increases by an amount $\delta a$ (Fig. 10.8). As the crack increases its length, the stress in the rock next to the crack relaxes (Fig. 10.9), that is, the rock next to the crack becomes less highly stressed than it was before crack propagation and thus loses elastic energy. This elastic energy could be the strain energy plus the work (or potential energy) if the walls or edges of the elastic body could move as the crack extended. However, when the walls are fixed, as here (Fig. 10.8), the elastic energy is the strain energy of the body.

For this case, the change in elastic strain energy $\delta U_0$ due to the extension of the crack by an amount $\delta a$ is given by

$$\delta U_0 = Gt\delta a \tag{10.1}$$

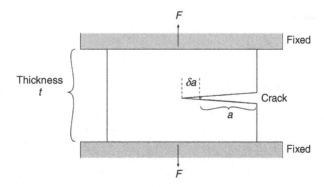

**Fig. 10.8**   Model for fracture propagation. The elastic plate containing the crack has a thickness $t$ and fixed (non-moving) boundaries.

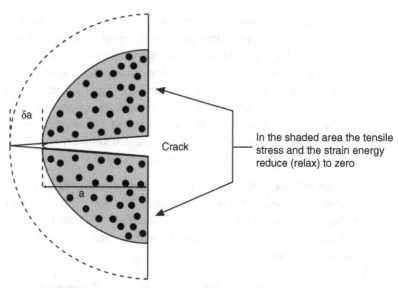

Fig. 10.9 Circular (or semi-circular) zone around a crack relaxes the associated tensile stress and thus loses its stored elastic strain energy.

where $G$ is the energy absorbed by unit area of crack extension, the crack area being $t\delta a$. The force $F$ (the load) generates stress $\sigma$ inside the rock specimen, producing strain $\varepsilon$. For uniaxial tension, as discussed here, the elastic strain energy per unit volume of rock specimen equals half the product of stress and strain, that is,

$$U_0 = \frac{1}{2}\sigma\varepsilon \qquad (10.2)$$

The elastic strain energy equals the area under the stress–strain curve in Fig. 10.1. Using the one-dimensional Hooke's law (Chapter 4),

$$\varepsilon = \frac{\sigma}{E} \qquad (10.3)$$

Equation (10.2) can be rewritten as

$$U_0 = \frac{1}{2} \times \sigma \times \frac{\sigma}{E} = \frac{\sigma^2}{2E} \qquad (10.4)$$

Now assume that as the crack grows, all the elastic strain energy in the semi-circular, dotted region in Fig. 10.9 is totally relaxed and thus reduces to zero. The area $A$ of the semicircle is

$$A = \frac{1}{2}\pi a^2 \qquad (10.5)$$

The strain energy relaxation is thus equal to

$$U_0 = \frac{1}{2}\pi a^2 t \frac{\sigma^2}{2E} \qquad (10.6)$$

Thus, when the crack extends by length $\delta a$ then the change in elastic strain energy $\delta U$ becomes

$$\delta U_0 = \frac{dU_0}{da}\delta a = \frac{\sigma^2}{2E}\frac{2\pi at}{2}\delta a \qquad (10.7)$$

Using Eqs. (10.1, 10.7), we get

$$Gt\delta a = \frac{\sigma^2}{2E}\frac{2\pi at}{2}\delta a \qquad (10.8)$$

Rearranging and solving for $G$, Eq. (10.8) reduces to

$$G = \frac{\sigma^2 \pi a}{2E} \qquad (10.9)$$

Here $G$ denotes the energy needed for crack extension, that is, for the onset of crack propagation. This is similar to Griffith's original result, but his more rigorous derivation gives the correct relation, namely

$$G = \frac{\sigma^2 \pi a}{E} \qquad (10.10)$$

Equation (10.10) is the correct formula for the energy release rate $G$ associated with crack propagation. This is the plane-stress formula, whereas the corresponding plane-strain formula is

$$G = \frac{\sigma^2 \left(1 - \nu^2\right) \pi a}{E} \qquad (10.11)$$

For a facture to propagate, the energy release rate of the rock much reach a certain minimum critical value. This critical value is the **material toughness** and is denoted by $G_c$. Thus, material toughness is the critical energy release rate for a crack to propagate. Rewriting Eq. (10.11), using the material toughness $G_c$ and solving for the tensile stress $\sigma$, we obtain the minimum tensile stress needed to initiate fracture propagation as

$$\sigma = \left(\frac{EG_c}{\pi a}\right)^{1/2} \qquad (10.12)$$

for plane stress. The corresponding formula for plane strain is

$$\sigma = \left(\frac{EG_c}{\pi(1 - \nu^2)a}\right)^{1/2} \qquad (10.13)$$

The plane-stress Eq. (10.12) may be rewritten as

$$\sigma\sqrt{\pi a} = \sqrt{EG_c} \qquad (10.14)$$

The term $\sigma\sqrt{\pi a}$ is the **stress intensity factor**, denoted by $K$. For a fracture to propagate, the stress intensity factor much reach a certain minimum critical value. The critical stress intensity factor for a fracture to propagate is the **fracture toughness**, denoted by $K_c$. We can solve Eq. (10.14) to obtain the fracture toughness, thus:

$$K_c = \sigma\sqrt{\pi a} \qquad (10.15)$$

# 10.4 Toughness of rock

The energy release rate $G$ and the stress intensity factor $K$ have been determined for many rock specimens or samples in the laboratory (Appendix E.2; Atkinson and Meredith, 1987). Typical values for the critical stress intensity factor, that is, fracture toughness, $K_c$, for mode I testing in most solid rock specimens at room temperature and pressure are 0.5–3 MPa m$^{1/2}$. Only rarely do rock specimens yield toughnesses as high as 5–20 MPa m$^{1/2}$ (Appendix E.2). Even at high temperatures and pressures, these values do not change much. For example, the laboratory fracture toughness of basaltic rocks (lava flows) from Iceland and Italy at a confining pressure as high as 30 MPa and temperature of 30–600°C, is between 1.4 and 3.8 MPa m$^{1/2}$ (Appendix E.2; Balme *et al.*, 2004).

To put these values into geological context, 30 MPa corresponds to crustal depths of 1.2–2 km. Temperatures of 600°C at that depth would normally only be reached close to magma chambers or recent igneous intrusions such as sills or dykes. At the high pressure of 60–100 MPa, some limestone and sandstone rock samples, though, have a fracture toughness of about 5 MPa m$^{1/2}$ (Appendix E.2; Atkinson and Meredith, 1987).

The energy release rate values of laboratory rock specimens under mode I testing are also low. The highest value quoted by Atkinson and Meredith (1987) is 1580 J m$^{-2}$, which is for a sandstone specimen normal to bedding. The lowest energy release rate value is 15 J m$^{-2}$, which is for a limestone specimen. The highest value quoted by Fourney (1983), for a basalt specimen, is 2298 J m$^{-2}$, whereas the lowest, for a limestone specimen, is about 8 J m$^{-2}$. However, most energy release rate values listed by these authors for mode I testing range between about 20 and 400 J m$^{-2}$ (Appendix E.2).

Similar results for the material toughness of rock samples under mode I loading are obtained through theoretical calculations. Using typical Young's moduli and Poisson's ratios for laboratory rock samples, the material toughness ranges between 2.5 and 900 J m$^{-2}$, with a typical value of about 100 J m$^{-2}$. Much higher material toughness values, however, are obtained for mode II and mode III loading. For example, Li (1987) provides a table for mode II testing, at various confining pressures, where most of the rock samples yield material toughness of the order of 0.01 MJ m$^{-2}$ (Appendix E.2).

Material toughness and fracture toughness have been determined experimentally for numerous materials. Ashby and Jones (2005) give typical laboratory values for many materials. A few typical values, all of which are for mode I loading, are given below:

- Water ice: material toughness, $G_c$, about 3 J m$^{-2}$; fracture toughness, $K_c$, about 0.2 MPa m$^{1/2}$.
- Pure ductile metals: material toughness, $G_c$, as high as 1 MJ m$^{-2}$; fracture toughness, $K_c$, as high as 200 MPa m$^{1/2}$.
- Man-made composites: material toughness, $G_c$, as high as 0.05–0.1 MJ m$^{-2}$; fracture toughness, $K_c$, as high as 50–100 MPa m$^{1/2}$.
- Mild steel: material toughness, $G_c$, about 0.1 MJ m$^{-2}$; fracture toughness, $K_c$, as high as 100 MPa m$^{1/2}$.

- Common wood, a natural composite: material toughness, $G_c$, 8–20 kJ m$^{-2}$ if measured perpendicular, but 0.5–2 1 kJ m$^{-2}$ if measured parallel, to the grain in the wood. Note that the units here are kJ m$^{-2}$, not MJ m$^{-2}$. Fracture toughness, $K_c$, 11–13 MPa m$^{1/2}$ if measured perpendicular, but 0.5–1 MPa m$^{1/2}$ if measured parallel, to the grain in the wood. As is explained below, the layering, such as the grain in the wood, increases the toughness or resistance to fractures that propagate perpendicular to the layers, but has less effects on fractures that propagate parallel with the layers.

All these materials, except water ice, have toughnesses that are generally much higher than those of typical rocks. All the results for $K$ and $G$ (and $K_c$ and $G_c$) are based on small laboratory rock specimens. One reason why fracture mechanics has not been applied much to rock fractures in nature is that these estimates are generally so low as to be unrealistic. When estimating the in-situ $K_c$ and $G_c$ as applicable to outcrop-scale or crustal-scale fractures, such as mineral veins, dykes, and seismogenic faults, a different approach is needed; the values obtained in the laboratory must be up-scaled. In subsequent sections, we derive formulas that allow realistic, outcrop-value estimates to be made of $K$ and $G$. But before that is possible, we must first discuss the concepts of process zone and damage zone and their implications for fracture development. Both concepts, and in particular the concept of a damage zone, are discussed in more detail in later chapters (Chapters 14 and 16). Here, we shall merely introduce them as a basis for up-scaling the toughness for rock fractures in the field.

## 10.5  Core, damage zone, and process zone

Field observations show that fault zones normally consist of two main structural units, a fault core and a fault damage zone (Fig. 10.10). The core takes up most of the fault displacement and is also referred to as the fault-slip zone. In major fault zones, the core may be as thick as tens of metres. It includes many small faults and fractures, but its most distinctive features are breccias, gouge, and other cataclastic rocks. The core rock is thus commonly crushed and altered into a soft material that can fail as brittle only during seismogenic faulting. As the core develops, its cavities and fractures become gradually filled with secondary minerals, but during fault slip the core has a granular-media structure at the millimetre or centimetre scale.

In major fault zones, the fault damage zone may be as thick as several kilometres (Chapters 14 and 16). The damage zone, also referred to as the transition zone, also contains breccias, but its characteristic features are fractures. The number of fractures in the damage zone increases, commonly in an irregular manner (Gudmundsson *et al.*, 2009) towards its contact with the core. The result is a general decrease in the effective Young's modulus towards the core (Fig. 10.10(d)). In an active fault, the fault gouge and breccia of the core itself would also normally have a very low Young's modulus, similar to that of clay, weak sedimentary rocks, or pyroclastic rocks such as tuff. As the core and the damage zone change and generally become thicker with time (Fig. 10.10), so do their mechanical properties, in particular their Young's moduli (Gudmundsson *et al.*, 2009).

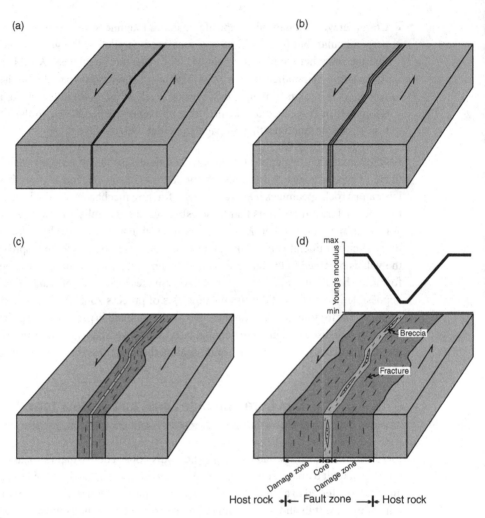

(a)

(b)

(c)

(d)

**Fig. 10.10**  Development of the damage zone and the fault core as the associated fault zone evolves. Both the core and the damage zone gradually become thicker (a–d) over time. The internal structure of the fault zone, and the associated variation in Young's modulus, are indicated in part (d).

As the fault displacement increases so does the thickness of the fault zone, so that there will be gradually thicker zones of brecciated and fractured fault rocks around the fault plane (Fig. 10.10). Because the fault rocks are normally soft in comparison with the host rocks, the stiffness of an active fault zone generally decreases with time.

It follows from Eqs. (10.10) and (10.11) that the energy release rate increases when Young's modulus decreases. That is to say, the material toughness is generally higher for fracture propagation in fault zones with comparatively low Young's modulus. This means that more energy is needed to propagate a fracture within a comparatively compliant fault zone than in the stiff host rock; the fault zone is, in comparison with the brittle host rock,

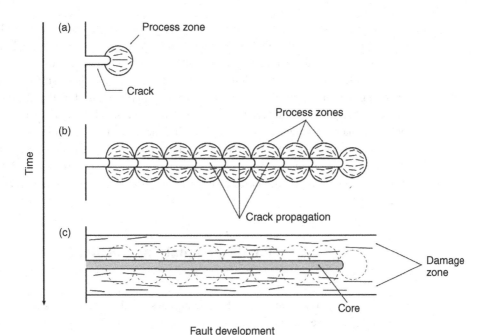

Fault development

**Fig. 10.11**  The process zone at the tip of a fracture at a shallow depth is primarily related to microcracking (a). As the fault zone evolves (Fig. 10.10), the process zone gradually becomes a part of the core and the damage zone (b and c).

comparatively tough. And the fault-zone toughness is partly related to the process of fracture extension, namely to the process zone or process region around the fracture tip at any instant during its propagation.

The concepts of a damage zone and process zone are by some regarded as identical. For rocks, particularly for faults, they are, however, better regarded as distinct. For rocks, the process zone or process region may be defined as follows (Figs. 10.2 and 10.11):

- The **process zone** is a zone or region of, mostly, microcracking at the tip of a fracture. The process zone corresponds to the plastic zone at a fracture tip in metals. In metals, the plastic deformation at the fracture tip is primarily through the movement of dislocations. At great depths in major fault zones, the process zone in rocks is also, partly at least, a plastic region. However, at shallower depths with low temperatures and pressures, there is little plastic deformation at the fracture tip but rather microcracking. It is thus primarily this zone of microcracking which is the process zone in rocks at shallow depths.

The process zone exists at the tip of a propagating fracture at any instant (Figs. 10.2 and 10.11). Because of all the microcracks in the process zone, not only is the Young's modulus of the zone comparatively small, but in addition the potential energy for fracture propagation is partly dissipated on many of the microcracks in the zone. This means that for each advancement of the main fracture tip, which requires energy, there will also be formation of, or slip on existing, numerous microcracks. The formation of and slip on the microcracks absorbs energy, thereby increasing the material toughness of the rock.

For a fault zone, the process zone gradually becomes a part of the core and the damage zone (Fig. 10.11). As the fault zone continues to grow and increase its total displacement, both the core and the damage zone gradually become thicker (Fig. 10.10). This increasing thickness may increase the volume of material where energy is dissipated through micro-cracking and/or plastic deformation before any major fault slip, thereby increasing the material toughness of the fault zone. The damage zone and the core may thus be one reason why the material toughness of a major fault zone, as estimated below (Example 10.3), is so much higher than that estimated for laboratory rock samples using single cracks. Before we can estimate the material toughness of outcrop-scale fractures such as dykes and fault zones we must discuss the basic equations for stress intensity factors and energy release rates.

## 10.6  Basic equations of $K$ and $G$

Here we summarise some of the basic equations for the stress intensity factor $K$ and for the energy release rate $G$ for various loading conditions. The equations are needed in the subsequent sections of this chapter. The loadings considered are uniform tensile stress $\sigma$ (minus sign omitted here), uniform shear stress $\tau$, and uniform fluid overpressure $p_o$. Young's modulus is, as always, denoted by $E$ and Poisson's ratio by $\nu$. Half the crack strike dimension is denoted by $a$. The subscripts I–III for $K$ and $G$ denote the mode of crack displacement.

We start with the formulas for the stress intensity factors. For the loading conditions given, the mode I stress intensity factors for (driving) tensile stress $\sigma$ (Eq. 10.15) is

$$K_I = \sigma (\pi a)^{1/2} \tag{10.16}$$

the mode II stress intensity factor for uniform driving shear stress $\tau_d$ is

$$K_{II} = \tau_d (\pi a)^{1/2} \tag{10.17}$$

the mode III stress intensity factor for uniform driving shear stress $\tau_d$ is

$$K_{III} = \tau_d (\pi a)^{1/2} \tag{10.18}$$

and the mode I stress intensity factor for uniform driving pressure (overpressure) $p_o$ is

$$K_I = p_o (\pi a)^{1/2} \tag{10.19}$$

In many applications of stress intensity factors, the situation is more complex than indicated by these equations. For example, the applied stress (or overpressure) may vary along the fracture, that is, it may be non-uniform. Also, Eqs. (10.16)–(10.19) apply to a central crack in an infinite elastic plate. If the plate is of a finite size, or if the crack is located at one of the edges of the plate, then a correction is needed through a geometric factor. Using this correction, Eq. (10.16) becomes

$$K_I = Y\sigma (\pi a)^{1/2} \tag{10.20}$$

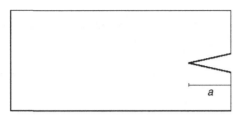

Fig. 10.12 Edge crack (part-through crack) of length $a$ in an elastic body.

where $Y$ is the geometric factor and $a$ is the full length of an edge crack (Fig. 10.12) or the half length of a central crack (Fig. 10.13). Similar geometric corrections can be made for Eqs. (10.17)–(10.19).

Here we provide some stress-intensity formulas for cracks in finite, elastic bodies. Many more results are given by Tata *et al.* (2000). Again, the stress is uniform for all these formulas. Some of the formulas are exact, others are approximations. All are widely used in practical applications.

The first solution is for a central crack of strike dimension (length) $2a$ in an elastic sheet or a very thin plate of finite width $W$ (Fig. 10.13). The model I stress intensity factor for constant tensile stress $\sigma$ is

$$K_{\mathrm{I}} = \sigma (\pi a)^{1/2} \left( \frac{W}{\pi a} \tan \frac{\pi a}{W} \right)^{1/2} \tag{10.21}$$

This equation is suitable for many rock fractures. For example, through-the-thickness tension fractures at the surface of a lava flow (Chapter 9). The tension fracture is then supposed to extend through the lava flow, or a flow unit of a pahoehoe flow, and meet with an essentially free surface at the contact between the top flow/unit and the next one (Fig. 9.13).

Many fractures and faults consist of segments (Figs. 12.3–12.5), some of which are offset whereas others are in line, that is, collinear. Consider an array of regularly spaced collinear cracks in an infinite elastic plate subject to remote tensile stress $\sigma$. Each crack has an equal strike dimension $2a$ and the distance between the centres of each pair of cracks is $d$ (Fig. 10.14). The model I stress intensity factor is then given by the formula

$$K_{\mathrm{I}} = \sigma (\pi a)^{1/2} \left( \frac{d}{\pi a} \tan \frac{\pi a}{d} \right)^{1/2} \tag{10.22}$$

which is analogous to Eq. (10.21), the only difference being that the distance $d$ is substituted for the plate width $W$. In fact, in the original derivation of Eq. (10.21) the solution for an infinite array of collinear cracks, Eq. (10.22), was used (Sanford, 2003). Equation (10.22) is very useful for analysing segmented fractures and faults and for assessing whether they are likely to link up during tensile loading.

A part-through crack, an edge crack, is also important in the analysis of rock fractures (Figs. 9.1, 9.5, and 10.12). This applies to cracks located in a semi-infinite plate. A typical example would be a fracture that is small in relation to the fissure swarm within which it is located and extends partly into the surface lava flow (Figs. 1.1 and 9.13).

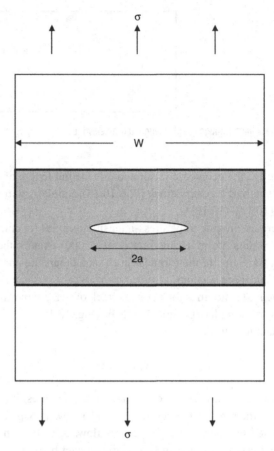

**Fig. 10.13** Mode I fracture of strike dimension or length $2a$ is subject to tensile stress $\sigma$ as the only loading. The width of the elastic plate containing the crack is denoted by $W$.

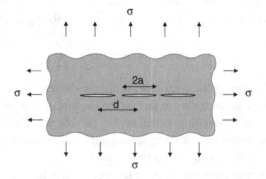

**Fig. 10.14** Set of collinear cracks, each of length (strike dimension) $2a$ and with the centres separated by the distance $d$, subject to biaxial tensile stress $\sigma$.

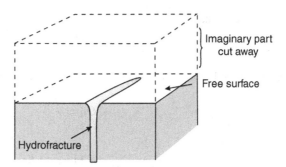

When a mode I crack, here a hydrofracture, reaches a free surface, for example the Earth's surface, it expands and increases its aperture. This is one reason why feeder-dykes may increase their thicknesses on approaching and reaching the Earth's surface (Figs. 1.3, 1.6; Geshi *et al.*, 2010).

For an edge crack of dip dimension $a$ (the depth to which the crack extends from the free surface), assuming that the dip dimension of the crack is small in relation to the total thickness of the hosting layer, and subject to tensile stress $\sigma$, the mode I stress intensity factor is

$$K_I = 1.12\sigma(\pi a)^{1/2} \tag{10.23}$$

The equation for an edge crack (Eq. 10.23) differs from that for a through crack (Eq. 10.16) only in the factor 1.12. The 12% difference in $K_I$ between these crack geometries is attributable to the effects of the free surface. The edge-crack geometry is obtained by cutting the plate containing the through crack through the middle of that crack. It follows that the edge crack opening is larger because it is less restrained than the through crack (Fig. 10.15). The through crack opens into a flat ellipse, whereas the opening of the edge crack is more wedge-shaped, or triangular (Geshi *et al.*, 2010).

Penny-shaped (circular) interior cracks (Chapter 9) are widely used to model rock fractures. If the crack is of radius $a$ and located in an infinite, elastic body subject to remote tensile stress $\sigma$, the mode I stress intensity factor is given by

$$K_I = \sigma(\pi a)^{1/2}\left(\frac{2}{\pi}\right) = 2\sigma\left(\frac{a}{\pi}\right)^{1/2} \tag{10.24}$$

If the penny-shaped crack is loaded by internal fluid overpressure, rather than external tension, the mode I stress intensity factor is given by

$$K_I = 2p_0\left(\frac{a}{\pi}\right)^{1/2} \tag{10.25}$$

Clearly, Eq. (10.25) is entirely analogous to Eq. (10.24) when the overpressure $p_o$ is substituted for the tensile stress $\sigma$.

The energy release rates can also be obtained for various loadings and crack geometries. The energy release rate can be expressed in terms of stress intensity factors. Here we present some of the fundamental results. For a mode I crack, the energy release rate $G_I$ is given by the formulas

$$G_I = \frac{K_I^2}{E} \tag{10.26}$$

for plane stress, and

$$G_{\mathrm{I}} = \frac{\left(1 - v^2\right) K_{\mathrm{I}}^2}{E} \tag{10.27}$$

for plane strain. For a mode II crack, the energy release rate $G_{\mathrm{II}}$ is given by the formulas

$$G_{\mathrm{II}} = \frac{K_{\mathrm{II}}^2}{E} \tag{10.28}$$

for plane stress, and

$$G_{\mathrm{II}} = \frac{\left(1 - v^2\right) K_{\mathrm{II}}^2}{E} \tag{10.29}$$

for plane strain. For a mode III crack, the energy release rate $G_{\mathrm{III}}$ is given by the formula

$$G_{\mathrm{III}} = \frac{(1 + v) K_{\mathrm{III}}^2}{E} \tag{10.30}$$

Many fractures, particularly those that are deflected or otherwise have irregular paths, propagate, at least during parts of their paths, as mixed-mode fractures. For a mixed-mode crack, the total energy release rate $G_{\mathrm{total}}$ is given by the formulas

$$G_{\mathrm{total}} = G_{\mathrm{I}} + G_{\mathrm{II}} + G_{\mathrm{III}} = \frac{1}{E}\left[ K_{\mathrm{I}}^2 + K_{\mathrm{II}}^2 + K_{\mathrm{III}}^2(1 + v) \right] \tag{10.31}$$

for plane stress, and

$$G_{\mathrm{total}} = G_{\mathrm{I}} + G_{\mathrm{II}} + G_{\mathrm{III}} = \frac{(1 - v^2)K_{\mathrm{I}}^2}{E} + \frac{(1 - v^2)K_{\mathrm{II}}^2}{E} + \frac{(1 + v)K_{\mathrm{III}}^2}{E} \tag{10.32}$$

for plane strain.

## 10.7 Toughness in terms of fracture displacements

To apply the principles of fracture mechanics to rock fractures, particularly at the scale of typical outcrops, it is helpful to rewrite the formulas for energy release rate and stress intensity factors in terms of some quantities that are easily measured in outcrops. The strike and dip of a fracture, of course, are normally measurable in the field, and so, sometimes, are the strike and dip dimensions. It is rare that both dimensions can be measured directly, but one (for example, the strike dimension or surface trace length) can often be measured and the other (for example, the dip dimension from geodetic and seismic studies) can often be inferred. When the strike and dip dimension are known, then the controlling dimension can be estimated.

Apart from the controlling dimension, perhaps the most important measurable quantity associated with a fracture in the field is its displacement. For an extension fracture, the displacement is reflected in its aperture, if an open fracture, and in its thickness, if a filled fracture such as a mineral vein or a dyke. For a shear fracture, the displacement of a dip-slip fault, for example, is reflected in the total throw or, in a single earthquake, the slip. The displacement is not always measurable. The aperture may be either very small, for instance, or very large and partly related to collapse. In both cases it may be difficult to measure it accurately. Similarly, the lack of clear marker layers (or marker horizons) in many rocks, particularly large plutons, makes it difficult, or practically impossible, to determine the displacement on a shear fracture.

For many rock fractures, however, both the controlling dimension and the displacement can be measured or inferred. These data can then be used to estimate the toughness of the rocks that host the fractures. In this section, we derive equations where the energy release rate is presented in terms of controlling dimensions and fracture displacements. These equations are useful for estimating the toughness of outcrop rocks, as they are important for understanding fluid-filled reservoirs of various types. The results can also be used to provide crude estimates of the effective toughness of large-scale structures, such as fault zones and volcanoes.

## 10.7.1 Estimates of $G_I$

The derivation uses the example of hydrofractures, but the results are equally applicable to tension fractures. For a hydrofracture with an overpressure $p_o$, the stress intensity factor is given by Eq. (10.19):

$$K_I = p_0(\pi a)^{1/2}$$

And, for a plane-strain condition, the energy release rate is given by Eq. (10.27):

$$G_I = \frac{(1 - v^2)\, K_I^2}{E}$$

Substituting Eq. (10.19) for $K_I$ in Eq. (10.27), we obtain an expression for the energy release rate of a hydrofracture in terms of its fluid overpressure:

$$G_I = \frac{P_0^2(1 - v^2)\pi a}{E} \tag{10.33}$$

Equation (10.33) is often helpful when estimating the likely energy release rate associated with the propagation of a hydrofracture with a known overpressure. However, in the field, a more useful expression can be obtained from the aperture or thickness of the hydrofracture. Because the aperture/thickness normally varies as in a flat ellipse (Chapter 9), little error is involved if the average of the measured apertures/thicknesses is taken as the maximum aperture/thickness of the hydrofracture.

From Eq. (9.16) we have

$$P_0 = \frac{\Delta u_I E}{2L\left(1 - v^2\right)} \tag{10.34}$$

which we substitute for $P_o$ in Eq. (10.33) to get

$$G_I = \frac{\left[\frac{\Delta u_I E \pi}{2(1-v^2)L}\right]^2}{E} \frac{(1 - v^2)\pi a}{E} \tag{10.35}$$

Using $2a$ for $L$ and simplifying, Eq. (10.35) reduces to

$$G_I = \frac{E \Delta u_I^2 \pi}{16(1 - v^2)a} \tag{10.36}$$

for plane strain, and

$$G_I = \frac{E \Delta u_I^2 \pi}{16a} \tag{10.37}$$

for plane stress. Thus, Eqs. (10.36) and (10.37) make it possible to determine the energy release rate $G_I$ of an extension fracture such as a hydrofracture provided we can estimate the controlling dimension of the fracture, $2a$, as well as its (maximum) aperture, $\Delta u_I$.

## 10.7.2  Estimates of $G_{II}$

For a mode II crack subject to a constant driving shear stress $\tau_d$, the stress intensity factor is given by Eq. (10.17):

$$K_{II} = \tau_d (\pi a)^{1/2}$$

And, for a plane-strain condition, the energy release rate is given by Eq. (10.29):

$$G_{II} = \frac{(1 - v^2) K_{II}^2}{E}$$

Substituting Eq. (10.17) for $K_{II}$ in Eq. (10.29), we obtain an expression for the energy release rate of a mode II crack in terms of the driving shear stress:

$$G_{II} = \frac{(1 - v^2) \left[ \tau_d (\pi a)^{1/2} \right]^2}{E} = \frac{\tau_d^2 (1 - v^2) \pi a}{E} \tag{10.38}$$

Using Eq. (9.27) with $2a$ substituted for $L$, we have

$$\Delta u_{II} = \frac{4 \tau_d (1 - v^2)a}{E} \tag{10.39}$$

Solving for $\tau_d$, we get

$$\tau_d = \frac{E \Delta u_{II}}{4(1 - v^2)a} \tag{10.40}$$

Substituting Eq. (10.40) for $\tau_d$ in Eq. (10.38), we get

$$G_{II} = \frac{\left[ \frac{E \Delta u_{II}}{4(1 - v^2)} \right]^2}{E} \frac{(1 - v^2) \pi a}{E} \tag{10.41}$$

which reduces to

$$G_{II} = \frac{E \Delta u_{II}^2 \pi}{16(1 - v^2)a} \tag{10.42}$$

for plane strain, and

$$G_{II} = \frac{E \Delta u_{II}^2 \pi}{16a} \tag{10.43}$$

for plane stress. Using Eqs. (10.42) and (10.43) we can determine the energy release rate $G_{\mathrm{II}}$ of a fault that can be modelled by a mode II crack. This applies to many dip-slip faults, and some strike-slip faults. Some dip-slip faults, and many strike-slip faults, however, are better modelled as mode III cracks.

### 10.7.3  Estimates of $G_{\mathrm{III}}$

For a mode III crack subject to a constant driving shear stress $\tau_d$, the stress intensity factor is given by Eq. (10.18):

$$K_{\mathrm{III}} = \tau_{\mathrm{d}}(\pi a)^{1/2}$$

And, for a plane-strain condition, the energy release rate is given by Eq. (10.30):

$$G_{\mathrm{III}} = \frac{(1 + v)\,K_{\mathrm{III}}^2}{E}$$

Substituting Eq. (10.18) for $K_{\mathrm{III}}$ in Eq. (10.30), we obtain an expression for the energy release rate of a mode III crack in terms of the driving shear stress:

$$G_{\mathrm{III}} = \frac{\tau_{\mathrm{d}}^2(1 + v)\pi a}{E} \tag{10.44}$$

Using Eq. (9.26) with $a$ substituted for $R$, we have

$$\Delta u_{\mathrm{III}} = \frac{4\tau_{\mathrm{d}}(1 + v)a}{E} \tag{10.45}$$

Solving for $\tau_d$, we get

$$\tau_{\mathrm{d}} = \frac{E\,\Delta u_{\mathrm{III}}}{4(1 + v)a} \tag{10.46}$$

Substituting Eq. (10.46) for $\tau_d$ in Eq. (10.44), by analogy with Eq. (10.41), we get

$$G_{\mathrm{III}} = \frac{E\,\Delta u_{\mathrm{III}}^2\pi}{16(1 + v)a} \tag{10.47}$$

Thus, we have established formulas (Eqs. (10.36), (10.37), (10.42), (10.43), and (10.47)) that allow us to calculate the energy release rate, and estimate the material toughness (and, thereby also, the fracture toughness), of different types of rock fractures as measured in the field. We shall use these results in some examples in this chapter, as well as in subsequent chapters, to quantify processes associated with fracture displacements.

## 10.8  Summary

- Material toughness is a measure of the energy absorbed by a material, such as rock, during fracture propagation. The greater the material toughness of a rock, the greater the energy that must be used to propagate a fracture through the rock; that is, the more fracture resistant is the rock. Material toughness has units of J m$^{-2}$.

- Fracture toughness is a measure of the critical stress intensity at the fracture tip needed for fracture propagation. It has units of MPa m$^{1/2}$. Fracture toughness and material toughness are related and are both a measure of the resistance of a rock (or other solid materials) to fracture propagation.
- Measurements of toughnesses of laboratory specimens normally yield small values. Common fracture-toughness values are 0.5–3 MPa m$^{1/2}$, whereas the material-toughness values are 20–400 J m$^{-2}$ for extension fractures. Field estimates, however, yield much larger values. For example, regional dykes yield fracture-toughness values of the order of 500 MPa m$^{1/2}$, and the corresponding material-toughness values are of the order of 10 MJ m$^{-2}$. Similar in-situ toughnesses are obtained for shear fractures, that is, faults.
- Layered rock masses are generally more resistant to fracture propagation than non-layered rock masses. This follows partly because the fractures tend to become deflected at the contacts between the layers, so that their propagation paths become more irregular (zigzag) and longer than paths of fractures that propagate in approximately homogeneous, isotropic rock masses. Increasing path length and propagation in different modes (parallel and perpendicular to contacts, for example) requires greater energy; thus, layered rocks absorb more energy than non-layered during fracture propagation and have thus normally greater material toughness.

## 10.9  Main symbols used

| | |
|---|---|
| $A$ | area |
| $a$ | half length (strike dimension) of a crack |
| $\delta a$ | increase in crack length |
| $d$ | distance between the centres of a pair of collinear cracks |
| $E$ | Young's modulus (modulus of elasticity) |
| $G$ | strain energy release rate during crack extension (propagation) |
| $G_c$ | critical strain energy release rate, material toughness |
| $G_{tot}$ | total strain energy release rate during crack extension (propagation) |
| $G_I$ | strain energy release rate during the extension of a mode I crack |
| $G_{II}$ | strain energy release rate during the extension of a mode II crack |
| $G_{III}$ | strain energy release rate during the extension of a mode III crack |
| $K$ | stress intensity factor |
| $K_c$ | critical stress intensity factor or plane-stress fracture toughness; for modes I, II, and III the plane-strain fracture toughnesses are denoted by $K_{Ic}$, $K_{IIc}$, and $K_{IIIc}$, but these are not used in the book |
| $K_I$ | stress intensity factor for a mode I crack |
| $K_{II}$ | stress intensity factor for a mode II crack |
| $K_{III}$ | stress intensity factor for a mode III crack |
| $p_0$ | fluid overpressure in a fracture |
| $t$ | thickness of plate or specimen containing a crack |

| | |
|---|---|
| $U_0$ | strain energy density (strain energy per unit volume) |
| $\delta U_0$ | increase in elastic strain energy |
| $\Delta u_{\mathrm{I}}$ | maximum aperture (twice the opening displacement) of a mode I crack |
| $\Delta u_{\mathrm{II}}$ | maximum aperture of a mode II crack |
| $\Delta u_{\mathrm{III}}$ | maximum aperture of a mode III crack |
| $W$ | finite width of a plate with a crack |
| $Y$ | geometric correction factor for stress-intensity factors |
| $\varepsilon$ | normal strain |
| $\nu$ | Poisson's ratio |
| $\pi$ | 3.1416 . . . |
| $\sigma$ | tensile loading, tensile stress |
| $\tau_d$ | driving shear stress associated with fault slip or displacement |

## 10.10 Worked examples

### Example 10.1

#### Problem

A wide but thin glass plate is loaded by a tensile stress of 40 MPa. If the material toughness of the glass is 10 J m$^{-2}$ and its Young's modulus is 70 GPa, find the maximum size of a crack that can exist in the glass without fracture propagation and failure.

#### Solution

Since the glass plate is wide and thin, the loading is in plane stress. From Eq. (10.9) we have

$$G = \frac{\sigma^2 \pi a}{E}$$

Rewriting this equation and solving for the half-length of the crack, we get

$$a = \frac{EG}{\pi \sigma^2}$$

Here Young's modulus $E = 70$ GPa, $G$ is the critical energy release rate $G_c$ for crack propagation, that is, the material toughness 10 J m$^{-2}$, and $\sigma$ is the tensile stress, 40 MPa. Using the present equation, we get the maximum half-length of the crack as

$$a = \frac{EG_c}{\pi \sigma^2} = \frac{7 \times 10^{10}\,\mathrm{Pa} \times 10\,\mathrm{Jm}^{-2}}{3.1416 \times (4 \times 10^7\,\mathrm{Pa})^2} = 0.00014\mathrm{m} = 0.14\,\mathrm{mm}$$

This is half the length of the crack. The total length is thus 0.000 28 m = 0.28 mm.

## Example 10.2

### Problem

For a single crystal of quartz the measured fracture toughness is 0.32 MPa m$^{1/2}$ and the material toughness 1 J m$^{-2}$. A laboratory specimen of granite, of which typically 20–40% is quartz, has a fracture toughness of 1.74 MPa m$^{1/2}$ and a material toughness of 56 J m$^{-2}$. Young's modulus of the quartz crystal is 100 GPa, and that of the granite specimen 60 GPa. Both are subject to tensile stress of 10 MPa.

(a) Calculate the maximum crack length that the quartz crystal can contain before failure, using first the fracture toughness formulation and then the material toughness formulation. Assume that the loading is in plane stress.
(b) Do the same as in (a) for the granite specimen.
(c) Explain the difference between the toughnesses of quartz and granite.

### Solution

(a) From Eq. (10.15) we have

$$K_c = \sigma (\pi a)^{1/2}$$

Rewriting and solving for the half-length of the crack $a$, we get

$$a = \frac{K_c^2}{\sigma^2 \pi} = \frac{(3.2 \times 10^5 \text{Pa} \times \text{m}^{0.5})^2}{(1 \times 10^7 \text{Pa})^2 \times 3.1416} = 0.0003 \text{ m} = 0.3 \text{ mm}$$

This is half the crack length, so that the total length $2a = 0.6$ mm.
Using material toughness ($G_c$), from Eqs. (10.12) and (10.14) we have

$$\sigma (\pi a)^{\frac{1}{2}} = (E G_c)^{\frac{1}{2}}$$

When each side of Eq. (10.14) is squared, we can solve for $G_c$ and obtain

$$G_c = \frac{\sigma^2 \pi a}{E} \qquad (10.48)$$

which we give a new equation number because although it is identical with Eq. (10.10), Eq. (10.48) explicitly includes the material toughness $G_c$. Rearranging and solving for the half-length of the crack $a$, we get

$$a = \frac{E G_c}{\pi \sigma^2} = \frac{1 \times 10^{11} \text{ Pa} \times 1 \text{Jm}^{-2}}{3.1416 \times (1 \times 10^7 \text{ Pa})^2} = 0.0003 \text{ m} = 0.3 \text{ mm}$$

The total length is thus $2a = 0.6$ mm, that is, exactly the same (as it should be) as when we used the fracture-toughness formulation. Thus, for the given loading conditions, the maximum length of a crack that the crystal can contain without fracturing is 0.6 mm.

(b) We proceed as in (a) so that we first use Eq. (10.15) in a rewritten form to calculate the crack half-length, thus:

$$a = \frac{K_c^2}{\sigma^2 \pi} = \frac{(1.74 \times 10^6 \text{Pa} \times \text{m}^{0.5})^2}{(1 \times 10^7 \text{ Pa})^2 \times 3.1416} = 0.01 \text{ m} = 10 \text{ mm}$$

The total length $2a$ is thus 20 mm.
From Eq. (10.48) solved for $a$ we have

$$a = \frac{EG_c}{\pi\sigma^2} = \frac{6 \times 10^{10}\text{Pa} \times 56\text{Jm}^{-2}}{3.1416 \times (1 \times 10^7\text{Pa})^2} = 0.01\text{m} = 10\text{mm}$$

So, again, the total length $2a$ is 20 mm.

(c) Thus, the results for the granite indicate that the maximum length of a crack that the rock can contain without failure, for the given loading conditions and properties, is about 50 times larger than a quartz crystal can contain for the same loading conditions. This is in general agreement with the observation that rock specimens composed largely of a certain crystal (here quartz) typically have an order of magnitude greater material toughness $G_c$ than the crystal itself. There are several reasons for this, the main ones including:

- The process zone in the rock specimen is larger and takes up much more energy than the one in the crystal.
- Because of the grain boundaries in the rock specimen, the crack path is much more complex than in the crystal, so that in the specimen the crack propagates partly in a mixed mode. Mixed mode implies that more energy is needed to propagate the crack than would be required in pure mode I propagation (Eq. 10.32), and also a longer path, and thus a greater material toughness than is needed for pure mode I propagation in the crystal.

## Example 10.3

### Problem

A dextral strike-slip fault 50 km long is hosted by a crustal segment with an average dynamic Young's modulus of 85 GPa and a Poisson's ratio of 0.25. During seismogenic slip on the fault, the entire 20-km-thick seismogenic layer is ruptured. The stress drop (driving stress) associated with the slip is 3 MPa.

(a) Calculate the energy release rate for the slip. Assume that the fault dissects the entire seismogenic layer (Chapter 9).
(b) Calculate as in (a) but assume here that the bottom of the seismogenic layer is in the roof of a fluid magma chamber, such as in a rift zone or a large collapse caldera (Chapter 9).

### Solution

(a) For a strike-slip fault that gradually changes into a ductile shear zone with depth (Figs. 9.15 and 9.19), the depth or dip dimension is the controlling dimension, so that the appropriate crack model is mode III. From Eq. (10.47) we have

$$G_{III} = \frac{E\Delta u_{III}^2 \pi}{16(1 + v)a}$$

Here all the parameters are known except for the surface slip, $\Delta u_{\mathrm{III}}$. From Eq. (10.45) the surface slip is calculated as follows:

$$\Delta u_{\mathrm{III}} = \frac{4\tau_{\mathrm{d}}(1+v)a}{E} = \frac{4 \times 3 \times 10^6 \, \mathrm{Pa} \times (1+0.25) \times 20000 \, \mathrm{m}}{8.5 \times 10^{10} \, \mathrm{Pa}} = 3.53\mathrm{m}$$

From Eq. (10.47) we then get the strain energy release rate as

$$G_{\mathrm{III}} = \frac{E\Delta u_{\mathrm{III}}^2 \pi}{16(1+v)a} = \frac{8.5 \times 10^{10} \, \mathrm{Pa} \times (3.53\mathrm{m})^2 \times 3.1416}{16 \times (1+0.25) \times 20000 \, \mathrm{m}} = 8.3 \times 10^6 \mathrm{Jm}^{-2}$$

Alternatively, we could use Eq. (10.44), in which case we do not need to calculate the surface slip, and obtain the energy release rate as

$$G_{\mathrm{III}} = \frac{\tau_{\mathrm{d}}^2(1+v)\pi a}{E} = \frac{(3 \times 10^6 \, \mathrm{Pa})^2 \times (1+0.25) \times 3.1416 \times 20000 \, \mathrm{m}}{8.5 \times 10^{10} \, \mathrm{Pa}}$$

$$= 8.3 \times 10^6 \mathrm{Jm}^{-2}$$

This gives the same result, as it should do.

(b) If the bottom of the seismogenic layer coincides with the top of a fluid magma chamber, it follows that there is a free surface at the bottom and at the top of the crustal segment hosting the fault. For this condition, the dip dimension is effectively infinite, and the smaller effective dimension, and thus the controlling dimension, is the strike dimension of the fault. In this case, a mode II crack is the most appropriate model for the fault. Since the crustal segment considered is 20 km thick (the thickness of the crust above the magma chamber) but more than 50 km wide (the strike dimension of the fault), it is much wider than it is thick, so that a plane-stress formulation is suitable. From Eq. (10.43) we then get the strain-energy release rate as

$$G_{\mathrm{II}} = \frac{E\Delta u_{\mathrm{II}}^2 \pi}{16a} = \frac{8.5 \times 10^{10} \, \mathrm{Pa} \times (3.53 \, \mathrm{m})^2 \times 3.1416}{16 \times 25000 \, \mathrm{m}} = 8.3 \times 10^6 \mathrm{Jm}^{-2}$$

Here we notice that since the strike dimension $2a = 50\,000$ m, then $a = 25\,000$ m. Alternatively, the strain-energy release rate could be calculated directly from Eq. (10.38) without the use of the estimated surface slip, thus:

$$G_{\mathrm{II}} = \frac{\tau_{\mathrm{d}}^2(1-v^2)\pi a}{E} = \frac{(3 \times 10^6 \, \mathrm{Pa})^2 \times 3.1416 \times 25000 \, \mathrm{m}}{8.5 \times 10^{10} \, \mathrm{Pa}} = 8.3 \times 10^6 \mathrm{Jm}^{-2}$$

Thus, the results are very consistent. However, this is so in this case primarily because the thickness of the seismogenic layer, that is, the dip dimension of the fault (20 km) is very similar to half the strike dimension of the fault (25 km). By contrast, if the strike dimension was 100 km instead of 50 km, while the dip dimension was the same (20 km), then the strain-energy release rate would be different depending on whether the fault was modelled as a mode II or a mode III crack. Generally, however, the mode III model is the most appropriate except when, as in (b) above, there is a free surface at the top and bottom of the crustal segment hosting the fault.

**Example 10.4**

Problem

At the level of exposure, a typical strike dimension/thickness ratio of regional dykes in Iceland is about 1000. This means that a dyke 10 km long is typically about 10 m thick. This also agrees roughly with the estimated thickness of 9 m for the 11-km-long feeder formed during the Krafla Fires in North Iceland (Figs. 3.4–3.6). Consider the 6-m-thick dyke in Fig. 8.1. We assume that at its depth of exposure its strike dimension is about 6 km, whereas the strike dimension is likely to increase at deeper crustal levels (Gudmundsson, 1990). Regional dykes are mostly derived from magma reservoirs located at depths of 10–20 km below the surface of the active rift zone at their time of emplacement. Assume that the strike dimension of the dyke in Fig. 8.1 is 6 km at the depth of exposure, and that Young's modulus at this depth was 10 GPa and Poisson's ratio 0.25 at the time of dyke emplacement. Calculate the energy release rate and the stress-intensity factor at this crustal depth during dyke emplacement.

Solution

We assume here that the dyke formed in a single injection. For a dyke that has a strike dimension of 6 km and a dip dimension of at least 10 km, the correct model is plane-strain mode I crack. From Eq. (10.36) the energy release rate is

$$G_{\mathrm{I}} = \frac{E \Delta u_{\mathrm{I}}^2 \pi}{16(1 - v^2)a} = \frac{1 \times 10^{10}\mathrm{Pa} \times (6\mathrm{m})^2 \times 3.1416}{16(1 - 0.25^2) \times 3000\,\mathrm{m}} = 2.43 \times 10^7 \mathrm{Jm}^{-2} = 24.3\,\mathrm{MJ\,m}^{-2}$$

Rewriting Eq. (10.27) in the form

$$K_{\mathrm{I}} = \left( \frac{E G_{\mathrm{I}}}{1 - v^2} \right)^{1/2} = \left( \frac{1 \times 10^{10}\mathrm{Pa} \times 2.43 \times 10^7 \mathrm{Jm}^{-2}}{1 - 0.25^2} \right)^{1/2}$$

$$= 4.93 \times 10^8\,\mathrm{Pa\,m}^{1/2} = 493\,\mathrm{MPa\,m}^{1/2}$$

Both the value of $G_{\mathrm{I}}$ and $K_{\mathrm{I}}$ are several orders of magnitude larger than the corresponding laboratory values, for reasons discussed in the text.

# 10.11 Exercises

10.1 Define material toughness. Give its units and indicate its relation to the concepts of crack-extension force and energy release rate.

10.2 Define fracture toughness. Give its units and indicate its relation to stress-intensity factor. Provide a formula that shows how fracture toughness and material toughness are related.

10.3 What is the process zone of a propagating fracture? How does the process zone relate to the fault core and damage zone of a fault zone?

10.4  Write the basic (plane stress where appropriate) equations (modes I–III) for (a) stress-intensity factors and (b) energy release rates for loading by tension, shear, and fluid overpressure.

10.5  What is the geometric factor $Y$ and when is it needed?

10.6  A thin plate loaded in tension (plane stress) has a Young's modulus of 100 GPa and a material toughness of 20 J m$^{-2}$. On inspection, it is found that the plate contains a crack with a total length (strike dimension) of 0.8 mm. How much tensile stress can the plate sustain before the crack begins to propagate?

10.7  The fracture toughness of a rock specimen is 2 MPa m$^{1/2}$. If the specimen is subject to tensile (plane) stress of 15 MPa, what is the maximum length of a crack that the specimen can contain before failure?

10.8  Estimate the critical energy release rate, the material toughness, for the specimen in Exercise 10.7 if its Young's modulus is 70 GPa.

10.9  A 150-km-long sinistral strike-slip fault is hosted by a crustal segment with an average dynamic Young's modulus of 100 GPa and a Poisson's ratio of 0.27. During seismogenic fault slip the entire seismogenic layer, 25 km thick, is ruptured. The fault zone gradually changes into a shear zone below the lower boundary of the seismogenic layer. If the stress drop during the seismogenic faulting is 5 MPa, calculate the energy release rate associated with the slip.

10.10  A 14-m-thick regional dyke has a strike dimension of 20 km and a depth of origin of 9 km below the present exposure at 1 km below the original surface of the rift zone. This means that the dip dimension of the dyke from the present level of exposure to the top of the source magma reservoir is 9 km. The Young's modulus of the host rock at the level of exposure is 10 GPa. If the dyke formed in a single magma injection, calculate the associated energy release rate.

# References and suggested reading

Anderson, T. L., 2005. *Fracture Mechanics: Fundamentals and Applications*, 3rd edn. London: Taylor & Francis.

Ashby, M. F. and Jones, D. R. H., 2005. *Engineering Materials 1*, 3rd edn. Amsterdam: Elsevier.

Atkins, A. G. and Mai, Y.W., 1985. *Elastic and Plastic Fracture*. Chichester: Horwood.

Atkinson, B. K. (ed.), 1987. *Fracture Mechanics of Rock*. London: Academic Press.

Atkinson, B. K. and Meredith, P. G., 1987. Experimental fracture mechanics data for rocks and minerals. In: Atkinson, B. K. (ed.), *Fracture Mechanics of Rock*. London: Academic Press, pp. 477–525.

Balint, D. S. and Hutchinson, J. W., 2001. Mode II edge delamination of compressed films. *ASME Journal of Applied Mechanics*, **68**, 725–730.

Balme, M. R., Rocchi, V., Jones, C., Sammonds, P. R., Meredith, P. G., and Boon, S., 2004. Fracture toughness measurements on igneous rocks using a high-pressure, high-temperature rock fracture mechanics cell. *Journal of Volcanology and Geothermal Research*, **132**, 159–172.

Broberg, K. B., 1999. *Cracks and Fracture*. New York: Academic Press.

Carmichael, R. S., 1989. *Practical Handbook of Physical Properties of Rocks and Minerals*. Boca Raton, Boston, MA: CRC Press.

Chawla, K. K., 1998. *Composite Materials: Science and Engineering*, 2nd edn. Berlin: Springer-Verlag.

Dahlberg, T. and Ekberg, A., 2002. *Failure, Fracture, Fatigue*. Studentlitteratur, Lund, Sweden.

Delaney, P. and Pollard, D. D., 1981. Deformation of host rocks and flow of magma during growth of minette dikes and breccia-bearing intrusions near Ship Rock, New Mexico. *US Geological Survey Professional Paper*, **1202**.

Fourney, W. L., 1983. Fracture control blasting. In: Rossmanith, H. P. (ed.), *Rock Fracture Mechanics*. New York: Springer-Verlag, pp. 301–319.

Geshi, N., Kusumoto, S., and Gudmundsson, A., 2010. The geometric difference between non-feeders and feeder dikes. *Geology*, **38**, 195–198.

Griffith, A. A., 1920. The phenomena of rupture and flow in solids. *Philosophical Transactions of the Royal Society of London*, **A221**, 163–198.

Griffith, A. A., 1924. Theory of rupture: In: Biezeno, C. B. and Burgers, J. M. (eds.), *Proceedings of the First International Congress on Applied Mechanics*. Delft: Waltman, pp. 55–63.

Gudmundsson, A., 1990. Emplacement of dikes, sills, and magma chambers at divergent plate boundaries. *Tectonophysics*, **176**, 257–275.

Gudmundsson, A., 2004. Effects of Young's modulus on fault displacement. *Comptes Rendus Geoscience*, **336**, 85–92.

Gudmundsson, A., 2009. Toughness and failure of volcanic edifices. *Tectonophysics*, **471**, 27–35.

Gudmundsson, A., 2010. Deflection of dykes into sills and magma-chamber formation. *Tectonophysics* (on line October 2009).

Gudmundsson, A., Simmenes, T. H., Larsen, B., and Philipp, S. L., 2009. Effects of internal structure and local stresses on fracture propagation, deflection, and arrest in fault zones. *Journal of Structural Geology*, doi:10.1016/j.jsg.2009.08.013.

Hull, D. and Clyne, T. W., 1996. *An Introduction to Composite Materials*, 2nd edn. Cambridge: Cambridge University Press.

Jin, Z. H. and Johnson, S. E., 2008. Magma-driven multiple dike propagation and fracture toughness of crustal rocks. *Journal of Geophysics Research*, **113**, B03206.

Janssen, M., Zuidema, J., and Wanhill, R., 2004. *Fracture Mechanics*, 2nd edn. Abingdon, UK: Spon Press.

Kobayashi, T., 2004. *Strength and Toughness of Materials*. Berlin: Springer-Verlag.

Larsen, B., Grunnaleite, I., and Gudmundsson, A., 2009. How fracture systems affect permeability development in shallow-water carbonate rocks: an example from the Gargano Peninsula, Italy. *Journal of Structural Geology*, doi:10.1016/jsg.2009.05.009

Li, V. C., 1987. Mechanics of shear rupture applied to earthquake zones. In: Atkinson, B. K. (ed.), *Fracture Mechanics of Rock*. London: Academic Press, pp. 351–428.

Li, H. X. and Xiao, X. R., 1995. An approach on mode-I fracture toughness anisotropy for materials with layered microstructures. *Engineering and Fracture Mechanics*, **52**, 671–683.

Nasseri, M. H. B. and Mohanty, B., 2008. Fracture toughness anisotropy in granitic rocks. *International Journal of Rock Mechanics and Mining Science*, **45**, 167–193.

Nasseri, M. H. B., Mohanty, B., and Young, R. P., 2006. Fracture toughness measurements and acoustic activity in brittle rocks. *Pure and Applied Geophysics*, **163**, 917–945.

Newhall, C. G. and Dzurisin, D., 1988. Historical unrest of large calderas of the world. *US Geological Survey Bulletin*, **1855**.

Oehler, J. F., de Vries, B. V. W., and Labazuy, P., 2005. Landslides and spreading of oceanic hot-spot and arc basaltic edifices on low strength layers (LSLs): an analogue modelling approach. *Journal of Volcanology and Geothermal Research*, **144**, 169–189.

Parker, A.P., 1981. *The Mechanics of Fracture and Fatigue*. London: Spon.

Paterson, M. S. and Wong, T. W., 2005. *Experimental Rock Deformation –the Brittle Field*, 2nd edn. Berlin: Springer-Verlag.

Rice, J. R., 2006. Heating and weakening of faults during earthquake slip. *Journal of Geophysical Research*, **111**, B05311.

Rice, J. R. and Cocco, M., 2007. Seismic fault rheology and earthquake dynamics. In: Handy, M. R., Hirth, G., and Horius, N. (eds.), *Tectonic Faults: Agents of Chance on a Dynamic Earth*. Cambridge, MA: The MIT Press, pp. 99–137.

Rivalta E. and Dahm, T., 2006. Acceleration of buoyancy-driven fractures and magmatic dikes beneath the free surface. *Geophysical Journal International*, **166**, 1424–1439.

Rossmanith, H. P. (ed.), 1983. *Rock Fracture Mechanics*. New York: Springer-Verlag, pp. 301–319.

Rossmanith, H. P. (ed.), 1997. *Fracture Research in Retrospect: An Anniversary Volume in Honour of George R. Irwin's 90th Birthday*. Rotterdam: Balkema.

Rubin, A. M. and Pollard, D. D., 1987. Origins of blade-like dikes in volcanic rift zones. *US Geological Survey Professional Paper*, **1350**, pp. 1449–1470.

Sanford, R. J., 2003. *Principles of Fracture Mechanics*. Upper Saddle River, NJ: Prentice-Hall.

Schultz, R. A., 1995. Limits on strength and deformation properties of jointed basaltic rock masses. *Rock Mechancis and Rock Engineering*, **28**, 1–15.

Shah, S. P., Swartz, S. E., and Ouyang, C., 1995. *Fracture Mechanics of Concrete: Applications of Fracture Mechanics to Concrete, Rock, and Other Quasi-Brittle Materials*. New York: Wiley.

Shulka, A. (ed.), 2006. *Dynamic Fracture Mechanics*. London: World Scientific.

Sneddon, I. N. and Lowengrub, M., 1969. *Crack Problems in the Classical Theory of Elasticity*. New York: Wiley.

Tada, H., Paris, P. C., and Irwin, G.R., 2000. *The Stress Analysis of Cracks Handbook*, 3rd edn. New York: American Society of Mechanical Engineers.

Tibaldi, A., Bistacchi, A., Pasquare, F. A., and Vezzoli, L., 2006. Extensional tectonics and volcano lateral collapses: insights from Ollague volcano (Chile-Bolivia) and analogue modelling. *Terra Nova* **18**, 282–289.

Tvergaard, V. and Hutchinson, J. W., 2008. Mode III effects on interface delamination. *Journal of the Mechanics and Physics of Solids*, **65**, 215–229.

Wooller, L., de Vries, B. V. W, Murray, J. B., Rymer, H., and Meyer, S., 2004. Volcano spreading controlled by dipping substrata. *Geology*, **32**, 573–576.

# 11 Field analysis of extension fractures

## 11.1 Aims

What measurements can we make in the field so as to be able to use the analytical framework developed in earlier chapters? In particular, how do we analyse extension fractures in the field? The main aims of this chapter are to:

- Give field examples of the main types of extension fractures.
- Illustrate how the most important field measurements of extension fractures are made.
- Provide typical field data on tension fractures.
- Provide typical field data on joints.
- Provide typical field data on mineral veins.
- Provide typical field data on dykes.
- Provide data on man-made hydraulic fractures.
- Illustrate the connection between extension fractures and faults.
- Outline the general methods of field measurements of fractures.

## 11.2 Types of extension fractures

Tension fractures are formed by absolute tension, so that they are mostly confined to the shallow parts of the crust (Chapter 8). Hydrofractures are partly or entirely generated by an internal fluid overpressure and can form at any crustal depth if there is high fluid pressure. Tension fractures include many large tension fractures in rift zones, as well as many joints. Hydrofractures are fluid-driven rock fractures that include many joints, mineral veins, and dykes. Man-made fractures injected under fluid overpressure into reservoir rocks to increase their permeabilities, commonly referred to as hydraulic fractures, are also hydrofractures. In many hydrofractures, for example fractures formed by gas, oil, and ground water, the fluid may disappear after the fracture has formed. Other hydrofractures, such as dykes, are driven open by fluids that solidify in the fracture once it is formed.

The main types of extension fractures are as follows:

- Tension fractures.
- Many joints.
- Many mineral veins.

- All dykes.
- All inclined sheets.
- All sills.
- All artificial hydraulic fractures.

In the following sections, we consider how to measure these various types of hydrofractures in the field, and how the analytical framework developed in the earlier chapters can be used to throw light on the physical processes and parameters associated with the fracture development.

## 11.3 Tension fractures

The mechanics and depth of propagation of tension fracture are discussed in Chapter 8. These fractures form when the tensile stress, that is, negative $\sigma_3$, reaches the tensile strength of the rock (Eq. 8.1). In situ, most solid rocks have tensile strengths 0.5–6 MPa, so that the tensile stress involved in their formation is generally in this range.

Consider a **tension fracture** in a rift zone (Fig. 11.1). As indicated, the attitude and aperture of the fracture are comparatively easy to measure. The aperture varies along the

**Fig. 11.1**    Tension fracture in the Holocene rift zone of Southwest Iceland. The fracture aperture is as much as 10 m. The fracture is partly filled with ground water, but the water uses the fracture only as a channel and did not participate in its formation.

strike dimension of the fracture, but several measurements near its central part yield an average that can be used when modelling the fracture. The maximum aperture of this fracture is about 10 m, its strike dimension is around 2000 m, and it is located in a 300-m-thick Holocene pahoehoe lava flow. When calculating the likely tensile stress associated with the formation of this fracture, there are two main factors that must be considered, namely:

- The controlling dimension of the fracture.
- Young's modulus and Poisson's ratio of the host rock.

The controlling dimension is likely to be the strike dimension. This conclusion is based on the following observations. First, the surface lava flow hosting the fracture contains numerous open contacts between flow units, many of which may have arrested the downward propagation of the fracture and act as a free surface at the lower edge of the fracture (Fig. 9.13). Second, the pahoehoe lava flow is underlain by thick layers of hyaloclastite, basaltic breccias, of mechanical properties dissimilar to those of the pahoeheo lava flow. This contact is likely to be weak or open, as is common for such contacts in the Quaternary lava pile in Iceland, and would thus be likely to act as a sort of free surface at the lower edge of the fracture. Thus, the 1100 m surface length of the fracture, its strike dimension, may be assumed to be its controlling dimension.

As Poisson's ratio of most solid rocks is between 0.2 and 0.3, a standard value is taken as 0.25 (Appendix D). Small variations in Poisson's ratio do not significantly affect the results, which is why we use the value of 0.25, as in most of the calculations and models in this book. However, Young's modulus can vary widely. In a typical basaltic lava pile, such as in Iceland, Young's modulus may vary by three orders of magnitude, from the compliant soil and scoria layers between the lava flows, to the dense and stiff central parts of the lava flows. For Young's Holocene pahoehoe lava flows, the static Young's modulus may be between 1 and 10 GPa, the average being close to 5 GPa (Gudmundsson, 1988).

Using a through-crack model with the strike dimension as the controlling dimension (Chapter 9), and taking the Young's modulus of the hosting lava flow as 5 GPa and its Poisson's ratio as 0.25, the tensile stress associated with the fracture formation (Fig. 11.1) is estimated at 13 MPa (Example 11.1). This estimated tensile stress value is higher than typical in-situ tensile strengths of solid rocks, which is 0.5–6 MPa, although occasional in-situ values may be as high as 9 MPa, such as at a depth of 9 km in the KTB drill hole in Germany (Amadei and Stephansson, 1997). The highest laboratory tensile strengths exceed 20 MPa and some are close to 30 MPa (Appendix E; Jumikis, 1979; Amadei and Stephansson, 1997).

Other surface tension fractures, however, may be modelled either as through cracks or as part-through cracks (Fig. 11.2). In general, to model a tension fracture at the surface, the following parameters must be estimated: its strike dimension, its likely dip dimension, its aperture, and the host-rock properties. The tension fracture in Fig. 11.2 is also hosted by a several-hundred-metres thick Holocene pahoehoe lava flow resting on soft Quaternary sediments. For both the fractures (Figs. 11.1 and 11.2) the strike dimensions and apertures are easily determined. However, to decide if the fractures are likely to be through cracks or

**Fig. 11.2** Tension fracture, a fissure (opening about 1 m), in a Holocene pahoehoe lava flow in the Krafla Fissure Swarm in North Iceland (Chapter 3). The open tension fracture was subsequently used as a part of a path for lava to the surface during the 1975–1984 Krafla Rifting Episode.

part-through cracks, we must have information on the geology of the areas. In particular, the thickness of the lava flows hosting the tension fractures must be estimated as well as the nature of the layers on which the lava flows rest. If those layers, as in these cases (Figs. 11.1 and 11.2), are of very different mechanical properties (here soft sediments and altered, soft hyaloclastites) from those of the lava flows, then the contact between the lava flows and the layers are likely to act as a sort of free surface, and the appropriate model for the fractures is a through crack.

# 11.4 Joints

A **joint** is a rock fracture with no visible displacement parallel with, but a small displacement normal to, the fracture plane (Fig. 11.3). Joints are thus primarily extension fractures. If there is clear displacement parallel with the fracture plane, the fracture is regarded as a shear fracture, that is, as a fault. However, the plane-parallel displacement may exist, even if it is not visible in the field. Hence the qualifying word 'visible' in the definition of a joint.

**Fig. 11.3**    Part of a 30-m-thick sill in the Quaternary lava pile of Southwest Iceland. The person provides a scale.

There are two main reasons why there may be no visible plane-parallel displacement in an outcrop even though the fracture is a shear fracture. First, the displacement may be too small to be seen by the naked eye in the field. Such a displacement might be detectable under the microscope, but in fracture studies in the field it is obviously not practical to sample each joint that is a suspected shear fracture for analysis under the microscope. Second, the host rock may lack suitable marker layers or horizons. Many joints occur in massif rocks, such as intrusions (Fig. 11.3) or thick lava flows (Figs. 11.4 and 11.5), where suitable marker layers are absent. For example, many columnar joints show evidence of subsequently generated shear stresses based on slickensides and minerals, although the rock normally lacks layering or other markers that could indicate the shear displacement.

Thus, some fractures classified as joints are not necessarily pure extension fractures, but rather hybrid fractures and best modelled as mixed-mode cracks. Cross-cutting relationships, however, indicate that many, and perhaps most, joins are primarily extension fractures and therefore should, to a first approximation, be modelled as such. Here we focus on joints as extension fractures.

Like all extension fractures, joints are either tension fractures or hydrofractures. Our definition of a joint implies that the opening displacement normal to the fracture plane is 'small', which would normally mean maximum openings of a few centimetres and generally much less than those of large tension fractures (fissures) in rift zones (Figs. 8.2, 11.1, and 11.2). Thus, in this book tension fractures in rift zones with apertures exceeding about 5 cm are not classified as joints. A good example of joints formed as extension fractures are

**Fig. 11.4**    Quaternary lava flow in South Iceland. The lower part forms regular columnar joints, whereas the upper part forms irregular, cube-jointed structure. The person provides a scale.

**Fig. 11.5** Holocene volcanic plug or neck (about 30 m tall) in Northeast Iceland. The irregular columnar jointing indicates that ground water in the crater cone to which this plug was a feeder affected the cooling.

columnar (cooling) joints in lava flows and intrusions (Figs. 11.3 and 11.4). Columnar joints form by cooling and shrinkage of the intrusion. Ideally, the columns should be hexagonal in plan view (Jaeger, 1968; DeGraff and Aydin, 1987), but more commonly they are pentagons (with five sides) or tetragons (with four sides). As the intrusion cools and shrinks, the extension fractures form.

External fluids play a role in the formation of columnar joints. This is clear when considering the formation of cube-jointed lavas (Fig. 11.4) and the irregularities in the joints in many lava flows and intrusions such as plugs (Fig. 11.5). It is generally accepted that the cube-jointed tops of otherwise regularly jointed lava flows are related to a more rapid cooling because of flow of water on top of the cooling lavas. Also, water is supposed to be partly responsible for the irregularities in many columnar joints in plugs and other intrusions close to the surface (Fig. 11.5).

During cooling of lava flows and intrusions the fluid involved in the development of columnar joints is normally vapour or steam. This follows because the temperatures at which columnar joints start to form is estimated at 60% of the original temperature of the magma (Jaeger, 1961). For a basaltic magma at about $1100\,°C$, 60% of the temperature would be $600–700\,°C$, in which case the fluid cannot be liquid water but must be steam or vapour. Steam is of course known to participate in the cooling of lava flows and intrusions, as is well demonstrated by high-temperature geothermal fields, most of which are related to cooling intrusions or magma chambers at depth.

We may thus conclude that columnar joints are extension fractures, partly driven by fluid overpressure and, therefore, hydrofractures. Because columnar joints are so common in lava flows and intrusions, they are often the primary weaknesses from which larger fractures develop. For example, many tension fractures and normal faults in rift zones develop from columnar joints (Figs. 9.7 and 11.6). Similarly, dykes use the columnar joints as weaknesses, and the linking up of those joints that are ahead of the dyke tip eventually leads to the formation of the dyke fracture and determines its propagation path. Once formed, columnar joints act as large-scale versions of Griffith cracks from which most large fractures, both extension fractures such as dykes and shear fractures such as normal faults, develop (Chapters 13 and 14).

There are other types of contractional fractures, most of which are also pure extension fractures. Some contractional fractures are pervasive and important sources of permeability in reservoirs and, like columnar joints, offer weaknesses from which other, tectonic fractures may develop. The contractional fractures include mud cracks (desiccation cracks), commonly wedge shaped and forming polygonals, but mostly restricted to thin near-surface layers. Mud cracks are thus generally of little importance for fluid flow and permeability in reservoirs. There are also syneresis fractures, generated by spontaneous dewatering and volume reduction of sediments. These fractures are commonly isotropically distributed, with high frequency, and occur throughout sedimentary rocks such as shales, siltstones, limestones, dolomites, and sandstones. Also included are mineral phase-change fractures resulting from host-rock volume reduction due to mineral phase changes, such as in the clay or carbonate constituents of a sedimentary rock; for example, during dolomitisation where the phase change of calcite to dolomite leads to a molar volume reduction of around 13%.

For an illustration of field studies of joints, consider the central joint in Fig. 11.7. How would we measure and model this joint? We note the following points:

• The joint goes through the hosting layer (here limestone).
• There is clear evidence of ductile deformation at the lower tip of the joint and, therefore, presumably at its upper end. This means that the lower and, presumably, the upper margins of the joint were in contact with essentially free surfaces.

**Fig. 11.6**    Tension fracture in the Holocene pahoehoe lava flow in North Iceland. Tension fractures, like other tectonic fractures in lava flows, develop from columnar joints. For this reason, the original shape of a tension fracture is often quite irregular.

- The joint is confined to the stiff (limestone) layer and does not penetrate the ductile layers below and (presumably) above. The joint is thus layerbound/stratabound.
- The joint is a pure extension fracture: there is no evidence of any vertical displacement nor (based on observations) any horizontal displacement. The only displacement is a small opening normal to the plane of the fracture.

**Fig. 11.7**    Joints cutting through a limestone layer at Kilve, on the Somerset coast, Southwest England. The soft deformed layer below the lower tip of the central joint shows that, when the joint formed, the layer behaved as ductile.

From these points we model the joint as follows: it is primarily a mode I through crack where the strike dimension is the controlling dimension (Fig. 11.7). Then we need to decide whether the joint formed as a tension fracture or as a hydrofracture. If it was formed by tension, then it must have formed at very shallow depths (Chapters 7 and 9). That is to say, its depth of origin below the surface must have been a few hundred metres or less for it to be a tension fracture (Chapter 8). However, the following evidence indicates that it is in fact a hydrofracture:

1. The hosting layer was stiff while the layers above and below were ductile. The limestone layer cannot have been stiff at the time of the deposition of the sediments, only much later during burial. Clearly, therefore, if the fracture formed as a tension fracture, it must have done so during erosion and uplift of the area, that is, at depth levels similar to those of the outcrop today.

2. The joint has essentially a constant aperture along its dip dimension (height). By contrast, a tension fracture extending from a free surface is normally not with constant aperture along its dip dimension; its aperture or opening tends to decrease with depth. This is well known from studies of tension fractures in rift zones and is seen in sedimentary basins as well.

3. Today the shale layers in between the limestone layers are stiff. Earlier in the history of this outcrop, however, the shale layers were soft (compliant), presumably to considerable

depths. This is indicated by the ductile behaviour of the shale during, and presumably following, the joint formation. This means that the joint is unlikely to have formed close to the surface. In other words, the joint is most likely to have formed at a considerable depth and thus to have been driven open by fluids.

4. This conclusion is further supported by there being some minerals in the joint. Minerals are deposited from geothermal fluids, indicating much higher geothermal gradients than presently in the area.

We therefore conclude that the joint was most likely formed as a hydrofracture. In fact, most joints are believed to be formed in this way (Secor, 1965). In layered rocks, stratabound joints may commonly be modelled as mode I through crack with a controlling strike dimension.

Many joints form systems. Some systems consist of parallel joints, others of orthogonal joints (Fig. 11.8). In Fig. 11.8 the joins are located in gneiss with marker layers, so that it can be demonstrated that the joints are, as seen in the outcrop, with no fracture-parallel displacements and thus pure extension fractures. The origin of such a system is not entirely clear. In this particular case, the horizontal joints have, partly at least, developed along contacts or interfaces in the layered gneiss. These contacts may have opened as exfoliation fractures, running parallel with the free surface of the eroded gneiss body. By contrast, the vertical joints, also orthogonal, are most likely the result of tensile stresses. There is no evidence of their being generated by fluid pressure in this particular system. Fluid pressure cannot, of course, be excluded, because gas and water, for example, might not leave any traces of their transport in the joints. But in this case, tensile stresses related to erosion and uplift is a likely cause of joint formation.

**Fig. 11.8**    Three sets of mutually orthogonal joints in gneiss on the island of Oygaarden, off the coast of Bergen, West Norway. The person provides a scale.

Joints are very common in carbonate rocks, in particular in limestone layers. As we saw earlier (Fig. 11.7), the joints are commonly largely confined to the stiff layers and only comparatively few of the joints in the stiff layers cut through the more compliant shales (Fig. 11.9). Field studies of such joint systems normally focus on the following points:

- The frequency of joints along a given measuring profile in a typical limestone layer.
- The proportion of joints in the limestone layer that also cut through the shale layer above, or below, or both.
- The mechanical type of fracture generating the joints in the different layers. That is, are all the joints extension fractures or are some shear fractures. In the latter case, are the shear fractures mostly, or entirely, confined to certain layers; for example, the shale layers. Some inclined (shear) joints occur in the shale layers in Fig. 11.9.
- The spacing of joints in these layers.
- The attitude and the apertures of the joints in these layers.
- The variation in attitude of joints between different layers.

In many carbonate rocks, there are vuggy (vesicle) fractures, characterised by an opening that is either a wide ellipse or a circle (Fig. 11.10). Their shapes are largely due to dissolution, commonly related to acid fluids, of the rock matrix surrounding the original fracture.

**Fig. 11.9**    Outcrop of limestone and shale layers in the Bristol Channel, Wales. Most of the fractures are confined to the limestone layers (light grey). Those fractures that pass through the shale layers (dark grey) tend to be inclined shear fractures.

**Fig. 11.10**   Vuggy fractures cutting through a limestone layer at Kilve, on the Somerset coast, Southwest England. (a) View northwest, all the vuggy fractures are confined to a single limestone layer. (b) Close-up of two of the vuggy fractures in (a).

The acid fluid is in disequilibrium with the walls of the initial fracture, so that dissolution may occur in a narrow zone surrounding the fracture. Vuggy fractures are commonly associated with the development of karst. The three-dimensional geometry of these fractures is normally a sphere or an oblate ellipsoid. Because of this shape, they have a high fracture stiffness in compression and rarely close, even at great depths. It follows that formation of vuggy fractures increases the permanent porosity and, if the fractures are interconnected, the permeability of the host reservoir.

## 11.5 Mineral veins

**Mineral veins** are those fractures that are partly or completely filled with secondary minerals such as quartz, zeolites, and calcite (Figs. 11.11–11.14). They are common in all main rock types, that is, igneous, sedimentary and metamorphic, and indicate that there were geothermal fields in the vicinity of the rock at the time of vein formation. Fractures completely filled with minerals (Figs. 11.11–11.14) are barriers to transverse fluid flow, but fractures only partly filled with minerals may contribute significantly and positively to the permeability parallel with the fracture. The fracture-parallel permeability is partly due to channels within the mineralised fracture, and partly due to the mineral aggregates increasing the stiffness of the fracture. Increased stiffness increases the fracture resistance to closure under normal compressive stresses. Partial filling is either due to an original lack of complete mineralisation in the fracture, or to partial dissolution of minerals in a filled fracture.

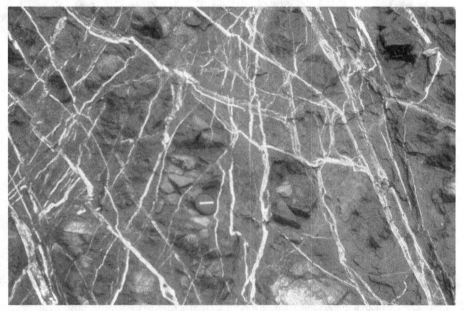

**Fig. 11.11**    Mineral veins in a basaltic lava flow of the damage zone of the Husavik-Flatey Fault in North Iceland. View vertical, the lens cap (cover) provides a scale.

**Fig. 11.12**    Two sets of mineral veins in a limestone layer at Kilve, on the Somerset coast, Southwest England. Some of these veins belong to the set that formed the joint/vein in Fig. 11.7. Oblique view, the measuring tape is 1 m long.

Some veins are shear fractures but many, and presumably most, are extension fractures, even those that are located within the damage zones of faults. For example, studies of cross-cutting relationships of several hundred mineral veins in the lava flows of the damage zone of the Husavik-Flatey Fault (Fig. 11.11), a transform fault in North Iceland, indicate that more than 80% of the mineral veins are extension fractures (Gudmundsson *et al.*, 2002). Similar studies have been made of mineral veins in sedimentary rocks, such as in the Bristol Channel of Britain. Based on cross-cutting relationships (Fig. 11.12) and the orientation of the vein fibres in relation to the fracture walls, the great majority of the veins are extension fractures (Philipp, 2008). It should be emphasised that these results do not mean that all the veins are extension fractures – there are many that are shear fractures – only that the great majority are extension fractures. In this book we therefore model mineral veins as extension fractures.

Consider the mineral vein in Fig. 11.13. This vein is located in a limestone layer in the Bristol Channel (the Somerset coast) in Britain. The dip dimensions of many veins in the limestone layers are confined to those layers; commonly the veins are arrested at the contacts between the limestone and shale layers (Fig. 11.14). The vein in Fig. 11.12 has clear tips, so that its strike dimension or length can be measured. In addition, its thickness, or palaeo-aperture, can be measured along its length. There are three questions to consider when measuring the vein thickness:

- What is the general maximum thickness and how can it be best estimated?
- How does the thickness vary along the length (strike dimension) of the vein?
- Is the vein multiple and, if yes, how does that affect the overpressure calculations?

**Fig. 11.13**   Mineral vein with both lateral ends exposed in a limestone layer at Kilve, on the Somerset coast, Southwest England. The measuring tape is 2 m long.

The maximum thickness is used to calculate the fluid overpressure during the formation of the fracture. Normally, we make three measurements of the thickness at those points along the vein where eye inspection indicates that it has the greatest thickness. The average of these measurements is used as the aperture in Eq. (9.16) and related equations for estimating the fluid overpressure (Example 9.1).

The thickness variation along veins varies considerably. However, veins that are non-restricted laterally (Fig. 11.13), that is, do not meet with other veins, fractures, or free surfaces, commonly have roughly elliptical thickness variations (Chapter 13). This is similar to what would be expected in a host rock that behaves as roughly isotropic in mechanical properties. While such a roughly elliptical variation in thickness/aperture is not uncommon in hydrofractures when measured at the surfaces of rock layers such as limestone or basaltic lavas, they are much less common when the hydrofracture apertures are measured in vertical

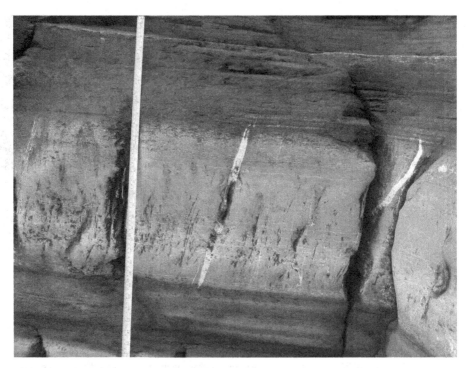

**Fig. 11.14** Vertically arrested mineral vein at contracts between a limestone layer, which hosts the vein, and adjacent shale layers. The height (length) of the measuring tape is 75 cm.

sections. This follows because in the vertical sections the layers that the hydrofracture cuts through commonly have widely different mechanical properties (Figs. 11.9 and 11.14). Consequently, the local stresses, in particular the normal stresses on the fracture, vary from one layer to another, and so does the overpressure (Chapter 13).

If the vein is multiple, it must have formed during many episodes of water injection and circulation. Then, clearly, the present thickness was at no time the aperture of the hydrofracture. However, many factures are known to grow as self-similar structures (Turcotte, 1997). And for those hydrofractures that grow as self-similar, the aspect (length/thickness) ratio remains constant during their evolution. Thus, for self-similar veins, the present aspect ratio is an indication of the fluid overpressure at the time of fracture formation. In addition, many veins appear to have formed in a single fluid injection (Philipp, 2008), in which case the present thickness is presumably a rough indication of the aperture at the time of fracture formation. The same applies to many other hydrofractures, such as dykes.

## 11.6 Dykes

All fissure eruptions are supplied with magma through fractures (Figs. 9.14, 9.15, and 11.2). Consequently, a primary physical condition for many and perhaps most eruptions is that a

Fig. 11.15 Subvertical dyke cutting through an inclined sheet, both being basaltic, in Ardnamurchan, Scotland. The cross-cutting relationship indicates that the dyke is younger than the sheet and that the dyke is a pure extension fracture. View vertical, the steel tape on the dyke is about 6 cm in diameter.

magma-driven fracture, a hydrofracture, is able to propagate to the surface. Magma-driven fractures, frozen or fluid, are referred to as sheet intrusions or sheets. Depending on its dip and intersection with the host rock, a sheet-like intrusion is referred to as a dyke, an inclined sheet (a cone sheet), or a sill. Often these terms are used more loosely in that the name dyke is given to a vertical or close-to-vertical sheet, the name sill to a horizontal or close to horizontal sheet, and the name inclined sheets to those that dip somewhere between steeply dipping (subvertical) and gently dipping (subhorizontal) intrusions.

Field measurements of dykes are similar to those of tension fractures and, in particular, mineral veins. In the field it is usually easy to determine the strike, dip, and thickness of a dyke (Figs. 1.3, 4.1, 9.8, and 11.16). When a dyke can be followed laterally, it is seen that the thickness varies along its strike dimension (Fig. 11.16). Commonly, the outcrop is limited to a vertical section, in which two or three thickness measurements are usually made so as to get the representative dyke thickness. Many dykes are formed by multiple injections. In this case the overpressure estimates (Eqs. 9.16 and 9.17) are approximately correct so long as the dyke grows in a self-similar way. There is, in fact, evidence that some dykes at least grow as self-similar structures.

At shallow depths of about 1 km below the free surface, the aspect ratios of many regional dykes are around 1000. Many and perhaps most regional dykes are likely to originate in deep-seated magma reservoirs (Gudmundsson, 1995). What is not so obvious is if the dykes should be treated as through cracks or as part-through cracks (Chapter 9). The argument for treating a regional dyke as a through crack, even if it is not known to be a feeder, is

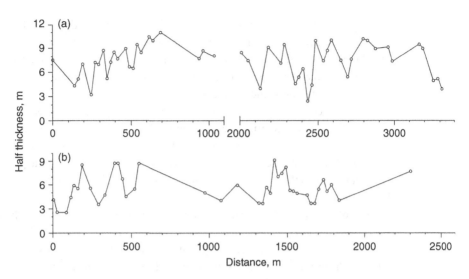

Aperture variation (measured as half thickness) in two segments of regional, basaltic dykes in East Iceland. (a) This dyke segment cuts through a basaltic lava pile with numerous red, soft soil and scoria layers in between the lavas. Many of the thickness changes are related to stiffness differences between the lava flows and the soil/scoria layers. (b) This dyke segment cuts fewer scoria/soil layers, which may be one reason for the smaller variation in aperture/thickness.

Part of the segmented Laki Fissure, seen cutting through the hyaloclastite mountain Laki, South Iceland. Aerial view north, this fissure which erupted in 1783 is 27 km long and in its central part, seen here, associated with a narrow graben.

shown by the way many dykes terminate vertically. Many dykes end vertically at weak contacts between layers (Chapter 9). While these contacts are strictly not free surfaces, they may be close to acting as such at shallow depths. This is particularly the case at the time of dyke emplacement when the contacts have not sealed or healed because of subsequent compaction, alteration, and secondary mineralisation.

There are clear arguments in favour of treating regional non-feeders as part-through cracks. Since the dykes originate at the solid–fluid boundary of the source reservoir, it follows that its lower edge meets with a free surface. However, a non-feeder dyke becomes arrested at a certain depth in the crust. It is thus, by definition, a part-through crack, particularly when the upper tip is not at a contact but inside a rock body such as a thick lava flow (Chapter 9).

One potential problem in modelling extension fractures in general, and dykes in particular, is that they tend to be segmented. A good example of a segmented dyke is provided by the feeder of the 1783 Laki eruption in Southeast Iceland (Thordarsson and Self, 1993). The total length of the volcanic fissure is about 27 km and the feeder dyke is estimated with a surface thickness of about 10 m. The volcanic fissure is composed of at least ten main segments (Fig. 11.17). Another segmented feeder is the 6000-year-old feeder to the Sveinar-Randarholar volcanic fissure in North Iceland (Figs. 1.3 and 1.6). This feeder is very rare in that its connection with the crater row is exposed in the vertical walls of the canyon Jokulsa a Fjollum (Chapter 1).

There exist analytical solutions for the opening displacements of segmented extension fractures such as these feeder dykes (Chapter 9; Sneddon and Lowengrub, 1969). The results show that as the distance between the nearby tips of the segmented cracks decreases in relation to the lengths of the cracks, the segments approach the behaviour of a single, continuous fracture.

Detailed comparison between feeders and non-feeders indicates that their geometries are very different and reflect different magmatic pressures (Geshi *et al.*, 2010). It is clear that feeders propagate as non-feeders until they reach the surface to supply magma to an eruption. Therefore, the geometric difference between feeders and non-feeders is primarily a reflection of the feeders reaching the surface. A non-feeder largely keeps its original emplacement geometry. By contrast, once a feeder reaches the surface and starts issuing magma, its overpressure gradually falls. Thus, during the course of the eruption, the thickness of a feeder gradually decreases to balance the horizontal stress in the host rock.

## 11.7 Inclined sheets

The analysis of inclined sheets is very similar to that of subvertical dykes (Figs. 11.15 and 11.18). The main differences between the sheets and the dykes is that the sheets are generally less steeply dipping and also thinner. For example, in Iceland regional Tertiary dykes are commonly 3–5 m thick, and may be as thick as 50–60 m, whereas the inclined sheets are normally 0.5–1 m thick, with a maximum thickness of 12–14 m (Gudmundsson, 1995).

Fig. 11.18 Cross-cutting sheets and dykes in the Quaternary lava pile in Southwest Iceland. The cross-cutting relationships indicate that all the intrusions are extension fractures and that the local stress field of the associated magma chamber changed significantly during their emplacement. This can be inferred from the fact that while subvertical dyke on the left cuts an inclined sheet (and is thus younger than the sheet), the subvertical dyke to the right is cut by (and thus older than) some of the inclined sheets. The hammer is about 30 cm long.

The local stresses in a sheet swarm also fluctuate more than in a regional dyke swarm (Chapter 8). This follows because the stress field controlling the emplacement of sheets is largely controlled by the geometry of the associated shallow magma chamber, which may change much during the evolution of the sheet swarm (Gudmundsson, 2006). By contrast, most regional dykes are injected from deep-seated magma reservoirs and the associated stress fields are largely generated by the regional plate pull associated with spreading. Thus, the stress field controlling the emplacement of regional dykes is comparatively stable, whereas the local stress field controlling the emplacement of the inclined sheets is variable. As a consequence, and also because the sheet intensity is much higher than the dyke intensity, cross-cutting relationships are much more common in sheet swarms than in regional dyke swarms.

Many inclined sheets thus cross-cut (Figs. 11.15 and 11.18). Almost all the cross-cutting relationships indicate that the sheets (and dykes) are pure extension fractures. From this we can infer that the sheets are hydrofractures, that is, are entirely generated by the internal magmatic overpressure. Dykes that cut sheets (Fig. 11.15 and 11.18) and lava flows (Figs. 1.5, 8.1, and 9.8) support the conclusion that most dykes are also pure extension fractures. Because the sheets are inclined, and generally much thinner than the regional dykes, the volumetric flow rates are usually much less than those of the regional dykes (Chapter 17).

# 11.8 Sills

In contrast with dykes and inclined sheets, which are mostly discordant to the host-rock layers, that is, cut the layers, sills tend to be mostly concordant, that is, parallel with the layers (Figs. 10.4, 10.5, 11.3, and 11.19). Field measurements are made in the same ways as those of dykes and inclined sheets. Because most sills are gently dipping (Fig. 11.19), it is often difficult to measure their dips very accurately from a distance. It is often best to measure the dip close to or at (depending on magnetic effects) either the bottom or top of the sill. However, because both the bottom and top are somewhat wavy or undulating, several measurements are needed to obtain a reasonably accurate average. In these respects, strike, dip, and thickness measurements of sills face the same difficulties as accurate field measurements of other flat-lying strata such as many lava flows.

Since the lava flows are cut by dykes and inclined sheets, there is normally no difficulty in distinguishing these intrusions from lava flows. Some sheets, however, at the outer margins of a sheet swarm may be very gently dipping and change into concordant, sill-like intrusions. Such sheets can be distinguished from lava flows using the criteria for sills.

It is usually comparatively easy to distinguish between lava flows and sills. The following criteria can be used:

- The sills are characterised by well-developed columnar joints that are, because of the slow cooling, normally much better developed than in lava flows (Fig. 11.3).

**Fig. 11.19**    Basaltic sill in East Iceland. The sill is about 30 m thick and exposed at a depth of about 1200 m below the original top of the lava pile.

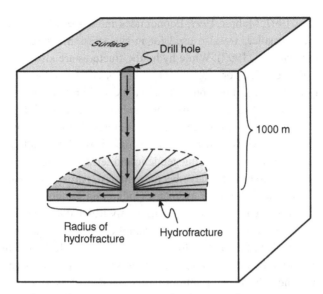

Fig. 11.20   Example 11.2. Diagram of one-half of a horizontal hydraulic fracture, a hydrofracture or, here, a water sill.

- Sills have chilled selvages (glassy margins) at their upper and lower contacts with the host rocks.
- There are normally no scoria layers associated with sills, as are common at the top and bottom of aa lava flows.
- There are normally much fewer and smaller vesicles in sills than in otherwise similar lava flows.

Most sills form when a propagating dyke meets an interface such as a contact where the mechanical properties and local stresses become unfavourable for dyke propagation, but are still favourable for sill formation. The exact mechanics of sill formation are discussed in Chapter 13. For sills that are small in relation to their depth below the free surface, the circular interior crack is a good model for the sill. If, however, the radius of the sill becomes similar to, or larger than, the depth to the sill (the overburden thickness), then a different mechanism comes to play (Gudmundsson, 1990). In this case, the magmatic pressure in the sill may cause up-doming (concave bending) of the layers above. If the bending is large, and the sill continues to receive magma (through dykes) while liquid, it may grow into a shallow magma chamber. Many shallow magma chambers presumably develop from a sill or sills, sometimes generating a laccolith.

## 11.9 Man-made hydraulic fractures

Man-made or artificial hydrofractures are those generated when fluid is injected under over-pressure into a reservoir rock to increase its permeabilities. The resulting hydrofractures are commonly referred to as hydraulic fractures (Valko and Economides, 1995; Charlez, 1997;

Yew, 1997; Mahrer, 1999; Economides and Nolte, 2000). These types of fractures, although much smaller, are also used for in-situ stress measurements in drill holes (Amadei and Stephansson, 1997). While hydraulic fractures are also used to generate artificial geothermal reservoirs (for example, enhanced or engineered geothermal systems, or hot-dry-rock reservoirs), the most common technique uses pressure pulses, resulting in existing fractures slipping and/or opening, in order to enhance the permeability and thereby form the reservoir.

Since most man-made hydraulic fractures are made at considerable depth, they cannot be observed directly. However, some hydraulic fractures made at comparatively shallow depths have been monitored using geodetic and seismic methods. In addition, many deep hydraulic fracture experiments have been monitored, particularly using seismic methods and fluid-pressure and flow-rate estimates (Valko and Economides, 1995; Economides and Nolte, 2000). There have also been many analytical and numerical models of hydraulic fractures.

Generally, the mechanics of hydraulic fracture propagation appears to be exactly analogous to that of natural hydrofractures. A remarkable feature of hydraulic factures, also seen in natural hydrofractures, is that their paths may rotate so as to become or continue to be perpendicular to the minimum principal compressive stress, $\sigma_3$ (Valko and Economides, 1995). This means that if a hydraulic fracture injected from a drill hole is for some reason initially not perpendicular to $\sigma_3$ (for example, because it uses a weakness that is oblique to $\sigma_3$), it may rotate after a short propagation distance to become perpendicular to $\sigma_3$. Similar observations are commonly made of sheet intrusions such as dykes. They may use weaknesses, perhaps joints or segments of steeply dipping normal faults, for some parts of their paths but soon rotate into a path that is perpendicular to $\sigma_3$. Thus, the general trend of all hydrofractures, including hydraulic fracture, is perpendicular to $\sigma_3$.

In present technology, a hydraulic fracture need not be injected horizontally as a blade-like fracture (Fig. 13.28). While horizontal hydraulic fractures are still the most common, particularly in the petroleum industry, a hydraulic fracture can, depending on the local stress field, have any attitude. For example, in hydraulic fracture experiments for man-made geothermal reservoirs, it is sometimes useful to make the drill hole inclined or close to horizontal and inject subvertical or inclined hydraulic fractures from the upper part of the subhorizontal drill hole.

Since all hydraulic fractures originate from drill holes they are normally modelled as part-through cracks. This applies in particular to hydraulic fractures that are injected horizontally into a specific target layer (Fig. 13.28), such as is common in the petroleum industry, to increase the permeability of that layer. Some hydraulic fractures, however, may become arrested at discontinuities such as faults or contacts. If these discontinuities are very weak, such as active faults with high fluid pressure or contacts at shallow depths, the hydraulic fractures may more appropriately be modelled as through cracks.

## 11.10  Field measurements of fractures

So far, we have primarily focused on the measurements of individual fractures. Commonly, however, field measurements include hundreds or thousands of fractures, many of which belong to one or more sets (Figs. 11.8–11.12). It is assumed that the reader is familiar with standard field techniques in geology, so that this is just a summary as it relates to fractures for ease of reference. More details are found in standard engineering mechanics handbooks and field geology manuals (McClay, 1987; Priest, 1993; Maley, 1994; Hudson and Harrison, 1997).

In the field, fractures are usually studied in traverses (profiles) or along scan lines. We refer to profiles (or traverses) when dealing with long distances and large fractures. In particular, dykes, inclined sheets (and large faults) are studied in profiles, commonly hundreds or thousands of metres long and extending along the coast lines of fjords and bays, as well as along river channels and valleys. Smaller fractures, such as found in individual quarries and outcrops are normally measured along scan lines. These are mostly with lengths of metres or tens of metres.

The difference is not only as regards size, the scan lines being much shorter, but also as regards orientation. The profiles have a general overall trend; for example, at right angles to the main trend of a certain fracture set to be studied. However, in detail the profiles tend to be quite irregular in shape. This follows because the coastline or the river channel that is followed is, in detail, rather irregular in shape. By contrast, a scan line usually follows a stretched measuring tape (tape measure) and can be made quite straight. In both cases, however, most of the fractures are projected onto the profile or scan line so as to measure their attitude and apertures/thicknesses. Figures 11.4, 11.8, 11.9, and 11.12 show examples of fine sites for measuring fractures along scan lines. The outcrops in Figs. 11.4 and 11.8 are very suitable for horizontal and vertical scan lines, whereas the outcrop in Fig. 11.12 is suitable for two horizontal scan lines, at right angles. The fracture set in Fig. 11.8 would be best measured in two horizontal and one vertical scan lines, all at right angles to each other.

When measuring fractures along a profile or a scan line, the principal measurements include the following:

1. **Attitude or orientation**. This means measuring the strike and dip, or the dip and the dip direction, of the fracture.
2. **Displacement**. For an extension fracture, this means measuring the aperture, if the fracture is open, or the thickness if the fracture is filled; for example, a dyke or a mineral vein. For a shear fracture, a fault, this means measuring the fault-parallel displacement or, alternatively, the vertical and (less commonly) the horizontal components of the displacement, that is, the throw and heave, respectively.
3. **Trace length** or just **length**. This is the length of the line of intersection of the fracture plane with the rock surface where the scan line or the profile is made. Depending on the orientation of the profile or scan line, which can be along a vertical section (cliffs, say, of a lava flow) or along a horizontal section (the flat plane of a limestone layer, for example), the trace length may be either the dip dimension (vertical section, Fig. 11.14) or strike dimension (lateral section, Fig. 11.13). For large fractures, such as many regional dykes, the strike dimension cannot be measured in the profile but must be obtained either by following the dyke along its length in the field (Fig. 11.16), or from aerial photographs or satellite images. Based on certain geometric assumption, the trace length, when measureable, can be used to infer the size of the fracture. For example, the fracture

may be inferred to be a part-through crack, and then often modelled as half-circular or half-elliptical in shape. Alternatively, the fracture may modelled as a through crack, in which case it is commonly assumed to be of similar strike or dip dimension inside the rock body as on its surface. For dykes, however, there is theoretical and observational evidence that the strike dimension increases with increasing crustal depth, whilst the thickness (or aperture) decreases (Gudmundsson, 1990; Geshi *et al.*, 2010).

4. **Type of fracture**. This means deciding whether the fracture is a joint, a mineral vein, a dyke, an inclined sheet or a sill, a dip-slip fault (normal, reverse, or thrust), a strike-slip fault (dextral or sinistral), or an oblique-slip fault.

5. **Infill material**. This means deciding if the fracture contains minerals (and then of what type), fault gouge, or fault breccias.

6. **Slickensides**. Here the focus is on the pitch of the striations and the sense of last movement along the fault.

7. **Fracture frequency**. This is the number of fractures measured or estimated in a unit length along the scan line or profile, or in a unit area, or unit volume of rock. This concept is sometimes referred to as **fracture intensity**.

8. **Fracture spacing**. This is the reciprocal of fracture frequency. In a general sense, fracture spacing is simply the distance between one fracture and another. Commonly, however, we are dealing with fracture sets, in which case fracture spacing may also be defined as the distance between two parallel or subparallel fractures of a given set. This is elaborated below.

When measuring fractures, we try to make at least two profiles or scan lines, at roughly right angles (Figs. 11.4, 11.8, 11.9, and 11.12). This applies both to each rock surface of an outcrop such as a quarry, where scan lines are used, and to larger areas containing a dyke swarm, for example where long profiles are used. Also, the three-dimensional view of the fracture sets can be obtained by making profiles and scan lines at different elevations in the studied areas. For horizontal fractures, vertical profiles or scan lines are used. These can be measured either directly, or in the case of vertical walls, such as in quarries, from photographs or lidar models.

Thus, for a three-dimensional presentation of a fracture network where the outcrop contains horizontal and vertical rock surfaces, we would normally try to obtain two scan lines on the vertical surface (for example, a cliff face), and two on the horizontal surface (Fig. 11.8). Each scan line should, if possible, dissect at least 30 fractures, so as to get a statistically significant specimen, and preferably 50–100 fractures.

The attitude of the rock fractures is always measured where the fracture, or its projection, dissects the scan line or profile. The trend and pitch of the scan line itself are also measured. For long, irregular profiles, the general trend and location above sea level is indicated.

Displacement and trace length (dip or strike dimension) are subject to limitation in that values of these parameters that are smaller or larger than certain numbers are omitted from the data sets. The sizes of the values omitted depend on the individual data sets being considered as well as on the available outcrops and aerial photographs or satellite images used. The term **truncated** normally means that strike dimensions, dip dimensions, or displacements smaller than certain threshold values are not recognised or recorded and thus omitted. The term censored means that values larger than certain threshold values are unobtainable and therefore omitted.

An example of truncation is that many fractures in any particular study are microscopic and thus obviously not seen and measured in an outcrop study. Furthermore, when studying tectonic fractures in a profile, columnar joints would normally be omitted. This follows because many of the tectonic fractures use the joints as weaknesses. The joints are thus not regarded as of tectonic origin but rather as the initial weaknesses or Griffith flaws in the lava pile from which the tectonic fractures develop (Figs. 11.3, 11.4, and 11.6). As an example of censoring, it is normally not possible to measure the lengths of fractures that extend beyond the field of vision. The field of vision, of course, depends on the fracture set and the type of study; it is, for example, different when working in a limited outcrop such as a quarry or working on an aerial photograph or satellite image study.

The definitions above are normally precise enough for most fracture studies. More precise definitions exist of some of these terms. For example, if values smaller than certain threshold values are omitted from the data set, it is referred to as **trimming**. Also, if no record is kept of the number of these omitted values, the specimen is referred to as truncated at the threshold value, whereas if the number of omitted values is recorded, the specimen is referred to as censored at the threshold value. If trace lengths, such as strike dimensions or other values, larger than a certain value are omitted, it is referred to as **curtailment**. If the number of omitted values is recorded, it is **truncation**, but if no record is kept, it is **censoring**.

The surface trace length of a fracture in an outcrop can be measured either as its total trace length or as its **semi-trace length**. The trace length is the linear distance between the end points of the fracture trace in the outcrop. The semi-trace length is the distance from the point of intersection between the profile (the scan line or the measuring tape) to the end of the fracture trace. When the profile is near the edge of the exposure, it is only possible to measure the semi-trace length on one side of the profile.

## 11.11 Summary

- Extension fractures form in a direction that is perpendicular to the minimum principal compressive (maximum tensile) stress, $\sigma_3$. Extension fractures can thus be used as stress indicators, that is, the fracture planes were, at their time of fracture formation, perpendicular to the direction of $\sigma_3$ and thus parallel with the direction of the other principal stresses, $\sigma_2$ and of $\sigma_1$.
- Extension fractures are mechanically of two main types: tension fractures and fluid-driven fractures or hydrofractures. Tension fractures form when the minimum principal stress is tensile, that is, $\sigma_3$ is negative. On an outcrop scale, tectonic stress conditions of this kind can only occur at shallow depths, normally at depths less than about 1 km. By contrast, hydrofractures can occur at any crustal depth; the only condition is that the total fluid pressure at the time of fracture initiation is larger than the minimum principal compressive stress plus the tensile strength (cf. Eq. 8.3).

- Tension fractures occur as gaping, large-scale fractures in rift zones worldwide. They are particularly common at divergent plate boundaries of the ocean ridges and in their on-land extensions as in Iceland. Some joints are tension fractures, particularly those formed at shallow depths, such as during erosion and uplift. Columnar joints formed in solidifying igneous bodies, such as lava flows, dykes, and sills, are partly pure tension fractures formed by thermal stresses and shrinkage of the igneous body, and partly fluid-driven, where the fluid is mainly vapour or steam.
- Hydrofractures, or fluid-driven fractures, are very common in all tectonic settings. They include many, and presumably most, joints and mineral veins and all dykes, sills, and inclined sheets. In addition, hydrofractures include artificial or man-made hydraulic fractures, such as are injected from drill holes to measure the local stresses or, on a much larger scale, to increase the permeability in fluid-filled reservoirs of various types.
- To model fracture systems, several basic fracture parameters are normally measured in well-exposed traverses (profiles) or along scan lines. These include attitude (strike and dip), displacement (including opening of extension fractures), and the trace length (length) of the intersection of the fracture with the rock surface where the profile is measured. Additional parameters that are commonly measured include fracture type (vein, joint, dyke, and so forth), infill material (if any), fracture frequency, and fracture spacing.
- There are several restrictions used when dealing with data on fracture sets. Displacement and trace-length values smaller than certain threshold values (which depend on the outcrop and aerial images available as well as the rock type) are truncated, that is, are not measured. Similarly, very large values of the same type cannot be measured because of the limitation of the outcrop or image and are thus censored (omitted).

## 11.12 Worked examples

### Example 11.1

#### Problem

The tension fracture in Fig. 11.1 has a total maximum opening (aperture) of 10 m, a strike dimension of 1100 m, and is hosted by a Holocene pahoehoe lava flow with a Young's modulus of 5 GPa and a Poisson's ratio of 0.25. If the tension fracture was formed in a single rifting episode, find the tensile stress that could have formed the fracture.

#### Solution

The thick lava flow hosting the fracture rests partly on late Quaternary sediments, some of which are quite soft. If we assume that the sediments are so soft that they deform as ductile, and not penetrated by the fracture, then we may regard the bottom, in addition to the surface, of the lava flow as roughly free surfaces. On that assumption, a plane-strain through crack mode I model is appropriate for the tension fracture. From Eq. (9.12) we then get the tensile stress associated with the fracture formation as follows:

$$-\sigma = \frac{\Delta u_I E}{2L(1-v^2)} = \frac{10 \text{ m} \times 5 \times 10^9 \text{Pa}}{2 \times 2000 \text{ m} \times (1-0.25^2)} = 1.3 \times 10^7 \text{Pa} = 13 \text{ MPa}$$

This is higher than typical in-situ tensile strength of rocks, 0.5–6 MPa (the maximum being about 9 MPa, Amadei and Stephansson, 1997). This is, however, much less than the maximum measured laboratory tensile strength, which is as high as 20–30 MPa (Jumikis, 1979; Myrvang, 2001). Hydraulic fracturing on 1 m$^3$ blocks of granite yields tensile strengths as high as 13.5 MPa (Rummel, 1987), similar to the above value.

---

### Example 11.2

Problem

Water under 2 MPa overpressure with reference to $\sigma_3$ (which is vertical) is pumped into a drill hole. At the crustal depth of 1000 m the water forms a horizontal hydraulic fracture which propagates radially from the hole and forms a circular hydrofracture (Fig. 11.20). Young's modulus of the host rock is 10 GPa and its Poisson's ratio is 0.25. If $1 \times 10^5$ litres are pumped into the hydrofracture, make a crude estimate of its radius.

Solution

From Eq. (9.20) we have the volume of fluid in a penny-shaped hydrofracture, namely

$$V_f = \frac{16(1-v^2)p_o a^3}{3E}$$

To find the radius, we rewrite Eq. (9.20) and solve for the radius $a$ as follows:

$$a = \left(\frac{3EV_f}{16(1-v^2)p_0}\right)^{1/3} = \left(\frac{3 \times 1 \times 10^{10} \times 100 \text{ m}^3}{16(1-0.25^2) \times 2 \times 10^6 \text{Pa}}\right)^{1/3} = 46.4 \text{ m}$$

The thing to remember here is that the volume $V_f$ is given in litres, which is not an SI unit. One litre was originally defined as one kilogram of pure water at 4 °C and standard atmospheric pressure, and is approximately equal to 1000 cm$^3$ or $10^{-3}$ m$^3$. To change it into cubic metres, which is an SI unit, we multiply with $10^{-3}$, thus:

$$1 \times 10^5 \text{ litres} \times 1 \times 10^{-3} = 100 \text{ m}^3$$

Since the calculated radius of the hydraulic sill-like fracture is about 46 m, it is clearly much smaller than the depth of the fracture, 1000m, so that the assumption we made, namely that the fracture is located in an infinite elastic crust or body, is valid. By contrast, if the fracture radius was about ten times larger, in which case its diameter would be similar to its depth below the surface, then the free-surface effects would come into play and affect the opening and volume of the fracture for the given elastic properties and overpressure (Example 11.3).

## Example 11.3

### Problem

When the radius of a horizontal hydrofracture, such as a hydraulic fracture (Fig. 11.20) or a sill (Fig. 11.19), becomes large in its relation to its depth below the Earth's surface, the fracture starts to bend (dome) the layers above. This is, for example, seen in the formation of laccoliths and many magma chambers (Gudmundsson, 1990). Usually, if the sill is of a depth similar to its diameter, an upward deflection or doming of the overburden is expected and the following equation from elasticity for the bending of a circular plate is used (Ugural, 1981; Solecki and Contant, 2003). The maximum deflection $u_{max}$ of the roof of the sill is then given by

$$u_{max} = \frac{p_0 a^4}{64 D_f} \left( \frac{5 + v}{1 + v} + \frac{4d^2}{(1 - v^2)a^2} \right) \tag{11.1}$$

where $p_0$ is the excess fluid pressure in the sill, $a$ is the radius of the sill, $v$ is Poisson's ratio of the host rock, $d$ is the depth to the top of the sill, and $D_f$ is the flexural rigidity of the host rock, that is, the layers above the sill and given by

$$D_f = \frac{E d^3}{12(1 - v^2)} \tag{11.2}$$

where $E$ is the Young's modulus of the host rock. For a layered rock, the effective flexural rigidity is commonly used. Then it is assumed that the layers can slip along their contacts, which reduces the rigidity (Reddy, 2004). In the present example, however, the sill is supposed to be at a very shallow depth (30 m), so that normally only one layer would form its roof, in which case the general flexural rigidity as presented by Eq. (11.2) should be used.

A circular hydrofracture (a water sill) with a radius of 50 m is injected at a depth of 30 m below the Earth's surface. The other values are as in Example 11.2, namely the fluid excess pressure is 2 MPa, the Young's modulus of the host rock is 10 GPa, and its Poisson's ratio is 0.25.

(a) Calculate the maximum aperture or opening of the hydrofracture using Eq. (11.1).
(b) Do the same for the hydrofracture in (a) if it was injected at a depth of 1000 m below the Earth's surface.

### Solution

(a) To use Eq. (11.1), we need to know the flexural rigidity, $D_f$, which we obtain from Eq. (11.2) as follows:

$$D_f = \frac{E d^3}{12(1 - v^2)} = \frac{1 \times 10^{10} \text{Pa} \times (30 \text{ m})^3}{12 \times (1 - 0.25^2)} = 2.4 \times 10^{13} \text{N m}$$

Using this result, from Eq. (11.1) we get the maximum opening displacement as

$$u_{max} = \frac{p_0 a^4}{64 D_f} \left( \frac{5 + v}{1 + v} + \frac{4d^2}{(1 - v^2)a^2} \right)$$

$$= \frac{2 \times 10^6 \text{Pa} \times (50 \text{ m})^4}{64 \times 2.4 \times 10^{13} \text{N m}} \left( \frac{5 + 0.25}{1 + 0.25} + \frac{4 \times (30 \text{ m})^2}{(1 - 0.25^2) \times (50 \text{ m})^2} \right)$$

$$= 0.0467 \text{ m}$$

This is the opening displacement and thus half the total opening ($2u_{max}$), about 0.093 m.

(b)  Since the fracture radius is a small fraction of its depth below the Earth's surface, the appropriate formula for calculating the total opening or aperture is Eq. (9.19) which gives

$$\Delta u_I = \frac{8 p_0 (1 - \nu^2) a}{\pi E} = \frac{8 \times 2 \times 10^6 \text{Pa} \times (1 - 0.25^2) \times 50 \text{ m}}{3.1416 \times 1 \times 10^{10} \text{Pa}} = 0.024 \text{ m}$$

Clearly, the total opening is much smaller for a penny-shaped hydrofracture at great depth, for a given excess pressure and elastic constants, than for the same fracture at very shallow depth (in comparison with its radius). And the reason is that when the fracture is at a shallow depth, the effects of the free surface result in the fluid excess pressure bending the layers above, that is, the layers that form the roof of the fracture. This is, in fact, one of the reasons why a magmatic sill may, on reaching a certain critical radius in relation to its depth below the surface, change into a laccolith and, commonly, a shallow magma chamber (Pollard and Johnson, 1973; Gudmundsson, 1990).

### Example 11.4

#### Problem

A columnar joint in a lava flow (Figs. 11.4 and 11.6) is opened by geothermal water with an overpressure of 4 MPa. The lava flow hosting the joint is 5 m thick and has a Young's modulus of 10 GPa and a Poisson's ratio of 0.25. If the strike dimension of the join is 0.5 m, find its maximum aperture.

#### Solution

Columnar joints in aa lava flows normally extend through the thicknesses of the lava flows (Fig. 1.6), although in exceptional situations they are confined to a part of the lava flow (Fig. 11.4). Here we assume that the joint extends through the lava flow, so that the dip dimension of the joint is 5 m. Since the strike dimension is 0.5 m, it follows that the strike dimension is the smaller of the two and thus the controlling dimension. The correct model is thus that of a mode I through crack opened by an internal fluid overpressure.

From Eq. (9.14) we get the aperture as follows:

$$\Delta u_I = \frac{2 p_0 (1 - \nu^2) L}{E} = \frac{2 \times 4 \times 10^6 \text{Pa} \times (1 - 0.25^2) \times 0.5 \text{ m}}{1 \times 10^{10} \text{Pa}}$$

$$= 0.000\,375 \text{ m} \approx 0.4 \text{ mm}$$

Many mineral veins are between 0.1 and 1 mm in thickness, so that this value is similar to that of typical veins in geothermal fields.

# 11.13 Exercises

11.1 Why is the controlling dimension of a tension fracture more likely to be its strike dimension than its dip dimension?

11.2 What information is needed to model a tension fracture at the Earth's surface?

11.3 What are the main field characteristics of a joint? Give field examples of joints.

11.4 What is the most widely considered mechanism of formation of joints at crustal depths of several kilometres or more?

11.5 What is a mineral vein? What are the main field characteristics of mineral veins?

11.6 What is a multiple vein and how does its being multiple affect the way that the vein is modelled?

11.7 What are the main field characteristics of dykes?

11.8 Describe and explain the main geometric difference between feeder-dykes and non-feeders.

11.9 What is the main geometric difference between dykes and inclined sheets? How do these differences relate to the local stresses controlling their emplacement mechanism and magma sources?

11.10 What are the main field characteristics of sills and how can they be used to distinguish between sills and lava flows?

11.11 What are man-made hydraulic fractures, how are they made, and what is the purpose of making them?

11.12 When measuring fractures along a scan line or profile, what are the principal measurements to be made?

11.13 As regards fracture measurements, what does the term truncated normally mean?

11.14 As regards fracture measurements, what does the term censored normally mean?

11.15 During a rifting episode, a new tension fracture is formed. Its strike dimension is 500 m and its maximum aperture (opening) 0.7 m. It is located in a pahoehoe lava flow with a Poisson's ratio of 0.25 and a Young's modulus of 5 GPa. Use a through-crack model to calculate the tensile stress associated with the fracture formation. How does the tensile stress compare with in-situ estimates of tensile strength?

11.16 Use a through-crack model to estimate the maximum aperture of mode I fracture in a host rock with a Young's modulus of 7 GPa and a Poisson's ratio of 0.27, if the tensile stress opening the fracture is 6 MPa and the fracture is 50 m long.

11.17 Magma is injected at 1500 m depth in a lava pile to form a circular sill 500 m in radius with a maximum thickness of 2 m. If the average Young's modulus of the host rock is 10 GPa and its Poisson's ratio is 0.26, calculate the magma overpressure associated with the sill formation.

11.18 A circular sill is formed at 1000 m depth in a lava pile with an average Young's modulus of 10 GPa and a Poisson's ratio of 0.25. The radius of the sill is 2000 m and the estimated magma overpressure during sill formation is 20 MPa. Calculate the maximum thickness of the resulting sill.

# References and further reading

Acocella, V. and Neri, M., 2003. What makes flank eruptions? The 2001 Etna eruption and its possible triggering mechanism. *Bulletin of Volcanology*, **65**, 517–529.

Acocella, V. and Neri, M., 2009. Dike propagation in volcanic edifices: overview and possible developments. *Tectonophysics*, **471**, 67–77.

Acocella, V., Neri, M., and Scarlato, P., 2006. Understanding shallow magma emplacement at volcanoes: orthogonal feeder dikes during the 2002–2003 Stromboli (Italy) eruption. *Geophysical Research Letters*, **33** (17): art. no. L17310 September 9, 2006.

Amadei, B. and Stephansson, O., 1997. *Rock Stress and its Measurement*. London: Chapman & Hall.

Baer, G., 1991. Mechanisms of dike propagation in layered rocks and in massive, porous sedimentary rocks. *Journal of Geophysical Research*, **96**, 11911–11929.

Bell, F. G., 2000. *Engineering Properties of Soils and Rocks*, 4th edn. Oxford: Blackwell.

Bonafede, M. and Rivalta, E., 1999. On tensile cracks close to and across the interface between two welded elastic half-spaces. *Geophysical Journal International*, **138**, 410–434.

Carmichael, R. S., 1989. *Practical Handbook of Physical Properties of Rocks and Minerals*. London: CRC Press.

Charlez, P. A., 1997. *Rock Mechanics, Volume 2: Petroleum Applications*. Paris: Editions Technip.

DeGraff, J. M. and Aydin, A., 1987. Surface-morphology of columnar joints and its significance to mechanics and direction of joint growth. *Geological Society of America Bulletin*, **99**, 605–617.

Delaney, P. and Pollard, D. D., 1981. Deformation of host rocks and flow of magma during growth of minette dikes and breccia-bearing intrusions near Ship Rock, New Mexico. *US Geological Survey Professional Paper*, **1202**.

Economides, M. J. and Nolte, K. G. (eds.), 2000. *Reservoir Stimulation*, 3rd edn. New York: Wiley.

Engelder, T., 1993. *Stress Regimes in the Lithosphere*. Princeton, NJ: Princeton University Press.

Engelder, T. and Geiser, P., 1980. On the use of regional joint sets as trajectories of paleostress fields during the development of the Appalachian Plateau, New York. *Journal of Geophysical Research*, **85**, 6319–6341.

Geshi, N., Kusumoto, S., and Gudmundsson, A., 2010. The geometric difference between non-feeders and feeder dikes. *Geology*, **38**, 195–198.

Gudmundsson, A., 1988. Effect of tensile stress concentration around magma chambers on intrusion and extrusion frequencies. *Journal of Volcanology and Geothermal Research*, **35**, 179–194.

Gudmundsson, A., 1990. Emplacement of dikes, sills and magma chambers at divergent plate boundaries. *Tectonophysics*, **176**, 257–275.

Gudmundsson, A., 1995. Infrastructure and mechanics of volcanic systems in Iceland. *Journal of Volcanology and Geothermal Research*, **64**, 1–22.

Gudmundsson, A., 2006. How local stresses control magma-chamber ruptures, dyke injections, and eruptions in composite volcanoes. *Earth-Science Reviews*, **79**, 1–31.

Gudmundsson, A., Fjeldskaar, I., and Brenner, S.L., 2002. Propagation pathways and fluid transport in jointed and layered rocks in geothermal fields. *Journal of Volcanology and Geothermal Research*, **116**, 257–278.

Harrison, J. P. and Hudson, J. A., 2000. *Engineering Rock Mechanics. Part 2: Illustrative Worked Examples*. Oxford: Elsevier.

Hudson, J. A. and Harrison, J. P., 1997. *Engineering Rock Mechanics. An Introduction to the Principles*. Oxford: Elsevier.

Jaeger, J. C., 1961. The cooling of irregularly shaped igneous bodies. *American Journal of Science*, **259**, 721–734.

Jaeger, J. C., 1968. Cooling and solidification of igneous rocks. In: Hess, H. H. and Poldervaart, A. (eds.), *Basalts, Vol. 2*. New York: Interscience, pp. 503–536.

Jaeger, J. C., Cook, N. G. W., and Zimmerman, R. W., 2007. *Fundamentals of Rock Mechanics*, 4th edn. Oxford: Blackwell.

Jumikis, A. R., 1979. *Rock Mechanics*. Clausthal, Germany: Trans Tech Publications.

Mahrer, K. D., 1999. A review and perspective on far-field hydraulic fracture geometry studies. *Journal of Petroleum Science and Engineering*, **24**, 13–28.

Maley, T. S., 1994. *Field Geology Illustrated*. Boise, ID: Mineral Land Publications.

McClay, K. R., 1987. *The Mapping of Geological Structures*. Milton Keynes: Open University Press.

Myrvang, A., 2001. *Rock Mechanics*. Trondheim: Norway University of Technology (NTNU), (in Norwegian).

Neri, M. and Acocella, V., 2006. The 2004–2005 Etna eruption: implications for flank deformation and structural behaviour of the volcano. *Journal of Volcanology and Geothermal Research*, **158**, 195–206.

Philipp, S. L., 2008. Geometry and formation of gypsum veins in mudstones at Watchet, Somerset, SW England. *Geological Magazine*, **145**, 831–844.

Pollard, D. D. and Aydin, A., 1988. Progress in understanding jointing over the past century. *Geological Society of America Bulletin*, **100**, 1181–1204.

Pollard, D. D. and Fletcher, R. C., 2005. *Fundamentals of Structural Geology*. Cambridge: Cambridge University Press.

Pollard, D. D. and Johnson, A. M., 1973. Mechanics of growth of some laccolithic intrusions in the Henry Mountains, Utah, II. Bending and failure of overburden layers and sill formation. *Tectonophysics*, **18**, 311–354.

Priest, S. D., 1993. *Discontinuity Analysis for Rock Engineering*. London: Chapman & Hall.

Reddy, J. N., 2004. *Mechanics of Laminated Composite Plates and Shells*, 2nd edn. London: CRC Press.

Rivalta E. and Dahm, T., 2006. Acceleration of buoyancy-driven fractures and magmatic dikes beneath the free surface. *Geophysical Journal International*, **166**, 1424–1439.

Rubin, A. M., 1993. Tensile fracture of rock at high confining pressure – implications for dike propagation. *Journal of Geophysical Research*, **98**, 15919–15935.

Rubin, A. M., 1995. Propagating magma-filled cracks. *Annual Reviews of Earth and Planetary Sciences*, **23**, 287–336.

Rummel, F., 1987. Fracture mechanics approach to hydraulic fracturing stress measurements. In: Atkinson, B. (ed.), *Fracture Mechanics of Rock*. London: Academic Press, pp. 217–239.

Schön, J. H., 2004. *Physical Properties of Rocks: Fundamentals and Principles of Petrophysics*. Amsterdam: Elsevier.

Secor, D. T., 1965. The role of fluid pressure in jointing. *American Journal of Science*, **263**, 633–646.

Sneddon, I. N. and Lowengrub, M., 1969. *Crack Problems in the Classical Theory of Elasticity*. New York: Wiley.

Solecki, R. and Conant, R.J., 2003. *Advanced Mechanics of Materials*. Oxford: Oxford University Press.

Thordarson, T. and Self, S., 1993. The Laki (Skaftar Fires) and Grimsvotn eruptions in 1783–1785. *Bulletin of Volcanology*, **55**, 233–263.

Turcotte, D. L., 1997. *Fractals and Chaos in Geology and Geophysics*, 2nd edn. Cambridge: Cambridge University Press.

Ugural, A. C., 1981. *Stresses in Plates and Shells*. New York: McGraw-Hill.

Valko, P. and Economides, M. J., 1995. *Hydraulic Fracture Mechanics*. New York: Wiley.

Yew, C. H., 1997. *Mechanics of Hydraulic Fracturing*. Houston, TX: Gulf Publishing.

# Field analysis of faults

## 12.1 Aims

Faults are measured in a very similar way as extension fractures. The main aims of this chapter are to:

- Give field examples of typical faults.
- Illustrate how the most important field measurements of faults are made.
- Provide typical field data on dip-slip faults.
- Provide field data on strike-slip faults.
- Explain and discuss oblique-slip faults.

## 12.2 Dip-slip faults

Dip-slip faults are those where the displacement is parallel with the dip of the fault (Fig. 12.1). They include normal faults, reverse faults, and thrust faults (Figs. 8.3 and 8.8). The block above the fault plane is the **hanging wall**, the block below the fault plane the **footwall** (Fig. 12.1). Depending on the sense of slip and the fault dip, the main types of dip-slip faults are normal faults, reverse faults, thrusts, and overthrusts. More specifically, when the hanging wall moves down relative to the footwall, the fault is a **normal fault**. When the hanging wall moves up relative to the footwall, the fault is a **reverse fault**.

When the dip of a fault is less than 45° it is referred to as a **low-angle fault**. A low-angle normal fault is just referred to as such, but sometimes as a lag fault. However, low-angle reverse faults have special names. In general, if the dip of the reverse fault is less than 45° (for some the angle should be 30°), it is referred to as a **thrust** or a **thrust fault**. A thrust fault with very large displacements, sometimes set at 5 km or more, is commonly referred to as an **overthrust**.

The words fault and a fault zone are used interchangeably in this text. There is, however, a slight difference between these terms. A **fault** is any planar discontinuity or a fracture across which the rock displacement is primarily parallel with the fracture plane. A **fault zone** is a tabular body with many faults and fractures as well as one or more cores. The faults and fractures of the zone may constitute the damage zone of a main fault, or they may be individual faults of more or less equal sizes that happen to be concentrated within a certain region. Until recently, a fault zone was often regarded as a zone of many parallel faults,

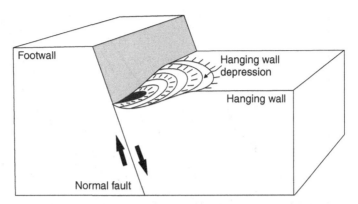

Fig. 12.1  Dip-slip fault, a normal fault, with the location of the hanging wall and the footwall indicated. There is commonly a hanging-wall depression somewhere along the central part of a major normal fault (cf. Fig. 1.10).

usually of similar sizes. But in recent years the term fault zone has been more widely used for one major fault with numerous smaller faults (and other fractures) in its damage zone, and with one or more well-defined cores. Most uses in this book refer to the second definition.

It is important to distinguish between a fault zone and a shear zone. A fault zone is primarily a structure formed by brittle deformation. By contrast, in this book a **shear zone** is a structure formed primarily by ductile deformation.

The horizontal and vertical components of the displacement are referred to as the **heave** and the **throw**, respectively (Fig. 1.10). The term heave is not used much in practice, but the use of the term throw is very common. In fact, in field studies it is normally the throw or vertical displacement that is measured rather than the true displacement. This follows because the displacement is commonly measured using a marker layer or horizon on either side of the fault, and the vertical displacement of this layer is normally the easiest to measure. In fact, the fault plane itself is commonly too poorly exposed to allow the true displacement to be measured.

The strike and dip of a dip-slip fault are measured as for extension fractures (Chapter 11). The throw on a dip-slip fault can be measured in various ways. The exact method used depends on the outcrop and the size of the displacement. If the displacement is small and there is a clear marker layer or horizon, such as a contact, then the measurement is normally made with a measuring tape (Fig. 12.2). The throw is easily measured when there is a clear marker horizon, and since the fault plane is clear, the strike and dip can be measured as well. It should be noted, however, that most fault planes are somewhat undulating and often more accurate average strike and dip measurements are obtained from a distance than from using any particular part of the fault plane itself for placing the compass.

When the displacement is tens of metres or more, there are many methods of measuring the throw. One is to use an altimeter. This may give reasonably accurate measurements, particularly in areas with comparatively stable weather so that the air pressure does not change much during the time when the measurements are being made (usually a whole day). They are less suitable, however, in areas where there are common depressions with changes in air pressure.

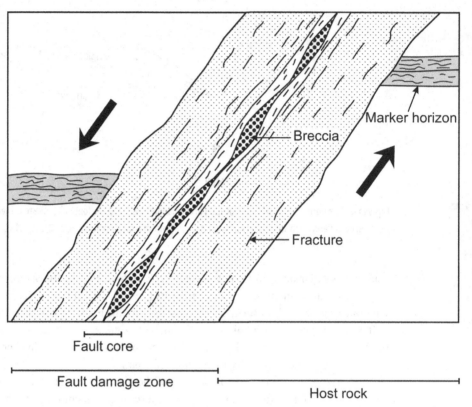

**Fig. 12.2** Marker horizons or marker layers, that is, contacts or layers easily identified on both sides of a fault make it possible to determine the fault slip quite accurately. Also indicated are the fault core and the damage zone of a typical dip-slip fault.

Another method is to use GPS instruments. Handheld GPS give reasonably accurate locations and can be used, particularly for the estimate of large displacements. The accuracy of GPS is increasing every year, so that these will be used more and more as time goes on. The third method is to use oblique photographs take from the ground or from helicopters or aircrafts. Then there must be one or more layers or intrusions with known thicknesses, or the total height of the cliff or mountain where the fault is located must be known, and from the marker layers the displacement can be estimated. This method can be very accurate for subvertical sea cliffs, such as in many parts of Greenland and Norway and Hawaii and other oceanic islands, but is less accurate when the faults are located in cliffs and mountain sides that are sloping much less than vertical.

The fourth method mentioned here is the use of handheld clinometers for levelling. This method is based on locating a marker layer in the footwall of the fault and then climbing up, while levelling, to the marker layer in the hanging wall (Fig. 12.2). When the mountain slope is not too steep, so that it is possible to climb it without special equipment, this method can yield quite accurate results. This method can also be used to measure the thicknesses of layers if they form vertical cliffs.

So far we have discussed dip-slip faults as seen at depth, that is, at crustal levels where all the principal stresses are compressive. Recall from Chapter 7 that this depth is generally a few hundred metres or less. At shallower depths, and at the surface of an active rift zone, one of the principal stresses, $\sigma_3$, may be tensile during the slip on a normal fault. If the tensile strength of the rock hosting the normal fault at the surface is close to zero, the fault may maintain its dip right to the surface (Figs. 1.10 and 12.2). For such a fault, the measurements of the displacement are straightforward and can be made in the field using levelling techniques of various types and accuracies, or from micrometers using stereoscopic technique on images, or directly from digital elevation maps.

If the tensile strength of the surface host rock is significant, say a few megapascals, the surface part of the normal fault may be an open, mixed-mode fracture (Fig. 9.10). Large-scale faults and tension fractures of this kind are referred to as **fissures**, hence the name **fissure swarm** (Figs. 12.3 and 12.4). Good examples of an open, gaping, normal fault are shown in Figs. 9.10 and 12.3. In these cases, the opening and the vertical displacement can be measured either from images, digital elevation maps, or in the field using various electronic techniques (such as lidar). The displacement can also be measured using a levelling technique, described above, and a measuring tape.

Normal faults show a number of interesting structures associated with their development. Some of the well-known structures may be discussed briefly as follows.

**Relay ramps**. These are sloping layers (ramps) where the displacement along a fault segment is relayed or passed on or transferred to another segment (Figs. 12.3 and 12.4).

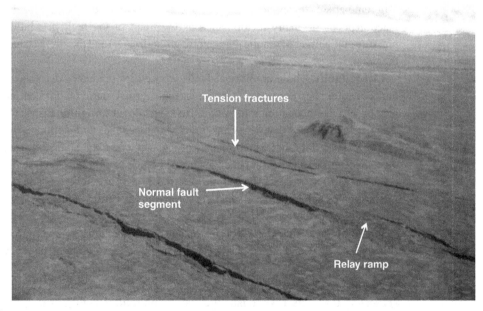

**Fig. 12.3**  Part of the Krafla Fissure Swarm, North Iceland (Figs. 3.4–3.6). Aerial view east, fissure swarms like this one, located in a Holocene pahoehoe lava flow, contain numerous tension fractures and normal faults, many of which have relay ramps.

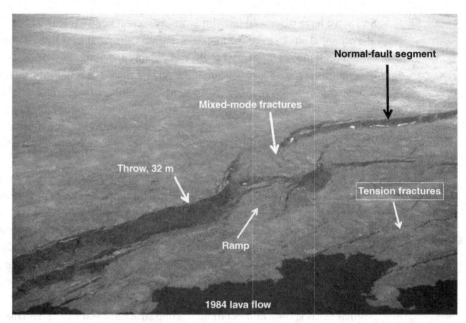

**Fig. 12.4** Part of the main western boundary fault, a normal fault, in the Krafla Fissure Swarm, North Iceland (Figs. 3.4–3.6). The faults link up through curved, hook-shaped fractures, and in-between the main fault segments there are mixed-mode fractures, partly tensile and partly shear (cf. Figs. 12.5 and 12.7).

The ramps are a direct consequence of how faults commonly develop from initially offset smaller fractures and fault segments (Fig. 12.5). They may be conduits of fluids between different parts of the associated fault zone. However, whether such a fluid transport takes place depends on many factors, such as the trend of the fault in relation to the fluid-pressure gradient (or hydraulic gradient for ground water) and the stresses that concentrate in the ramp.

**Listric faults**. These are normal faults whose dips gradually decrease with depth; the faults therefore become concave upwards (Fig. 12.6). A listric fault is commonly accompanied by an accommodation fold, referred to as a **rollover anticline**. Listric faults are mostly associated with rocks that behave as largely homogeneous and isotropic.

**Ramps and flats**. In layered rocks, particularly where the layers have widely different mechanical properties, abrupt changes in fault dip are common. Some segments may then coincide with layer contacts, and are referred to as **flats**, whereas other segments form inclined paths across the layers and are referred to as **ramps**. Flats and ramps occur in all types of dip-slip faults.

**Fault linkage**. Many faults grow by linkage of smaller fault segments into larger segments (Figs. 12.4, 12.5, and 12.7; Chapter 14). The basic mechanics of fault linkage are well understood and depend in detail on the offset and underlap of the nearby fault segments in addition to host-rock properties and loading. For a pair of nearby fault segments in a vertical section, **offset** is the lateral distance and **underlap** the vertical distance between their nearby ends or tips (Fig. 12.7). If the underlap is negative, the configuration is referred to as **overlap**.

**Fig. 12.5** View southwest along the Almannagja Fault, the western boundary fault of the Thingvellir Graben (Figs. 3.13 and 3.14). The major segments curve towards each other and link roughly where the road crosses the fault.

The fractures that eventually link the fault segments are of two types: **hook-shaped fractures** and **transfer faults** (Fig. 12.8). The hook-shaped fractures are ideally extension fractures, but may develop into normal faults. If the linked fault segments are open (gaping) fractures at the surface, or filled with a fluid, then the angle between the hook-shaped fracture and the fault segment should be about 90°. This follows because then there is no shear stress acting (laterally) parallel with the fault, that is, its open part is a free surface. Otherwise, the angle should be less than 90°. Transfer faults are, as the name implies, shear fractures and form along high shear-stress zones. Before linking has occurred, either through hook-shaped fractures or transfer faults, the offset fault segment is sometimes regarded as being **soft-linked**. Similarly, once linking of the segments has occurred, the offset segments are sometimes referred to as being **hard-linked**. It should be understood, however, that long before the actual linkage occurs, the segments start to function as a single fault.

## 12.3  Strike-slip faults

Strike-slip faults are those where the displacement is primarily parallel with the strike of the fault (Fig. 12.9). They are also sometimes referred to as wrench faults or transcurrent faults. Strike-slip faults are normally steeply dipping, and many are close to vertical.

Fig. 12.6 Listric fault in young sedimentary rocks (late Quaternary tillites) in the Husavik-Flatey Fault in North Iceland (Fig. 12.16). The person provides a scale.

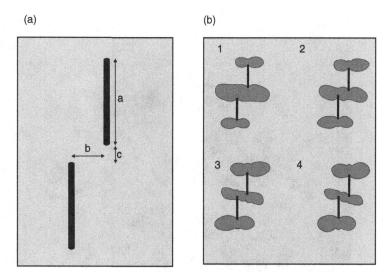

(a) Two fractures of length *a*, offset *b*, and underlap *c* subject to tensile loading. (b) Schematic presentation of the tensile and shear stresses between the nearby tips of the fractures (in different configurations, 1–4) where transfer/transform faults would tend to develop (cf. Gudmundsson, 2007).

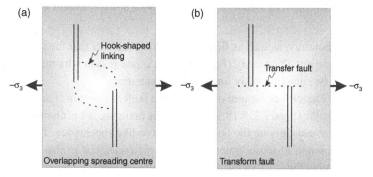

Offset fracture/fault segments link primarily through (a) hook-shaped, curved fractures (Figs. 12.4 and 1.5) and through (b) transfer/transform faults (Fig. 12.16).

Depending on the direction of movement of the fault block opposite the observer, there are two main types of strike-slip faults. If the opposite fault block (the fault block on the far side of the fault) moves to the right, the fault is termed **dextral** or **right-lateral**. If the opposite fault block moves to the left, the fault is termed **sinistral** or **left-lateral** (Figs. 7.6 and 12.10).

In many areas of strike-slip faulting, there are pairs of dextral and sinistral faults making an angle of about 60°. These pairs are referred to as **conjugate faults**. The angle between them is the conjugate angle, which is bisected by the maximum principal compressive stress, $\sigma_1$ (Figs. 7.6 and 12.10; Chapter 7). Conjugate faults are common among strike-slip faults, but they occur among all types of faults and shear fractures in general. Experiments in the

Fig. 12.9 Sinistral (conjugate) part of one of the earthquake faults (mainly dextral) which slipped during the June 2000 earthquakes in the South Iceland Seismic Zone. Aerial view, the measured fault slip is about 1 m.

laboratory, as well as field observations, indicate that the faults in the conjugate pair often form simultaneously (Fig. 7.3). This is one reason why many conjugate joints are thought to be simultaneously formed shear fractures.

The surface rupture of a strike-slip fault has certain characteristics that are easily recognisable (Figs. 7.7, 9.18, and 12.9). The main one is a pronounced en echelon arrangement of segments, giving the fault trace a wave-like appearance. This characteristic is particularly common in loose, granular surface materials, but is also seen in Holocene basaltic lava flows (Figs. 7.7 and 9.18). The fracture segments generated at the surface during slip on a strike-slip fault also have a special arrangement. Thus, for a dextral strike-slip fault these fractures are left-stepping, that is, where one segment of a pair ends, the next appears to the left of the previous one (Fig. 9.18). Similarly, for a sinistral strike-slip fault, the associated fractures are right-stepping.

Among the most conspicuous features associated with strike-slip faults are pull-aparts and push-ups. **Pull-aparts** are structures that open up due to local tension, referred to as transtension, along faults with irregular strike (Fig. 12.11). Large pull-aparts may become filled with water, forming sag-ponds or basins (Fig. 12.12). These are surface, or near-surface features, but at greater depths where absolute tension is unlikely the associated relative tension may help generate abnormally thick hydrofractures, as is often seen in mineral veins (Fig. 12.13). Part of the volcanism associated with many transform faults may be related to pull-apart structures. This means that the extension in the pull-aparts reduces the minimum principal compressive stress, $\sigma_3$, so as to allow dykes to be injected.

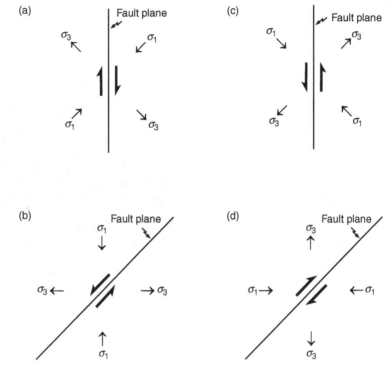

**Fig. 12.10**   Main faults in the South Iceland Seismic Zone are conjugate strike-slip faults, where the north-trending faults are dextral (a) and the northeast-trending faults sinistral (b). However, there are also some north-trending sinistral segments (c), as well as northeast-trending dextral segments (d) (cf. Figs. 7.7, 9.18, and 12.9).

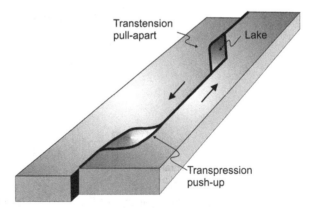

**Fig. 12.11**   Sinistral strike-slip faults, of irregular shape, give rise to regions of transtension, where pull-apart structures may develop, and regions of transpression, where push-up structures may develop.

A well-known pull-apart basis is the Dead Sea on the boundary between Israel and Jordan. The Dead Sea is related to a sinistral strike-slip fault, the Dead Sea Transform Fault, which marks the boundary between the western part of the Arabian Plate and the northern part of the African Plate. The Dead Sea is just one of several pull-apart basins along the transform

**Fig. 12.12**    Pull-apart basin, a sag pond, on the Tjornes Peninsula, along the Husavik-Flatey Fault, North Iceland (Fig. 12.16).

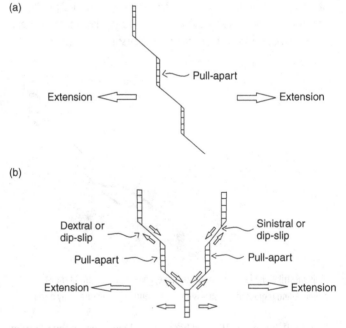

**Fig. 12.13**    Pull-apart structures are not restricted to strike-slip faults but can also develop along dip-slip faults. Such structures may explain unusual thickness variations in mineral veins. This section can be viewed either as vertical (for a dip-slip fault) or horizontal (for a strike-slip fault).

**Fig. 12.14**  Small push-up (about 20 cm high) in grass/soil along a dextral fault segment that slipped during the June 2000 earthquakes in South Iceland.

fault, some of which reach many tens of kilometres in length and up to ten kilometres in width.

**Push-ups** are structures generated by local compression, referred to as transpression, along faults with irregular strike (Fig. 12.11). These structures vary in size from centimetres to hundreds of metres or more. Consider first small push-ups that were generated during strike-slip faulting (Fig. 12.14). This is a very good example of push-up in a soft material, that is, in soil and grass. These were generated during M6.5 earthquakes in the South Iceland Seismic Zone in 2000. A larger example is provided by push-ups in the Holocene lava flows of the South Iceland Seismic Zone (Fig. 12.15). This push-up is ridge-shaped and belongs to the dextral fault seen in Figs. 7.7 and 9.18. The end of the ridge (Fig. 12.15) is a 4-m-high hill of broken Holocene pahoehoe lava flow. This push-up is related to a Holocene earthquake of magnitude of at least 7.1–7.3. The estimated slip, partly based on the size of this push-up, is just over two metres.

Push-ups can be much larger. As an example of a large-scale push-up, consider the Tjornes Peninsula in North Iceland (Fig. 12.16). The northern peninsula has been slowly rising, uplifted for at least 1–2 Ma. The total uplift, as estimated from marine sediments exposed on land, is as much as 600 m. This uplift is presumably partly due to the transpressional effects of the Husavik-Flatey Fault, a dextral transform fault. Thus, the peninsula, long regarded as a horst, may be viewed as a large push-up uplifted for millions of years in response to transpression generated by the Husavik-Flatey Fault. As a result of the uplift, the fault pattern on the peninsula is not that of a simple rifting but rather much more complex. This follows because some of the faults are rift-related, whereas others are related to the transpression and uplift (Gudmundsson, 2007).

Pull-aparts and push-ups are not restricted to strike-slip faults. These structures are commonly seen in vertical sections along normal faults, for example Fig. 12.13. Many veins are

**Fig. 12.15**    Part of a push-up ridge along a dextral segment of the fault shown in Figs. 7.7 and 9.18.

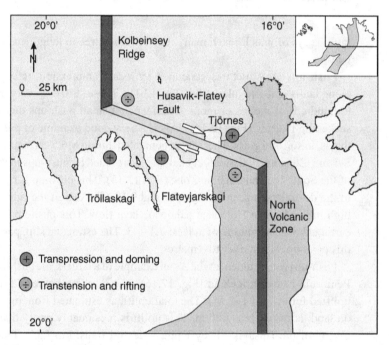

**Fig. 12.16**    The Husavik-Flatey Fault, a transform fault partly exposed on land in North Iceland, has caused an uplift of the peninsulas of Tjornes and Flateyjarskagi (Gudmundsson, 2007).

unusually thick in certain layers. The thick parts tend to be along regions of transtension, that is, they are pull-apart structures. Similarly, transpression regions, while not so conspicuous, also occur along dip-slip faults. During repeated slips on a fault, it develops into a fault zone where pull-aparts and transpression areas gradually become a part of the core and the damage zone and thus often disappear. Many of these features, as regards dip-slip faults, are therefore most easily recognisable on faults with small displacements (Fig. 12.13).

## 12.4 Slickensides and oblique-slip faults

The sense of movement along fault planes can often be inferred from slickensides. **Slicken-sides** are fault surfaces that have been polished by frictional sliding (Angelier, 1984, 1990). The surfaces are often very smooth, even shiny or mirror-like. Slikensides commonly form on the host rock of the fault, but sometimes within part of the core, primarily in gouge.

To infer the direction of fault slip, we measure the **slickenlines.**These are mostly straight lines on the fault surface. They are generated by irregularities or asperities on one fault wall that project into the other fault wall, usually resulting in **striations** or **stria**. The striations thus indicate the general type of fault slip, that is, whether the last slip on the fault was primarily dip-slip or strike-slip. To decide if the dip-slip fault was normal or reverse, or if the strike-slip fault was sinistral or dextral, additional information is needed. The most accurate information found on the slickensides are **Riedel shears**, that is, surface fractures or cavities that project into one of the fault walls. They usually make an acute angle with the fault plane in the direction that the missing wall (the wall causing the Riedel-shear fracture) was moving. From this movement, the sense of slip can be fully determined.

Slickensides, marker layers, and other indicators show that many faults are neither pure dip-slip nor pure strike-slip but rather **oblique-slip**. On some faults the deviation from pure dip-slip or strike-slip is small. For example, many normal faults in Iceland show stria that plunge about 80°, so that the strike-slip component is minor. Other faults, however, show stria plunging about 45°, indicating that the strike-slip and the dip-slip components were equal.

Many faults also show evidence of having slipped in various directions during their evolution. This is commonly seen in the slickensides. On a slickenside there may be many different sets of stria. One set, for example, may indicate a pure dip-slip, another a pure strike-slip, and the third an oblique-slip. Normally, the younger stria, that is, the later slips, are the clearest and by looking at the different sets of stria, how weathered they are and dissected by later sets, we can infer crudely the slip-history of the fault and its relation with the evolution of the local stress field.

A clear example of changes in the slip history of faults occurs in the South Iceland Seismic Zone (Chapter 16). Many strike-slip faults in this zone show evidence of changing their sense of slip (Gudmundsson, 2007). They may start as dextral strike-slip faults and later be reactivated as sinistral strike-slip faults, or vice versa. Other examples of the chance in the sense of slip on faults occurs among some dip-slip faults. In particular, normal faults that are subsequently subject to a horizontal tectonic compressive stress perpendicular to the strike of the faults may result in some reverse movement along the faults. Good examples of this occur in many sedimentary basins where, during inversion, many normal faults develop some reverse slips. Also, in rift zones, some normal faults may develop reverse slips when overpressured dykes become emplaced in their vicinity. Many examples of this are known from the lava pile in Iceland. A fine example is seen in Fig. 1.6, where a dyke with an overpressure of, say, 10 MPa, has generated a reverse slip on one of the boundary faults of a nearby graben.

# 12.5 Summary

- If the slip (displacement) is parallel with the dip of the fault, it is known as a dip-slip fault. The block above the fault plane is the hanging wall; the block below the fault plane is the footwall. If the hanging wall moves down relative to the footwall, the fault is a normal fault. If the hanging wall moves up relative to the footwall, the fault is a reverse fault. If the dip of a reverse fault is less than 45° (some put the critical dip as low as 30°) it is referred to as a thrust or a thrust fault. The horizontal and vertical components of the fault displacement are referred to as the heave and the throw.
- A single fault is a fracture across which the displacement of the rocks on either side is primarily parallel with the fracture plane. A fault zone is a tabular rock body with many faults and fractures, commonly with one or more zones of breccias and gouge. The faults and fractures of the tabular body may constitute the damage zone of the main fault, whereas the breccia and gouge constitute the core of the main fault. A fault zone is primarily formed through brittle deformation (fracturing), whereas its extension into the deeper, warmer part of the crust is a shear zone, primarily formed through ductile deformation (plastic flow).
- Swarms of large tension fractures and normal faults are known as fissure swarms. The sloping layers (ramps) where the displacement along a fault segment is relayed or transferred to another segment are known as relay ramps. When the dip of a (normal) fault gradually decreases with depth, it is known as a listric fault. Fault segments that coincide with layer contacts are known as flats, the steeper segments (dissecting the layers) as ramps, the overall fault structure as flats and ramps. For a pair of nearby fault segments in a vertical section, offset is the lateral distance and underlap the vertical distance between their nearby tips. A negative underlap is known as overlap.
- Many faults grow through segment linkage. The two main types of linkage are hook-shaped (primarily) extension fractures, and transfer faults. Mechanically similar linkages at mid-ocean ridges are referred to as overlapping-spreading centres and transform faults, respectively.
- The above summary refers primarily to dip-slip faults. If the slip (displacement) is parallel with the strike of the fault, it is known as a strike-slip fault. If the block on the far side of the fault has moved to the left, the fault is known as sinistral or left-lateral; if the block has moved to the right, the fault is known as dextral or right-lateral. Strike-slip faults commonly occur in pairs as seen in lateral sections, one sinistral the other dextral, with an angle of about 60° between, known as conjugate faults (such pairs are also common in vertical sections among dip-slip faults). Local extension along a strike-slip fault may generate a pull-apart structure, whereas local compression along a strike-slip fault generates a push-up.

- Slickensides are fault surfaces that have been polished by frictional sliding. Slickenlines, striations or stria, on the fault surface form a part of the slickensides and can be used to infer the direction (and commonly also the sense) of slip. Sense of slip means whether, for example, a dip-slip fault is a reverse fault or a normal fault. Additional features used to infer the sense of slip are Riedel shears, that is, surface fractures or cavities that project into one of the fault walls (surfaces) at an acute angle to the fault surface. Systematic studies of slickensides and associated structures can be used to infer the palaeostresses associated with the fault development.

## 12.6 Worked examples

### Example 12.1

#### Problem

The fault in Fig. 12.9 is a conjugate, sinistral part of a 20-km-long mainly dextral strike-slip fault in the South Iceland Seismic Zone. The rupture occurred during an M6.5 earthquake in June 2000. The entire seismogenic layer, about 10 km thick, ruptured during the earthquake (Pedersen *et al.*, 2003). If the average dynamic Young's modulus of the seismogenic layer is 85 GPa and Poisson's ratio is 0.25, and the maximum surface displacement is 1 m, calculate the driving stress (or stress drop) associated with the fault slip.

#### Solution

From Eq. (9.26), the maximum total surface displacement $\Delta u_{\mathrm{III}}$ and the driving stress $\tau_d$ are related to the formula

$$\Delta u_{\mathrm{III}} = \frac{4\tau_d(1+v)R}{E}$$

This equation can be rewritten to find the driving shear stress $\tau_d$ as follows:

$$\tau_d = \frac{\Delta u_{\mathrm{III}}E}{4(1+v)R} = \frac{1\,\mathrm{m} \times 8.5 \times 10^{10}\mathrm{Pa}}{4 \times (1+0.25) \times 1 \times 10^4\mathrm{m}} = 1.7 \times 10^6\mathrm{Pa} = 1.7\,\mathrm{MPa}$$

This is a reasonable, although comparatively low, driving stress. Other estimates (Pedersen *et al.*, 2003) indicate that the maximum slip may have been 2–3 times the surface slip, in which case the driving stress would be correspondingly higher.

### Example 12.2

#### Problem

In palaeorift zones, such as in the Tertiary areas of East and West Iceland, one occasionally finds steeply dipping reverse faults. All these faults are located in deeply eroded (by glaciers) valleys and coastal areas. The level of erosion is commonly between 1 and 2 km. Assume that the state of stress before glacial erosion was lithostatic and that 1.5 km of the crust were removed by geologically rapid glacial erosion. The average density of the uppermost part of the crust is taken as 2600 kg m$^{-3}$ and its Poisson's ratio as 0.25. Estimate roughly the surface crustal stresses after the erosion and use the results to explain steeply dipping reverse faults.

### Solution

Before glacial erosion the state of stress was lithostatic so that $\sigma_1 = \sigma_2 = \sigma_3$. The lithostatic stress is equal to the overburden pressure, that is, to the vertical stress $\sigma_v$. To find the stress change due to the erosion, we compare the stresses before and after the erosion as follows
   Before erosion the stress at a depth of 1.5 km was, from Eq. (4.45),

$$\sigma_v = \rho_r g z = 2600 \text{ kg m}^{-3} \times 9.81 \text{ m s}^{-2} \times 1500 \text{ m} = 3.826 \times 10^7 \text{Pa} = 38 \text{ MPa}$$

During the erosion 1500 m of rock is removed, that is, the vertical stress at the new surface is reduced to zero, that is,

$$\sigma_v = \rho_r g z - \rho_r g z = 0 \text{ Pa}$$

The horizontal stress is also reduced due to the erosion, but by the value determined by Eq. (4.53), as follows:

$$\sigma_h = \sigma_v - \frac{\sigma_v}{m - 1} = 3.826 \times 10^7 \text{Pa} - \frac{3.826 \times 10^7 \text{Pa}}{4 - 1} = 2.55 \times 10^7 \text{Pa} = 25.5 \text{ MPa}$$

Horizontal stress of 25 MPa is certainly enough to generate reverse movements on favourably orientated faults. Such faults are mostly old normal faults that become reactivated by the horizontal compressive stress as reverse faults. The reverse faults tend to be very steeply dipping for the simple reason that most of the existing normal faults in the Tertiary areas of Iceland are very steeply dipping. The high horizontal compressive stresses generated by glacial erosion are also a principal reason for the formation of surface-parallel exfoliation joints (Fig. 11.8).
   Reactivation of normal faults as reverse faults is, of course, not confined to horizontal compression generated by glacial erosion. Horizontal compressive stresses generated by nearby overpressured dykes may be sufficiently high to cause reverse slip on existing normal faults (Fig. 1.6).

## 12.7 Exercises

12.1  For a dip-slip fault define hanging wall and footwall.
12.2  Explain the concepts of a normal fault, a reverse fault, a thrust fault, and an overthrust.
12.3  Define a fault, a fault zone, and a shear zone.
12.4  Define heave and throw.
12.5  What are a fissure and a fissure swarm?
12.6  What is a relay ramp?

12.7  What is a listric fault?

12.8  What are ramps and flats?

12.9  For a segmented fracture, define offset, underlap, and overlap.

12.10  For fracture/fault linkage, explain what is meant by a hook-shaped fracture, a transfer fault, soft-linked segments, and hard-linked segments.

12.11  Explain what is meant by a sinistral and a dextral strike-slip fault.

12.12  What is a pull-apart? Give examples.

12.13  What is a push-up? Give examples.

12.14  Explain slickensides, slickenlines, stria, and Riedel shears and how these can be used in fault analysis.

12.15  During seismogenic slip, a sinistral strike slip fault ruptures a 15-km-thick seismogenic layer with an average dynamic Young's modulus of 90 GPa and a Poisson's ratio of 0.25. The driving shear stress associated with the fault slip is estimated at 4 MPa. Calculate the surface displacement.

# References and suggested reading

Amadei, B. and Stephansson, O., 1997. *Rock Stress and its Measurements*. New York: Chapman & Hall.

Angelier, J., 1984. Tectonic analysis of fault slip data sets. *Journal of Geophysical Research*, **89**, 5835–5848.

Angelier, J., 1990. Inversion of field data in fault tectonics to obtain the regional stress 3. A new rapid direct inversion method by analytical means. *Geophysical Journal International*, **103**, 363–376.

Angelier, J. and Bergerat, F., 2002. Behaviour of a rupture of the 21 June 2000 earthquake in South Iceland as revealed in asphalted car park. *Journal of Structural Geology*, **24**, 1925–1936.

Angelier, J., Bergerat, F., and Homberg, C., 2000. Variable coupling across weak oceanic transform fault: Flateyjarskagi, Iceland. *Terra Nova*, **12**, 97–101.

Angelier, J., Bergerat, F., Bellou, M., and Homberg, C., 2004. Co-seismic strike-slip fault displacement determined from push-up structures: the Selsund Fault case, South Iceland. *Journal of Structural Geology*, **26**, 709–724.

Bergerat, F. and Angelier, J., 2003. Mechanical behaviour of the Arnes and Hestfjall Faults of the June 2000 earthquakes in Southern Iceland: inferences from surface traces and tectonic model. *Journal of Structural Geology*, **25**, 1507–1523.

Cartwright, J. A., Trudgill, B. D., and Mansfield, C. S., 1995. Fault growth by segment linkage: an explanation for scatter in maximum displacement and trace length data from the Canyonlands Grabens of SE Utah. *Journal of Structural Geology*, **17**, 1319–1326.

Clark, R. M. and Cox, S. J. D., 1996. A modern regression approach to determining fault displacement-length scaling relationships. *Journal of Structural Geology*, **18**, 147–152.

Cowie, P. A. and Scholz, C. H., 1992. Displacement-length scaling relationships for faults: Data synthesis and discussion. *Journal of Structural Geology*, **14**, 1149–1156.

Davis, G. H. and Reynolds, S. J., 1996. *Structural Geology of Rocks and Regions*, 2nd edn. New York: Wiley.

Dawers, N. H., Anders, M. H., and Scholz, C. H., 1993. Growth of normal faults: displacement-length scaling. *Geology*, **21**, 1107–1110.

Gudmundsson, A., 2007. Infrastructure and evolution of ocean-ridge discontinuities in Iceland. *Journal of Geodynamics*, **43**, 6–29.

Kanamori, H. and Anderson, D. L., 1975. Theoretical basis of some empirical relations in seismology. *Bulletin of the Seismological Society of America*, **65**, 1073–1095.

Kasahara, K., 1981. *Earthquake Mechanics*. New York: Cambridge University Press.

Moores, E. M. and Twiss, R. J., 1995. *Tectonics*. New York: W.H. Freeman.

Nicol, A., Watterson, J., Walsh, J. J., and Childs, C., 1996. The shapes, major axis orientations and displacement pattern of fault surfaces. *Journal of Structural Geology*, **18**, 235–248.

Park, R. G., 1997. *Foundation of Structural Geology*, 3rd edn. London: Routledge.

Pedersen, R., Jonsson, S., Arnadottir, T., Sigmundsson, F., and Feigl, K., 2003. Fault slip distribution of two June 2000 Mw 6.4 earthquakes in South Iceland estimated from joint inversion of InSAR and GPS measurements. *Earth and Planetary Science Letters*, **213**, 487–502.

Pollard, D. D. and Segall, P., 1987. Theoretical displacements and stresses near fractures in rock: with applications to faults, joints, veins, dikes and solution surfaces. In: Atkinson, B. (ed.), *Fracture Mechanics of Rock*. London: Academic Press, pp. 277–349.

Pollard, D. D. and Fletcher, R. C., 2005. *Fundamentals of Structural Geology*. Cambridge: Cambridge University Press.

Price, N. J. and Cosgrove, J. W., 1990. *Analysis of Geological Structures*. Cambridge: Cambridge University Press.

Rice, J. R., 1980. The mechanics of earthquake rupture. In: Dziewonski, A. M. and Boschi E. (eds.), *Physics of the Earth's Interior*. Amsterdam: North Holland, pp. 555–649.

Rippon, J. H., 1985. Contoured patterns of the throw and hade of normal faults in the Coal Measures (Westphalian) of north-east Derbyshire. *Proceedings of the Yorkshire Geological Society*, **45**, 147–161.

Sanderson, D. J. and Marchini, W. R. D., 1984. Transpression. *Journal of Structural Geology*, **6**, 449–458.

Schlische, R. W., Young, S. S., Ackermann, R. V., and Gupta, A., 1996. Geometry and scaling relations of a population of very small rift-related normal faults. *Geology*, **24**, 683–686.

Scholz, C. H., 1990. *The Mechanics of Earthquakes and Faulting*. New York: Cambridge University Press.

Schultz, R. A., 1997. Displacement-length scaling for terrestrial and Martian faults: implications for Valles Marineris and shallow planetary grabens. *Journal of Geophysical Research*, **102**, 12 009–12 015.

Twiss, R. J. and Moores, E. M., 2007. *Structural Geology*, 2nd edn. New York: W.H. Freeman.

van der Pluijm, B. A. and Marshak, S., 2004. *Earth Structure*, 2nd edn. London: W.W. Norton.

Willemse, E. J. M., Pollard, D. D., and Aydin, A., 1996. Three-dimensional analysis of slip distributions on normal fault arrays with consequences for fault scaling. *Journal of Structural Geology*, **18**, 295–309.

Yeats, R. S., Sieh, K., and Allen, C. R., 1997. *The Geology of Earthquakes*. Oxford: Oxford University Press.

# 13 Evolution of extension fractures

## 13.1 Aims

How do small extension fractures link to form larger fractures? More specifically, what main factors determine the path of an extension fracture? These topics are of great importance in understanding the development and maintenance of permeability in crustal rocks and reservoirs. The topic is also fundamental in understanding how volcanoes work. This follows because most volcanic eruptions are supplied with magma through dykes and inclined sheets, both of which are extension fractures. For assessing volcanic hazards and risks, we need to understand how dykes and sheets propagate to the surface, resulting in an eruption, or, alternatively, become arrested within layers or at layer contacts or other discontinuities. The main aims of this chapter are to present results on the:

- General development of tension fractures.
- General propagation and path-selection of hydrofractures.
- Conditions for hydrofracture deflection and arrest.
- Apertures of hydrofractures subject to constant fluid overpressure.
- Apertures of hydrofractures subject to varying fluid overpressure.
- Surface deformation related to arrested hydrofractures.

## 13.2 Development of tension fractures

Tension fractures develop in a similar way to hydrofractures. Most of the conclusions regarding hydrofracture propagation, deflection, and arrest, presented below, apply, with modifications, to tension fractures as well. The main differences are, first, that large tension fractures are restricted to shallow crustal depths, usually the uppermost 0.5–1 km of the crust (Chapters 7 and 8). Second, tension fractures are driven by external tensile stress, whereas hydrofractures are driven by internal fluid overpressure. While tension fractures are common close to and at the surface of rift zones, grabens, and other areas undergoing extension, hydrofractures are common everywhere in the brittle crust and upper mantle. Hydrofractures are much more important than tension fractures for transporting fluids in the crust. Thus, the focus in this chapter is on hydrofractures. But first we provide a brief overview of the development of tension fractures.

Fig. 13.1 Fracture opening in a Holocene pahoehoe lava flow in the rift zone of Iceland. View vertical, the fracture is formed by tensile stresses, and thus is essentially a tension fracture. However, because the columnar (hexagonal) joints (Chapter 11) function as weaknesses, part of the fracture follows the joints in directions that deviate from being perpendicular to $\sigma_3$. Such irregularities are common in newly formed, narrow tension and extension fractures, but less marked as the fracture-opening displacement increases. The general trend of the fracture is perpendicular to $\sigma_3$.

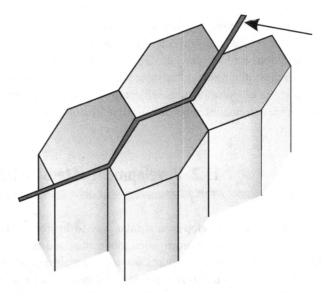

Fig. 13.2 Propagating tension-fracture path (the front indicated by an arrow) partly follows the columnar joints in the lava flow; hence the irregular or zigzag geometry seen in Fig. 13.1.

Like other fractures, tension fractures commonly start from small joints or other weaknesses that gradually link into larger fracture segments. Consider, for example, the evolution of tension fractures in a rift zone. They nucleate, that is, start their development, from columnar joints (Figs. 13.1–13.3) and from these expand laterally and vertically. As a consequence

Tension fracture, partly opened in a rifting episode in North Iceland in 1975–1984 in the Krafla Fissure Swarm (located in Figs. 3.4 and 3.6). View south, the person provides a scale. The fracture is still irregular, partly due to its growth along columnar joints, but, because of its much greater opening, the irregularity is much less noticeable than in Fig. 13.1. Photo: Sonja L. Philipp.

of the weaknesses being columnar joints, the original, small tension fractures are commonly irregular, that is, with a zigzag geometry (Figs. 13.1 and 13.2). During further rifting episodes, the small irregular tension fractures start to link laterally and vertically into larger segments.

In a lateral section, the general development of tension fractures is as indicated in Fig. 13.4. During rifting events, the tension fractures gradually become longer. When the fractures reach a certain length that is large in comparison with the distances between the nearby tips of each fracture pair, they start to act as a single fracture.

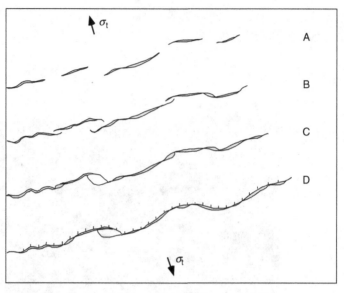

Fig. 13.4 General evolution of tension fractures into a normal fault in a rift-zone fissure swarm. The first stage is A (offset tension fractures), the last stage is D (where the original tension fractures have linked up into a major normal fault). As the tension fractures link up (A–C) and the resulting fracture increases in strike dimension (length), its dip dimension or depth also increases. When the depth reaches a critical limit, the developing tension fracture must change into a normal fault.

If the tension fractures in the set are collinear, or close to collinear, and equally spaced, then the opening displacement $u(x/a)$ of a set of tension fractures is given by (Sneddon and Lowengrub, 1969; Gray, 1992)

$$u(x/a) = \frac{4\sigma(1-\nu^2)d}{E\pi}\left[\ln\left\{\cos\left(\frac{\pi x}{2d}\right) + \left[\cos^2\left(\frac{\pi x}{2d}\right) - \cos^2\left(\frac{\pi a}{2d}\right)\right]^{1/2}\right\} - \ln\cos\left(\frac{\pi a}{2d}\right)\right]$$

$$(13.1)$$

where $\sigma$ is the tensile stress, $d$ is half the distance between the centres of a pair of fractures, $a$ is half the strike dimension of each fracture (all are assumed to be of equal strike dimension), $\nu$ is the Poisson's ratio of the host rock and $E$ is its Young's modulus (Fig. 13.5). Equation (13.1) yields an elliptical opening displacement (Fig. 13.5). As the distance $2d$ between the fractures decreases the fracture opening displacement approaches that of a single fracture of a length equal to the total length of all the fractures. Thus, for instance, when $a/d$, the ratio between the strike dimension of a fracture and its distance from a nearby fracture, increases from 0 to 0.9, the opening displacement $u$ increases by about 80%.

Since the maximum opening of the fracture set approaches that of a single fracture with a length of the total set, the tension fractures gradually function more and more as a single fracture until, eventually, they combine physically in to a single fracture (Fig. 13.4; Chapter 11). If rifting events continue, and the tension fractures reach a critical depth (Chapter 8), they change into a normal fault (Fig. 13.4; Chapter 12).

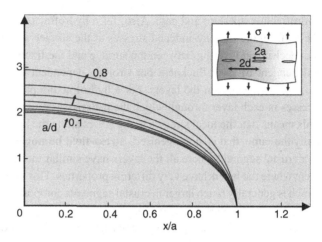

Fig. 13.5 Increase in opening displacement (half the aperture, vertical axis) of a set of tension fractures (subject to tensile stress $\sigma$) as the distance $2d$ between the centres of the cracks (each of length $2a$) decreases. When $a/d = 1.0$, the set has become a single fracture. The opening displacement is elliptical, as it should be for a constant tensile loading in a homogeneous, isotropic material. The arrows show the gradual increase in $a/d$ (data from Gray, 1992).

Some rift-zone tension fractures may form on the crests of anticlines above buried normal faults. Many, however, change into normal faults on reaching a critical depth, as indicated by the Griffith crack theory (Chapter 7). A tension fracture that gradually increases its depth during rifting events must eventually reach the critical depth at which it changes into a normal fault (Fig. 8.4).

## 13.3  Propagation of hydrofractures

The propagation of a hydrofacture, like that of other fractures, depends on local stresses in the rock layers or units along the potential fracture path. These local stresses, in turn, depend on the mechanical properties of these layers and units and their contacts and interfaces. Because hydrofractures transport fluids that affect the permeability of reservoirs, as well as magma, to the surface in volcanic eruptions, it is important to understand the main factors and processes that control hydrofracture propagation paths. In addition, hydrofractures presumably play an important role in triggering earthquakes. In Appendix E.2 there are some data on the physical properties of common crustal fluids. Here the focus is on those factors that control the geometry of a hydrofracture path. Particular attention is given to the conditions that determine whether the hydrofracture eventually reaches the Earth's surface or, alternatively, becomes arrested in a particular layer at depth.

In this chapter we present field data on hydrofracture paths, and numerical models to explain the paths. Generally, the modelling results indicate that the properties of the mechanical layers and their contacts largely control the local stresses, including hydrofracture-induced stresses ahead of a propagating hydrofracture and, thereby, the hydrofracture aperture and path. Also, for a hydrofracture propagating vertically towards the surface, the fracture-induced stresses at the surface at any instant depend strongly on the mechanical layering between the surface and the fracture tip. These results are largely independent of layer thickness but strongly dependent on the stiffness (Young's modulus) contrasts between the layers. For a hydrofracture path to reach the surface, the local stresses in each layer through which the path passes must favour hydrofracture formation. This means that the local stresses in the layers along the hydrofracture path must be essentially the same, that is, homogenised. **Stress-field homogenisation** is much easier to reach in a crustal segment where all the layers have similar mechanical properties than in a segment where the layers have very different properties. This relates also to material toughness, which is generally much larger in crustal segments composed of layers with widely different mechanical properties (Chapter 10).

Consider first the propagation path of a hydrofracture in a homogeneous, isotropic crustal segment. Since hydrofractures are extension fractures, they propagate in a direction that is parallel with the maximum principal compressive stress, $\sigma_1$. Thus, the likely path of a hydrofracture can be mapped approximately using the modelled trajectories of $\sigma_1$. For a simple two-dimensional, circular model of a fluid-filled reservoir with an internal fluid excess pressure as the only loading and located far from any free surfaces, the trajectories of $\sigma_1$ are radial from the reservoir (Fig. 13.6). Clearly, so long as there is sufficient fluid overpressure in the hydrofracture it should, theoretically, reach the surface.

Similar results are obtained for reservoirs close to the surface if, again, the internal excess pressure is the only loading (Fig. 13.7). If the only loading of a circular magma chamber (or another fluid source) is external tension, for example during a rifting episode, the trajectories

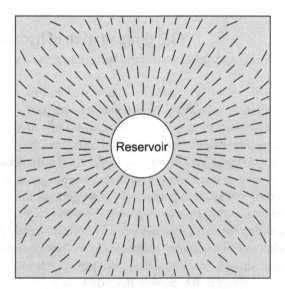

Fig. 13.6    Trajectories (ticks) of $\sigma_1$ around a reservoir of a circular cross-section subject to internal excess fluid pressure as the only loading. In this numerical model the excess pressure is 10 MPa, but the geometry of the stress trajectories does not depend on the magnitude of the pressure. Hydrofractures propagate parallel with the $\sigma_1$-trajectories.

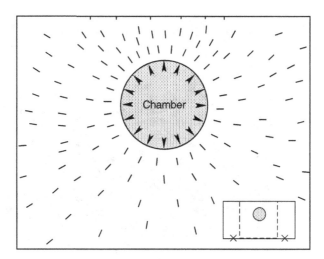

**Fig. 13.7**  Trajectories (ticks) of $\sigma_1$ around a reservoir of a circular cross-section subject to internal excess fluid pressure as the only loading. In this numerical model the excess pressure is 5 MPa. The main difference in the pattern of $\sigma_1$-trajectories in this numerical model and the one in Fig. 13.6 is because this reservoir is close to the free surface that modifies the stress field.

of $\sigma_1$ have a geometry that differs from that in the models above (Gudmundsson, 2006). The trajectories are steeper than in the previous models, partly because of the effects of the free surface, and partly because the $\sigma_1$-trajectories tend to be perpendicular to the horizontal tension, that is, to the direction of the minimum principal compressive stress, $\sigma_3$. For all these models, however, even if the reservoirs were of different shape, sufficiently buoyant hydrofractures would tend to reach the Earth's surface.

These simple models for a fluid source in a homogeneous, isotropic crustal segment give a general idea of the propagation paths of many hydrofractures. For example, analytical models for magma chambers in homogeneous, isotropic crustal segments have been used to map out the likely paths of inclined sheets in central volcanoes (Anderson, 1936; Gudmundsson, 2006). On these assumptions, an indication can sometimes be obtained as to the location of the associated shallow source magma chamber (Silver and Karson, 2009). Such indications are useful in a similar way to how Mogi models are useful for estimating the depth to the source of a surface deformation in an active volcano (Chapter 3). More refined considerations, however, take into account the effects of crustal layering, interfaces, and discontinuities on the potential paths of hydrofractures.

When layering is taken into account, the calculated geometry of the $\sigma_1$-trajectories may be very different from that of a homogeneous, isotropic (non-layered) model. For a layered crustal segment, the potential hydrofracture paths are therefore commonly very different from those inferred from the non-layered models above. Examples have already been given in Chapter 6 (Figs. 6.21 and 6.22). In these models, the part of the crust above the fluid source, a magma chamber, is composed of alternating thick and stiff layers and thin and compliant layers. Clearly, the $\sigma_1$-trajectories are very different from those in Figs. 13.6 and 13.7. For the circular model (Fig. 6.21) the most significant difference is that at the

**Fig. 13.8**  Deflected dyke at the contact between dissimilar rocks. The contact itself is composed of scoria. The dyke meets the contact, changes into a thin 8-m-long sill, and then continues its vertical propagation (cf. Figs. 13.9 and 13.11).

contacts between the middle and upper stiff and soft layers there is a 90° rotation of the $\sigma_1$-trajectories. This means that any vertical hydrofracture such as a dyke propagating from the source or chamber towards the surface would tend to be either deflected along, or arrested at, these contacts.

For the sill-like model (Fig. 6.22) the trajectories of $\sigma_1$ show abrupt changes at the contacts between layers. In this model, there is a 90° rotation of the $\sigma_1$-trajectories at each contact between the thin/compliant and the thick/stiff layers. Thus, hydrofractures injected from the sill-like source would tend to be deflected many times on their way towards the surface. For typical layered crustal segments, there would be many more layers, and it is very likely that the hydrofracture would be arrested at some of the contacts.

Deflection and arrest of this kind is very common for dykes (Fig. 13.8) and other hydrofractures (Fig. 13.9; Chapter 10). The deflection does not depend only on the local stress, but also on the rock properties and other factors, as discussed later in this chapter. Here, however, the focus is on how much changes in local stresses may contribute to the determination of the hydrofracture paths.

As examples of hydrofracture propagation in multi-layer segments, consider first a model with 30 layers above the fluid source (Fig. 13.10). Again, stress rotation indicates that the hydrofracture path would tend to become irregular, with many deflected segments, particularly in the central part of the segment. The hydrofracture would presumably become arrested at one of the contacts. Similar results are obtained when the layers are dipping, as is common (Gudmundsson, 2006). At many layer contacts there is a 90° rotation of the $\sigma_1$-trajectories, resulting in either hydrofracture deflection or arrest.

These simple two-dimensional numerical models show that in a layered crustal segment, the actual paths of the hydrofractures are likely to be irregular. Furthermore, they indicate that many hydrofractures become arrested and never propagate very far from

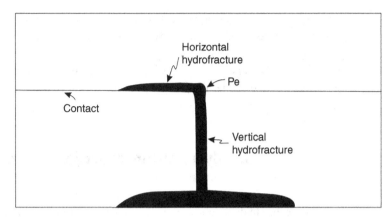

**Fig. 13.9**    Vertically propagating hydrofracture changes into a horizontal hydrofracture on meeting a contact. The hydrofracture is deflected in one direction along the contact (cf. Figs. 13.8 and 13.11).

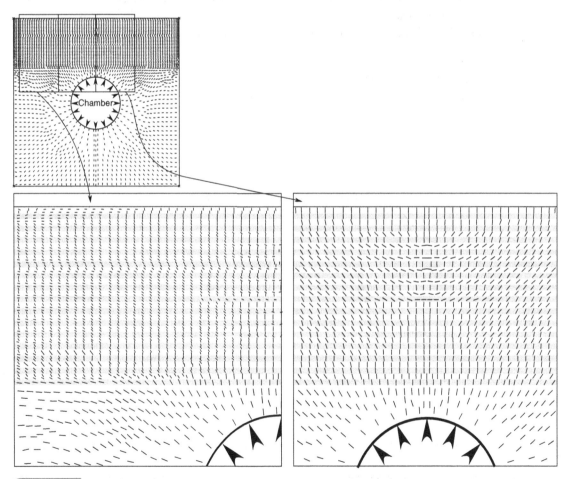

**Fig. 13.10**    Stress trajectories (ticks) of $\sigma_1$ in a 30-layer numerical model of a circular reservoir or a magma chamber. The layers alternate in stiffness between 1 GPa and 100 GPa (cf. Gudmundsson and Philipp, 2006).

their sources. This is, indeed, what is observed. Not only is it common to see sheets and dykes become arrested at contacts between dissimilar rocks, but the same applies to other types of hydrofractures such as mineral veins and man-made hydraulic fractures (Chapter 10). In order to understand the conditions of hydrofracture deflection and arrest, however, we must also consider the properties of the rock layers and the contacts or interfaces themselves.

## 13.4 Hydrofracture deflection and arrest

As we have seen, a hydrofracture in a layered rock mass or crustal segment commonly meets layers with local stresses that are unfavourable for the fracture propagation. Such layers are referred to as **stress barriers**. The concept simply means layers or rock units where the local stress is unfavourable for the propagation of that type of fracture, here a hydrofracture. When a propagating hydrofracture meets an interface or discontinuity such as a contact or an existing fracture, the hydrofracture may do one of the following (Fig. 13.11):

- become arrested so as to stop its propagation
- penetrate the contact
- deflect along the contact, in one or two directions.

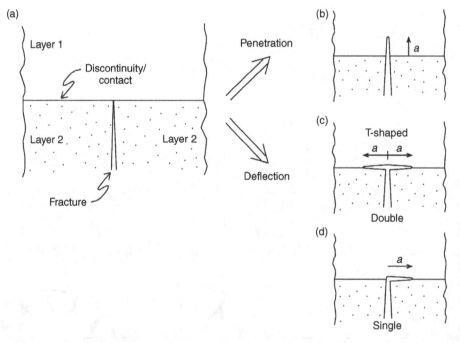

**Fig. 13.11**    On meeting a discontinuity such as a contact an extension fracture may (a) become arrested, (b) penetrate the layers above the contact, or become doubly (c) or singly (d) deflected along the contact. Modified from Hutchinson (1996).

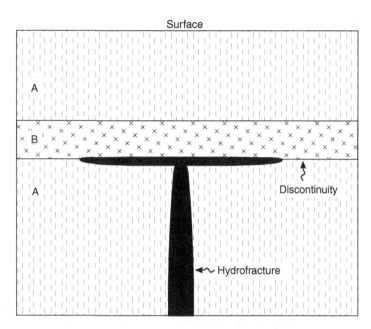

**Fig. 13.12**    Vertically propagating hydrofracture meets the lower contact between layers A and B and becomes doubly deflected (deflected in two directions) along the contact to form what is known as a T-shaped fracture in petroleum geology and geothermy and as a dyke–sill pair or contact in volcanology (cf. Fig. 17.7).

These three possibilities are well known from field observations of hydrofractures (Figs. 3.1, 6.20, 8.1, 13.8, 13.9, and 13.12). These scenarios can be explained in terms of three related factors:

1. the rotation of the principal stresses at the interface/discontinuity
2. an induced tensile stress ahead of the propagating hydrofracture tip, that is, the Cook–Gordon debonding or delamination mechanism
3. the material properties of the contact and the adjacent rock layers, particularly as regards elastic mismatch and material toughness.

We have already discussed the first factor, that is, the rotation of the principal stresses, and its effect on hydrofracture propagation. Here the focus is on the second and third mechanisms, namely induced stresses and material toughness (Chapter 10).

## 13.4.1 Induced stresses

Consider first a homogeneous, isotropic material. Then the tensile stress induced ahead of, and parallel with, a propagating hydrofracture, a mode I crack, is around 20% of the induced tensile stress ahead of and perpendicular to the crack (Fig. 13.13). The induced tensile stress may thus open an interface/discontinuity such as a contact ahead of the hydrofracture tip (Fig. 13.14) so long as the interface tensile strength is less than about 20% of the tensile strength of the adjacent rock layers in a direction perpendicular to the hydrofracture. For an

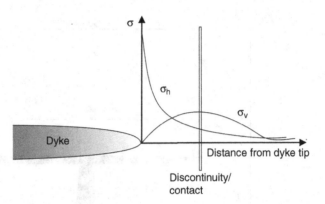

**Fig. 13.13**   Cook–Gordon debonding (delamination) mechanism.

in-situ rock tensile strength of 2–3 MPa (Chapter 7), an interface would tend to open if its tensile strength was 0.4–0.6 MPa. This latter value is very similar to the minimum in-situ tensile strength in basaltic rocks (Chapter 7), showing that such an opening, for example of contacts, is possible at shallow crustal depths (Fig. 13.8).

Dynamic crack-propagation experiments suggest that Cook–Gordon debonding is a common mechanism of **delamination** in composite materials (Xu *et al.*, 2003; Xu and Rosakis, 2003; Kim *et al.* 2006; Wang and Xu, 2006). In a geological context, these results throw light on the propagation of faults, since they are dynamic fractures. However, the results are also relevant to the semi-static propagation of hydrofractures. The experiments suggest that when a vertically propagating hydrofracture approaches an interface such as a contact, particularly in the uppermost 1–3 km of the crust, the hydrofracture-induced tensile stress may open up the contact before the hydrofracture tip reaches it. It is mainly the tensile strength of the contact itself which determines if the debonding takes place. On meeting an open contact (Fig. 13.14), a hydrofracture may change into a T-shape fracture (Figs. 13.11 and 13.12) or become offset along the contact (Figs. 13.8, 13.9, and 13.11), provided the stress field is favourable for such a path change. The experimental results (Xu *et al.*, 2003) support theoretical results (He and Hutchinson, 1989) in that on becoming deflected along a contact, a mode I crack such as a hydrofracture changes into a mixed-mode fracture.

### 13.4.2  Material toughness and elastic mismatch

The third mechanism for hydrofracture deflection/arrest at an interface depends on the difference in material toughness between the contact and the adjacent rock layers and the elastic mismatch between the layers.

From Eq. (10.31), the total strain energy release rate, $G_{\text{total}}$, in a mixed-mode plain-strain fracture propagation is

$$G_{\text{total}} = G_{\text{I}} + G_{\text{II}} + G_{\text{III}} = \frac{(1-\nu^2)K_{\text{I}}^2}{E} + \frac{(1-\nu^2)K_{\text{II}}^2}{E} + \frac{(1+\nu)K_{\text{III}}^2}{E} \qquad (13.2)$$

**Fig. 13.14**  Thick pahoehoe lava flows are composed of tens or hundreds of flow units, mostly 0.5–2 m thick, all of which have essentially the same mechanical properties. Some of the contacts between the flow units are weak and/or open. The photo shows a part of the wall of the western boundary fault of the Thingvellir Graben, Southwest Iceland (Figs. 1.5, 3.13, and 3.14).

where $G_{\mathrm{I-III}}$ are for the ideal crack-displacement modes I–III, $E$ is Young's modulus, $\nu$ is Poisson's ratio, and $K_{\mathrm{I-III}}$ are the stress intensity factors for modes I–III. If the rock layers on either side of a contact have the same mechanical properties, then the conditions for hydrofracture penetration (Figs. 13.11 and 13.15) on meeting the contact are that the

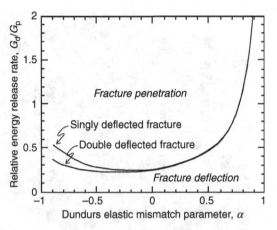

When a fracture meets an interface, the ratio of the strain energy release rate for fracture deflection $G_d$ to that of fracture penetration $G_p$ controls the fracture propagation. The ratio is shown here as a function of the Dundurs elastic mismatch parameter $\alpha$ (Eqs. 13.7 and 13.8). There is little difference in the elastic strain energy release rate for a single or double deflection. For negative values of $\alpha$, layer 2 (the fracture-hosting layer) is stiffer than layer 1 and there is little tendency to fracture deflection along the interface. However, as the stiffness of layer 1 increases in relation to that of layer 2, the tendency to fracture deflection along the interface greatly increases. If there is no Young's modulus mismatch across the contact/interface, then fracture deflection occurs only if the contact toughness, that is, the toughness for deflection $G_d$, is about 26% of the toughness of the material on the other side of the contact, $G_p$. However, when the mismatch increases (to the right), deflection will still occur even if the contact toughness $G_d$ becomes equal to or higher than the toughness $G_p$ of layer 1. Modified from He *et al.* (1994).

strain energy release rate for mode I crack, $G_p$, (with subscript p for penetration) reaches the critical value for crack extension, namely the material toughness of the layer, $\Gamma_L$ (with subscript L for rock layer). From Eq. (13.2) the conditions are

$$G_p = \frac{(1-\nu^2)K_I^2}{E} = \Gamma_L \qquad (13.3)$$

However, if the strain energy release rate reaches the material toughness of the discontinuity, $\Gamma_D$ (the superscript D denotes discontinuity), then the hydrofracture will deflect into the interface/discontinuity (Fig. 13.8, 13.9, and 13.11). As the hydrofracture propagates in mixed-mode (modes I and II) along the discontinuity (Hutchinson, 1996; Xu *et al.*, 2003; Wang and Xu, 2006), it follows from Eq. (13.3) that deflection (with subscript d for deflection) into the discontinuity occurs when

$$G_d = \frac{(1-\nu^2)}{E}(K_I^2 + K_{II}^2) = \Gamma_D \qquad (13.4)$$

Here the stress-intensity factors $K_I + K_{II}$ refer to the interface/discontinuity. From Eqs. (13.3) and (13.4) it follows that the hydrofracture penetrates the discontinuity if

$$\frac{G_d}{G_p} < \frac{\Gamma_D}{\Gamma_L} \qquad (13.5)$$

By contrast, the hydrofracture becomes deflected into the discontinuity if

$$\frac{G_d}{G_p} \geq \frac{\Gamma_D}{\Gamma_L} \tag{13.6}$$

Equations (13.3)–(13.6) are likely to control, partly at least, whether a hydrofracture penetrates or becomes deflected along a contact in a pile of mechanically similar layers such as of basaltic lava flows or flow units (Figs. 8.1 and 13.14).

In many rocks, however, there is an abrupt change in mechanical properties at contacts and other discontinuities (Figs. 2.4 and 6.20), resulting in an elastic mismatch. One measure of the mechanical change across an interface is provided by the Dundurs (1969) elastic mismatch parameters. Using these parameters, the conditions for hydrofracture penetration or deflection, indicated above in Eqs. (13.3)–(13.6), can be modified so as to include dissimilar layers. The two Dundurs parameters, $\alpha$ and $\beta$, may be presented as follows (Hutchinson, 1996; Freund and Suresh, 2003):

$$\alpha = \frac{E_1^* - E_2^*}{E_1^* + E_2^*} \tag{13.7}$$

$$\beta = \frac{1}{2} \frac{\mu_1(1 - 2\nu_2) - \mu_2(1 - 2\nu_1)}{\mu_1(1 - \nu_2) + \mu_2(1 - \nu_1)} \tag{13.8}$$

Here $\mu$ is the shear modulus (commonly denoted by $G$ but here by $\mu$ to avoid confusion with the energy release rate $G$), $\nu$ is Poisson's ratio, and $E^* = E/(1 - \nu^2)$ is the plain-strain Young's modulus. The subscript 2 is used for the moduli of the rock containing the hydrofracture, whereas the subscript 1 is used for the material on the other side of (here above) the interface (Fig. 13.11). The parameter $\alpha$ is a measure of mismatch in extensional or uniaxial stiffness and $\beta$ of volumetric or areal stiffness (Freund and Suresh, 2003).

The strain energy release rate (Chapter 10) associated with hydrofracture penetration into the layer above the interface is denoted by $G_p$ and that associated with hydrofracture deflection into the interface or discontinuity is $G_d$ (Fig. 13.15). By analogy with Eqs. (13.5) and (13.6), the hydrofracture is likely to penetrate the interface/discontinuity if the following inequality is satisfied:

$$\frac{G_d}{G_p} < \frac{\Gamma_D(\psi)}{\Gamma_L^1} \tag{13.9}$$

By contrast, hydrofracture deflection into the interface/discontinuity is likely to occur if the following condition is satisfied:

$$\frac{G_d}{G_p} \geq \frac{\Gamma_D(\psi)}{\Gamma_L^1} \tag{13.10}$$

Here the subscript for the material toughness is for layer 1 (Fig. 13.11) and $\psi$ is a measure of the relative proportion of mode II to mode I, namely

$$\psi = \tan^{-1}(K_{II}/K_I) \tag{13.11}$$

Thus, $\psi = 0°$ is for pure mode I and $\psi = \pm 90°$ for pure mode II (Chapter 9).

For a given hydrofracture-segment length $a$ (Fig. 13.11), the energy release rate depends only on the Dundurs parameter $\alpha$ ($\beta = 0$). Thus, the ratio $G_d/G_p$ (Eqs. 13.9 and 13.10) can be shown as a function of $\alpha$ (Fig. 13.15). When the ratio is below the curves deflection of a hydrofracture into the interface is favoured. By contrast, when the ratio is above the curves then vertical penetration of the hydrofracture through the interface and into layer 1 is favoured.

When stiffnesses of layers 1 and 2 are equal, in which case the Dundurs parameter $\alpha = 0$, then $G_d/G_p$ becomes equal to 0.26 (Fig. 13.15). This means that a hydrofracture would become deflected into a sill-like fracture along the interface/discontinuity only if the material toughness of the interface/discontinuity ($\Gamma_D$) is less than 26% of the material toughness of layer 1 ($\Gamma_L^1$) above the contact. This condition may sometimes be met, but is probably uncommon, which may partly explain why hydrofractures tend to penetrate piles of rocks with similar mechanical properties (Figs. 8.1 and 13.14).

The curves for the formation of a single-directed and a double-directed deflection (Figs. 13.9, 13.11, and 13.12) along the discontinuity are very similar for most values of the parameter $\alpha$ (Fig. 13.15). Thus, for practical purposes, the tendency for a hydrofracture to form a sill-like fracture through a single- or double-directed deflection is the same.

When $\alpha$ is negative, so that the stiffness of layer 1 is less than that of layer 2, there is much less tendency for a hydrofracture to deflect into a sill-like fracture along the interface/discontinuity than when $\alpha$ is positive. As the positive value of $\alpha$ increases, so that layer 1 becomes stiffer in relation to layer 2, there is a greatly increased tendency for a hydrofracture to deflect into the discontinuity (Fig. 13.15). A hydrofracture propagating through a soft pyroclastic or shale layer towards a stiff lava flow or limestone would therefore be more likely to deflect into the interface than a hydrofracture propagating from a stiff limestone towards a soft shale (other factors, such as stress rotation, also have effects).

This conclusion is supported by many experiments on fracture propagation and arrest at contacts between dissimilar layers (Kim *et al.*, 2006) and in geological analogue experiments (Kavanagh *et al.*, 2006). When the deflection is not possible because of the orientation of the principal stresses (Gudmundsson, 2010; Gudmundsson and Philipp, 2006) a hydrofracture propagating from a soft towards a stiff layer would tend to become arrested at the interface/contact. This process is often seen in the field (Figs. 3.1 and 6.20) where the interface/contact commonly acts as a trap and arrests the vertical hydrofracture propagation.

## 13.5 Hydrofracture aperture and overpressure

The volume of fluid transported per unit time through a fracture depends on its strike or dip dimension but primarily on its aperture (opening). The volumetric flow rate through a fracture is a function of the aperture in the third power, a relationship known as the 'cubic law' in hydrogeology and related fields. Because of this law, it is very important to understand the factors that determine the aperture variation of a fracture.

In an isotropic, homogeneous rock, an isolated hydrofracture driven open by a constant fluid overpressure develops an aperture that varies in size as a flat ellipse (Chapter 9).

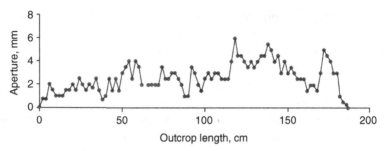

Fig. 13.16  Aperture (here, thickness) variation of a vein in gneiss in West Norway. The vein is close to 190 cm long, with a maximum aperture (thickness) of 6 mm, the fill being quartz.

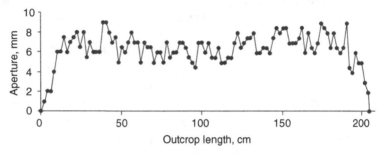

Fig. 13.17  Aperture (thickness) variation, measured as thickness of the mineral fill, of a vein in gneiss in West Norway. The vein, 205 cm long and with a maximum thickness of 9 mm, is made of quartz.

The apertures of some hydrofractures approach the ideal elliptical geometry (Chapter 9), but many show irregular aperture variations (Figs. 13.16 and 13.17). Many mineral veins with strike dimensions up to several metres show this kind of irregular aperture variation. And large tension fractures, normal faults, and dykes with strike dimensions up to many kilometres show similar variations. Because of the cubic law, large variations in fracture aperture may result in much of the fluid transport becoming confined to the parts of the fracture with the largest aperture, that is, in flow channelling (Tsang and Neretnieks, 1998). Flow channelling may be one reason for the formation of crater cones along volcanic fissures, and it is also well known from tunnelling and fractured reservoirs.

Hydrofracture geometry, in particular its aperture variation, is thus of great significance for understanding fluid transport in hydrofractures or extension fractures in general. In this section, some of the analytical techniques for understanding aperture variations are reviewed. This section is regarded as 'advanced' since it requires a command of mathematics that the rest of the book, for the most part, does not require. Two analytical approaches are used: polynomials and Fourier series to model the aperture variation. The results are then compared with numerical models.

It should be noted that the aperture variation can be attributed either to the intrinsic variation in fluid overpressure or directly to changes in the mechanical properties of the rock dissected by the hydrofracture. This follows because overpressure is defined as the total fluid pressure minus the horizontal stress on the fracture, which, for a hydrofracture, is the minimum principal compressive stress $\sigma_3$. In a heterogeneous and anisotropic rock the local $\sigma_3$ will vary even if the remote regional loading is constant and, as a consequence, the overpressure will vary. This is easiest to visualise in vertical section through horizontally layered rocks. Both in compression and extension, the stiff layers become highly stressed, whereas the soft layers largely maintain their state of stress (Fig. 6.4). It follows that the stiff layers increase their horizontal compressive stress during compression and decrease it during extension. For a hydrofracture with a constant total fluid pressure propagating through a layered rock with contrasting stiffnesses, such as in a sedimentary basin or a stratovolcano, the overpressure would change as follows. During extension the fluid over-pressure of the fracture would increase in the stiff layers, since they concentrate (relative) tensile stresses. By contrast, during compression, the overpressure would decrease in the stiff layers because they concentrate compressive stresses. Thus, mechanical layering and heterogeneities directly affect the fluid overpressure variation in hydrofractures.

### 13.5.1 Overpressure modelled by polynomials

Consider a two-dimensional crack, a line crack, located along the $x$-axis, with a length $2a$ and defined by $y = 0$, $-a \leq x \leq a$ (Fig. 13.18). The final shape of the mode I crack considered here is determined by the normal displacement $u = u_y (x, 0)$ of the crack walls in the direction of the $y$-axis. According to the boundary conditions, there is no displacement of the walls outside the tips of the crack, that is, $u = 0$ for $x > a$. The general solution for the normal displacement of one (the upper) crack wall in the direction of the $y$-axis (Sneddon, 1973; Maugis, 2000) is

$$u = \frac{4(1 - v^2)}{\pi E} \int_x^a \frac{t\,dt}{(t^2 - x^2)^{1/2}} \int_0^t \frac{p(x)dx}{(t^2 - x^2)^{1/2}} \qquad (13.12)$$

where $0 < t < a$. The second integral is also used when calculating the theoretical hydrofracture tip stresses. It is convenient to represent the second integral separately as

$$q(t) = \int_0^t \frac{p(x)dx}{(t^2 - x^2)^{1/2}} \qquad (13.13)$$

Here we consider first two types of fluid overpressure in the crack: constant and linearly varying from a maximum at the crack centre to its tips. For a constant overpressure, we have

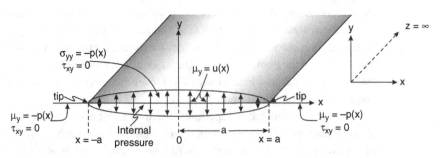

**Fig. 13.18**    Coordinate system of a hydrofracture subject to internal fluid overpressure as the only loading.

$p(x) = P_0$ and Eq. (13.13) gives $q(t) = P_0\pi/2$. Using this value for the second integral in Eq. (13.12), and solving the first integral, we obtain

$$u = \frac{2(1 - v^2)P_0}{E}(a^2 - x^2)^{1/2} \tag{13.14}$$

This is the plane-strain formula for a two-dimensional, elliptical through crack, a 'tunnel crack' (Chapter 9), whereas for plane stress, the term $(1 - v^2) = 1$. The shape of the crack is then reflected in its total opening, measured by its aperture $\Delta u = 2u$. From Eq. (13.14) it follows that a constant fluid overpressure in a two-dimensional mode I crack in a homogeneous, isotropic rock, opens the crack into a flat ellipse.

Consider next a through crack where the fluid overpressure $p(x)$ changes from a maximum value $P_0$ at the crack centre by a linear gradient $p_1 x$ to the crack tips, thus:

$$p(x) = P_0 + p_1 x \tag{13.15}$$

The crack opening related to the constant term $P_0$ is already given by Eq. (13.14). From Eqs. (13.13) and (13.15) we have $q(t) = p_1 t$, in which case Eq. (13.12) becomes

$$u = \frac{4(1 - v^2)p_1}{\pi E} \int_x^a \frac{t^2 dt}{(t^2 - x^2)^{1/2}} \tag{13.16}$$

which gives the opening displacement as

$$u = \frac{2p_1(1 - v^2)}{\pi E} \left[ a(a^2 - x^2)^{1/2} + x^2 \ln\left[a + (a^2 - x^2)^{1/2}\right] - x^2 \ln(x) \right] \tag{13.17}$$

Combining Eqs. (13.14) and (13.17), the aperture, $2u = \Delta u$, of a hydrofracture subject to the linear overpressure distribution presented by Eq. (13.15) becomes

$$\Delta u = \frac{4(1 - v^2)}{E} \left[ P_0 k + \frac{p_1}{\pi}\left(ak + x^2 \ln\frac{a + k}{x}\right) \right] \tag{13.18}$$

where

$$k = (a^2 - x^2)^{1/2} \tag{13.19}$$

The aperture as a function of $x$ is a smooth curve, not so dissimilar from that of an ellipse (Fig. 13.19). Thus, a constant overpressure and linearly varying overpressure for a hydrofracture in a homogeneous, isotropic rock yields a smoothly varying opening displacement profile. Generally, the opening shape compares well with the opening profile of linearly varying overpressure in the numerical models below.

## 13.5.2 Overpressure modelled by Fourier series

The overpressure distribution inside the two-dimensional fracture can be presented by various functions. The solutions for constant overpressure (Chapter 9) as well as polynomials (Section 13.5.1; Valko and Economides, 1995; Gudmundsson et al., 2002) have been widely

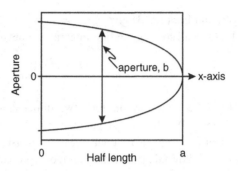

**Fig. 13.19** Variation in aperture of a hydrofracture.

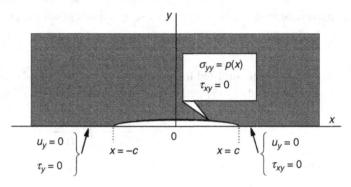

**Fig. 13.20** Crack model used for aperture variation in a hydrofracture of length $2c$ opened by its internal fluid overpressure, given by the even function $p(x) = p(-x)$. There are no stresses outside the lateral ends (tips) of the crack ($u_y = 0$, $\tau_{xy} = 0$) and the stress field is symmetric about the line $y = 0$.

used to analyse aperture variations in rock fractures. Here we consider a solution for the fracture opening displacement or aperture when the overpressure variation is given by Fourier cosine series.

The model is two-dimensional with the basic configuration as defined in Fig. 13.20. A mathematical crack is opened in a direction of the y-axis by a fluid overpressure given by the function $p(x)$. The opening displacement, $u_y(x)$, is given as follows (Valko and Economides, 1995):

$$u_y(x) = \frac{4(1 - \nu^2)}{E\pi} \int_x^c \frac{\xi g(\xi)}{\sqrt{\xi^2 - x^2}} d\xi \qquad (13.20)$$

where $0 \leq x \leq c$. The crack is located along the x-axis and c, E, and $\nu$ are half the length of the crack, the Young's modulus (stiffness), and the Poisson's ratio of the host rock, respectively, and $g(\xi)$ is defined as

$$g(\xi) = \int_0^\xi \frac{p(x)}{\sqrt{\xi^2 - x^2}} dx \qquad (13.21)$$

The opening displacement in the $y$-direction, $u_y(x)$, is half the aperture (Fig. 13.20). Most observations of aperture variations in rock factures are based on exposures at the Earth's surface, so that we take the crack surface length $2c$ to be the fracture strike dimension. Then the fracture dip dimension, that is, the depth or extension of the fracture from the surface and into the rock layers is assumed infinite. Outside the crack lateral ends, the tips, there is no displacement. Thus, the boundary conditions are (Fig. 13.20)

$$\tau_{xy} = 0, \quad \sigma_{yy} = p(x), \quad |x| \le c,$$
$$\tau_{xy} = 0, \quad u_y(x) = 0, \quad |x| > c \tag{13.22}$$

The boundary conditions at infinity are

$$\sigma_{xx} \to 0, \quad \sigma_{yy} \to 0, \quad \tau_{xy} \to 0, \quad \text{as} \quad \sqrt{x^2 + y^2} \to \infty, \quad y \ge 0, \tag{13.23}$$

for the half-plane $y \ge 0$ where, as usual, $\tau$ is the shear stress and $\sigma$ the normal stress.

If the overpressure is defined as an even function in the range of $-c \le x \le c$, the overpressure distribution $p(x)$ can be presented by Fourier cosine series as

$$p(x) = \frac{a_0}{2} + \sum_{n=1}^{\infty} a_n \cos n\omega x \tag{13.24}$$

where $n$ is the Fourier coefficient ($n = 1, 2, \ldots$) and $\omega$ is an angular frequency defined as $\omega = \pi/c$. By substituting Eq. (13.23) into Eq. (13.21), we get

$$g(\xi) = \frac{\pi}{4} a_0 + \frac{\pi}{2} \sum_{n=1}^{\infty} a_n J_0(n\omega\xi) \tag{13.25}$$

where $J_0(n\omega\xi)$ is a zero-order Bessel function of the first kind. Substituting $g(\xi)$ into Eq. (13.20), we obtain an equation for the normal opening-displacement of the crack $u_y(x)$ as

$$u_y(x) = \frac{2(1 - v^2)}{E} \left\{ \frac{a_0}{2} \sqrt{c^2 - x^2} + \int_x^c \frac{\xi \left[ \sum_{n=1}^{\infty} a_n J_0(n\omega\xi) \right]}{\sqrt{\xi^2 - x^2}} d\xi \right\} \tag{13.26}$$

The second term in this equation is difficult to integrate analytically; however, it can be solved by numerical integration (for example, by Gauss–Legendre numerical integration).

As an example of the use of these results, consider an irregular overpressure distribution in a hydrofracture (Fig. 13.21). Here, $x_N$ is equal to half the length or strike dimension of the crack, $c$. The interval $\Delta x (= x_{i+1} - x_i)$ may be constant but need not be so. The overpressure function, $p(x)$, can be described by the Fourier cosine series as

$$p(x) = \frac{x_N p_N + \sum_{i=1}^{N-1} (p_i - p_{i+1}) x_i}{c} + \frac{2}{\pi} \sum_{n=1}^{\infty} \sum_{i=1}^{N-1} \frac{(p_i - p_{i+1})}{n} \sin(n\omega x_i) \cos(n\omega x) \tag{13.27}$$

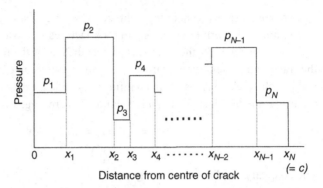

**Fig. 13.21** Step-like overpressure variation in the upper half of a hydrofracture. The overpressure varies abruptly from one part of the crack to the next, as is indicated by the symbols $p_1, p_2 \ldots$

Comparing Eqs. (13.23) and (13.26), the initial and general terms of the Fourier cosine series are

$$\frac{a_0}{2} = \frac{x_N p_N + \sum\limits_{i=1}^{N-1} (p_i - p_{i+1}) x_i}{c} \tag{13.28}$$

$$a_n = \frac{2}{\pi} \sum_{i=1}^{N-1} \frac{(p_i - p_{i+1})}{n} \sin(n\omega x_i) \tag{13.29}$$

Substituting the general terms given by Eqs. (13.27) and (13.28) into Eq. (13.24), the function $g(\xi)$ becomes

$$g(\xi) = \frac{\pi}{2} \frac{\left[ x_N p_N + \sum\limits_{i=1}^{N-1} (p_i - p_{i+1}) x_i \right]}{c} + \sum_{n=1}^{\infty} \sum_{i=1}^{N-1} \frac{(p_i - p_{i+1})}{n} \sin(n\omega x_i) J_0(n\omega\xi) \tag{13.30}$$

Finally, putting Eq. (13.29) into Eq. (13.20) gives the normal crack-opening displacement (half the aperture) for the irregular overpressure distribution as

$$u_y(x) = \frac{2(1 - v^2)}{E} \left\{ \frac{\left[ x_N p_N + \sum\limits_{i=1}^{N-1} (p_i - p_{i+1}) x_i \right]}{c} \sqrt{c^2 - x^2} \right.$$

$$\left. + \int_x^c \frac{\xi \left[ \sum\limits_{n=1}^{\infty} \sum\limits_{i=1}^{N-1} \frac{(p_i - p_{i+1})}{n} \sin(n\omega x_i) J_0(n\omega\xi) \right]}{\sqrt{\xi^2 - x^2}} d\xi \right\} \tag{13.31}$$

Figures 13.22 and 13.23 present examples of crack-opening displacement (half the aperture) as a function of irregular overpressure and different Young's moduli, yielding results

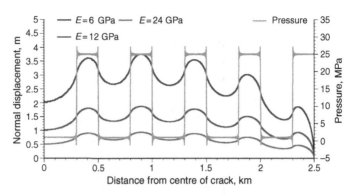

Fig. 13.22 Opening-displacement (half the aperture) of a hydrofracture where the overpressure varies abruptly between 1 MPa and 25 MPa in a host rock with different generalised Young's moduli (6 GPa, 12 GPa, and 24 GPa). This extreme overpressure variation may easily occur in layered rocks with abrupt changes in the mechanical properties between layers. The stiffness variation of the real rock is already included in the local $\sigma_3$ and, therefore, in the overpressure variation. Thus, when the overpressure variation has been calculated, a generalised, constant Young's modulus is given to the host rock. The largest apertures occur in the softest rock.

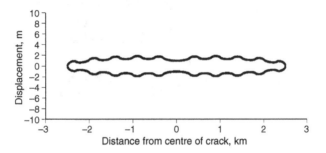

Fig. 13.23 Total aperture for a fracture hosted by rock with a generalised stiffness of 12 GPa. The loading and Poisson's ratio (0.25) are the same as in Fig. 13.22. In both models, the lengths of the fractures are arbitrary; for kilometre-long fractures, the apertures are in metres (as shown here), for metre-long fractures, the model apertures would be in millimetres as in Figs. 13.16 and 13.17.

formally similar to those in Figs. 13.16 and 13.17. In all these models the actual opening depends on the length chosen. If the modelled fracture length is in the order of kilometres the opening is in metres and thus suitable for many dykes. By contrast, if the length is considered metres, the opening is in millimetres (as in Figs. 13.16 and 13.17).

### 13.5.3 Numerical models of aperture variation

To compare with and add to the analytical solutions given above, we present here some numerical models on hydrofracture aperture variations. These models are particularly suitable for vertical hydrofractures in layered rocks with horizontal or gently dipping layers. However, the results can be generalised. The overpressure may be intrinsic and due to variation in the total fluid pressure. This may happen, for example, because of density changes in the fluid as it is transported through the crust, volume changes of the fracture itself,

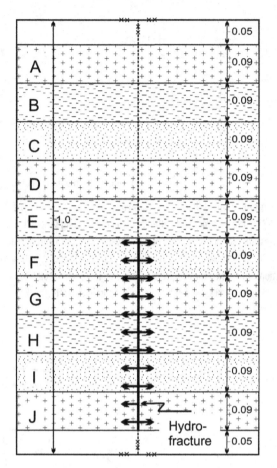

**Fig. 13.24**    Basic geometry and loading conditions and layering for the numerical models in Figs. 13.25–13.27.

variation in the supply of fluid from the source, as well as other factors. But the overpressure variation may also be a consequence of the mechanical layering itself in lateral and/or vertical sections, and to other rock heterogeneities, as explained above.

The general configuration of the models is presented in Fig. 13.24. The model is for a hydrofracture in mechanically layered rocks. The layering can be either in a horizontal plane (Fig. 1.11) or in a vertical plane. The latter is assumed here. The vertical dimension or height of the model is taken as a unit and all the layers have the same typical Poisson's ratio, 0.25. The model has ten layers of equal thickness; each layer is 0.09 units, the total height being 1.0. In the models the layering is as follows: the lowermost layer, J, is very stiff, with a Young's modulus of 100 GPa. The adjacent layer, I, is very compliant or soft, with a Young's modulus of 1 GPa. In ascending order, the third layer H, is moderately stiff, with a Young's modulus of 10 GPa. This three-layer sequence, that is, very stiff–very soft–moderately soft, is repeated up to the top of the model. Thus, the topmost layer, A, is very stiff, layer B moderately stiff, and layer C very soft, as is seen in the first model (Fig. 13.25).

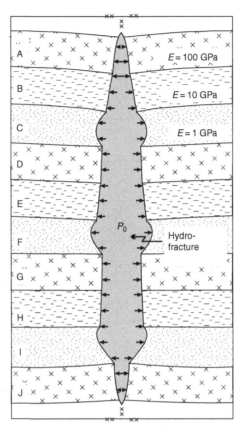

**Fig. 13.25** Numerical model of the aperture variation of a hydrofracture in a layered rock. The hydrofracture is subject to constant overpressure of 6 MPa.

The hydrofracture is static; it is confined to the layers indicated. In the model, the fracture tips, therefore, are not allowed to propagate up or down. The tips are fastened, as well as the model itself, using the conditions of no displacement (the displacement vectors being zero). The fluid overpressure varies within the fracture, depending on the local horizontal stresses in the adjacent layers, as explained above, and the location of the fluid front at any instant.

In the first model (Fig. 13.25) the fluid fills the entire fracture and has a constant overpressure of 6 MPa as the only loading. This model would be appropriate for many hydrofractures, for example for dykes injected laterally from vertical conduits in stratovolcanoes, and also for hydraulic fractures injected laterally into layered target petroleum reservoirs. The results (Fig. 13.25) clearly indicate that the apertures are largest in the soft layers and smallest in the stiff layers. This is as expected since the constant internal fluid overpressure is the only loading. Because of the comparatively large aperture variations, there would be a tendency to flow channelling during flow parallel with the strike dimension (into the page) of this fracture.

Hydraulic-fracture experiments (Chapter 7) indicate that, while the hydraulic fracture is propagating, the fracture tip is usually ahead of the fluid front (Warpinski, 1985). It is also

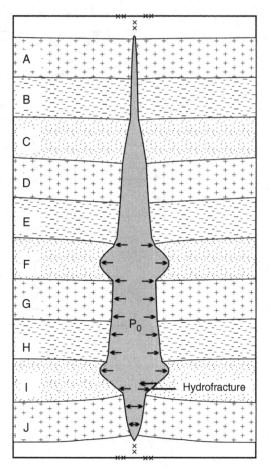

**Fig. 13.26** Numerical model of the aperture variation of a hydrofracture in a layered rock. The hydrofracture is subject to constant overpressure of 6 MPa, but only in its lower half.

known from field studies of rock fractures, such as joints formed through cooling (DeGraff and Aydin, 1987) and fluid pressure (Secor, 1965), that they grow in a semi-static way. This means that when the stresses at the fracture tip reach the rupture criteria, the fracture tip runs ahead of the loading front, becomes temporarily arrested on entering rock units with unfavourable stress conditions, and then resumes its propagation once the loading (due to cooling or build up of fluid pressure) has modified the stress conditions at the fracture tip so as to encourage further propagation.

Thus, in the second model (Fig. 13.26) the fluid fills only the lower half of the fracture, that is, the lowermost 5 layers dissected by the fracture. The fluid overpressure is constant, at 6 MPa, and applied only to the lower half of the fracture. Clearly, the aperture in both the soft layers, I and F, becomes large, which encourages subsequent flow channelling for flow parallel with the strike dimension of the fracture.

In the third model (Fig. 13.27) the fluid overpressure, again, is applied only to the lower half of the fracture. That is to say, the upper half of the fracture is not subject to any fluid

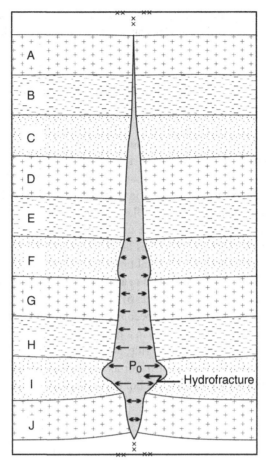

**Fig. 13.27**  Numerical model of the aperture variation of a hydrofracture in a layered rock. The hydrofracture is subject to overpressure only in its lower half. The overpressure varies from 10 MPa at the bottom of the fracture to 0 MPa in its central part.

overpressure. In this model, however, the fluid overpressure is not constant but rather varies linearly from 10 MPa at the bottom of the fracture to 0 MPa at the top of the fifth layer, layer F. This loading condition may be common during fracture propagation, either if the fracture in Fig. 13.27 is regarded as the entire fracture or, alternatively, if it is regarded as only the near-tip top part of a larger fracture. The overpressure values, 6 and 10 MPa, are chosen so as to correspond to the typical maximum tensile strengths of solid rocks. The results are clearly very different from those in Fig. 13.26 because in the present model only the aperture in layer I becomes large – but it is exceptionally large and thus favours flow channelling.

The results show some of the effects that mechanical layering may have on hydrofracture apertures. The layering is reflected in the aperture variation between the layers. The modelling can be either through abrupt overpressure variations between layers, as in the analytical models above, or through a constant or linearly varying overpressure, where the real variation comes through abrupt changes in the stiffnesses of the layers that the hydrofracture dissects. This latter approach is used in the numerical models.

## 13.6  Tip stresses and surface deformation

Hydrofractures propagate by advancing their tips when the associated tensile stresses exceed the tensile strength of the host rock (this criterion can also be formulated using fracture toughness rather than tensile strength). To calculate the tip stresses of a rock fracture in general, and a hydrofracture in particular, two approaches have normally been used. One is to consider the crack as a flat elliptic hole (Chapter 6), an approach frequently used in rock mechanics. The other is to model the rock fracture as a mathematical crack of zero thickness (Section 13.5). This latter approach is more common in fracture mechanics. Let us first calculate the potential tip stress of rock fractures, using both approaches, and then analyse briefly the deformation at the Earth's surface generated by arrested hydrofractures.

The elliptical hole (Chapter 6) is an appropriate model for many types of open fractures, mineral veins (before they become filled with vein material), and vuggy fractures. In addition, an elliptical hole is often a reasonable model for some tension fractures in rift zones, as well as for many dykes, sills, and inclined sheets.

If an **elliptical hole** of major axis $2a$ and minor axis $2b$ is subject to a constant fluid overpressure $P_0$, then from Eq. (6.8) the minimum principal compressive stress $\sigma_3$ at its tip is

$$\sigma_3 = -P_0 \left[ \frac{2a}{b} - 1 \right] \tag{13.32}$$

For filled hydrofractures such as mineral veins and dykes, as well as for many open and vuggy fractures, the tip may be exposed and measurable. Then the radius of curvature of the tip, $\rho_c = b^2/a$, may be used to calculate the crack-tip stress at the time of hydrofracture emplacement, using Eq. (6.6), as

$$\sigma_3 = -P_0 \left[ 2(a/\rho_c)^{1/2} - 1 \right] \tag{13.33}$$

Typical hydrofracture crack-tip tensile stresses based on the elliptical hole model may be calculated using results on 384 mineral-filled veins in a fault zone in North Iceland (Chapters 9 and 11). We define the aspect ratio of a vein as the smaller of the dip or strike dimensions of the vein divided by its maximum thickness or aperture. For these 384 veins modelled as elliptical holes, the average aspect ratio is $a/b = 400$. The estimated average fluid overpressure at the time of vein emplacement is $P_0 = 20$ MPa (Chapter 9), in which case Eq. (13.31) yields a crack-tip tensile stress of $\sim 1.6 \times 10^4$ MPa. Since the typical in-situ rock tensile strength is 0.5–6 MPa, the theoretical crack-tip tensile stresses are as much as four orders of magnitude greater than the in-situ tensile strength. These theoretical stresses may be regarded as effectively infinite in comparison with that strength.

Consider next a **mathematical-crack model**, as an appropriate model for many hydrofractures. In particular, narrow cracks with very thin, hair-like tips, such as many joints, are best analysed using a mathematical-crack model.

A two-dimensional line crack of length $2a$ along the $x$-axis can be defined by $y = 0$, $-a \leq x \leq a$ (Fig. 13.18). Because most hydrofractures are extension fractures (mode I cracks) where, by definition, the minimum principal compressive stress, $\sigma_3$, is the normal stress on the crack, it follows that $\sigma_y = \sigma_3$. The internal fluid overpressure is given by the even function $p(x) = p(-x)$, meaning that the pressure is the same on both walls of the hydrofracture. Inside the crack tips, the stress $\sigma_3(x, 0) = -p(x)$, for $0 \leq x \leq a$, whereas beyond the crack tips, for $x > a$, the stress $\sigma_3$ is given by (Sneddon, 1995; Maugis, 2000)

$$\sigma_3 = -\frac{2x}{\pi} \int_0^a \frac{t \, dt}{(x^2 - t^2)^{3/2}} \int_0^t \frac{p(x) dx}{(t^2 - x^2)^{1/2}} \tag{13.34}$$

where $0 < t < a$. The second integral is also used when considering the displacement, or shape, of hydrofractures and may be presented as shown in Eq. (13.13).

As in the numerical (Figs. 13.24–13.27) and analytical models of fracture shape, we consider two types of internal fluid overpressure distribution in the crack: a constant pressure and a pressure varying linearly from a maximum at the crack centre to zero at its tips. We believe that these pressure distributions are good approximations of those commonly occurring in natural hydrofractures.

For a constant overpressure, $p(x) = P_0$, the integral in Eq. (13.13) gives $P_0 \pi/2$. Substituting this into Eq. (13.33) and solving the first integral, the crack-tip tensile stress becomes

$$\sigma_3 = -P_0 x \left( \frac{1}{(x^2 - a^2)^{1/2}} - \frac{1}{x} \right) \tag{13.35}$$

indicating that when $x \to a$, in which case the tip of the hydrofracture is approached (from outside the tip), the theoretical tensile stress $\sigma_3$ becomes infinite, or $\sigma_3 \to -\infty$.

We next consider a mathematical crack (line crack), where the fluid overpressure $p(x)$ decreases linearly from a maximum value $P_0$ at the crack centre by a linear gradient $p_1 x$ to the arrested crack tips. The overpressure $p(x)$ is then given by Eq. (13.15). Using this overpressure gradient, from Eq. (13.13) we get $q(t) = p_1 t$, and then from Eq. (13.33)

$$\sigma_3 = \frac{2x p_1}{\pi} \int_0^a \frac{t^2 dt}{(x^2 - a^2)^{3/2}} \tag{13.36}$$

The solution to Eq. (13.35) is

$$\sigma_3 = -\frac{2 p_1}{\pi} \left[ \frac{xa}{(x^2 - a^2)^{1/2}} - x \arcsin \frac{a}{x} \right] \tag{13.37}$$

Equation (13.36) shows that, when the tip of the hydrofracture is approached, $x \to a$, then $\sigma_3 \to -\infty$. Equations (13.31), (13.32), and (13.36), show that, regardless of whether the hydrofracture is modelled as a mathematical crack or an elliptical hole, and subject to constant or linearly varying overpressure, the theoretical tip tensile stresses are normally so high that a continuous and buoyant hydrofracture should continue its propagation to the surface. Of course, plastic deformation and/or microcracking in the process zone of the propagating hydrofracture reduces the theoretical stress and keeps it finite (Chapter 10). Nevertheless, both modelling approaches indicate that even when the fluid overpressure is minor (fraction of a megapascal, for example), the theoretical tensile stress at the fracture tip is so high that, if this stress was the only controlling factor, all buoyant hydrofractures should reach the surface. This was also the conclusion of the numerical models for a homogeneous, isotropic crust (Figs. 13.6 and 13.7). However, most hydrofractures become arrested, as has been discussed in Section 13.4. And when they become arrested at comparatively shallow depths, the commonly high tensile (and shear) stresses induced by the hydrofractures may cause surface deformation. This deformation, in turn, can sometimes be used to infer the depth to the tip of the arrested hydrofracture. This is important for all hydrofractures approaching the Earth's surface, and particularly so for dykes because of their potential for generating fissure eruptions.

Commonly observed, arrested hydrofractures include sheet-like igneous intrusions such as dykes (Fig. 6.20) and mineral-filled veins (Fig. 11.14), as well as many joints (Figs. 11.9–11.10) and man-made hydraulic fractures. These latter are common in petroleum engineering where hydraulic fractures are injected under fluid overpressure into reservoir rocks to increase their permeabilities (Daneshy, 1978; Valko and Economides, 1995; Yew, 1997; Economides and Nolte, 2000). Commonly, these hydraulic fractures become arrested when their vertical tips enter layers of high fracture-perpendicular compressive stresses or meet with sharp contacts between mechanically contrasting layers (Section 13.4).

In the geothermal and petroleum industries, the aim is to keep the hydraulic fracture, injected to increase the permeability of a certain target layer or unit, within the target unit (Fig. 13.28). Commonly, the hydraulic fracture is contained within the target unit because

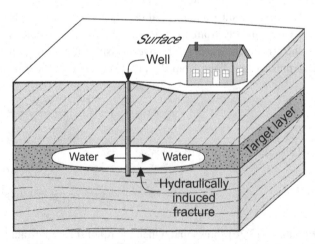

**Fig. 13.28**    Hydraulic fractures are used to increase permeability in target layers or units, as indicated schematically here.

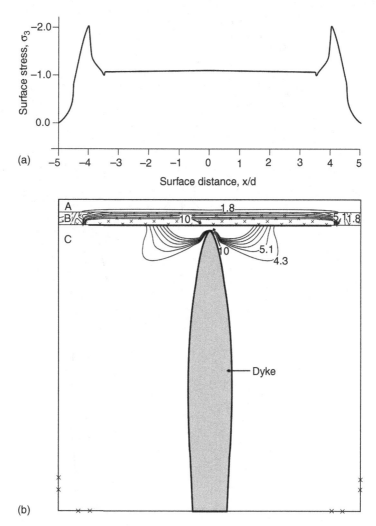

**Fig. 13.29** Numerical model of the surface stresses induced by a dyke arrested below a layered crustal segment.

there is an abrupt change in stiffness across the contacts between that unit and the adjacent layers. Hydraulic fractures are commonly injected at depths of several kilometres (except for some small-scale hydraulic stress measurements), and normally do not generate much surface deformation.

Many dykes, however, become arrested at comparatively shallow depths and may induce considerable stresses and deformation at the surface of the host volcano or rift zone. The exact magnitude and distribution of surface stresses and deformation, however, depends on the layering of the crust. Two numerical models (Figs. 13.29 and 13.30) show the significant difference in the induced surface tensile stresses of an arrested dyke for the same loading by different crustal layering. The results also show that inference as to the depth and location of the dyke, a so-called geodetic inversion, can be misleading when using homogeneous,

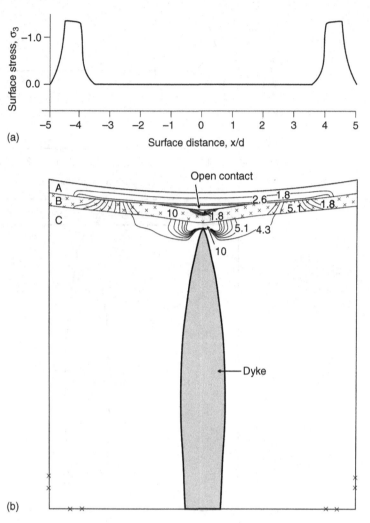

Fig. 13.30 Numerical model of the surface stresses induced by a dyke arrested below a layered crustal segment where the contact between two of the layers is weak (has a low tensile strength) and opens through a Cook–Gordon delamination or debonding mechanism.

isotropic half-space models when the actual crustal segment is strongly layered and with weak contacts, as is commonly the case in active rift zones and volcanoes.

## 13.7 Summary

- Extension fractures, both tension fractures and hydrofractures, form by the linking of smaller fractures, such as columnar joints, as well as contacts and other existing weaknesses in the rock. Narrow and newly formed extension fractures often have an irregular, zigzag geometry because part of the path is along hexagonal columnar joints. Tension fractures on an outcrop scale can only form close to the surface where there is absolute tension, such as during rifting episodes. As tension-fracture segments become closer to each other in relation to their controlling dimensions (strike dimensions at the surface of a rift zone, for example), the segments gradually function more and more as a single, segmented fracture.

- In a homogeneous, isotropic crust, all buoyant hydrofractures should reach the Earth's surface. Partly because of mechanical layering, however, most become arrested at various crustal depths. Hydrofracture arrest and/or deflection is common at contacts between mechanically dissimilar layers, particularly where there is an abrupt change in Young's modulus across a contact between layers. On meeting a contact, an extension fracture (normally a hydrofracture except at shallow depths) may (a) become arrested, (b) penetrate the contact, or (c) become deflected in one or two directions. Which of (a)–(c) happens depends on whether or not there is, at the contact, (i) rotation of the principal stresses, generating a stress barrier, (ii) debonding or delamination (opening) of the contact, or (iii) unfavourable material toughness and elastic mismatch between the layers and the contact material itself.

- A hydrofracture propagating thorough heterogeneous and anisotropic rocks (layered rocks, for example) develops fluid overpressure that depends on the local stresses in the hosting layers. The local stresses, in turn, depend on the loading conditions and stiffnesses of the layers. For example, during extension, the stiff layers concentrate (relative) tensile stresses and increase the effective overpressure of those parts of the hydrofracture that cut through the stiff layers, whereas during compression the stiff layers concentrate compressive stresses and reduce the hydrofracture overpressure. Overpressure variation in a hydrofracture can be modelled analytically in many ways: here we show modelling using polynomials and Fourier cosine series. Numerical models can also be used to model aperture variations of hydrofractures, subject to different overpressures, propagating in layered rocks. Large-aperture parts of hydrofractures normally transport fluids at greater volumetric flow rates and may result in flow channelling.

- Theoretical elastic tensile stresses at the tips of many hydrofractures are very high even if the fluid overpressure is very low. This follows from the aspect ratio of the fracture; in particular, if the fractures are modelled as mathematical cracks, the theoretical tensile stresses for any fluid overpressure at the tips approach infinity. Plastic deformation and/or microcracking in the process zone at the tip reduce the tensile stresses and keep them finite. Nevertheless, the high theoretical tensile stresses at hydrofracture tips would normally result in such fractures, if buoyant and in a homogeneous and isotropic crustal segment, reaching the Earth's surface.

- Most hydrofractures (and other rock fractures), however, become arrested at various depths, particularly at layer contacts. The arrest is attributed to the factors discussed above. An arrested hydrofracture may generate tensile stresses and deformation at the Earth's surface, such as in volcanoes and rift zones. The magnitude and location of the deformation and stress, however, depends on the mechanical layering of the crust. The layering must be known in order to use inversion of the surface deformation to infer the depth of the associated arrested hydrofracture.

# 13.8 Main symbols used

| | |
|---|---|
| $a$ | half length (strike dimension) of a crack |
| $2a$ | major axis of an elliptical hole |
| $2b$ | minor axis of an elliptical hole |
| $c$ | half surface length of a crack (depending on orientation, may be equal to $a$) |
| $d$ | distance between the centres of a pair of collinear cracks |
| $E$ | Young's modulus (modulus of elasticity) |
| $E_1*$ | plane-strain Young's modulus of layer 1 |
| $E_2*$ | plane-strain Young's modulus of layer 2 |
| $G_d$ | strain energy release rate for fracture deflection into a discontinuity |
| $G_p$ | strain energy release rate for fracture penetration of a discontinuity |
| $G_{tot}$ | total strain energy release rate during crack extension (propagation) |
| $G_I$ | strain energy release rate during the extension of a mode I crack |
| $G_{II}$ | strain energy release rate during the extension of a mode II crack |
| $G_{III}$ | strain energy release rate during the extension of a mode III crack |
| $g(\xi)$ | function defined in Eqs. (13.21), (13.24), and (13.29) |
| $J_0$ | zero-order Bessel function of the first kind |
| $K_I$ | stress intensity factor for a mode I crack |
| $K_{II}$ | stress intensity factor for a mode II crack |
| $K_{III}$ | stress intensity factor for a mode III crack |
| $k$ | parameter defined in Eq. (13.19) |
| $n$ | Fourier coefficient |
| $p_0$ | fluid overpressure in a crack |
| $p(x)$ | polynomial expression of the fluid overpressure in a crack |
| $p_1 x$ | linear gradient of fluid overpressure in a crack |
| $u$ | crack opening displacement (half the aperture) |
| $u_y(x)$ | crack opening displacement parallel with the $y$-axis |
| $\Delta u$ | maximum aperture (twice the opening displacement) of a hydrofracture |
| $x_N$ | half surface length of the crack (equal to $c$) |
| $\Delta x$ | interval $(x_{i+1} - x_i)$ |
| $\alpha$ | Dundurs uniaxial (extensional) elastic mismatch parameter |
| $\beta$ | Dundurs areal (volumetric) elastic mismatch parameter |
| $\Gamma_D$ | material toughness of a discontinuity or an interface (a contact, for example) |
| $\Gamma_L$ | material toughness of a layer (material toughness in general is denoted by $G_c$) |
| $\Gamma_L^1$ | material toughness of layer 1 |
| $\mu$ | shear modulus (denoted by $G$ in most other chapters) |
| $\mu_1$ | shear modulus of layer 1 |
| $\mu_2$ | shear modulus of layer 2 |
| $\nu$ | Poisson's ratio |
| $\nu_1$ | Poisson's ratio of layer 1 |
| $\nu_2$ | Poisson's ratio of layer 2 |
| $\pi$ | $3.1416\ldots$ |

$\rho_c$   radius of curvature
$\sigma$   tensile loading, tensile stress
$\sigma_3$   minimum compressive (maximum tensile) principal stress
$\tau_{ij}$   shear-stress component (for example, $\tau_{xy}$)
$\psi$   measure of the relative proportion of mode II to mode I
$\omega$   angular frequency

# 13.9 Worked examples

---

### Example 13.1

#### Problem

A joint in a 2-m-thick flow unit of a Pleistocene pahoehoe lava flow has a strike dimension of 1 m and extends through the flow unit. The Young's modulus of the flow unit is 10 GPa and its Poisson's ratio is 0.25. Geothermal water with an overpressure of 10 MPa enters and fills the joint to its tips. Calculate the theoretical tensile stress at the tips of the joint in a direction perpendicular to the strike dimension of the joint if the fluid overpressure is the only loading when the joint is modelled as:

(a) a mathematical crack
(b) an elliptical hole with a maximum total opening, or aperture, of 0.8 mm.

#### Solution

(a) Since the fluid overpressure $p_0$ is constant, the tip stress for a mathematical crack is given by Eq. (13.34) as

$$\sigma_3 = -p_0 x \left( \frac{1}{(x^2 - a^2)^{1/2}} - \frac{1}{x} \right)$$

Here the crack lies along the $x$ coordinate axis. Thus, when the crack tip (at $x = a$) from outside the crack (that is, from the right of the tip) then $x \to a$ and Eq. (13.34) approaches the form

$$\sigma_3 \to -1 \times 10^7 \, \text{Pa} \times 0.5 \, \text{m} \times \left( \frac{1}{[(0.5 \, \text{m})^2 - (0.5 \, \text{m})^2]^{1/2}} - \frac{1}{0.5 \, \text{m}} \right)$$

Since $(a^2 - a^2)^{1/2} = 0$ and $(1/0) = \infty$, that is, infinite, it follows that on approaching the tip of the crack, $-\sigma_3 \to \infty$ whatever the value of $p_0 a$. Thus, the theoretical tensile stress of a mathematical crack becomes infinitely high at its tip (that is, a singularity). This theoretical result is well established. And so is the fact that the tensile stress does not become infinite at the tip of a real crack: there is always microcracking or plastic deformation (dislocation glide) at the tip to keep the actual tensile stress finite, however narrow the crack tip. The size of the process zone at the crack tip where microcracking and/or plastic deformation takes place and keeps the tensile stress finite is a measure of how brittle the fracture is: if the zone is small, the fracture (and the associated process) is brittle; if the zone is large, the fracture is more ductile.

(b) When the fracture is modelled as an elliptical hole, Eq. (6.8) can be used to estimate the crack-tip tensile stress as follows:

$$\sigma_{max} = \sigma_3 = -p_0 \left( \frac{2a}{b} - 1 \right) = -1 \times 10^7 Pa \times \left( \frac{2 \times 0.5 \text{ m}}{4 \times 10^{-4} \text{ m}} - 1 \right)$$
$$= -2.5 \times 10^{10} Pa = -25 \text{ GPa}$$

Since for an elliptical hole, by definition, $b \neq 0$, that is, the semi-minor axis has a non-zero length, an elliptical hole model always gives a finite tensile stress at the crack tip. For a mathematical crack, however, the aperture or thickness is by definition zero so that, when applied to an elliptical hole, it would mean $b = 0$. Then, of course, the theoretical tensile stress would be infinite at the crack tips, whatever the internal fluid overpressure.

## Example 13.2

### Problem

A hydrofracture has a strike dimension of 200 m. The fluid overpressure decreases linearly from 5 MPa at the fracture centre to zero at the fracture tips. The host-rock Young's modulus and Poisson's ratio are 10 GPa and 0.25, respectively. Use a through-crack mode I model to find the maximum fracture aperture.

### Solution

We can obtain the aperture or total opening $\Delta u$ from Eq. (13.18), which is

$$\Delta u = \frac{4(1 - v^2)}{E} \left[ p_0 k + \frac{p_1}{\pi} \left( ak + x^2 \ln \frac{a + k}{x} \right) \right]$$

We have the following information. Poisson's ratio $v$ is 0.25, Young's modulus $E$ is $1 \times 10^{10}$ Pa, the overpressure in the centre of the fracture $p_0$ is $5 \times 10^6$ Pa, the half length of the fracture $a$ is 100 m, and the factor $k$ is given by Eq. (13.19) as $k = (a^2 - x^2)^{1/2}$. The overpressure gradient $p_1$ is obtained from the following consideration. The overpressure changes from $5 \times 10^6$ Pa in the centre to 0 Pa at the tips, and this changes takes place over the distance $a$, the half length of the fracture, namely 100 m. Thus the gradient is

$$p_1 = -\frac{p_0}{a} = \frac{-5 \times 10^6 Pa}{100 \text{ m}} = -50\,000 \text{ Pa m}^{-1}$$

The gradient is negative because the overpressure decreases towards the tips. The maximum aperture, $\Delta u_{max}$, occurs in the centre of the fracture because that is where the fluid overpressure is greatest. Since the fracture has its centre at the origin of the coordinate system, the centre is where $x = 0$. Substituting $x = 0$ into Eq. (13.18) we get

$$\Delta u_{max} = \frac{4(1 - v^2)}{E}\left[p_0 k + \frac{p_1 a k}{\pi}\right]$$

Putting $x = 0$ into Eq. (13.19), namely, $k = (a^2 - x^2)^{1/2}$, so that $k = (a^2)^{1/2} = a$, and substituting this for $k$, the last equation can be written as

$$\Delta u_{max} = \frac{4(1 - v^2)a}{E}\left[p_0 + \frac{p_1 a}{\pi}\right] \tag{13.38}$$

Putting in the values above, we get

$$\Delta u_{max} = \frac{4 \times (1 - 0.25^2) \times 100 \text{ m}}{1 \times 10^{10}\text{Pa}}\left[5 \times 10^6\text{Pa} - \frac{5 \times 10^4\text{Pa m}^{-1} \times 100 \text{ m}}{3.1416}\right]$$

$$= 0.13 \text{ m} = 13 \text{ cm}$$

Thus, these calculations indicate that the ratio of the strike dimension, 200 m, to that of the maximum aperture, 13 cm, is around 1500, a very reasonable value and similar to that of many natural hydrofractures such as dykes.

## Example 13.3

### Problem

A horizontal hydrofracture has a total internal fluid pressure of 5 MPa in its centre and 0 MPa at its lateral ends. The fracture is located at a depth of 100 m. The average density of the layer above the hydrofracture (its roof) is 2500 kg m$^{-3}$, its Young's modulus is 10 GPa and its Poisson's ratio 0.25. The fracture is a tunnel crack with a width (or length perpendicular to the direction of fluid flow) of 50 m. Find:

(a) the total opening (aperture) of the fracture in its centre
(b) the total opening (aperture) as a function of the distance from its centre
(c) the opening (aperture) at a distance of 10 m from the centre.

### Solution

(a) From Eq. (13.37) we have

$$\Delta u_{max} = \frac{4(1 - v^2)a}{E}\left[p_0 + \frac{p_1 a}{\pi}\right]$$

Here there are two unknowns: the overpressure in the horizontal fracture $p_0$ and the overpressure gradient $p_1$. We are only given the total fluid pressure, so we have to find the overpressure. For a horizontal (sill-like) hydrofracture to form, the minimum principal compressive stress $\sigma_3$ must be vertical (Chapters 8 and 11). The vertical stress on the water sill is obtained from Eq. (4.45):

$$\sigma_v = \sigma_3 = \rho_r g z = 2500 \text{ kg m}^{-3} \times 9.81 \text{ m s}^{-2} \times 100 \text{ m} = 2.45 \times 10^6 \text{ Pa} \approx 2.5 \text{ MPa}$$

Since the fluid overpressure $p_0$ is defined as the total fluid pressure minus the normal stress on the fracture (for an extension fracture the normal stress is $\sigma_3$), we get the fluid overpressure as follows:

$$p_0 = p_t - \sigma_3 = 5 \text{ MPa} - 2.5 \text{ MPa} = 2.5 \text{ MPa}$$

The gradient $p_1$ is obtained as in Example 13.2, namely

$$p_1 = -\frac{p_0}{a} = \frac{-2.5 \times 10^6 \text{Pa}}{25 \text{ m}} = -1 \times 10^5 \text{Pa m}^{-1}$$

Putting these values into Eq. (13.37) we get the maximum opening or aperture as

$$\Delta u_{max} = \frac{4 \times (1 - 0.25^2) \times 25 \text{ m}}{1 \times 10^{10} \text{Pa}} \left[ 2.5 \times 10^6 \text{Pa} - \frac{1 \times 10^5 \text{Pa m}^{-1} \times 25 \text{ m}}{3.1416} \right]$$

$$= 0.0153 \text{ m} = 15.3 \text{ mm}$$

(b) From Eq. (13.18) we have

$$\Delta u = \frac{4(1 - v^2)}{E} \left[ p_0 k + \frac{p_1}{\pi} \left( ak + x^2 \ln \frac{a + k}{x} \right) \right]$$

This equation is perhaps easier to use for the present purpose if we substitute the aperture $b$ for $\Delta u$ and show $b$ as a function of $x$, as follows:

$$b(x) = \frac{4(1 - v^2)}{E} \left[ p_0 k + \frac{p_1}{\pi} \left( ak + x^2 \ln \frac{a + k}{x} \right) \right] \qquad (13.39)$$

Substituting the appropriate values from Example 13.3, including the gradient $p_1$ into Eq. (13.38), we obtain

$$b(x) = \frac{4 \times (1 - 0.25^2)}{1 \times 10^{10} \text{Pa}} \times \left[ 2.5 \times 10^6 \text{Pa} \times [(25 \text{ m})^2 - x^2]^{1/2} - \frac{1 \times 10^5 \text{Pa}}{3.1416} \right.$$

$$\left. \times \left( 25 \text{ m} \times [(25\text{m})^2 - x^2)]^{1/2} + x^2 \ln \frac{25 \text{ m} + [(25 \text{ m})^2 + x^2]^{1/2}}{x} \right) \right]$$

This gives a smooth opening variation, not very dissimilar to that of an ellipse.

(c) On substituting $x = 10$ m in the equation above, and omitting the units during the calculations for convenience, we get

$$b(10) \approx 3.75 \times 10^{-10} \times \left[ 2.5 \times 10^6 \times (25^2 - 10^2)^{1/2} - 3.18 \times 10^4 \right.$$

$$\left. \times (25 \, [25^2 - 10^2]^{1/2} + 10^2 \times \ln \frac{25 + [25^2 - 10^2]^{1/2}}{10} \right]$$

$$\approx 3.75 \times 10^{-10} \times [5.7 \times 10^7 - 3.18 \times 10^4 \times (573 + 157)]$$

$$\approx 3.75 \times 10^{-10} \times 3.38 \times 10^7 = 0.0127 \text{ m} = 12.7 \text{ mm}$$

Thus, from the centre of the hydrofracture to a distance of 10 m from the centre, the total opening or aperture of the hydrofracture decreases from 15.3 mm to 12.7 mm, or by 2.6 mm.

## Example 13.4

### Problem

A regional vertical basaltic dyke in Iceland (similar to those in Fig. 14.8) is exposed at a crustal depth of 1 km below the surface of the rift zone at its time of formation. The dyke is traced laterally for 10 km, so that its strike dimension is 10 km. Its dip dimension is between 10–20 km, that is, it is assumed to come from a deep-seated reservoir in the lower crust or at the crust mantle boundary (Chapter 6). At its lateral ends, the overpressure must have been essentially zero. We assume that the overpressure varies linearly from 10 MPa in the centre to 0 MPa at the lateral ends. At a crustal depth of 1 km, the static Young's modulus is about 15 GPa and Poisson's ratio is 0.25. Calculate the maximum total opening (now thickness) of the dyke.

### Solution

We can obtain the maximum opening from Eq. (13.37). To do so, however, we must first find the gradient $p_1$. We proceed as in Example 13.2 to get the gradient

$$p_1 = -\frac{p_0}{a} = \frac{-1 \times 10^7 \text{Pa}}{5000 \text{ m}} = -2000 \text{ Pa m}^{-1}$$

Now we can calculate the maximum opening from Eq. (13.37) as follows:

$$\Delta u_{\max} = \frac{4(1 - v^2)a}{E} \left[ p_0 + \frac{p_1 a}{\pi} \right]$$

$$= \frac{4 \times (1 - 0.25^2) \times 5000 \text{ m}}{1.5 \times 10^{10}} \left[ 1 \times 10^7 \text{Pa} - \frac{2 \times 10^3 \text{Pa m}^{-1} \times 5000 \text{ m}}{3.1416} \right] = 8.5 \text{ m}$$

Many of the regional dykes, such as those seen in Fig. 14.8, have thicknesses between 5 m and 10 m, and some dykes in East Iceland have been traced laterally between 4 and 22 km. These results are thus very reasonable. However, the thickness/opening variation is normally not linear but rather more irregular and better described by Fourier series as in Section 13.5.2.

# 13.10 Exercises

13.1 Explain briefly the mechanism(s) by which extension fractures grow.

13.2 Explain how the maximum displacement on a segmented tension fracture changes as the distances between nearby fracture segments gradually decrease.

13.3 What is meant by stress-field homogenisation and how does it affect fracture propagation and arrest?

13.4 What is a stress barrier and how can it affect fracture propagation?

13.5 What three scenarios are possible when a propagating hydrofracture meets an interface or discontinuity (such as a contact) in a crustal segment? Briefly explain the three scenarios.

13.6 Explain what is mean by fracture-induced stresses. Give a geological example.

13.7 What is elastic mismatch and how can it affect extension-fracture propagation?

13.8 What is the general difference between aperture variation in a hydrofracture in a homogeneous and isotropic rock, on the one hand, and a layered rock, on the other?

13.9 Name and explain briefly two analytical methods for modelling overpressure variation in hydrofractures.

13.10 Name and explain briefly two models for the theoretical calculation of stresses at fracture tips.

13.11 Explain briefly how fracture-induced surface stresses (and deformation), such as those generated by dykes in volcanoes, depend on the mechanical layering of the associated crustal segment.

13.12 A 2-m-long (strike dimension) fracture, modelled as a mode I through-crack, in a host rock with a Young's modulus of 5 GPa and a Poisson's ratio of 0.27 is subject to a constant fluid overpressure of 5 MPa. Calculate the theoretical tensile stress at the fracture tip using the elliptical-hole model.

13.13 Repeat Exercise 13.12 using a mathematical-crack model. Comment on the difference between the results. Why are stress singularities not reached in nature? What happens?

13.14 A fluid-driven fracture is 50 m long (strike dimension) and can be modelled as a mode I through crack. The fluid overpressure in the fracture is 3 MPa at its centre and decreases linearly to zero at the fracture tips. The Young's modulus of the host rock is 5 GPa and the Poisson's ratio 0.27. Calculate the maximum fracture aperture.

13.15 A horizontal fluid-driven fracture is located at a depth of 60 m in a host rock with an average Young's modulus of 7 GPa, a Poisson's ratio of 0.27, and a density of 2400 kg m$^{-3}$. The fluid overpressure in the fracture centre is 10 MPa and 0 MPa at its ends. The fracture is a tunnel crack with a length perpendicular to the direction of fluid flow of 40 m. Calculate (a) the total opening (aperture) in the centre of the fracture, and (b) the aperture at 10 m from the centre.

# References and suggested reading

Anderson, E. M., 1936. The dynamics of formation of cone sheets, ring dykes and caldron subsidences. *Proceedings of the Royal Society of Edinburgh*, **56**, 128–163.

Anderson, T. L., 2005. *Fracture Mechanics: Fundamentals and Applications*, 3rd edn. London: Taylor & Francis.

Brebbia, C. A. and Dominquez, J., 1992. *Boundary Elements: an Introductory Course*. Southampton: Computational Mechanics Publications.

Broberg, K. B., 1999. *Cracks and Fracture*. New York: Academic Press.

Broek, D., 1978. *Elementary Engineering Fracture Mechanics*, 2nd edn. Leyden: Noord-hoff.

Chawla, K. K., 1998. *Composite Materials: Science and Engineering*, 2nd edn. Berlin: Springer-Verlag.

Cook, J. and Gordon, J. E., 1964. A mechanism for the control of crack propagation in all-brittle systems. *Proceedings of the Royal Society of London*, **A282**, 508–520.

Daneshy, A. A., 1978. Hydraulic fracture propagation in layered formations. *AIME Society of Petroleum Engineers Journal*, 33–41.

DeGraff, J. M. and Aydin, A., 1987. Surface-morphology of columnar joints and its significance to mechanics and direction of joint growth. *Geological Society of America Bulletin*, **99**, 605–617.

Dundurs, J., 1969. Edge-bonded dissimilar orthogonal wedges. *Journal of Applied Mechanics*, **36**, 650–652.

Economides, M. J. and Nolte, K. G. (eds.), 2000. *Reservoir Stimulation*. New York: Wiley.

Freund, L. B. and Suresh, S., 2003. *Thin Film Materials: Stress, Defect Formation and Surface Evolution*. Cambridge: Cambridge University Press.

Gray, T. G. F., 1992. *Handbook of Crack Opening Data*. Cambridge: Abingdon.

Gudmundsson, A., 2003. Surface stresses associated with arrested dykes in rift zones. *Bulletin of Volcanology*, **65**, 606–619.

Gudmundsson, A., 2006. How local stresses control magma-chamber ruptures, dyke injections, and eruptions in composite volcanoes. *Earth-Science Reviews*, **79**, 1–31.

Gudmundsson, A., 2010. Deflection of dykes into sills and magma-chamber formation. *Tectonophysics*, document doi:10.1016/j.tecto.2009.10.015.

Gudmundsson, A. and Philipp, S. L., 2006. How local stress fields prevent volcanic eruptions. *Journal of Volcanology and Geothermal Research*, **158**, 257–268.

Gudmundsson, A., Fjeldskaar, I., and Brenner, S. L., 2002. Propagation pathways and fluid transport of hydrofractures in jointed and layered rocks in geothermal fields. *Journal of Volcanology and Geothermal Research*, **116**, 257–278.

Gudmundsson, A., Kusumoto, S., Simmenes, T. H., Philipp, S. L., and Larsen, B., 2010. Opening-displacement variations and fluid flow rates in rock fractures. *Tectonophysics* (submitted).

He, M. Y. and Hutchison, J. W., 1989. Crack deflection at an interface between dissimilar elastic materials. *International Journal of Solids and Structures*, **25**, 1053–1067.

He, M. Y., Evans, A. G., and Hutchinson, J. W., 1994. Crack deflection at an interface between dissimilar elastic materials: role of residual stresses. *International Journal of Solids and Structures*, **31**, 3443–3455.

Hull, D. and Clyne, T. W., 1996. *An Introduction to Composite Materials*, 2nd edn. Cambridge: Cambridge University Press.

Hutchison, J. W., 1996. *Stresses and Failure Modes in Thin Films and Multilayers. Notes for a DCAMM Course*. Lyngby: Technical University of Denmark, pp. 1–45.

Kavanagh, J. L., Menand, T. and Sparks, R. S. J., 2006. An experimental investigation of sill formation and propagation in layered elastic media. *Earth and Planetary Science Letters*, **245**, 799–813.

Kim, J. W., Bhowmick, S., Hermann, I., and Lawn, B. R., 2006. Transverse fracture of brittle bilayers: relevance to failure of all-ceramic dental crowns. *Journal of Biomedical Materials Research*, **79B**, 58–65.

Kusumoto, S. and Gudmundsson, A., 2010. Displacement and stress fields around rock fractures opened by irregular overpressure variation given by Fourier cosine series. *Geophysical Journal International* (accepted with revision).

Maugis, D., 2000. *Contact, Adhesion and Rupture of Elastic Solids*. New York: Springer-Verlag.

Philipp, S. L., 2008. Geometry and formation of gypsum veins in mudstones at Watchet, Somerset, SW-England. *Geological Magazine*, **145**, 831–844.

Schön, J. H., 2004. *Physical Properties of Rocks: Fundamentals and Principles of Petrophysics*. Oxford: Elsevier.

Secor, D. T., 1965. The role of fluid pressure in jointing. *American Journal of Science*, **263**, 633–646.

Siler, D. L. and Karson, J. A., 2009. Three-dimensional structure of inclined sheet swarms: implications for crustal thickening and subsidence in the volcanic rift zone of Iceland. *Journal of Volcanology and Geothermal Research*, **188**, 333–346.

Sneddon, I. N., 1948. The distribution of stress in the neighbourhood of a crack in an elastic solid. *Proceedings of the Royal Society of London*, **A187**, 229–260.

Sneddon, I. N., 1995. *Fourier Transforms*. New York: Dover.

Sneddon, I. N. and Lowengrub, M., 1969. *Crack Problems in the Classical Theory of Elasticity*. New York: Wiley.

Spiegel, M. R., 1974. *Fourier Analysis with Applications to Boundary Value Problems*. New York: McGraw-Hill.

Sutton, A. P. and Balluffi, R. W., 1995. *Interfaces in Crystalline Materials*. Oxford: Oxford University Press.

Thouless, M. D. and Parmigiani, J. P., 2007. Mixed-mode cohesive-zone models for delamination and deflection in composites. In: Sørensen, B. F., Mikkelson, L. P., Lilhot, H., Goutianos, S., and Abdul-Mahdi, F. S. (eds.), *Proceedings of the 28th Risø International Symposium on Materials Science: Interface Design of Polymer Matrix Composites*, Roskilde, Denmark, pp. 93–111.

Tsang, C. F. and Neretnieks, I., 1998. Flow channeling in heterogeneous fractured rocks. *Reviews of Geophysics*, **36**, 275–298.

Valko, P. and Economides, M. J., 1995. *Hydraulic Fracture Mechanics*. New York: Wiley.

Wang, P. and Xu, L. R., 2006. Dynamic interfacial debonding initiation induced by an incident crack. *International Journal of Solids and Structures*, **43**, 6535–6550.

Warpinski, H. R., 1985. Measurement of width and pressure in a propagating hydraulic fracture. *Journal of the Society of Petroleum Engineers*, February, pp. 46–54.

Xu, L. R. and Rosakis, A.J., 2003. An experimental study of impact-induced failure events in homogeneous layered materials using dynamic photoelasticity and high-speed photography. *Optics and Lasers in Engineering*, **40**, 263–288.

Xu, L. R., Huang, Y. Y., and Rosakis, A. J., 2003. Dynamics of crack deflection and penetration at interfaces in homogeneous materials: experimental studies and model predictions. *Journal of the Mechanics and Physics of Solids*, **51**, 461–486.

Yew, C.H., 1997. *Mechanics of Hydraulic Fracturing*. Houston, TX: Gulf Publishing.

Zhang, X., Jeffrey, R. G., and Thiercelin, M., 2007. Deflection and propagation of fluid-driven fractures at frictional bedding interfaces: a numerical investigation. *Journal of Structural Geology*, **29**, 396–410.

# 14 Evolution of faults

## 14.1 Aims

How faults transport crustal fluids is important in many fields of earth sciences, such as petroleum geology, geothermal research, volcanology, seismology, and hydrogeology. In order to understand the permeability evolution and maintenance of a fault zone, its internal hydromechanical structure and associated local stresses and mechanical properties must be known. The internal structure and, therefore, the local stresses in turn depend to a large degree on how the fault zone develops through time. The principal aims of this chapter are to present the current understanding of:

- How faults initiate and grow.
- The formation and development of the fault core.
- The formation and development of the fault damage zone.
- Local stresses within fault zones and their effects on permeability.
- The evolution of fault slip.

## 14.2 Initiation of faults

### 14.2.1 General

Faults, like any other tectonic fractures, initiate from existing weaknesses in the rock. Most weaknesses are joints and contacts/interfaces of various types. The basic theory of fracture initiation, including that of faults, is due to Griffith (1924). Earlier, the Griffith theory was discussed in connection with rock failure (Chapter 7), fracture mechanics (Chapter 10), and the maximum depth of tension fractures (Chapter 8). Here we use Griffith theory to explain the initiation of shear fractures from existing joints and other weaknesses in the host rock.

Griffith suggested that in a biaxial stress field, that is, where two of the principal stresses are non-zero, fracture initiation would be most likely from the points of highest tensile stress on the walls or surfaces of existing flaws in the material, the Griffith cracks. For rocks, joints and contacts that may be regarded as analogous, on an outcrop scale, to Griffith cracks. Contacts and joints are common in many rocks, such as sedimentary rocks and many igneous rocks, particularly lava flows. As an example, columnar joins in lava flows offer weaknesses from which faults in a volcanic rift zone initiate (Figs. 11.3 and 11.4).

Consider a biaxial stress field with $\sigma_1$ the maximum principal compressive stress and $\sigma_3$ the minimum principal compressive stress. If $T_0$ is the tensile strength of the rock, then the two-dimensional Griffith criterion is (Chapter 7)

$$(\sigma_1 - \sigma_3)^2 = 8T_0(\sigma_1 + \sigma_3) \tag{14.1}$$

when $\sigma_1 > -3\sigma_3$. Compressive stress is regarded as positive and tensile stress as negative, hence the minus sign. Since we are here talking about magnitudes and $\sigma_3$ may be negative, we use the minus sign in order to obtain the absolute value of $\sigma_3$ and compare it with the absolute value of $\sigma_1$, which, by definition, is always positive. Equation (14.1) applies to the compressive tectonic regime. For the tensile regime, that is, when $\sigma_1 < -3\sigma_3$, the two-dimensional Griffith criterion becomes (Chapter 7)

$$\sigma_3 = -T_0 \tag{14.2}$$

This is the well-known simple criterion for tensile failure. It is effectively a tautology, since it simply identifies the tensile strength $T_0$ with that magnitude of the principal tensile stress $\sigma_3$ at which tensile failure occurs. Despite being essentially tautologous, this is a useful criterion, particularly in practical work on tensile failure of rock masses at shallow depths or at the Earth's surface.

Equations (14.1) and (14.2) may be combined into a single equation which represents the shear stress $\tau$ for shear failure, that is, fault slip, namely

$$\tau = [4T_0(\sigma_n + T_0)]^{1/2} \tag{14.3}$$

The curve of Eq. (14.3) is the well-known Mohr's envelope (Chapter 7) and the normal stress $\sigma_n$ is given by

$$\sigma_n = \frac{\sigma_1 + \sigma_3}{2} - \frac{\sigma_1 - \sigma_3}{2} \cos 2\alpha \tag{14.4}$$

Where $\alpha$ is the angle between $\sigma_1$ and the plane of shear failure.

The conditions of Eq. (14.1) apply to a stress regime that is primarily compressive. If $\alpha$ is the angle between the major axis of an elliptical flaw or weakness, such as a columnar joint, and the direction of $\sigma_1$, then the orientation of the elliptical crack or joint at which failure is initiated, that is, fault initiation, is given by (Chapter 7)

$$\cos 2\alpha = \frac{\sigma_1 - \sigma_3}{2(\sigma_1 + \sigma_3)} \tag{14.5}$$

If follows that the joint that is most likely to propagate is oblique to the directions of the principal stresses and is therefore a shear fracture (Chapter 7).

The local maximum tensile stress at the elliptical joint does not, however, occur at the very end, or tip, of the ellipse but rather at a certain distance from its tip (Fig. 14.1). Furthermore, the tension fractures that nucleate from the elliptical joint make an angle of roughly 70° with the major axis of the elliptical joint (Fig. 14.2; Nemat-Nasser and Hori, 1999).

It follows from these theoretical considerations that a single shear fracture, propagating from a joint or another type of Griffith flaw, should be unable to grow in its own plane and develop into a large-scale fault. What should happen, instead, is that tensile fractures propagate from the sites of maximum tensile stresses at the surfaces of the elliptical joints

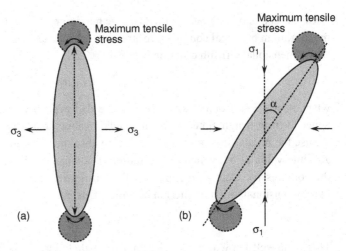

Fig. 14.1    During tilting of strata, original vertical joints (a) become inclined (b) and are no longer in a principal stress plane. As a consequence, the maximum tensile stress around the elliptical joint shifts away from its tips and becomes located somewhat to the side of the tips.

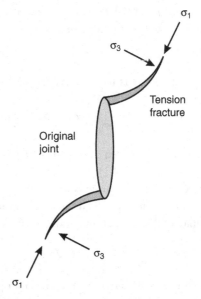

Fig. 14.2    Extension or tension fractures that originate from inclined joints propagate from the points of maximum tensile stress concentration (Fig. 14.1) and then curve so as to become parallel with $\sigma_1$ and perpendicular to $\sigma_3$.

along curved paths that gradually become parallel with the direction of the maximum principal compressive stress, $\sigma_1$, and thereby perpendicular to the maximum principal tensile stress, $\sigma_3$, and soon stop propagating (Fig. 14.2). These curving fractures commonly stop propagating because they lack the energy needed to advance the tip further (Chapter 10) or because of crack–crack interaction. That is, they enter regions where the driving stresses have already decreased, or relaxed, as a consequence of curved fractures propagating from the opposite direction from other inclined, elliptical joints, as discussed below. The conclusion that a single, inclined elliptical joint (or another type of fracture) is unable to propagate in its own plane so as to generate a large fault follows directly from Griffith's assumption that cracks are flat ellipses, and is supported by many experimental results (Nemat-Nasser and Hori, 1999; Paterson and Wong, 2005).

If instead of a simple, elliptical shear fracture, an inclined joint, there is a set of shear fractures, then a large fault may develop. This kind of development has been analysed in detail (Peng and Johnson, 1972; Nemat-Nasser and Hori, 1999), and the findings indicate that a fault can result from the linking up of smaller shear fractures (inclined joints) that initially form an array of offset, en echelon joints or other original flaws in the rock (Fig. 14.3).

This development of offset en echelon joints into faults is commonly seen in the field. For instance, such a development is commonly seen in igneous and sedimentary rocks (Figs. 14.3–14.6). Then the linking of the original joints often occurs along contacts between the layers. The process of joints linking into faults along contacts is discussed in detail through numerical models below, but first we will focus on certain analytical aspects of the process.

Equations (14.1) and (14.2) assume that the en echelon joints or cracks remain open during the loading to failure. That assumption may be valid for the tensile regime, and in particular during active extension in that regime, but many joints are likely to close in the compressive regime at deeper levels in the crust. This closure may be partial or complete. To take the effects of crack closure and associated friction into account during fault initiation from a set of en echelon joints, McClintock and Walsh (1962) modified the Griffith criterion into the form

$$\mu(\sigma_1 + \sigma_3 - 2\sigma_c) + (\sigma_1 - \sigma_3)(1 + \mu)^{1/2} = 4T_0 \left(1 - \frac{\sigma_c}{T_0}\right)^{1/2} \tag{14.6}$$

where $\sigma_c$ is the remote normal stress needed to close the elliptical joint and $\mu$ is the coefficient of sliding friction between the walls or surfaces of the partly or completely closed joint. If the joint or crack closes under very low remote stress, then a great simplification is obtained by assuming $\sigma_c = 0$. Then Eq. (14.6) reduces to the form

$$\mu(\sigma_1 + \sigma_3 - 2\sigma_c) + (\sigma_1 - \sigma_3)(1 + \mu)^{1/2} = 4T_0 \tag{14.7}$$

Equation (14.7) represents a linear relationship between the principal stresses at shear failure and can be written in the form

$$\tau = 2T_0 + \mu\sigma_n \tag{14.8}$$

where $\tau$ is the shear stress on the fault plane and $\sigma_n$ is the normal stress on the fault plane, given by Eq. (14.4). This is the **Modified Griffith criterion** (Chapter 7).

Equation (14.8) is almost identical to the empirical criterion for the formation of shear fractures or faults, namely the Coulomb criterion (Chapter 7). This criterion states that in a compressive stress regime shear failure or faulting occurs when the shear stress $\tau$ on the potential fault plane satisfies the equation

$$\tau = \tau_0 + \mu\sigma_n \tag{14.9}$$

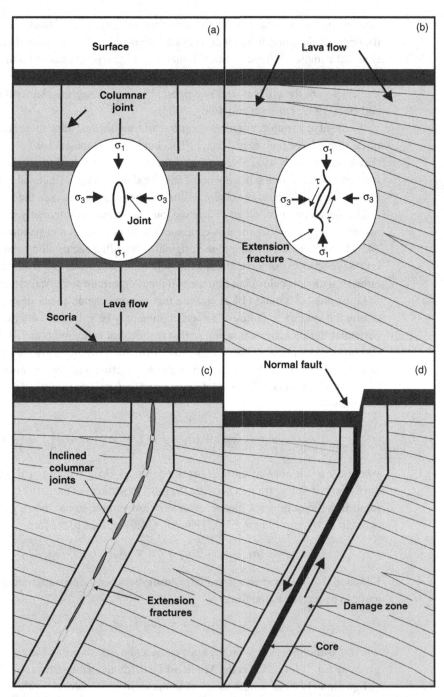

**Fig. 14.3** Schematic conceptual models as to how tilting of a pile of lava flow (or sedimentary strata), which originally, at the surface, have vertical columnar joints results in the joints becoming inclined and thus no longer in a principal stress plane. As the joints rotate, they become potential shear fractures and send out tension fractures that may link with other tension fractures, forming a set of interconnected fractures that eventually develops into a major shear fracture, that is, a fault (Gudmundsson, 1992).

**Fig. 14.4** Vertical cooling joints (some indicated by black lines) and horizontal contacts (some pointed at by white arrows) between flow units in a Holocene pahoehoe lava flow in the rift zone of Southwest Iceland. This is stage (a) in Fig. 14.3 before tilting occurs. When the joints become inclined, they may link through the indicated sets.

where $\tau_0$ is the cohesive shear strength or inherent shear strength of the rock and $\mu$ is the coefficient of internal friction. In Eq. (14.9), both $\tau_0$ and $\mu$ refer to unfractured rocks, but their exact physical meaning is somewhat obscure. By contrast, the corresponding terms $2T_0$ and $\mu$ in Eq. (14.8) have a clear physical meaning. Equation (14.8) may be regarded as the physical basis for the Coulomb criterion for faulting under predominantly compressive conditions (Brace, 1960).

## 14.2.2  Normal-fault initiation from tension fractures

In a rift zone, the shorter fractures in any fissure swarm tend to be pure tension fractures, whereas all the longer fractures are normal faults (Fig. 12.3 and 12.4). Consider, for example,

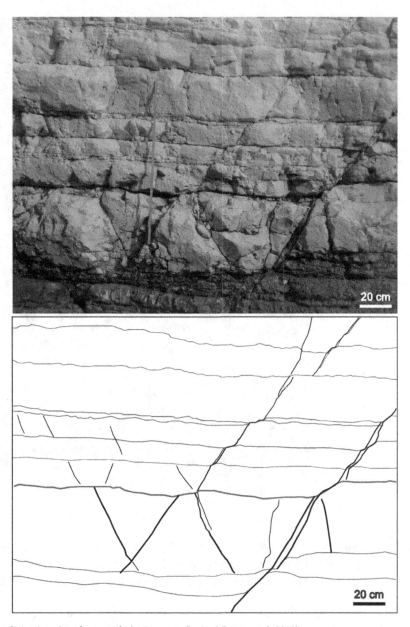

**Fig. 14.5**    Linking of joints into shear fractures, faults, in eastern England (Larsen *et al.*, 2010).

the Vogar Fissure Swarm located in a Holocene pahoehoe lava flow on the Reykjanes Peninsula in Southwest Iceland (Gudmundsson, 1987). The swarm contains about 120 fractures, 75% of which are tension fractures and 25% of which are normal faults. The average length of the tension fractures is 370 m, whereas the average length of the normal faults is 1990 m. All the normal faults change into tension fractures on approaching their tips. While many normal faults nucleate at depth, as discussed in Section 14.2.3, these results suggest that some of the normal faults may develop from tension fractures at the

**Fig. 14.6**   Another example of linking of joints into shear fractures, faults, in chalk in eastern England (Larsen *et al.*, 2010).

surface. More specifically, as the tension fractures gradually become longer during rifting episodes, they also become deeper (Chapter 9). When they reach a critical depth, they must, according to the Griffith criterion, change into normal faults (Chapter 7).

The failure criterion presented by Eq. (14.2) applies only when the inequality $\sigma_1 < -3\sigma_3$ is satisfied. Substituting $-T_0$ for $\sigma_3$ in this inequality, we obtain the inequality

$$\sigma_1 - \sigma_3 < 4T_0 \qquad (14.10)$$

which suggests that for Eq. (14.2) to apply, the stress difference $\sigma_1 - \sigma_3$ must be less than four times the tensile strength $T_0$. In a rift zone, or any other area undergoing extension, the maximum principal compressive stress is normally vertical and given by

$$\sigma_1 = \rho_r g z \qquad (14.11)$$

where $\rho_r$ is the rock density, $g$ the acceleration due to gravity, and $z$ the crustal depth. The maximum depth at which Eq. (14.2) applies is $\sigma_1 - \sigma_3 = 4T_0$, or $\sigma_1 = 4T_0 + \sigma_3$. Substituting $-T_0$ for $\sigma_3$, we obtain the vertical stress at this maximum depth $z_{max}$ of tensile fracture as

$$\sigma_1 = 3T_0 \qquad (14.12)$$

Using Eq. (14.11), the depth at which a downward propagating tension fracture in a rift zone must change into a normal fault is (Chapter 7)

$$z_{max} = \frac{3T_0}{\rho_r g} \qquad (14.13)$$

## 14.2.3  Fault initiation on sets of inclined joints

When subsidence takes place, such as in sedimentary basins and rift zones, the layers that constitute the upper part of the crust become tilted. The tilting is due to extension and piling up of layers on top of older layers, but rotation of the piles along major faults may also play a role. These layers normally contain numerous joints and contacts. Joints that were originally vertical become steeply dipping, and contacts that were originally horizontal become gently dipping (Fig. 14.7). Examples of this kind include joints and contacts in carbonate rocks of sedimentary basins (Figs. 14.5 and 14.6) and in lava piles (Fig. 14.8).

Consider the details of this process as seen in the lava pile of Iceland. Outside major topographic highs, such as large stratovolcanoes (central volcanoes), the surface of the rift zone is relatively flat (Figs. 12.3–12.5). When young lava flows pile up in the rift zone

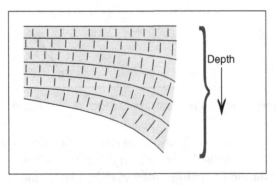

**Fig. 14.7**  Layers that are originally horizontal with vertical joints gradually increase their dips with depth, that is, become tilted. The associated joints then become gradually more shallowly dipping and may link into shear fractures.

**Fig. 14.8** Tilting of the Tertiary lava pile in Southeast Iceland. As the depth increases, the dip of the lava flow increases (as indicated by white arrows) by about one degree for every 150 m depth. Many regional dykes (some indicated), mostly 5–10 m thick, are seen dissecting the lava pile.

(Fig. 12.4), the older ones become buried and tilted (Figs. 6.2 and 14.8). The tilting is generally towards the axis of the rift zone. As a consequence of this tilting, the regional average dip of the lava pile increases by about one degree for every 150 m increase in depth in the pile (Fig. 14.8). It follows that the dips of the originally vertical joints decrease at the same rate with depth in the crust.

The normal state of stress in a flat rift zone, and in a sedimentary basin undergoing extension, is that $\sigma_1$ is vertical and $\sigma_3$ is horizontal and parallel with the axis of extension (spreading vector in a rift zone). It then follows that as soon as the joints become tilted with the lava flows, so that their dip becomes less than the original 90°, they are no longer parallel with $\sigma_1$ and are thus potential shear fractures. Many such inclined joints form an en echelon set of fractures from which faults may develop (Figs. 14.3–14.8).

The ideal cross-sectional shape of a joint is that of a flat ellipse (Figs. 14.1 and 14.2). The joint should start to propagate when the tensile stresses at its surface or walls reach the host-rock tensile strength. Let $\alpha$ be the angle between the major axis of the elliptical joint and the direction of $\sigma_1$. Then it follows from the conditions of Eq. (14.1) that the attitude of the joint when it starts to propagate is given by Eq. (14.5), which again shows that the joint most likely to propagate is oblique to $\sigma_1$ and thus a potential shear fracture. The maximum tensile stress at the surface of the joint does not occur at its tips (very ends) but rather at points at certain distances from the tips (Fig. 14.1). When the tensile stresses at these points reach the tensile strength of the host rock, tension fractures begin to propagate from these points of highest tensile-stress concentration. Close to and at these points of tensile-fracture nucleation, the propagating tensile fractures make an angle of about 70° to the major axis of the parent joint (Fig. 14.2). The propagation paths of the tension fractures, however, may be curved if they gradually become parallel with the direction of

$\sigma_1$. When the propagation path of the tension fractures becomes parallel with that of $\sigma_1$, the propagation stops.

These results suggest that, for the conditions of Eq. (14.1), a single small-scale shear fracture is normally unable to propagate in its own plane and develop into a large-scale fault. However, there are indications that under large confining pressures, that is, at great crustal depths, shear fractures may grow in their own plane into large-scale faults (Cox and Scholz, 1988; Lin and Parmentier, 1988; Shimada, 2000). It is then assumed that a process zone composed largely of mode I cracks develops at the expanding edge of the shear fracture, and that the fractures in this zone eventually link up into a single shear fracture, a fault. One possibility is that if the stress-intensity factor $K_I$ is close to zero, then a shear fracture can propagate in its own plane (Lin and Parmentier, 1988), but this conjecture remains to be supported by compelling data.

A more common situation for fault development, perhaps, is its growth from a set of en echelon fractures (Figs. 14.3–14.8). Experimental support for such sets developing into single shear fractures or faults is compelling (Cox and Scholz, 1988; Du and Aydin, 1991; Nemat-Nasser and Hori, 1999). As indicated above, such en echelon sets are bound to form during tilting of sedimentary or igneous layers in basins and rift zones (Fig. 14.7). Some sets dip towards the axis of the rift zone, or the centre of the sedimentary basin; others dip away from the axis, or towards the end margins of the sedimentary basin.

The depth of nucleation of faults on sets of en echelon cracks or joints can be estimated as follows. The shear stress $\tau$ on each of the en echelon fractures is given by

$$\tau = \frac{\sigma_1 - \sigma_3}{2} \sin 2\alpha \tag{14.14}$$

Using Eq. (14.11), and combining Eq. (14.14) with Eq. (14.3), yields the critical depth $z_c$ at which fractures of an en echelon system start to link up into a dip-slip (here normal) fault, as

$$z_c = \frac{2\left[4T_0(\sigma_n + T_0)\right]^{1/2}}{\rho_r g \sin 2\alpha} + \frac{\sigma_3}{\rho_r g} \tag{14.15}$$

where all the symbols are as defined above. Equation (14.15) is suitable for calculating the depth of fault nucleation only if all the en echelon fractures remain open (as elliptical joints, for example) during the process. This assumption is likely to be valid at comparatively shallow depths, and definitely in the tensile regime, such as close to the surface of a rift zone. In fact, the assumption may hold at considerable depths. For example, studies in many drill holes worldwide indicate that joints may be open, and conduct fluids, to depths of many kilometres.

Nevertheless, at considerable crustal depths many en echelon fractures may close during fault development. Fracture closure is taken into account in the modified Griffith criterion, Eq. (14.6). Some theoretical results suggest that the en echelon fractures in some sets may already close at very low stresses of $\sigma_c = 0-3T_0$ (Paterson and Wong, 2005). For $\sigma_c = 0T_0$, Eq. (14.6) reduces to Eq. (14.7) and, eventually, to Eq. (14.8). Combining Eqs. (14.14) and

(14.8) and proceeding as for Eq. (14.15), the critical depth when $\sigma_c = 0T_0$ becomes

$$z_c = \frac{4T_0 + 2\mu\sigma_n}{\rho_r g \sin 2\alpha} + \frac{\sigma_3}{\rho_r g} \tag{14.16}$$

During seismogenic faulting, there is normally a high fluid pressure on the fault plane. In fact, it is probable that no tectonic earthquakes ever occur without the involvement of fluids. When there is fluid pressure on the developing plane of an en echelon fracture, that is, the potential fault plane, then the term $2\mu\sigma_n$ becomes $2\mu(\sigma_n - p_t)$, where $p_t$ is the total fluid overpressure on the plane (Eq. 9.22). When the pressure $p_t$ approaches or equals the normal stress $\sigma_n$, then the term $2\mu(\sigma_n - p_t)$ approaches or equals zero (Chapter 9), in which case Eq. (14.16) reduces to

$$z_c = \frac{4T_0}{\rho_r g \sin 2\alpha} + \frac{\sigma_3}{\rho_r g} \tag{14.17}$$

Many normal faults may start their nucleation at depths not far below the tensile regime in a rift zone. In the tensile regime, $\sigma_3$ is likely to be about $T_0$, that is, a tensile stress of the order of several megapascals, during rifting episodes. Theoretically, $\sigma_3$ should change abruptly to zero or to a compressive stress at the boundary between the tensile and the compressive regime. However, in a typical layered crust, where the mechanical properties of the layers vary abruptly, there are commonly layers with $\sigma_3$ close to zero at depths well below the main boundary between the tensile and compressive regimes. We must therefore define the tensile regime not only as that part of the crust where $\sigma_3$ can become negative during rifting episodes, but where its magnitude $\sigma_3$ can reach the tensile strength of the rocks. That is to say, the tensile regime is that part of the crust where Eq. (14.2) can be satisfied during rifting episodes.

It follows from this definition of the tensile regime that in the compressive regime there may be layers where $\sigma_3$ is close to zero or even negative, but always less in magnitude than $T_0$. It is likely that many normal faults nucleate close to the boundary between the compressive and tensile regimes, partly because that is where friction is likely to start to have significant effects and also because the dip of the fracture sets is likely to change abruptly – from essentially vertical to inclined (Fig. 14.9). If $\sigma_3 = 0$, then Eq. (14.17) reduces to

$$z_c = \frac{4T_0}{\rho_r g \sin 2\alpha} \tag{14.18}$$

In Example 14.1 we use Eq. (14.18) to estimate roughly the depth of nucleation of normal faults on en echelon sets in the rift zone of Iceland. The results indicate that the nucleation could start at between about 800 m and 2100 m, depending on the tensile strength of the lava pile, and that a typical value could be about 1500 m. This does not mean, of course, that nucleation cannot start at greater depths. On the contrary, these values indicate the **minimum depth** at which the nucleation of normal faults could start, based on the simplified version of the criterion given in Eq. (14.17). If $\sigma_3$ is not zero, then the minimum depth of nucleation **increases**, and must be calculated from Eq. (14.17). For example, if $\sigma_3$ is 10 MPa then the depth would increase by about 400 m (Example 14.2).

**Fig. 14.9**  Closed normal fault in the eastern part of the Holocene Thingvellir Fissure Swarm in Southwest Iceland. Aerial view northeast, the central part of the fault has a vertical displacement of about 25 m.

These results, although very crude, indicate that in a rift zone or a sedimentary basin with existing joints, the nucleation of normal faults from inclined sets of joints could start at depths of 1–3 km or more. From these depths of nucleation, the rupture would propagate in all directions, up and down the inclined plane as well as horizontally, to form the fault.

Depending on the relative importance of friction and tensile stresses, that is, opening at and close to the surface, the near-surface profiles of normal faults have different geometries.

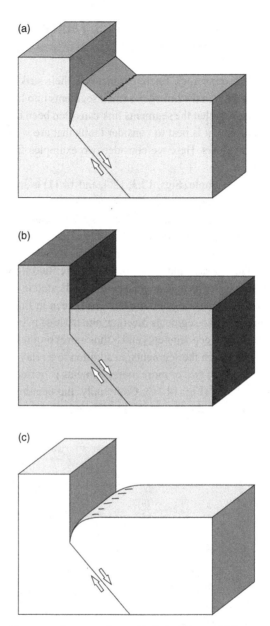

**Fig. 14.10**   Various geometric types of the near-surface parts of normal faults in a rift zone. (a) Some are open (gaping) or (b) closed, while (c) others show roll-over structures.

While many normal faults are open, gaping, at the surface (Figs. 1.5, and 12.3–12.5), others are essentially closed (Fig. 14.9). Reverse faults (Fig. 1.6) and strike-slip faults (Fig. 7.7) are normally closed. Some typical surface geometries observed of normal faults in rift zones are indicated in Fig. 14.10.

## 14.3 Fault growth

Faults, like other rock fractures, increase their strike and dip dimensions, that is, grow primarily through the linkage of small segments into larger ones (Figs. 12.4, 12.5, 12.7, and 13.4). The way that the segments link can often been clearly seen in developing faults (Fig. 12.7). For this, it is best to consider faults that are well exposed, such as at the surface of young lava flows. Here we consider two examples from the Holocene rift zone of North Iceland.

The first example (Figs. 12.3, 12.4, and 14.11) is an aerial view of several normal faults, all segmented, and tension fractures, also segmented, in a several-hundred-metre-thick Holocene pahoehoe lava flow in the Krafla Fissure Swarm (Figs. 3.4–3.6). Thus, all the fractures are less than 10 000 years old. The segments of the normal fault in the centre of Fig. 12.3 are particularly clear and marked by relay ramps. Such ramps characterise normal-fault growth and are often sites where fluid flow takes place. This is an example of a very simple segment growth, as is also indicated in Fig. 13.4.

A more complex segment interaction is seen in Figs. 12.4 and 14.11. In this case there are more than two segments meeting, and the area in between them is subject to mechanical interaction between segments and is thus more complex. For example, there are mixed-mode fractures between the segments, in addition to a relay ramp and other features.

The linking of two or more more segments is normally seen in the displacement profiles of the fractures (Fig. 14.12). Commonly, the opening and vertical displacement become close to, or actually, zero at the junction where the segments meet, when they are still not

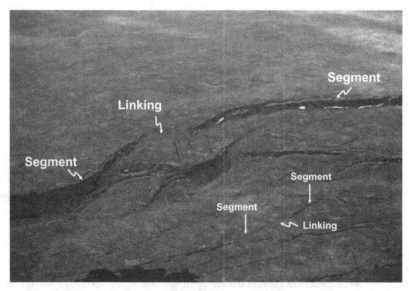

**Fig. 14.11**   Overlapping and linking segments of a major normal fault in the Holocene Krafla Fissure Swarm in North Iceland. Aerial view west (cf. Fig. 12.4)

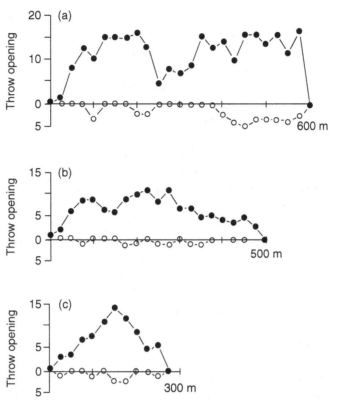

**Fig. 14.12**  Segments that have become linked are commonly recognisable in the opening and throw variations (a–c). In the top profile (a), it is clear that two former segments were linked where the opening shows a local minimum. The horizontal axis denotes segment length.

part of a single fault with a single profile. Later they become part of a larger segment and the irregularity in the displacement profile where the segments joined disappears.

The process of linking of fault segments into larger segments and faults has received much attention. There are four types of models used to analyse the process of segment linking: analytical, experimental, analogue, and numerical. Analytical models are very general and involve calculating the stress magnitudes and trajectories between nearby fracture segments (Chapter 13). Analytical models are used to calculate the stress intensity factors between the nearby tips of fracture segments, as well as the opening displacement of the fractures. The results indicate that when the nearby tips become close in relation to the dimensions of the segments, the segmented fractures begin to function as a single fracture. This means that the maximum displacement of the fractures, such as the opening of a segmented extension fracture composed of collinear segments spaced at short distances approaches that of a continuous fracture of the same length. Detailed analytical solutions of the displacements and stress intensity factors of segmented fractures are given by Sneddon and Lowengrub (1969), Melin (1983), Gray (1992), Broberg (1999), and Tada *et al.* (2000).

Experimental models on segment interaction and linkage involve methods such as photoelasticity (Dally and Riley, 1977). An early analysis of fracture-segment interaction was made by Lange (1968). As an example of a photoelastic analysis of segment interaction, we may consider the interaction between ocean-ridge segments. Transform faults connect many ocean-ridge segments and generally form along zones of high shear stress (Pollard and Aydin, 1984), in agreement with earlier photoelastic studies (Lange, 1968). Gudmundsson *et al.* (1993) made a photoelastic study of the mechanical interaction between the rift zone in North Iceland and the nearby segment of the Kolbeinsey Ridge, north of Iceland. They modelled the rift zone and the ridge segment as single extension fractures, so that the results are equally applicable to smaller segments of tension fractures and normal faults, where the fault segments are then regarded as being subject to horizontal tensile stress during a rifting episode.

The photoelastic material used had properties similar to that of rocks. That is, it had a Young's modulus of 3.4 GPa and Poisson's ratio of 0.37. These values are similar to surface rocks, such as the basaltic lava flows at the surfaces of the rift zone in North Iceland and the Kolbeinsey Ridge. The rift zone itself and the ridge segment were modelled as cuts, 5 cm long and with a 2 mm opening. A uniaxial tensile loading was applied in a direction parallel with the spreading vector in North Iceland, that is, in a direction of 73° to the trend of the cuts.

The results show that zones of high shear stress connect the nearby tips of the rift zone and the ridge segment (Fig. 14.13). These results thus support earlier models indicating that transform faults form along zones of high shear stress between the nearby tips of ridge segments (Pollard and Aydin, 1984). Also, in general, they support the results of Lange (1968) that between nearby offset segments of any extension fractures subject to tensile loading at a high angle to the segments, there will be zones of high shear stress between the nearby tips of the segments. For extension fractures and normal faults, the zones of high shear stress commonly result in the formation of transfer faults, which are small-scale versions of oceanic transform faults.

The results can also be used to understand the propagation paths of fault segments. From the photoelastic results, the stress trajectories can be calculated. The results (Fig. 14.14) show that the trajectories change their trends in the vicinity of, and particularly between, the nearby tips of offset segments. Normal faults and extension fractures propagate along the direction of the maximum principal compressive horizontal stress, which in this case would be $\sigma_2$ ($\sigma_1$ being vertical and $\sigma_3$ being perpendicular to $\sigma_2$). Consequently, as the faults or ridge segments propagate, their nearby ends would tend to curve towards each other, as indicated by the curved trends of the $\sigma_2$-trajectories. These results fit very well with the fault segments in Fig. 12.5. Whether the segments would eventually link along these curved paths, or propagate past each other, depends on many factors, such as the initial offset between the segments, their lengths in relation to the offsets, the thickness of the crustal segment within which they are located, and the type of loading to which they are subject. The two main ways of linking (Fig. 12.8), that is, through hook-shaped extension fractures and/or though transfer faults are quite common but, as said, the exact conditions that favour one mechanism of linking over the other are, as yet, not fully understood.

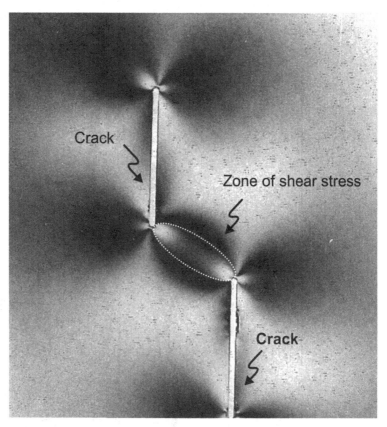

Fig. 14.13  Photoelastic experiment of two offset cracks subject to crack-perpendicular tensile stress as the only loading.

Other analogue models, in addition to the photoelastic ones, include experiments with granular media, in particular sand. Shear fractures in sand follow the same principles as in rocks, and the Coulomb criterion can be used for the compressive regime (Chapter 8). Tension fractures cannot, however, be modelled because dry sand has no cohesive or tensile strength. Earlier experiments used wet clay, that is, clay with a water film on its surface, to model tension fracture. Analogue experiments of this type are very useful in that they provide ideas and conceptual models as to how rock fractures may link into larger fractures (McClay and White, 1995; Acocella *et al.*, 2005; Acocella, 2008).

Numerical modelling of the linking of fracture segments into larger segments and major faults can be done in various ways. The models can be either two-dimensional or three-dimensional, and show the fault growing in lateral (along strike) and vertical (along dip) sections. Here we show very simple models to illustrate the principles, based on a study by Gudmundsson *et al.* (2003). We consider how sets of fractures link up when the underlap and offset between the adjacent tips of the original fractures vary (Fig. 14.15). The fracture segments are modelled as simple extension fractures, but may be regarded either as tension-fracture segments (which subsequently develop into normal faults), or as open segments of normal faults, such as are common in rift zones.

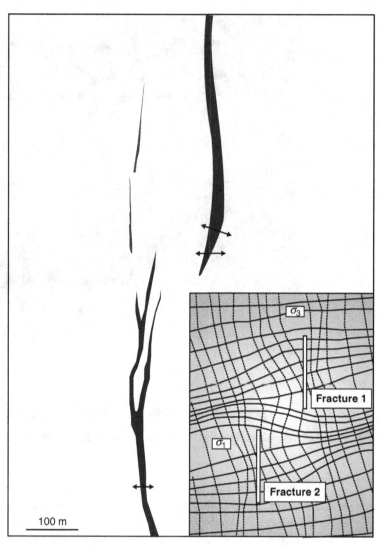

Fig. 14.14  Major fault segments curve towards each other and, eventually, link. This is as predicted by the stress trajectories (inset) calculated from the photoelastic experiment in Fig. 14.13.

Any variation in original underlap and offset between fracture segments may be attributed to different factors. Often, the variation is due to different arrangement of extension fracture sources or points of nucleation. The nucleation is normally related to stress raisers, areas of stress concentration, such as vesicles, pores, columnar joints, or other original weaknesses in the host rock that concentrate stresses and develop into fractures and fault segments. In all the models, the vertical distance between the nearby fracture tips (the underlap) is denoted by $c$, the horizontal distance (the offset) between the tips by $b$, and the (constant) original length of individual fracture segments by $a$. All the fracture segments are subject to a horizontal far-field tensile stress of 6 MPa, which is equal to the maximum in-situ tensile strength of common solid rocks.

(a)                                                          (b)

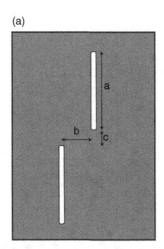                    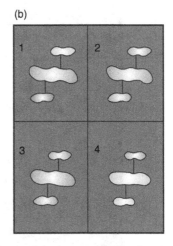

Fig. 14.15    Schematic numerical modelling results on the shear-stress concentration between two offset fractures of length (strike dimension) *a*, offset *b*, and underlap *c*. (a) Model configuration. (b) Stress results (cf. Fig. 12.7).

The results (Fig. 14.15) show zones of high stresses, stress-concentration zones, between the nearby tips of the fracture segments. For the variation in underlap *c*, only the shear stress is shown here, but zones of high tensile stress coincide roughly with the shear-stress zones. The numerical models suggest that so long as the underlap between the nearby tips of the fracture segments is significantly larger than zero, that is, is positive, the zones of high stress concentration between the nearby segment tips remain large and are likely to develop transfer faults (Fig. 12.8). However, once the underlap becomes, first, close to zero and, subsequently, negative, that is, changes into an overlap (not shown in Fig. 14.15), the stress-concentration zone shrinks.

These results are in agreement with the photoelastic study presented above (Fig. 14.13), which shows that once the underlap changes to an overlap, that is, when *c* becomes negative, the stress concentration zone between the nearby tips decreases in size. These results indicate that the most favourable configuration of offset extension fractures for the development of interconnecting transverse faults between the nearby ends of fault or fracture segments is that of a significant underlapping. This conclusion is supported by the geometry of the en echelon fracture set in Fig. 14.16. Here transverse faults have developed between some of the fault segments while there was a significant underlapping of the original segments.

When the offset *b* decreases so that the original fracture segments are closer to being collinear, the shear stress intensity decreases but the tensile stress intensity increases (Gudmundsson *et al.*, 2003). Thus, collinear fracture segments have the greatest probability of in-plane propagation, which would lead to linking up (through extension fractures) of the nearby ends. This is in agreement with the analytical studies summarised above (Sneddon and Lowengrub, 1969). Thus, the original configuration of the fractures in Figs. 14.13 and 14.15 favours shear stress concentration and transverse faults, whereas collinear extension fractures favour tensile stresses and the growth of extension-fracture segments into a single, segmented large fracture. In both cases, however, the loading results in stress concentrations that favour the linking of the fracture segments. The linking results either

Aerial view west, segmented part of the Almannagja Fault in of the Thingvellir Fissure Swarm in Southwest Iceland. Two of the segments have already linked through a transfer fault, and the other segments appear to be developing such a link.

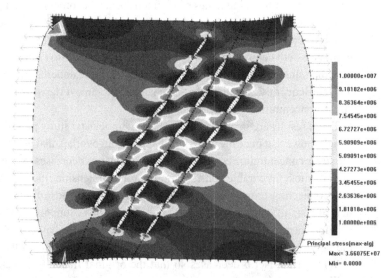

Numerical model results (www.beasy.com) on the linking of fractures in a fissure swarm. The high stresses between many of the fracture tips indicate a tendency to the formation of transfer faults.

in a single fracture with the offset segments linked through transverse faults, or in a single fracture composed of collinear segments, depending on the configuration of the original fracture segments.

Numerical modelling of this kind can be extended to parts of, or whole, fracture and fault swarms. One such example is shown in (Fig. 14.17). In this example a set of fractures and normal faults, a part of a fissure swarm (cf. Figs. 3.13, 3.14, and 13.4), is subject to tensile loading, such as would be common during rifting episodes at divergent plate boundaries. The results show that there are stress-concentration zones between the nearby tips of many of the fractures. Consequently, in many of these zones, transfer faults would be likely to develop and connect the nearby fracture tips. Thus, this is one way of forming highly permeable, interconnected sets of fractures. But this is also an indication, on a much smaller scale, of how fractures in a damage zone or a fault zone may form an interconnected network. Such networks, with generally a very high permeability, are common (Figs. 11.11 and 11.12).

When the strike dimension $L$ of a dip-slip fault is its controlling dimension, the ratio of the strike dimension to the mode III fault displacement during a single slip event is given by (Chapter 9)

$$\frac{L}{\Delta u_{\text{III}}} = \frac{E}{2\tau_{\text{d}}(1 + \nu)} \tag{14.19}$$

where $\Delta u_{\text{III}}$ is the displacement on a dip-slip fault, $E$ is the in-situ Young's modulus and $\nu$ the Poisson's ratio of the rock dissected by the fault, and $\tau_d$ is the shear stress driving the fault slip. Similarly, for a strike-slip fault where the controlling dimension is the dip dimension $R$, the relationship between the slip and the controlling dimension is given by

$$\frac{R}{\Delta u_{\text{III}}} = \frac{E}{4\tau_{\text{d}}(1 + \nu)} \tag{14.20}$$

where $\Delta u_{\text{III}}$ is the slip on the strike-slip fault.

These results show that the ratio between the controlling dimension of the fault and the displacement during a particular slip is, for given mechanical properties and driving stress, constant. That is, as the fault grows through the linking of segments, the displacement during individual slip events is likely to increase.

For many faults, the controlling dimension is the trace length at the surface (Chapter 9). When considering the relationship between faults in comparatively homogeneous and isotropic rock layers, such as thick basaltic pahoehoe lava flows, the fault displacement or throw may be approximately a linear function of the fault strike length (Fig. 14.18). In the particular case of 26 normal faults in the Holocene rift zone of Iceland (Fig. 14.18), the correlation between maximum vertical displacement $\Delta u_{\text{III}}$ and length $L$ is given approximately by the equation

$$\Delta u_{\text{III}} = 0.0028L - 0.4114 \tag{14.21}$$

The high linear correlation coefficient, $r = 0.91$, indicates that some 83% of the variation in vertical displacement can be explained in terms of the variation in the fault strike length. While the linear correlation is thus high, it should be emphasised that all these faults are young, that is, Holocene in age, and located in rocks with essentially uniform mechanical properties, namely several-hundred-metres-thick pahohehoe lava flows.

Similar linear relations have been obtained for single slips on faults during many large earthquakes (Scholz, 1990). However, attempts to find universal scaling laws between faults

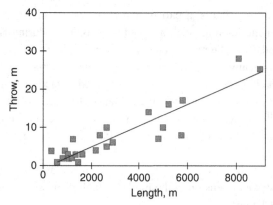

Fig. 14.18 Linear correlation between throw or displacement on 26 normal faults and their lengths (strike dimensions) in the Holocene rift zone of Iceland.

and their displacements have largely failed. First, many of the relationships are non-linear and show a large scatter. Second, when the relationships are linear, they vary between fracture populations, within a population, and also on individual faults and fault segments. Although roughly linear relations between length and displacement commonly exist for individual faults over short periods of time, over longer periods of time the correlation between rupture length and displacement has a large scatter, commonly by an order of magnitude or more. These observations suggest that some of the properties of the rock determining the fault displacement may be highly variable.

When we take a closer look at Eqs. (14.19) and (14.20), we see that the displacement also depends on the Young's modulus, $E$, of the rock. Alternatively, the displacement for a given controlling dimension is inversely proportional to $E$. Other things being equal, as $E$ gets lower, the displacement increases. The displacement is also directly proportional to the rock's Poisson's ratio, $\nu$, and of course the shear stress $\tau_d$ driving the fault displacement. Poisson's ratio does depend on the rock type under consideration, but its range is generally small compared with that of the Young's modulus, and for most solid crustal rocks Poisson's ratio is between 0.10 and 0.35 (Appendix D.1).

As discussed in Chapter 9, one approximate measure of the driving shear stress is the stress drop in earthquakes. For most large interplate earthquakes, the stress drop is around 3 MPa, but for intraplate earthquakes it is around 10 MPa. The average stress is around 6 MPa (Kanamori and Anderson, 1975). Thus, the driving shear stress for seismogenic faulting is relatively constant and, as indicated above, the same applies to the Poisson's ratio of the rocks hosting the faults.

By contrast, the Young's modulus, $E$, can vary widely between rocks and inside fault zones. Perhaps the most important effect on the magnitude of the field Young's modulus, particularly at comparatively shallow depths in tectonically active areas, is the fracture frequency of the rock mass. It is well known that the Young's modulus of a rock mass is normally less than that of a laboratory sample of the same type of rock. This difference is mainly attributed to fractures and pores in the rock mass which do not occur in the small laboratory samples. With an increasing number of fractures, in particular in a direction perpendicular to the loading, the ratio $E_{is}/E_{la}$ ($E$ in situ /$E$ laboratory) shows a rapid decay. Thus, increasing fracturing normally decreases the Young's modulus, and so does the presence of gouge and breccia in the core and damage zone of an active fault zone (Fig. 10.10).

It is thus likely that the attempts to find universal scaling laws for fault displacement versus fault length have failed because they do not take into account the changes in the mechanical properties, in particular the Young's modulus (stiffness), of the fault zone as it evolves. It follows from Eqs. (14.19) and (14.20) that the Young's modulus affects fault displacement both spatially and temporally. The spatial effect occurs when the trace of a fault at a given time dissects host rocks of different stiffnesses. The temporal effect occurs when the stiffness of the fault zone itself changes. During the evolution of an active fault zone, the effective Young's modulus of its damage zone and fault core normally decreases. To explore this effect of the Young's modulus on fault development, we must consider further the evolution of the fault damage zone and core.

## 14.4 Fault damage zone and core

On maps, images, and from a distance, fault zones appear as lineaments (Fig. 14.19). And this is how they are commonly viewed: as lineaments with little or no internal structure and heterogeneity. As a consequence, fault zones are commonly modelled as single, elastic

**Fig. 14.19**   View west, two parallel fault zones seen as lineaments (marked by arrows) dissecting layers of limestone and shale in the Bristol Channel, Kilve, Somerset coast, Southwest England.

**Fig. 14.20**  Fault core and part of the damage zone of the Husavik-Flatey Fault in North Iceland (Gudmundsson, 2007). View west, the 10-m-thick core (see the person for scale) strikes N62°W and is mainly composed of breccias, whereas the damage zone is characterised by tilted lava flows (dipping 40°NW, black arrows) and fractures of various sizes and types, including mineral-vein networks.

cracks or dislocations. As indicated above, such simple crack models are very useful for understanding the growth of faults through segment linkage and fault–fault interaction. They can also be used to analyse fault effects on regional stresses. These models, however, are less useful for understanding the local stresses around and within the fault zone itself. These local stresses largely control the slip and fracture development and thus the permeability of the fault zone. It follows that the internal mechanical structure of the fault zone must be considered with a view of understanding its potential for slip and associated fluid transport properties.

Many field studies of well-exposed fault zones show that they normally consist of two main hydromechanical units or zones: a fault core and a fault damage zone (Chapter 10). The core takes up most of the fault displacement. It is also referred to as the fault-slip zone (Bruhn *et al.*, 1994). The core contains many small faults and fractures, but its characteristic features are breccias and cataclastic rocks. Commonly, the core rock is crushed and altered into a porous material (Fig. 14.20). The core rock behaves as ductile or quasi-brittle except at very high strain rates, such as during seismogenic faulting. In many fault cores, there are numerous mineral veins and amygdales commonly spaced at centimetres or millimetres, which form dense networks. Before the associated hydrofractures become sealed with secondary minerals, that is, while the network consists of fractures that are subject to slip and thus active in transporting fluids, the network gives the core a granular-media structure at the millimetre or centimetre scale. This granular-media structure is one additional reason for modelling the core as a porous medium.

While the field description in this chapter of the fault core and damage zone focuses on large fault zones, it should be emphasised that the same units are seen in much smaller fault zones. In fact, laboratory experiments on small rock samples may produce very similar units, that is, a thin core and a thicker damage zone where the frequency of fractures changes irregularly, but generally decreases, with increasing distance from the core (Shimada, 2000).

In major fault zones, the thickness of the core is commonly from several metres to a few tens of metres (Fig. 14.20; Gudmundsson et al., 2009). Some very large faults zones, such as many oceanic transform faults, may, however, develop several fault cores, with damage zones in between (Faulkner et al., 2006; Gudmundsson, 2007). The permeability of a fault core is generally very low during most of the interseismic (non-slip) period. The core may thus act as a barrier to fluid flow, except during periods of high strain rates such as are associated with fault slip.

The damage zone consists partly of lenses of breccias and other heterogeneities, and partly of sets of extension fractures, and to a lesser degree, shear fractures. Many, and presumably most, of the extension fractures are hydrofractures that eventually become filled with minerals, that is, mineral veins. The fractures and faults make the damage zone generally much more permeable than the fault core. For instance, laboratory measurements indicate that hydraulic conductivities in the damage zone are as much as several orders of magnitude greater than those of either the fault core or the host rock (Evans et al., 1997; Seront et al., 1998).

The boundary between the damage zone and the host rock is normally diffuse. In the host rock, also referred to as the protolith (Bruhn et al., 1994), the number of fractures is normally less than in the damage zone. Although the boundaries between the fault core and the damage zone are sharper than those between the damage zone and the host rock, all these boundaries vary along the length of the fault and change, in time and space, with the evolution of the fault zone (Fig. 10.10).

The hydromechanical properties of the fault core, the damage zone, and the host rock have been studied much in recent years. The main results are as follows. Normally, the damage zone is the main conduit for water flow along a major fault zone. Laboratory measurements of small samples indicate that the damage zone has a permeability as much as 10 000 times higher than that of the core or the host rock. Evans et al. (1997) suggest that while the laboratory samples from the core yield hydraulic conductivity values that may not be much lower than the bulk in-situ values, the laboratory values of hydraulic conductivity for the damage zone are likely to be considerably lower than the corresponding bulk in-situ values. This difference is because the larger, highly conductive fractures that are common in the damage zone are not represented in the small, relatively non-fractured laboratory samples. This suggestion is supported by reported in-situ measurements giving fault-zone permeabilities as much as 1000 times greater than the maximum laboratory values of Evans et al. (1997). It follows that the in-situ permeability difference between the core and the damage zone may be even greater than the cited laboratory values would indicate.

## 14.5  Local stresses in fault zones

A fault zone is in many ways analogous to, and may be modelled as, an elastic inclusion (Fig. 6.3). As defined in Chapter 6, an elastic inclusion has elastic properties that differ from those of the host material, that is, the matrix to which the inclusion welded. This definition of an elastic inclusion follows that in the classical elasticity and rock mechanics literature (Eshelby, 1957; Savin, 1961; Jaeger *et al.*, 2007). In micromechanics the word inhomogeneity is used for a body with elastic properties that differ from those of the matrix, and the word inclusion is used for a body that has the same elastic properties but is a misfit with the surrounding matrix and thus strained (Nemat-Nasser and Hori, 1999; Qu and Cherkaoui, 2006). However, inclusions can be both homogeneous (having the same elastic properties as the matrix) and inhomogeneous (having different elastic properties from the matrix), as the concept is used in the solid-mechanics literature (Asaro and Lubarda, 2006), and this is the definition used here: namely, an elastic inclusion denotes a material body with different elastic properties or different strains from that of the larger hosting body.

The elastic properties of the damage zone and core, which in a fault zone constitute the inclusion, are normally different from those of the host rock (the matrix). This applies particularly to active fault zones. Since active, or potentially active, fault zones are of greatest interest here, we assume, unless stated otherwise, that the fault-zone inclusion has mechanical properties different from those of the host rock.

The presence of a fault zone as an elastic inclusion modifies the regional stress field and generates a local stress field that operates both within the inclusion and in its vicinity (Chapter 6). During any loading through regional stresses, displacements, or fluid pressure in the crust, the mechanical responses of the rocks constituting the fault-zone inclusion differ from those of the host rock. For example, if the rocks that constitute the fault-zone inclusion are stiffer (with a higher Young's modulus) than the host rocks that constitute the matrix, then the inclusion takes on most of the loading and becomes subject to either relative tensile stresses (if the loading is tension) or compressive stresses (if the loading is compression). If, however, the fault rocks forming the inclusion are more compliant or softer than the host rock forming the matrix, then most of the loading is taken up by the host rock. It follows then that the host rock will develop local high tensile or compressive stresses depending on whether the loading is tension or compression. The loading also results in shear stresses in the host rock and in the inclusion.

Each fault zone is composed of two principal mechanical units, a core and a damage zone. These units have widely different mechanical properties (Fig. 14.21), which, thereby, affects the local stresses within and around the fault-zone inclusion. The damage zone itself, and more rarely the core, is commonly composed of subzones with different mechanical properties. The variation in mechanical properties of the subzones is partly due to their different fracture frequencies (Fig. 14.22). It follows that the local stresses within a fault-zone inclusion may be expected to vary not only between the host rock and the fault zone, or between the core and the damage zone, but also within the damage zone itself.

The difference in stiffness between the host rock and the fault zone, and how this affects the local stresses of fault zones, is considered in Fig. 14.23. This model, based on a normal fault zone in Vaksdal, close to Bergen in West Norway (Fig. 14.22; Gudmundsson *et al.*, 2010), divides the fault zone into four main subzones. The fault plane itself is in the centre and modelled as an internal, compliant elastic spring. Based on estimates from open fractures (Gudmundsson and Brenner, 2003), the stiffness of the spring is assumed to be 6 MPa m$^{-1}$.

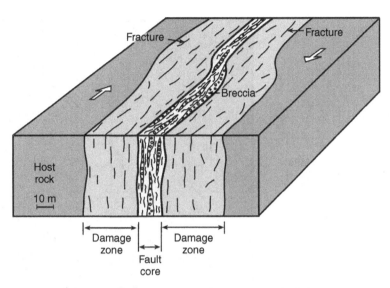

**Fig. 14.21**   Schematic illustration of a fault core and fault damage zone of a sinistral strike-slip fault.

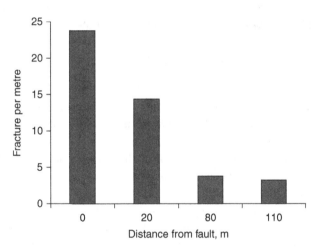

**Fig. 14.22**   Variation in fracture frequency with distance from the core of a fault in Vaksdal, West Norway.

Adjacent to the fault plane is the fault core, whose Young's modulus is taken as 0.1 GPa. This value is based on typical Young's moduli of unconsolidated rocks, as well as in-situ measurements from various fault cores worldwide. For these types of rocks, common stiffnesses are between 0.1 and 1 GPa (Appendix D.1; Hoek, 2000; Schön, 2004). The Young's modulus of the inner damage zone is taken as 1 GPa, that of the outer damage zone as 10 GPa, and that of the host rock as 50 GPa. These values reflect the decreasing number of fractures (Fig. 14.22) with increasing distance from the inner damage zone to the host rock. The rock itself is gneiss. Taking into account the well-known effects of fractures and other cavities on the Young's modulus (Section 4.4; Farmer, 1983; Priest, 1993; Nemat-Nasser

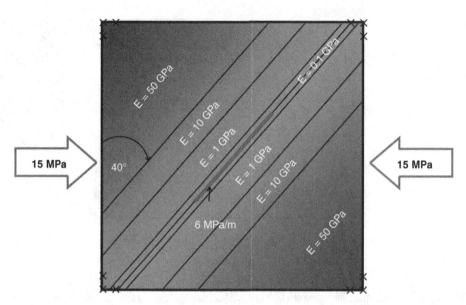

Fig. 14.23 Set up of the model in Fig. 14.24 largely based on the internal structure of the fault zone in Fig. 14.22. The loading is E–W compressive stress of magnitude 15 MPa, as inferred from stress data from Norway. The fault plane has a stiffness of 6 MPa/m, the core a Young's modulus of 0.1 GPa, the inner damage zone 1 GPa, the outer damage zone 10 GPa, and the host rock 50 GPa. The model is fastened in the corners (indicated by crosses) to avoid rigid-body rotation and translation (Gudmundsson *et al.*, 2010).

and Hori, 1999; Sadd, 2005), the Young's modulus is low for the highly fractured inner damage zone (Fig. 14.22), somewhat higher for the less-fractured outer damage zone, and highest for the normally fractured host rock.

The results (Fig. 14.24) indicate that, due to the Young's modulus inside the fault zone being lower than outside it, for the given loading conditions the intensity of the von Mises shear stresses in the fault zone is less than in the host rock. This numerical result may seem surprising since, when generalised, fault slip is mostly confined to the fault zone rather than the host rock. However, the von Mises shear stresses reach the typical stress drops/driving stresses for seismogenic fault slip, mostly 1–12 MPa. Slip would thus tend to occur in the fault zone simply because it already contains a weak fault plane and, most likely, a much higher pore-fluid pressure than the host rock. It is well known that tectonic earthquakes are usually related to zones of high fluid pressure, so that, using the Modified Coulomb criterion (Eq. 14.8; Chapter 7), the driving shear stress for seismogenic fault slip, $\tau$, becomes

$$\tau = 2T_0 + \mu(\sigma_n - p_t) \tag{14.22}$$

where $T_0$ is the tensile strength of the rock, $\mu$ is the coefficient of internal friction, $\sigma_n$ is the normal stress on the fault plane, and $p_t$ is the total fluid pressure on the fault plane at the time of slip. When the fluid pressure approaches or equals the normal stress, the term $\mu(\sigma_n - p_t)$ approaches or equals zero and the driving shear stress for slip becomes $2T_0$. Since the in-situ tensile strength of rocks is commonly in the range 0.5–6 MPa (Appendix E), it follows that,

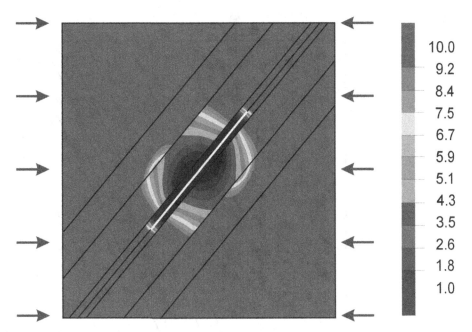

**Fig. 14.24** Numerical-model results showing the von Mises shear stress concentration (in MPa) around the fault zone in Fig. 14.23. The white line represents the fault plane. The fault core and inner damage zone have comparatively low shear stress, 1–5 MPa, whereas the outer damage zone has comparatively high shear stress.

for high-fluid-pressure fault zones, the driving shear stress for slip should be 1–12 MPa. Although the low Young's modulus in the damage zone and core yields comparatively low shear stresses in many active fault zones, they tend to slip because of the existing weak fault plane (or planes), the high fluid pressure (and thus low friction), and the low effective normal stress on the fault plane.

As we have seen, there is commonly a significant difference in mechanical properties between the core and the damage zone, as well as between the various subzones of the damage zone itself. In the next model the fault zone is divided into five subzones (Fig. 14.25). The central subzone represents the core with a Young's modulus of 1 GPa. This stiffness is similar to that of many compliant or soft breccias and unconsolidated rocks (Appendix D.1; Hoek, 2000; Schön, 2004). The next subzone (two because of symmetry about the core) represents the inner part of the damage zone with a Young's modulus of 5 GPa. This stiffness is similar to that of many fractured rocks. The final subzones represent the outer part of the damage zone with a stiffness of 10 GPa. This value corresponds to that of many fractured rocks where the fractures are not very dense (Gudmundsson *et al.*, 2010). The rock hosting the fault-zone inclusion is with a Young's modulus of 40 GPa. This value is similar to that of many near-surface solid rocks (Appendix D.1; Bell, 2000; Nilsen and Palmström, 2000).

We consider the case where the loading is extension and oblique to the fault (Fig. 14.25). This loading would be appropriate for a reverse fault in a vertical section, and a dextral

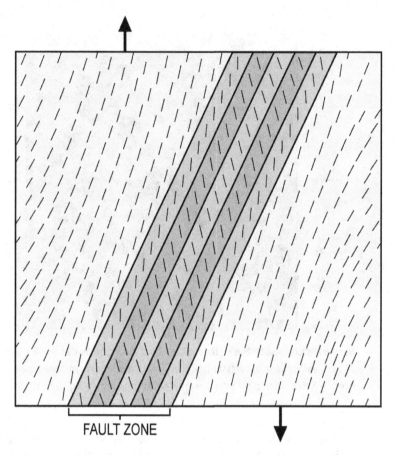

FAULT ZONE

Numerical-model results showing the stress trajectories of $\sigma_3$ in a fault zone composed of a core with a Young's modulus of 1 GPa, an inner damage zone with a Young's modulus of 5 GPa, and an outer damage zone of Young's modulus of 10 GPa located in a host rock with a Young's modulus of 40 GPa. The fault zone is subject to oblique loading, a tensile stress of 5 MPa. The trends of $\sigma_3$ differ between the core and the damage zone and the damage zone and the host rock, as well as between the subzones of the damage zone. This model demonstrates the variation in the local stresses within a typical 'layered' fault zone.

strike-slip fault in a lateral section. The oblique loading combined with the decrease in stiffness through the damage zone and to the core results in rotation of the principal stresses. Such rotations of the principal stresses within a fault zone are inferred from observations in the field. They are, for example, indicated by differences in the trends of damage-zone fractures close to the core (and the fault plane) and at greater distance from the core (Fig. 14.25).

The fault-zone inclusion thus develops a local stress field that controls its mechanical behaviour, slip, and permeability. The mechanical layering inside the fault-zone inclusion implies that the local stresses vary between its subzones so that there will rarely be uniform stresses over large parts of a fault zone. Since stress-field homogenisation over extensive parts of a fault zone is presumably a necessary condition for major earthquake ruptures (Gudmundsson and Homberg, 1999), the conditions needed for a simultaneous fault slip in a large part of, or the entire, fault zone are rarely reached. Thus, most fault slips, both along the main fault itself as well as along faults in the damage zone, remain small. Variations in local stresses and rock properties inside the fault-zone inclusion imply that stress fields favouring a particular type of fracture propagation (such as a normal fault) are usually only reached within a comparatively small region within the fault zone, such as along a section of part of a single subzone. When a fault that originates or slips in that particular section of the subzone attempts to propagate beyond the stress-homogenised region of its origin, the fracture enters regions that, normally, have local stresses that are unfavourable to the propagation of that particular fracture. Consequently, the fracture propagation becomes deflected and, commonly, arrested.

## 14.6 Fracture deflection and arrest

Most fractures in fault-zone inclusions propagate for only very short distances before they become arrested. This applies also to crustal fractures in general (Chapter 13). We have already mentioned that when a propagating fracture meets an interface or a discontinuity, such as a weak or open contact between dissimilar rocks or an earlier fracture, the fracture may (Fig. 13.11) (1) become arrested so as to stop its propagation, (2) penetrate the discontinuity, or (3) become deflected along the discontinuity, in one or two directions.

These three scenarios are well known from field observations and can be understood in terms of three related factors, namely: (i) the induced tensile stress ahead of the propagating fracture tip, (ii) rotation of the principal stresses at the discontinuity, and (iii) the material toughness or critical energy release rate of the discontinuity in relation to that of the adjacent rock layers. Fracture deflection and arrest has already been discussed in detail in Chapter 13, where the theory was applied to the propagation of extension fractures. Here we shall use the same theory, with some modifications, and apply it to the propagation, deflection and arrest of fractures in the damage zones of major fault zones.

The Cook–Gordon and stress rotation (stress barrier) mechanisms cause many fractures in heterogeneous rocks in general, and fault zones in particular, to become deflected and/or arrested at discontinuities and contacts between mechanically dissimilar rocks. But the difference in material toughness between the interface/discontinuity/contact and the adjacent rock layers is also of great importance. The total strain energy release rate $G_{\text{total}}$ in a mixed-mode fracture propagation is given by Eq. (13.2):

$$G_{\text{total}} = G_{\text{I}} + G_{\text{II}} + G_{\text{III}} = \frac{(1 - v^2)K_{\text{I}}^2}{E} + \frac{(1 - v^2)K_{\text{II}}^2}{E} + \frac{(1 + v)K_{\text{III}}^2}{E} \qquad (14.23)$$

where $G_{\text{I–III}}$ are the values of the material toughness for the ideal crack-displacement modes I–III, $E$ is the Young's modulus (compliance or stiffness), $v$ is Poisson's ratio, and $K_{\text{I–III}}$ are the associated stress intensity factors. The critical value of the stress intensity $K_c$ denotes the fracture toughness. Equation (14.23) applies to plane-strain conditions. By their nature and loading, fractures that become deflected into discontinuities or interfaces (contacts) are generally of a mixed mode (Hutchinson, 1996; Xu *et al.*, 2003).

If the subzones on either side of the core have the same mechanical properties, the condition for an extension fracture to penetrate a contact between subzones or the core and the damage zone (Fig. 14.25 ) is that the strain energy release rate $G_p$, (with subscript p for penetration) reaches the critical value for fracture extension, namely the material toughness of the layer, $\Gamma_L$ (with subscript L for rock layer). In parts of some fault zones this may be approximately the case. From Eq. (14.23) we then have

$$G_p = \frac{(1 - \nu^2)K_I^2}{E} = \Gamma_L \tag{14.24}$$

The fracture will kink at or deflect into the contact or discontinuity if the strain energy release rate reaches the material toughness of the discontinuity itself, $\Gamma_D$ (with superscript D for discontinuity). The fracture propagates in a mixed mode (mode I and II) along the contact or discontinuity (Hutchinson, 1996; Xu et al., 2003; Wang and Xu, 2006) so that, from Eq. (14.23), the fracture deflects into the discontinuity when (subscript d for deflection)

$$G_d = \frac{(1 - \nu^2)}{E}(K_I^2 + K_{II}^2) = \Gamma_D \tag{14.25}$$

Here the stress-intensity factor $K_I + K_{II}$ is for the discontinuity. From Eqs. (14.24) and (14.23) it follows that an extension fracture penetrates the discontinuity when

$$\frac{G_d}{G_p} < \frac{\Gamma_D}{\Gamma_L} \tag{14.26}$$

Similarly, the fracture becomes deflected into the discontinuity when

$$\frac{G_d}{G_p} \geq \frac{\Gamma_D}{\Gamma_L} \tag{14.27}$$

Equations (14.24)–(14.27) are likely to control, at least partly, whether a fracture penetrates or becomes deflected along a discontinuity such as a contact in a fault zone, if the adjacent parts of that contact have essentially the same mechanical properties. Commonly, however, the properties within a fault zone, such as between subzones or between the core and the damage zone, change abruptly across the contact. An abrupt change of this type is known as elastic mismatch. The primary measure of size or magnitude of the mismatch across a contact or discontinuity in general are the Dundurs (1969) elastic mismatch parameters. The two Dundurs parameters, $\alpha$ and $\beta$, are given (He and Hutchinson, 1989; Hutchinson, 1996; Freund and Suresh, 2003) by the equations (Chapter 13)

$$\alpha = \frac{E_1^* - E_2^*}{E_1^* + E_2^*} \tag{14.28}$$

$$\beta = \frac{1}{2}\frac{\mu_1(1 - 2\nu_2) - \mu_2(1 - 2\nu_1)}{\mu_1(1 - \nu_2) + \mu_2(1 - \nu_1)} \tag{14.29}$$

where $\mu$ is shear modulus, $\nu$ is Poisson's ratio, and the plain strain Young's modulus is $E^* = E/(1-\nu^2)$. The subscript 2 is used for the modulus of the rock hosting the propagating fracture and subscript 1 for the material on the other side (the far side with respect to the fracture tip) of the discontinuity or contact.

The following analysis is based on work by He and Hutchinson (1989) and He *et al.* (1994). The strain energy release rate (Chapter 10) associated with fracture penetration into the layer above the discontinuity, $G_p$, is given by

$$G_p = \frac{1-\nu_1}{2\mu_1} K_I^2 = \frac{1-\nu_1}{2\mu_1} c^2 k_1^2 a^{1-2\lambda} \tag{14.30}$$

where $a$ is the length of fracture penetration (Fig. 13.11), $k_1$ is an amplitude factor, proportional to the loading, which in this case is the driving stress or fluid overpressure for fracture propagation, and $\lambda$ is a real and $c$ a non-dimensional complex-valued function, both of which depend on the Dundurs parameters given in Eqs. (14.28) and (14.29). When the propagating fracture becomes deflected into the contact or discontinuity, the associated strain energy release rate, $G_d$, is given by

$$G_d = [(1-\nu_1)/\mu_1 + (1-\nu_2)/\mu_2] (K_I^2 + K_{II}^2)/(4\cosh^2 \pi\varepsilon) \tag{14.31}$$

where

$$K_I^2 + K_{II}^2 = k_1^2 a^{1-2\lambda} \left[ |d|^2 + |e|^2 + 2Re(de) \right] \tag{14.32}$$

where $d$ and $e$ are non-dimensional complex-valued functions that depend on the Dundurs parameters given in Eqs. (14.28) and (14.29). The ratio $G_d/G_p$, which is independent of $k_1$ and the fracture-segment length $a$ (Fig. 13.11), is given by

$$\frac{G_d}{G_p} = \frac{1-\beta^2}{1-\alpha} \times \frac{|d|^2 + |e|^2 + 2Re(de)}{c^2} \tag{14.33}$$

By analogy with Eqs. (14.26) and (14.27), the propagating fracture is likely to penetrate the discontinuity or contact between the dissimilar layers or subzones in the fault zone if

$$\frac{G_d}{G_p} < \frac{\Gamma_D(\psi)}{\Gamma_L^1} \tag{14.34}$$

Conversely, the propagating fracture is likely to become deflected into the discontinuity or contact, and commonly arrested in the contact, if

$$\frac{G_d}{G_p} \geq \frac{\Gamma_D(\psi)}{\Gamma_L^1} \tag{14.35}$$

The subscript for the material toughness is for layer 1 (Fig. 13.11) and $\psi$ is a measure of the relative proportion of mode II to mode I, namely, $\psi = \tan^{-1}(K_{II}/K_I)$, so that $\psi = 0°$ is for pure mode I and $\psi = \pm 90°$ for pure mode II (Chapter 13).

For a given fracture-segment length $a$ (Fig. 13.11), the energy release rate depends on $\alpha$ (assuming that $\beta = 0$), and the ratio $G_d/G_p$ (Eqs. 14.33–14.35) can be plotted as a function of $\alpha$ (Fig. 13.15). In the area below the curves, the ratio $G_d/G_p$ favours the deflection of a propagating fracture into the discontinuity or contact. In the area above the curves, the ratio favours the vertical penetration of the propagating fracture through the discontinuity and into layer 1. Also, when the stiffnesses of layers 1 and 2 are equal, the Dundurs parameter $\alpha = 0$ and $G_d/G_p= 0.26$. When this is the case, the propagating fracture deflects into the discontinuity or contact only when the material toughness of the discontinuity ($\Gamma_D$) is less than 26% of the material toughness of layer 1 ($\Gamma_L^1$). This latter condition is probably uncommon, which may partly explain why fractures tend to penetrate layered rocks, rather than become deflected or arrested at the layer contacts, where all the layers have similar mechanical properties (Chapter 13).

Figure 13.15 shows that when $\alpha$ is negative, so that the stiffness of layer 1 is less than that of layer 2, there is little tendency for the propagating fracture to be deflected into the discontinuity. When the positive value of $\alpha$ increases, so that layer 1 becomes much stiffer than layer 2 (Eq. 14.28), there is a greatly increased tendency for a propagating fracture to deflect into the discontinuity or contact between subzones within the damage zone or between the core and the damage zone. This means that, in terms of this model, it is normally easier for fractures to propagate from stiffer layers into more compliant or softer layers than in the opposite direction; hence fault zones grow, partly at least, by fractures propagating from the host rock into the outer damage zone, and from the outer damage zone into the inner one. (Other geological factors, however, such as stress rotation and Cook–Gordon debonding also affect the ease of propagation of rock fractures through contacts.) Because fracture propagation in the opposite direction tends to become arrested, the elastic mismatch between the host rock and the outer damage zone is likely to be one reason why fault zones remain comparatively thin in comparison with their strike and dip dimensions.

## 14.7  Evolution of fault slip

The formation of the core and the damage zone results in a gradual decrease in the effective Young's modulus (stiffness) towards the core, that is, the fault plane (Fig. 14.26). During the fault-zone evolution, the core and, particularly, the damage zone increase in thickness (Fig. 10.10). Numerical and analytical results show that when the thickness of the damage zone increases then, for a given fault geometry and loading conditions, fault slips in individual, equal-size ruptures increase. Thus, a fault zone may gradually accumulate larger and larger displacements while maintaining essentially constant dimensions. These theoretical results are supported by field observations.

The numerical models in Figs. 14.27–14.29 show the effects on fault slip of increasing the damage-zone thickness. With modifications, the basic results are applicable to any type of fault. In the models, however, the faults considered are sinistral strike-slip faults and dip-slip faults, idealised as vertical. The damage zone is divided into several mechanical units, each with a stiffness estimated from its fracture frequency as in Fig. 14.22. The model boundaries running parallel with, and far from, the fault are fastened (fixed) using the conditions of no displacement (Fig. 14.27). A typical stress drop of 5 MPa is used as the driving stress. Each model has an area of $2600 \times 2400$ length units, a fault-plane length (strike dimension) of 400 length units, and an opening (thickness of the fault core) of 0.2 length units. A new mechanical unit (of 10 length units) is added to the damage zone on either side of the fault, to simulate an increase in the damage-zone thickness between each model run – ten in total (Fig. 14.28). The fault dimensions and the loading are constant; the thickness of the damage zone is the only parameter that changes between model runs.

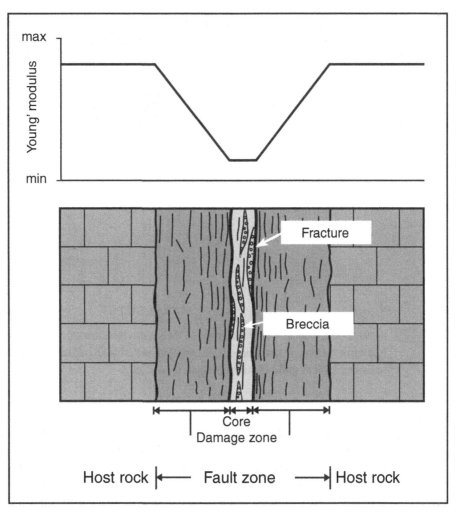

Fig. 14.26 Schematic illustration of a typical fault zone consisting of two main mechanical units: a comparatively thin core and a much thicker damage zone. The effective Young's modulus (stiffness) decreases from the host rock to the boundary between the core and the damage zone.

While earthquake rupture is a dynamic process, the final fault slip during and after an earthquake involves postseismic deformation that is partly related to the fault adjusting to the static stiffness of the rock. For a given rock unit the dynamic stiffness, particularly for fractured and porous rocks at shallow depths, may be as much as ten times the static modulus (Chapter 4), so to take the postseismic slip into account, we use the static stiffnesses. We scale the static laboratory measurements on small samples, commonly 1.5–5 times the field stiffnesses of the same rocks, to effective stiffnesses using the methods in Chapter 4. In particular, taking into account that as the fracture frequency in the damage zone increases, particularly in a direction perpendicular to the loading, the ratio between $E$ in situ and $E$ in the laboratory decays rapidly. Typical laboratory values of stiffness for igneous rocks,

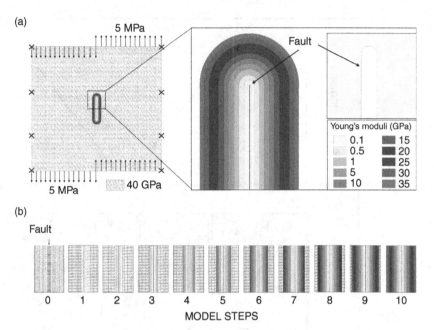

**Fig. 14.27** Numerical-model results on fault displacement. (a) The boundary conditions and the variation in Young's modulus (stiffness) in the core and damage zone. (b) The damage-zone thickness is gradually increased in ten steps. Step 10 corresponds to the mechanical units and stiffnesses shown in (a).

such as basalts, are thus scaled to lower values in order to take into account the effects of fractures in the damage zones. In the final fault-model geometry (step 10 in Fig. 14.27) the stiffnesses of the damage-zone units or subzones gradually decrease from 35 GPa in the outermost part of the damage zone, at the contact with the host rock, to 0.5 GPa next to the core. The highest value, 35 GPa, is chosen so as to be somewhat lower than the host-rock stiffness of 40 GPa. The core stiffness, 0.1 GPa, is based on the typical static stiffnesses of unconsolidated rocks as well as in-situ measurements from various fault cores worldwide.

When the damage-zone thickness gradually increases, the maximum slip or displacement during a particular rupture on the fault also increases (Fig. 14.28). Thus, the fault slip generated during a particular earthquake, including the postseismic slip, gradually increases with increasing total damage-zone thickness. For a seismogenic fault of constant rupture (trace) length, the ratio of the maximum displacement to the rupture length should therefore decrease with time (Fig. 14.28). The fault–displacement curves, however, remain geometrically similar throughout the gradually increasing displacement. Thus, irrespective of its size, the displacement shape or profile remains similar: a smooth curve with a maximum displacement at the centre and minimum displacements at the lateral tips of the fault (Fig. 14.29).

Let us now apply analytical results (Chapter 9) to understand better the gradual changes in fault displacement as the stiffness of the fault zone gradually decreases. For a dip-slip fault modelled as a mode III through crack (Chapter 9), the ratio of the strike dimension $L$

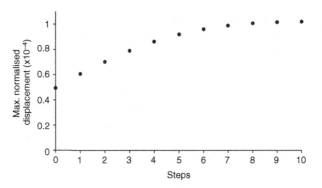

**Fig. 14.28** Maximum normalised displacement (MND) in the fault centre in each of the ten steps (Fig. 14.27). Here MND = $10^4$ (MD/FL), where MD is the maximum displacement and FL the fault length, both expressed in model length units (LU).

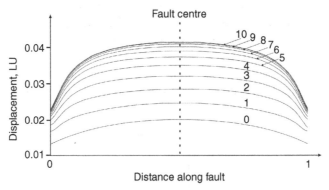

**Fig. 14.29** Displacement curves along the length of the fault in each of the ten steps.

to a single-event displacement (slip) in an earthquake (Eq. 14.19) is

$$\frac{L}{\Delta u_\text{d}} = \frac{E}{2\tau_\text{d}(1 + v)} \tag{14.36}$$

Here $\Delta u_\text{d}$ is the dip-slip fault displacement, $E$ the effective stiffness and $v$ the Poisson's ratio of the rock dissected by the fault, and $\tau$ is the shear stress driving the fault slip.

Similarly, for a strike-slip fault modelled as a mode III crack, where the dip dimension $R$ is the controlling one, we have from Eq. (14.20):

$$\frac{R}{\Delta u_\text{s}} = \frac{E}{4\tau_\text{d}(1 + v)} \tag{14.37}$$

where $\Delta u_\text{s}$ is the slip on the strike-slip fault, and the other symbols are as in Eq. (14.36).

Equations (14.31) and (14.32) show that the fault displacement is directly proportional to the rock Poisson's ratio $v$ and the driving shear stress $\tau_\text{d}$ generating the displacement. Poisson's ratio depends on the type of rock, but its range is generally small compared with that of Young's modulus; for most crustal rocks Poisson's ratio is between 0.2 and 0.3 (Chapter 4). The stress drop or driving shear stress in most earthquakes also has a remarkably narrow range and does not depend upon the rupture size or the earthquake magnitude. Most stress drops are between 1 and 12 MPa, and commonly 3–6 MPa (Kanamori and Anderson, 1975).

Thus, for a given stress drop and Poisson's ratio, the ratio of the controlling dimension $L$ to the displacement or slip $\Delta u$ during a particular earthquake depends primarily on the stiffness of the rock. Alternatively, for a given controlling dimension and driving stress the displacement increases as the fault-zone stiffness decreases, in agreement with the numerical results above.

The present theoretical results are also supported by direct observations. Field measurements show that as a fault evolves, not only does its damage zone and core become thicker (Figs. 10.10 and 14.27), but its displacement $\Delta u$ becomes larger in proportion to its trace length $L$ so that the $L/\Delta u$ ratio becomes smaller (Fig. 14.30). The data in Fig. 14.30 are similar to other data sets suggesting that the $L/\Delta u$ ratio becomes smaller as a fault-zone evolves. The numerical models (Figs. 14.27–14.29) show that the displacement depends much on the stiffness of the damage zone. When the comparatively soft damage zone thickens (Figs. 10.10 and 14.27) as the fault-zone evolves, then, from Eqs. (14.31) and (14.32), the $L/\Delta u$ ratio decreases (Fig. 14.30).

The length/total displacement ratios of major faults are usually much smaller than the rupture length/individual slip ratios during earthquakes in the same fault zone. The

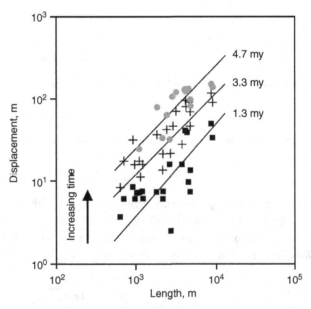

**Fig. 14.30** When a fault zone evolves its displacement $u$ becomes larger in proportion to its trace length $L$ so that the $L/\Delta u$ ratio becomes gradually smaller, as for these 23 Late Miocene normal faults from the Timor Sea of Western Australia. Millions of years from the onset of faulting are indicated: filled boxes (1.3 my), crosses (3.3 my), and grey circles (4.7 my) (data from Walsh *et al.*, 2002.)

length/total displacement ratios are commonly of the order of $10^1$ to $10^2$, whereas rupture length/individual slip ratios are normally of the order of $10^3$ to $10^4$. These ratios thus differ by a factor of 10–100.

The differences between instantaneous slip during an earthquake rupture and the long-term displacement in the same fault zone can be partly explained as being due to different effective stiffnesses. During an earthquake rupture, the dynamic stiffness controls the instantaneous dynamic surface slip on the fault. The long-term static displacement on the same fault, however, is determined by the static stiffness. Thus, the size of the aseismic slip following a seismic slip depends primarily on the difference between the rock dynamic and static stiffnesses. In this view, the aseismic slip in a fault zone gradually brings the rupture length/individual slip ratio closer to the length/total displacement ratio. Aseismic slips are very common on faults: it is estimated that in many earthquake areas, such as in subduction and transform zones, around 50% of the slip is aseismic.

These considerations indicate that at least part of the aseismic slip is related to a postseismic gradual increase in displacement so as to adjust to a value determined by the static stiffness. It follows that as the damage zone of an active fault becomes thicker and the stiffness lower, then the postseismic slip following an earthquake may gradually increase.

## 14.8 Summary

- All faults originate from weaknesses in the rocks, commonly from joints, pores, contacts, or other stress raisers. The basic theory for fault initiation is Griffith theory, which can be applied to both the tensile and the compressive regimes. If the original weakness from which a fault develops, say a columnar joint in a lava flow and modelled as an ellipse, is oblique to the principal stresses, the points of maximum tensile stress occur not at the vertices (the tips) of the ellipse but rather at certain distances from the tips. From these points, extension fractures propagate along curved paths that eventually become parallel with $\sigma_1$ and tend to link up the columnar joints into a large-scale shear fracture.
- Normal faults in a rift zone may initiate either from large-scale tension fractures or from sets of inclined (columnar) joints. When a normal fault originates from a tension fracture it follows from Griffith theory that the tension fracture can only reach a certain critical depth, at which point it must change into a normal fault. Depending on the in-situ tensile strength and density, this critical depth is commonly about 0.5 km and generally in the range 0.2–0.8 km. A normal fault that initiates on sets of inclined joints may form at any depth and propagate upward and downwards (as well as laterally) from the place of initiation. The minimum depth of initiation, however, is generally between 800 m and 2100 m.
- Faults grow, that is, increase their strike and dip dimensions, primarily through linkage of smaller segments into larger ones. There are primarily two mechanisms by which linkage occurs: the formation of hook-shaped fractures and the development of transfer faults.

Hook-shaped fractures are curved fractures that link nearby segments. They are initially extension fractures but may develop into faults. Large-scale versions of hook-shaped fractures at mid-ocean ridges are referred to as overlapping spreading centres. Transfer faults are comparatively straight shear fractures that connect nearby segments of a fault. Large-scale versions of transfer faults at mid-ocean ridges are referred to as transform faults.

- All fault zones develop two primary hydromechanical zones or units parallel with the fault plane: a core and a damage zone. The core consists primarily of breccias and gouge and can become very stiff in an inactive fault (because of healing and sealing). In an active fault zone, however, the core is normally soft, that is, compliant, and thus has a comparatively low Young's modulus. The damage zone, by contrast, consists primarily of fractures of various types that vary in frequency with distance from the core. The contact between the damage zone and the core is normally sharp, whereas that between the damage zone and the host rock (also referred to as the protolith) is normally diffuse. The damage zone is the main conduit for fluids in a fault zone, except during and shortly after fault slip when the permeability of the core may, temporarily, greatly increase as all the pores and cavities in its granular medium become interconnected along the slip plane.

- Fault zones may be regarded as elastic inclusions, composed of a core and a damage zone, which itself is commonly composed of several subzones with different elastic properties (different stiffnesses). All inclusions composed of materials with properties that differ from the hosting matrix (here, the host rock) modify the stress field around them and generate local stresses inside themselves. For fault zones, these local stresses, rather than the regional host-rock stresses, determine both when the fault slips and the associated permeability.

- Most fractures in a fault zone propagate for only short distances before they become arrested, commonly on meeting other discontinuities such as fractures and contacts. When a fracture meets a discontinuity or an interface, there are three possible scenarios: the fracture may (i) become arrested at the discontinuity, (ii) penetrate the discontinuity, or (iii) become deflected into the discontinuity. These three scenarios have strong effects on the development of fault zones over time, particularly with regards to changes in the thickness of the fault zone, primarily the thickness of the damage zone. The scenarios can be understood in terms of three related factors: (a) the induced tensile stress ahead of a propagating fracture (which may open discontinuities), (b) rotation of the principal stresses at discontinuities (generating stress barriers), and (c) the material toughness of the discontinuity in relation to that of the adjacent rock layers.

- During the evolution of an active fault zone, its core and damage zone gradually increase in thickness. The final fault slip, during and after an earthquake, involves postseismic slip, which is partly the result of the fault adjusting to the static stiffness of the core and the damage zone. The overall result is that, for an active fault of a constant strike dimension but increasing thickness of the core and damage zone, the fault slip in subsequent ruptures gradually increases. It follows that the ratio between the fault slip and rupture length in a particular earthquake tends to increase with time. This conclusion is supported by field observations.

# 14.9 Main symbols used

| | |
|---|---|
| $a$ | length of fracture penetration |
| $c$ | function in Eq. (14.30) |
| $d$ | function in Eqs. (14.32) and (14.33) |
| $E$ | Young's modulus (modulus of elasticity) |
| $E_1^*$ | plane-strain Young's modulus of layer 1 |
| $E_2^*$ | plane-strain Young's modulus of layer 2 |
| $e$ | function in Eqs. (14.32) and (14.33) |
| $G_d$ | strain energy release rate for fracture deflection into a discontinuity |
| $G_p$ | strain energy release rate for fracture penetration of a discontinuity |
| $G_{tot}$ | total strain energy release rate during crack extension (propagation) |
| $G_I$ | strain energy release rate during the extension of a mode I crack |
| $G_{II}$ | strain energy release rate during the extension of a mode II crack |
| $G_{III}$ | strain energy release rate during the extension of a mode III crack |
| $g$ | acceleration due to gravity |
| $K_I$ | stress intensity factor for a mode I crack |
| $K_{II}$ | stress intensity factor for a mode II crack |
| $K_{III}$ | stress intensity factor for a mode III crack |
| $k_1$ | amplitude factor in Eq. (14.30) |
| $L$ | strike dimension of a dip-slip fault (as controlling dimension) |
| $p_t$ | total fluid pressure |
| $R$ | dip dimension of a strike-slip fault (as controlling dimension) |
| Re | real part of a complex number |
| $T_0$ | tensile strength |
| $\Delta u_d$ | total dip-slip fault displacement |
| $\Delta u_s$ | total strike-slip fault displacement |
| $\Delta u_{III}$ | total displacement on a mode III crack |
| $z_c$ | the critical depth at which $\sigma_c = 0$ |
| $z_{max}$ | maximum depth of tension fracture (at which it changes into a normal fault) |
| $\alpha$ | angle between a fault plane and $\sigma_1$ |
| $\alpha$ | Dundurs uniaxial (extensional) elastic mismatch parameter |
| $\beta$ | Dundurs area (volumetric) elastic mismatch parameter |
| $\Gamma_D$ | material toughness of a discontinuity or an interface (a contact, for example) |
| $\Gamma_L$ | material toughness of a layer (material toughness in general is denoted by $G_c$) |
| $\Gamma_L^1$ | material toughness of layer 1 |
| $\lambda$ | real number (for homogeneous materials, 1/2) |
| $\mu$ | coefficient of internal friction |
| $\mu$ | shear modulus (denoted by $G$ in most other chapters) |
| $\mu_1$ | shear modulus of layer 1 |
| $\mu_2$ | shear modulus of layer 2 |
| $\nu$ | Poisson's ratio |

$\nu_1$     Poisson's ratio of layer 1
$\nu_2$     Poisson's ratio of layer 2
$\rho_r$     rock or crustal density
$\sigma_c$     normal stress for closure of elliptical cracks (joints)
$\sigma_n$     normal stress (mostly used in connection with a slip or fault planes)
$\sigma_1$     maximum compressive principal stress
$\sigma_2$     intermediate compressive principal stress
$\sigma_3$     minimum compressive (maximum tensile) principal stress
$\tau$     shear stress
$\tau_d$     driving shear stress associated with fault slip (or displacement)
$\tau_0$     inherent shear strength (or cohesion or cohesive strength)
$\psi$     measure of the relative proportion of mode II to mode I

## 14.10  Worked examples

### Example 14.1

#### Problem

During a rifting episode, the minimum compressive principal stress $\sigma_3$ is reduced to zero at a certain depth in the rift zone of Iceland. The average density of the uppermost 2 km of the crust in the rift zone of Iceland is about 2500 kg m$^{-3}$ (Gudmundsson, 1988). If a typical normal-fault dip in the rift zone is 75° (Forslund and Gudmundsson, 1992), find the minimum depth of initiation of a typical normal fault on a set of en echelon columnar joints in the tilted lava pile.

#### Solution

For the conditions given, Eq. (14.18) can be used to calculate the depth. Since the dip of the developing fault is supposed to be 75°, it follows that it makes an angle of 15° to $\sigma_1$ if, as is normal, $\sigma_1$ is vertical in the rift zone. The minimum depth of fault nucleation then follows from Eq. (14.18) as follows:

$$ z_c = \frac{4T_0}{\rho_r g \sin 2\alpha} = \frac{4 \times (0.5 - 6) \times 10^6 \, \text{Pa}}{2500 \, \text{kg m}^{-3} \times 9.81 \, \text{m s}^{-2} \times \sin 30°} = 163 - 1957 \, \text{m} $$

A tensile strength of 0.5 MPa is, however, rare and a more typical value, based on fracture studies in the fissure swarms, is about 3–4 MPa, in which case the depth would be about 1000–1300 m. Thus, a likely minimum depth of initiation of typical normal faults in the rift zone could be 1–1.3 km and, for high-strength lava flows, close to 2 km.

## Example 14.2

### Problem

Repeat Example 14.1 but assume that $\sigma_3$ is not zero but rather 10 MPa.

### Solution

Since the minimum principal stress is not zero, then Eq. (14.17) is the appropriate one and gives the critical minimum depth for normal-fault nucleation as

$$z_c = \frac{4T_0}{\rho_r g \sin 2\alpha} + \frac{\sigma_3}{\rho_r g} = \frac{4 \times (0.5 - 6) \times 10^6 \, \text{Pa}}{2500 \, \text{kg m}^{-3} \times 9.81 \, \text{m s}^{-2} \times \sin 30°}$$

$$+ \frac{1 \times 10^7 \, \text{Pa}}{2500 \, \text{kg m}^{-3} \times 9.81 \, \text{m s}^{-2}} = (163 \, \text{m} - 1957 \, \text{m}) + 408 \, \text{m}$$

$$\approx 570 - 2360 \, \text{m}$$

Thus the minimum depth for the nucleation of a normal fault on sets of inclined columnar joints is about 400 m deeper when $\sigma_3 = 10$ MPa than when $\sigma_3 = 0$ MPa. Again, the depth range depends on the tensile strength of the crust, and the shallow depth of 570 m is unlikely to be realistic since that depth assumes the lowest possible tensile strength, namely 0.5 MPa, whereas estimates from tension fractures in the rift zone of Iceland indicate that the in-situ tensile strength is at least a few megapascals.

## Example 14.3

### Problem

Show what Example 14.1 implies for the ratio between the tensile strength of the rocks at the time of fault initiation using the Griffith criterion and the maximum principal compressive stress, and also the depth at which that strength is reached.

### Solution

Since $\sigma_3$ is zero rather than negative, the normal-fault nucleation may be assumed to take place in the compressive regime. Equation (14.1) is

$$(\sigma_1 - \sigma_3)^2 = 8T_0(\sigma_1 + \sigma_3)$$

In Example 14.1 we have $\sigma_3 = 0$, so that from Eq. (14.1) we get

$$(\sigma_1 - 0)^2 = 8T_0(\sigma_1 + 0), \text{ so that } \sigma_1^2 = 8T_0\sigma_1 \text{ or } T_0 = \frac{1}{8}\sigma_1$$

If we take the maximum in-situ tensile strength as 6 MPa, then $\sigma_1$ has to reach 6 MPa $\times$ 8 = 48 MPa. Using Eq. (14.11) and a rock density of 2500 kg m$^{-3}$, that magnitude would be reached at a depth of

$$z = \frac{4.8 \times 10^7 \text{Pa}}{2500 \text{ kg m}^{-3} \times 9.81 \text{ m s}^{-2}} = 1957 \text{ m}$$

This is the depth that we calculated in Example 14.1 for the maximum tensile strength, as it should be. Similarly, if we use the minimum tensile strength of 0.5 MPa, then $\sigma_1$ has to reach 0.5 MPa $\times$ 8 = 4 MPa. For the given density, this happens at a depth of 163 m, as calculated in Example 14.1 above.

### Example 14.4

#### Problem

A 20-m-thick Holocene lava flow overlaying soft Holocene sediments is at the surface of a seismic zone characterised by strike-slip faulting (Fig. 9.18). The lava flow is subject to a biaxial horizontal stress field with a minimum horizontal principal tensile stress of 3 MPa and a maximum principal compressive stress of unknown magnitude. The in-situ tensile strength is 3 MPa. Use the Griffith criterion for failure in the compressive regime to determine the maximum principal compressive stress $\sigma_1$ at failure.

#### Solution

At the surface of the lava flow, one of the principal stresses is atmospheric (0.1 MPa or 1 bar) and regarded as zero. Equation (14.1) gives

$$(\sigma_1 - \sigma_3)^2 = 8T_0(\sigma_1 + \sigma_3)$$

which can be rewritten in the form

$$(\sigma_1 - \sigma_3)^2 - 8T_0(\sigma_1 + \sigma_3) = 0$$

Here we can use megapascals since no basic SI units enter the calculations. In Eq. (14.1) the tensile stresses are negative, so that a tensile stress of 3 MPa is written −3MPa and in the first parenthesis as −(−3 MPa). However, the tensile strength is always given as a positive value, so that a tensile strength of 3 MPa has a positive sign. The term for the tensile strength in the equation is thus −8 × 3 MPa and not −8 × (−3 MPa). This may sound confusing, but is simply the way these terms are used in rock mechanics, structural geology, and related fields. For convenience we do not write the units (MPa) in the following calculations. Substituting 3 for $T_0$ and −3 for $\sigma_3$, Eq. (14.1) gives

$$(\sigma_1 - (-3))^2 - 8 \times 3 \times (\sigma_1 - 3) = 0$$

or

$$\sigma_1^2 + 6\sigma_1 + 9 - 24\sigma_1 + 72 = 0$$

Simplifying,

$$\sigma_1^2 - 18\sigma_1 + 81 = 0$$

All the numbers above are in megapascals. This is a quadratic equation of the form

$$ax^2 + bx + c = 0$$

The quadratic equation has the general solution

$$x = \frac{-b \pm (b^2 + 4ac)^{1/2}}{2a}$$

Here we have $x = \sigma_1$ and $a = 1$, $b = -18$, and $c = 81$. Putting these values and symbols into the general solution of the quadratic equation, we get

$$\sigma_1 = \frac{18 \pm ((-18)^2 - 4 \times 1 \times 81)^{1/2}}{2 \times 1} = \frac{18 \pm 0}{2} = 9$$

Since all the numbers are in megapascals, the value of $\sigma_1$ is 9 MPa. Thus, for this strike-slip regime we have $\sigma_1 = 9$ MPa, $\sigma_2 = 0$ MPa, and $\sigma_3 = -3$ MPa. The intermediate stress is the one that is vertical.

The value of $\sigma_1$, 9 MPa, is very reasonable for the uppermost part of a strike-slip regime, such as seen in Fig. 9.18. Since this seismogenic zone is in Iceland and located between two overlapping spreading centres (Fig. 12.8; Gudmundsson, 2007), one way to easily generate a $\sigma_1$-value of 9 MPa and, in fact, much higher through (feeder) dyke injections in the nearby rift zones (Gudmundsson, 2000). Overpressured dykes may generate high horizontal compressive stresses in their vicinity (Figs. 1.6 and 2.13).

## 14.11 Exercises

14.1 Explain briefly what happens to a tension fracture in a rift zone as it propagates to greater and greater depths.

14.3 Joints in lava flows and sedimentary rocks that are vertical at the surface commonly become inclined at depth. Explain how this may happen.

14.3 Describe briefly the initiation of a normal fault on a set of inclined joints in a lava pile. At what crustal depth would the nucleation of joints into a fault be likely to start?

14.4 Describe the main mechanisms of fault growth through segment linkage. Indicate the relevant results from experimental, analytical, and numerical models.

14.5 What is the general relation between the displacement during a particular slip event on a fault and its controlling dimension?

14.6 What is the relationship between displacement during a particular slip event on a fault and the Young's modulus of the fault rock? What is the implication of this relationship for scaling laws between fault displacement and fault length (strike dimension)?

14.7 Define and describe briefly the main hydromechanical units of a typical fault zone. Which of these zones controls fault-zone permeability during non-slipping periods?

14.8 What are the analogies between a fault zone and an elastic inclusion?

14.9 Explain how a fault zone develops a local stress field that controls its mechanical behaviour and, thereby, its permeability.

14.10 Why is it commonly easier for a fault zone to grow by fractures propagating from the host rock and into the outer part of its damage zone rather than from the damage zone into the host rock?

14.11 Which mechanical conditions allow a mature fault zone to accumulate larger and larger displacements while maintaining essentially constant strike and dip dimensions?

14.12 Explain how the postseismic, static slip following the dynamic slip associated with an earthquake in a fault zone may gradually increase as the fault zone evolves.

14.13 During extension in a rift zone composed of lava flows with columnar joints, $\sigma_3$ is reduced to zero (or negative values, absolute tension) to a certain depth. The rock density to that depth is 2400 kg m$^{-3}$, the in-situ tensile strength is 3 MPa, and the normal fault that develops during the extension has a dip of 70°. Calculate the minimum depth of normal-fault nucleation on a set of columnar joints in the tilted lava pile based on these data.

14.14 Repeat Exercise 14.13 with $\sigma_3 = 5$ MPa.

14.15 Repeat Exercise 14.13 with a host-rock density of 2500 kg m$^{-3}$, in-situ tensile strength of 6 MPa, fault dip of 73°, and $\sigma_3 = 30$ MPa.

# References and suggested reading

Acocella, V., 2008. Transform faults or overlapping spreading centers? Oceanic ridge interaction revealed by analogue models. *Earth and Planetary Science Letters*, **265**, 379–385.

Acocella, V., Morvillo, P., and Funiciello, R., 2005. What controls relay ramps and transfer faults within rift zones? Insights from analogue models. *Journal of Structural Geology*, **27**, 397–408.

Amadei, B. and Stephansson, O., 1997. *Rock Stress and its Measurement*. London: Chapman & Hall.

Asaro, R. J. and Lubarda, V. A., 2006. *Mechanics of Solids and Materials*. Cambridge: Cambridge University Press.

Bell, F. G., 2000. *Engineering Properties of Soils and Rocks*, 4th edn. Oxford: Blackwell.

Brace, W. F., 1960. An extension of the Griffith theory of fracture to rock. *Journal of Geophysical Research*, **65**, 3477–3480.

Broberg, K. B., 1999. *Cracks and Fracture*. New York: Academic Press.

Bruhn, R. L., Parry, W. T., Yonkee, W. A. and Thompson, T., 1994. Fracturing and hydrothermal alteration in normal fault zones. *Pure and Applied Geophysics*, **142**, 609–644.

Carmichael, R. S., 1989. *Practical Handbook of Physical Properties of Rocks and Minerals*. London: CRC Press.

Cartwright, J. A., Trudgill, B. D., and Mansfield, C. S., 1995. Fault growth by segment linkage – an explanation for scatter in maximum displacement and trace length data from the Canyonlands Grabens of SE Utah. *Journal of Structural Geolology*, **17**, 1319–1326.

Cox, S. J. D. and Scholz, C. H., 1988. On the formation and growth of faults: an experimental study. *Journal of Structural Geology*, **10**, 413–430.

Dally, J. W. and Riley, W. F., 1977. *Experimental Stress Analysis*, 2nd edn. New York: McGraw-Hill.

Du, Y. and Aydin, A., 1991. Interaction of multiple cracks and formation of echelon crack arrays. *International Journal for Numerical and Analytical Methods in Geomechanics*, **15**, 205–218.

Dundurs, J., 1969. Edge-bonded dissimilar orthogonal wedges. *Journal of Applied Mechanics*, **36**, 650–652.

Eshelby, J. D., 1957. The determination of the elastic field of an ellipsoidal inclusion, and related problems. *Proceedings of the Royal Society of London*, **A241**, 376–396.

Evans, J. P., Forster, C. B., and Goddard, J. V., 1997. Permeability of fault-related rocks, and implications for hydraulic structure of fault zones. *Journal of Structural Geolology*, **19**, 1393–1404.

Farmer, I., 1983. *Engineering Behaviour of Rocks*, 2nd edn. London: Chapman & Hall.

Faulkner, D. R., Mitchell, T. M., Healy, D., and Heap, M. J., 2006. Slip on 'weak' faults by the rotation of regional stress in the fracture damage zone. *Nature*, **444**, 922–925.

Forslund, T. and Gudmundsson, A., 1992. Structure of Tertiary and Pleistocene normal faults in Iceland. *Tectonics*, **11**, 57–68.

Freund, L. B. and Suresh, S., 2003. *Thin Film Materials: Stress, Defect Formation and Surface Evolution*. Cambridge: Cambridge University Press.

Gray, T. G. F., 1992. *Handbook of Crack Opening Data*. Cambridge: Abingdon.

Griffith, A. A., 1920. The phenomena of rupture and flow in solids. *Philosophical Transactions of the Royal Society of London*, **A221**, 163–198.

Griffith, A. A., 1924. Theory of rupture: In: Biezeno, C. B. and Burgers, J. M. (eds.), *Proceedings of the First International Congress on Applied Mechanics*. Delft: Waltman, pp. 55–63.

Gudmundsson, A., 1987. Geometry, formation and development of tectonic fractures on the Reykjanes Peninsula, southwest Iceland. *Tectonophysics*, **139**, 295–308.

Gudmundsson, A., 1988. Effect of tensile stress concentration around magma chambers on intrusion and extrusion frequencies. *Journal of Volcanology and Geothermal Research*, **35**, 179–194.

Gudmundsson, A., 1992. Formation and growth of normal faults at the divergent plate boundary in Iceland. *Terra Nova*, **4**, 464–471.

Gudmundsson, A., 2000. Dynamics of volcanic systems in Iceland. *Annual Review of Earth and Planetary Sciences*, **28**, 107–140.

Gudmundsson, A., 2007. Infrastructure and evolution of ocean-ridge discontinuities in Iceland. *Journal of Geodynamics*, **43**, 6–29.

Gudmundsson, A. and Homberg, C., 1999. Evolution of stress fields and faulting in seismic zones. *Pure and Applied Geophysics*, **154**, 257–280.

Gudmundsson, A., Brynjolfsson, S., and Jonsson, M. T., 1993. Structural analysis of a transform fault-rift zone junction in North Iceland. *Tectonophysics*, **220**, 205–221.

Gudmundsson, A., Gjesdal, O., Brenner, S.L. and Fjeldskaar, I., 2003. Effects of growth of fractures through linking up of discontinuities on groundwater transport. *Hydrogeology Journal*, **11**, 84–99.

Gudmundsson, A., Simmenes, T. H., Larsen, B., and Philipp, S. L., 2010. Effects of internal structure and local stresses on fracture propagation, deflection, and arrest in fault zones. *Journal of Structural Geology*, document doi:10.1016/j.jsg.2009.08.013.

Haimson, B. and Rummel, F., 1982. Hydrofracturing stress measurements in the Iceland research drilling project drill hole at Reydarfjordur, Iceland. *Journal of Geophysical Research*, **87**, 6631–6649.

He, M. Y. and Hutchison, J. W., 1989. Crack deflection at an interface between dissimilar elastic materials. *International Journal of Solids and Structures*, **25**, 1053–1067.

He, M. Y., Evans, A. G., and Hutchinson, J. W., 1994. Crack deflection at an interface between dissimilar elastic materials: role of residual stresses. *International Journal of Solids and Structures*, **31**, 3443–3455.

Hoek, E., 2000. *Practical Rock Engineering*, available at http://www.rockscience.com.

Hudson, J. A. and Harrison, J. P., 1997. *Engineering Rock Mechanics. An Introduction to the Principles*. Oxford: Elsevier.

Hutchison, J. W., 1996. *Stresses and failure modes in thin films and multilayers. Notes for a DCAMM Course*. Lyngby: Technical University of Denmark, pp. 1–45.

Jaeger, J. C., Cook, N. G. W., and Zimmerman, R. W., 2007. *Fundamentals of Rock Mechanics*, 4th edn. Oxford: Blackwell.

Jumikis, A. R., 1979. *Rock Mechanics*. Clausthal, Germany: Trans Tech Publications.

Kanamori, H. and Anderson, D. L., 1975. Theoretical basis of some empirical relations in seismology. *Bulletin of the Seismological Society of America*, **65**, 1073–1095.

Kasahara, K., 1981. *Earthquake Mechanics*. New York: Cambridge University Press.

Lange, F. F., 1968. Interaction between overlapping parallel cracks: a photoelastic study. *International Journal of Fracture Mechanics*, **4**, 287–294.

Larsen, B., Grunnaleite, I., and Gudmundsson, A., 2009. How fracture systems affect permeability development in shallow-water carbonate rocks: an example from the Gargano Peninsula, Italy. *Journal of Structural Geology*, doi:10.1016/jsg.2009.05.009

Larsen, B., Gudmundsson, A., Grunnaleite, I., Sælen, G., Talbot, M. R., and Buckley, S., 2010. Effects of sedimentary interfaces on fracture pattern, linkage, and cluster formation in peritidal carbonate rocks. *Marine and Petroleum Geology*, doi: 10.1016/j.marpetgeo.2010.03.011.

Lin, J. and Parmentier, E. M., 1988. Quasi-static propagation of a normal fault: a fracture mechanics model. *Journal of Structural Geology*, **10**, 249–262.

McClay, K. R. and White, M. J., 1995. Analog modelling of orthogonal and oblique rifting. *Marine and Petroleum Geology*, **12**, 137–151.

McClintock, F. A. and Walsh, J. B., 1962. Friction on Griffith cracks in rocks under pressure. In: *Proceedings of the Fourth US National Congress on Applied Mechanics, Vol. 2*. New York: American Society of Mechanical Engineers, pp. 1015–1021.

Melin, S., 1983. Why do cracks avoid each other? *International Journal of Fracture*, **23**, 37–45.

Myrvang, A., 2001. *Rock Mechanics*. Trondheim: Norway University of Technology (NTNU), (in Norwegian).

Nemat-Nasser, S. and Hori, M., 1999. *Micromechanics. Overall Properties of Heterogeneous Materials*, 2nd edn. Amsterdam: Elsevier.

Nilsen, B. and Palmström, A., 2000. *Engineering Geology and Rock Engineering*. Oslo: Norwegian Soil and Rock Engineering Association (NJFF).

Paterson, M. S. and Wong, T. W., 2005. *Experimental Rock Deformation –the Brittle Field*, 2nd edn. Berlin: Springer-Verlag.

Peng, A. and Johnson, A. M., 1972. Crack growth and faulting in cylindrical specimens of the Chelmsford Granite. *International Journal of Rock Mechanics and Mining Sciences*, **9**, 37–86.

Pollard, D. D. and Aydin, A., 1984. Propagation and linkage of oceanic ridge segments. *Journal of Geophysical Research*, **89**, 10017–10028.

Priest, S. D., 1993. *Discontinuity Analysis for Rock Engineering*. London: Chapman & Hall.

Qu, J. and Cherkaoui, M., 2006. *Fundamentals of Micromechanics of Solids*. New Jersey: Wiley.

Sadd, M. H., 2005. *Elasticity: Theory, Applications, and Numerics*. Amsterdam: Elsevier.

Savin, G. N., 1961. *Stress Concentration Around Holes*. New York: Pergamon.

Scholz, C. H., 1990. *The Mechanics of Earthquakes and Faulting*. New York: Cambridge University Press.

Schön, J. H., 2004. *Physical Properties of Rocks: Fundamentals and Principles of Petrophysics*. Amsterdam: Elsevier.

Seront, B., Wong, T. F., Caine, J. S., Forster, C. B., and Bruhn, R. L., 1998. Laboratory characterisation of hydromechanical properties of a seismogenic normal fault system. *Journal of Structural Geology*, **20**, 865–881.

Shimada, M., 2000. *Mechanical Behaviour of Rocks Under High Pressure Conditions*. Rotterdam: Balkema.

Sheorey, P. R., 1997. *Empirical Rock Failure Criteria*. Rotterdam: Balkema.

Sneddon, I. N. and Lowengrub, M., 1969. *Crack Problems in the Classical Theory of Elasticity*. New York: Wiley.

Tada, H., Paris, P. C., and Irwin, G. R., 2000. *The Stress Analysis of Cracks Handbook*, 3rd edn. New York: American Society of Mechanical Engineers.

Walsh, J. J., Nicol, A., and Childs, C., 2002. An alternative model for the growth of fault. *Journal of Structural Geology*, **24**, 1669–1675.

Wang, P. and Xu, L. R., 2006. Dynamic interfacial debonding initiation induced by an incident crack. *International Journal of Solids and Structures*, **43**, 6535–6550.

Xu, L. R., Huang, Y. Y., and Rosakis, A. J., 2003. Dynamics crack deflection and penetration at interfaces in homogeneous materials: experimental studies and model predictions. *Journal of the Mechanics and Physics of Solids*, **51**, 461–486.

# Fluid transport in rocks – the basics

## 15.1 Aims

To understand fluid transport in the crust in general, and in rock fractures such as faults and hydrofractures in particular, some of the basic principles of fluid transport in porous and fractured media must be given. These principles include Darcy's law for flow in porous media and the cubic law for flow in fractures. Only the basic definitions and formulas that are needed in this and subsequent chapters are given. There are many excellent books on hydrogeology and fluid flow in porous media where the details can be found. The main aims of this chapter are to:

- Explain Darcy's law for fluid flow in porous media.
- Explain the cubic law for fluid flow in fractures.
- Present the model for fluid flow in a single set of fractures.
- Present models for fluid flow in orthogonal sets of fractures.
- Present some basic general results on fracture-related permeability.

## 15.2 Fluid transport in porous and fractured rocks

Fluid flow in crustal rocks is through pores, through fractures, or, most commonly, through both pores and fractures. How fluids are transported is of great academic and applied interest; understanding all kinds of fluid reservoirs depends on knowing and, preferably, being able to predict their permeabilities and fluid-transport or hydromechanical properties. Fluid flow in the crust is primarily through porous or fractured media, or a mixture of both (see Appendix E.2 for the physical properties of some crustal fluids).

Fluid transport in porous media, such as sedimentary rocks and unconsolidated rocks (for example, gravel, sand, and silt), derives originally from Darcy's experiments. The theoretical framework is very well established and tested and forms the basis of hydrogeology and related fields (de Marsily, 1986; Fetter, 1994; Domenico and Schwartz, 1998; Deming, 2002). The basic results have also, with appropriate modifications, been applied with considerable success to fluid transport in porous crustal rocks and to magma transport in the mantle. There exist numerous analytical solutions, as well as powerful numerical programs, for analysing and predicting the flow of fluids in porous media (Bruggeman, 1999).

By contrast, fluid transport in fractured rocks is still not well understood. There are several reasons for this. One is that rock fractures are, in contrast to circular or elliptical pores, normally very sensitive to changes in fluid pressure or rock stress. Thus, compressive stresses perpendicular to the fracture plane tend to close the fracture or, at least, reduce its aperture, whereas increasing the fluid pressure in a fracture tends to open the fracture and increase its aperture. The aperture of a fracture may also vary abruptly between rock layers with different mechanical properties, partly because the local stresses, such as the compressive stress perpendicular to the fracture, in such layers are likely to change abruptly. Variations in rock-fracture apertures are very important because large-aperture fractures tend to dominate the fluid transport in any set of fractures. Thus, in a set of tens or hundreds of fractures, perhaps less than 10% of the fractures contribute significantly to the fluid transport through that set.

In this chapter the focus is on fluid transport in fractures: in individual fractures, single sets of fractures, and in orthogonal sets of fractures. The models are analytical and are based on various simplifying assumptions. For understanding various concepts related to fluid transport in fractures, particularly as regards ground-water flow, it is necessary, however, to provide a brief outline of the elements of fluid transport in porous media, primarily as regards Darcy's law.

## 15.3  Darcy's law

One of the basic principles that we apply to ground-water flow or other crustal fluid flow is that its velocity, that is, the distance travelled during some time interval, is in proportion to some driving force. This principle leads to Darcy's law, the basic law for fluid transport in a porous medium. Darcy's law may be stated thus: the velocity of fluid flow is proportional to the hydraulic gradient. The law may be represented by the following equation:

$$q = \frac{Q}{A} = -K\frac{\Delta h}{\Delta L} = -K\frac{\partial h}{\partial L} \tag{15.1}$$

where $q$ is the specific discharge (or discharge velocity), $Q$ is the volume of flow in unit time, $A$ is the cross-sectional area normal to the flow, $K$ is the coefficient of permeability, $\Delta h$ is the difference of the total head (the difference in water levels in the manometers in Darcy's experiments), and $\Delta L$ is the length of flow in the sand (in the experiment). The minus sign is to indicate that $q$ is in the direction of decreasing total head, but this sign is commonly omitted. We can clarify these concepts as follows:

- $q$ is also known as the volumetric flow rate per unit surface area. It has units of metres per second, or m s$^{-1}$.
- $Q$ is also known as the **volumetric flow rate**. It has units of cubic metres per second, or m$^3$ s$^{-1}$.
- $A$ is the cross-sectional area and has units of square metres, that is, m$^2$. In the original experiments of Darcy, where he let water flow through a cylinder filled with sand, $A$ was the cross-sectional area through which the water flowed.

- $K$ is primarily used in hydrogeology, is known as the **hydraulic conductivity**, and has units of m s$^{-1}$. The minus sign in Eq. (15.1) is to indicate that the flow is in the direction of decreasing total head, that is, decreasing water level in the manometers (open tubes) in Darcy's experiments (water wells in nature). Commonly, the minus sign is omitted in the equations presenting Darcy's law.
- $\Delta h = h_1 - h_2$ is the difference in water levels in the manometers, that is, the difference in total head. It has units of metres, or m.
- $\Delta L$ is the distance that the water flows in the sand in Darcy's experiment. It has units of metres, or m.
- $\Delta h / \Delta L$ or $\partial h / \partial L$ is the **hydraulic gradient**. It has no units, that is, is a pure number, because it is length over length. $\partial h / \partial L$ is the partial derivative used in various versions of Darcy's law and applies to very small distances $\partial L$.

Darcy's law, although initially derived from experiments on water flowing through sand, is valid for most **granular media** so long as the flow is **laminar**. Laminar means that the fluid particles in a stream move along straight, parallel paths in layers or laminae. The particles retain the same relative positions in successive cross-sections through the stream. The viscosity of the fluid suppresses any tendency to turbulent conditions. The flow is also known as viscous or **streamline flow**. In particular, Darcy's law is generally valid for sediments and for porous, solid rocks, such as scoria layers between lava flows (Fig. 15.1). It can also be used as a crude approximation for some densely fractured rocks, particularly if the fractures have various trends and form a well-interconnected pore space.

The physical meaning of Darcy's law is that the flow is driven towards regions of **lower potential energy** and not, except in the special case of horizontal flow, from higher to lower pressure.

**Fig. 15.1** Flow of ground water through young lava flows can be modelled using Darcy's law, in particular flow through the highly porous scoria parts in between the main, massive parts of the flows. View west, ground water flows through a Holocene basaltic aa lava flow into a glacial river in West Iceland.

Darcy's law can be presented in various forms. Commonly, the letter $i$ is used for the hydraulic gradient $\Delta h / \Delta L$. Perhaps the best known of the simple forms of Darcy's law are the following:

$$q = Ki \tag{15.2}$$

and

$$Q = KiA \tag{15.3}$$

where $K$ is the hydraulic conductivity, $i$ is the hydraulic gradient, and $A$ is the cross-sectional area perpendicular to the flow.

When Darcy's law is applied to fluids other than water, the hydraulic connectivity (which refers to water) is normally not used, but rather a more general term known as permeability. Qualitatively, **permeability** is the ease with which fluid can move through rock. Permeability is **not** the same as **porosity**. A **porous** rock is one that contains holes or cavities (including fractures). The pores may contain fluids such as water, oil, or gas. The pores may be (i) interconnected or disconnected, (ii) with any kind of distribution within the rock, and (iii) of round, equidimensional shape, or fracture-like. The **degree of connectivity** between the pores determines the permeability of the rock. The rock may be highly porous, with a high total porosity, but with a low permeability. Examples include many limestones. Clay is also a very porous material but normally has a very low permeability. By contrast, the rock may have a low porosity but a high permeability. Examples include many crystalline rocks with well-connected fractures.

There are **two measures** or concepts of permeability. These are:

1. Coefficient of permeability or **hydraulic conductivity**. This is the coefficient $K$ in Darcy's law when it is written in a form such as in Eqs. (15.1)–(15.3). It depends on the properties of the rock and the fluid that flows through the rock and represents the ability of rock to **transmit water**. The units are m s$^{-1}$.
2. **Intrinsic permeability**. This is a measure of the material properties allowing the flow of **any fluid**. Normally, intrinsic permeability has the symbol $k$. It has the dimensions of area so its units are m$^2$. The units of $k$ are known as **darcy**. One darcy, 1 d $\approx 10^{-8}$ cm$^2 = 10^{-12}$ m$^2$.

The hydraulic conductivity $K$ and the intrinsic permeability $k$ are related through the equation

$$K = \frac{N\rho_w g d}{\mu_w} = \frac{k\rho_w g}{\mu_w} \tag{15.4}$$

where $N$ is a dimensionless shape factor relating to the geometry of the flow channels in the rock, $\rho_w$ is the density of water, $g$ is the acceleration due to gravity, $d$ is the mean diameter of the grains (of the sand used in Darcy's experiments), and $\mu_w$ is the viscosity of water. We see from Eq. (15.4) that the intrinsic permeability may be expressed as

$$k = Nd^2 \tag{15.5}$$

where the symbols are as defined for Eq. (15.4). Clearly, $k$ depends only on the properties of the rock, not on the properties of the fluid. Hydraulic conductivity is normally used in hydrogeology when dealing with a single fluid phase, namely water. Intrinsic permeability is normally used in other fields, such as in petroleum geology, where there are several fluids with different properties (and thus with different $K$ values), namely oil, gas, and water.

A more general version of Darcy's law, for isotropic hydraulic conductivity $K$, may be given as

$$\bar{q} = K \operatorname{grad} h = K \nabla h \tag{15.6}$$

where $\bar{q}$ is the specific discharge (a vector), the minus sign being omitted, and grad $h$ is the gradient of the scalar field $h$ given by

$$\operatorname{grad} h = \nabla h = \bar{i}\frac{\partial h}{\partial x} + \bar{j}\frac{\partial h}{\partial y} + \bar{k}\frac{\partial h}{\partial z} \tag{15.7}$$

where $\nabla$ is del, or nabla, and $\bar{i}$, $\bar{j}$, and $\bar{k}$ are unit vectors in the directions of the coordinate axes $x$, $y$, and $z$, respectively. The gradient $\nabla h$ represents the rate of change of $h$. The direction of $\nabla h$ coincides with that in which $h$ changes fastest; namely for isotropic $K$, the direction of ground-water flow.

## 15.4 The cubic law

Fractures in a rock contribute to its porosity and, particularly, to its permeability. One reason for the importance of fractures for fluid flow (ground water, geothermal water, oil and gas) is that one or a few fractures, even narrow ones, may largely control the flow. In fact, in a low-porosity rock, the fracture-related permeability may entirely control the fluid transport. This is, for example, common in many fractured reservoirs, such as in some carbonate rocks.

The flow in conduits that are either elliptical in shape (elliptical openings) or have parallel walls are often, with some modifications, applicable to flow in fractures. Fluid flow along a single fracture (Fig. 15.2) is most commonly modelled as flow between parallel plates (Fig. 15.3). It is then generally assumed that the walls of the fracture are smooth (Figs. 15.3 and 15.4). The model is based on a well-known special solution to the Navier–Stokes equation for the flow of a viscous fluid. It is widely used in hydrogeology, where it is known as the **cubic law**. For horizontal plates and flow, the volumetric flow rate $Q$ (Fig. 15.3) is given by

$$Q = \frac{(P_1 - P_2)\,b^3}{12\mu_f}\frac{W}{L} \tag{15.8}$$

where $P_1 - P_2$ is the pressure difference over the length $L$, $b$ is the aperture or opening of the fracture (the separation between the parallel plates), $L$ is the length of the plates parallel with the flow direction (indicated by an arrow and $Q$), $\mu_f$ is the dynamic viscosity of the fluid (when water, then sometimes with subscript $w$), $W$ is the width of plates perpendicular to flow, that is, the dimension of fracture perpendicular to the flow direction (and **not** the thickness or opening of the fracture). Equation (15.8) is for horizontal flow where acceleration due to gravity, $g$, does not enter, nor the density of the fluid, $\rho_f$.

**Fig. 15.2**  Flow of groundwater along a tension fracture in the Holocene rift zone of Southwest Iceland. View northeast, the maximum aperture of the fracture is as much as 10 m. The walls of fractures in lava flows can be reasonably smooth, but often have some irregularities (notches and jogs), as seen here.

Let us now generalise the results in Eq. (15.8) to flow in a fracture of any orientation, with aperture (opening) $b$ and width $W$ in a direction that is perpendicular to the flow direction. The cross-sectional area of the fracture perpendicular to the flow is thus $A = bW$. Using Eq. (15.3), the volumetric flow rate $Q$, that is, the volume of fluid flowing per unit time through the fracture, is then given by

$$Q = \frac{\rho_f g W b^3}{12\mu_f} i \tag{15.9}$$

or, using the gradient $\nabla h$ of the hydraulic head (Eq. 15.7), in the form

$$Q = \frac{\rho_f g W b^3}{12\mu_f} \nabla h \tag{15.10}$$

where $\rho_f$ is the fluid density, $g$ is the acceleration due to gravity, $\mu_f$ is the dynamic (absolute) viscosity of the fluid, and $\nabla h$ is the gradient of the hydraulic head $h$ (grad $h$), which can be expressed in terms of heads, when dealing with ground-water flow, as

$$\nabla h = \operatorname{grad} h = \operatorname{grad}\left(z + \frac{P}{\rho_w g}\right) \tag{15.11}$$

where $P$ is the fluid pressure at the bottom of a drill hole or water well (piezometer), located at a certain elevation $z$ (**elevation head**) above a reference level (usually sea level) and other symbols are as defined above. The term $P/(\rho_w g)$ is known as the **pressure head** and the sum

$$h = z + \frac{P}{\rho_w g} \tag{15.12}$$

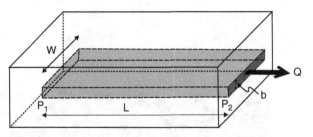

**Fig. 15.3** Modelling horizontal flow between parallel plates. For horizontal flow it is the pressure gradient, here the pressure difference $P_1 - P_2$ along the length of the flow, $L$, which drives the flow. $W$ is the width of the plates in a direction perpendicular to the flow, $b$ is the aperture (separation between the plates), and $Q$ is the volumetric flow rate.

as the **total head** for ground-water flow. The total head $h$ can be further explained as follows:

(a) The total head is the sum of the elevation of the base of the piezometer ($z$) and the length (the height) of the water column in it.
(b) The elevation of the water level in a piezometer is a measure of the total head.
(c) The total head is also the potential energy of the fluid per unit weight, but it is measured in metres.

It is understood that $Q$ is a function of both location and time. Crustal fluids flow in the direction of decreasing total head, as given by Eq. (15.12); thus, Eq. (15.10) commonly has a minus sign before its right-hand term, but often this sign is omitted, as we do here.

Equations (15.9) and (15.10) are valid for any fracture with smooth, parallel walls, independent of its orientation. They have the same form as Darcy's law in Eq. (15.3), where, in the fracture version, the area $A$ is given by $bW$. This indicates that the equation for flow between parallel, smooth plates (Eq. 15.9) may be written schematically as

$$Q = Cb^3 \Delta h \qquad (15.13)$$

which shows that $Q$ is proportional to the cube of the fracture aperture, hence the name **cubic law**. Notice that $\Delta h$ is the **change in hydraulic head** and not the hydraulic gradient. In Eq. (15.13) the constant $C$ is

$$C = \frac{\rho_f g}{12\mu_f} \frac{W}{L} \qquad (15.14)$$

for **straight flow**, and

$$C = \frac{2\pi \rho_f g}{12\mu_f} \ln\left(\frac{R}{r}\right) \qquad (15.15)$$

for **radial flow**, such as from a well or drill hole, where $R$ is the outer and $r$ the inner radii of the flow.

The hydraulic conductivity for the fracture flow is given by

$$K = \frac{\rho_f g b^2}{12\mu_f} \qquad (15.16)$$

The cubic law is the basic law of fluid transport in a rock fracture. It can be generalised so as to include fracture sets of various types, as in Sections 15.5 and 15.6. Here we make the following comments on the cubic law:

(a) The cubic law for fluid transport in rock fractures is particularly useful for fractures with reasonably smooth walls. It is generally less suitable for fractures with **rough walls**, or where the aperture varies greatly along the fracture strike and dip dimensions. Because of potential channelling of the fluid flow along the large-aperture parts of a fracture, the aperture variation is of great importance. Also, the surface roughness of natural fractures affects their hydraulic conductivity, particularly when the fractures are subject to stresses.

(b) **Flow channelling** is a preferred flow path in rock fractures, the path of least resistance, where most of the flow goes (cf. Tsang and Neretnieks, 1998). Then the fluid uses only a part, or parts, of a fracture as a channel(s). Flow channelling may occur at various scales and is related to (i) irregularities and general variations in the fracture aperture, (ii) fracture frequencies or densities, and (iii) connectivity, that is, how well the fractures are connected. Because the aperture of a fracture in an elastic rock depends on the local stress field and fluid overpressure, then so does flow channelling.

(c) The aperture-size distribution in most rock masses is a power law, a log-normal law, or some other law with a negative exponent (cf. Turcotte, 1997). Because of the cubic law, the relatively few large-aperture (wide open) fractures may largely control the overall permeability of the rock mass.

(d) The cubic law may not hold for natural fractures subject to normal stresses exceeding 20 MPa. Overburden pressure is 20 MPa at 700–800 m depth. Normal stress at that depth may close part of a fracture. In fact, some estimates indicate that at 20 MPa between 10% and 20% of the surface area of mode I (extension) fracture may be closed (cf. Lee and Farmer, 1993). However, in very deep drill holes, such as the KTB drill hole in Germany (drilled to a depth of 9 km), there were open fractures conducting fluids at depths of many kilometres (Amadei and Stephansson, 1997).

(e) The effect of shear stress on the hydraulic conductivity of natural fractures is complex and depends on the associated normal stress and the surface roughness. Shear deformation results in dilatancy (Chapter 5). As the normal stress increases at greater depths, however, it may eliminate any increase in permeability caused by shear-stress induced dilatancy.

## 15.5 Fluid transport in a single set of fractures

We shall consider only two main types of sets. For most complex sets simple analytical expressions are not available so that numerical or analogue models must be used. Here we consider a single set of smooth, planar fractures or discontinuities, all being parallel and also parallel with the direction of fluid flow (Fig. 15.4). The set consists of a single family of $m$ fractures, that is, the number of fractures is denoted by $m$. All the fractures have the

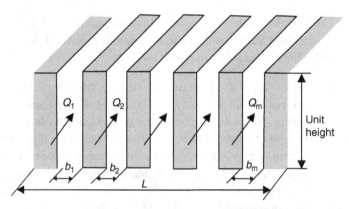

**Fig. 15.4**    Fluid flow along a single set of vertical fractures. The profile (scan-line) is of length $L$ and unit height. The fracture apertures are denoted by $b_1$, $b_2$, etc.

same aperture and strike parallel with the direction of ground-water flow. If $L$ is here the length of a profile measured in a section perpendicular to the fracture strike and the height of the measured section (the outcrop) is taken as a unit, which could also be the thickness of the layer under consideration, then we have

$$Q = \frac{\rho_f g m b^3}{12 \mu_f} i \tag{15.17}$$

$$K_f = \frac{\rho_f g m b^3}{12 \mu_f L} \tag{15.18}$$

where $Q$ is the volumetric flow rate, $\rho_f$ is the fluid density, $g$ is the acceleration due to gravity, $b$ is the aperture of the fractures (all are of equal size), $i$ is the hydraulic gradient, $\mu_f$ is the dynamic viscosity of the fluid (usually water), $K_f$ is the hydraulic conductivity of the fractured rock, $m$ is the number of fractures dissected by profile, and the profile is assumed to be of unit height.

The porosity of the fractured rock, $\phi$, can be defined as

$$\varphi = \frac{mb}{L} \tag{15.19}$$

where $m$ is the number of parallel fractures, $b$ is the aperture of the fractures, and $L$ is the length of the profile, which again is assumed to be of unit height.

Equations (15.17) and (15.18) assume that all the fractures have the same aperture, $b$. If the fractures are of different aperture, then the equations are modified as follows:

$$Q = \frac{\rho_f g M}{12 \mu_f} \left( \sum_{i=1}^{m} b_i^3 \right) i \tag{15.20}$$

and

$$K_f = \frac{\rho_f g}{12 \mu_f L} \left( \sum_{i=1}^{m} b_i^3 \right) \tag{15.21}$$

where all the symbols have the same meanings above, and $\sum$, the Greek capital letter sigma, refers to a summation procedure (Example 15.3).

## 15.6  Fluid transport in two orthogonal sets of fractures

Consider now a rock body dissected by two sets of smooth, planar fractures or discontinuities, meeting at right angles. That is, we have two orthogonal sets of fractures, both sets trending parallel with the direction of fluid flow (Fig. 15.5). The discontinuities need not be fractures. For example, the model is applicable when there is one set of vertical fractures and one set of horizontal open bedding or layer contacts and a horizontal fluid flow direction (Fig. 15.6).

Both sets strike parallel with the direction of the fluid flow. The lengths of the profiles are $L_1$ and $L_2$. The cross-sectional area normal to the flow direction is then $L_1 \times L_2$. If the spacing or number, $m$, of horizontal and vertical fractures is the same, and all are of equal aperture $b$, then

$$Q = \frac{\rho_f g 2 m b^3 M}{12 \mu_f} i \qquad (15.22)$$

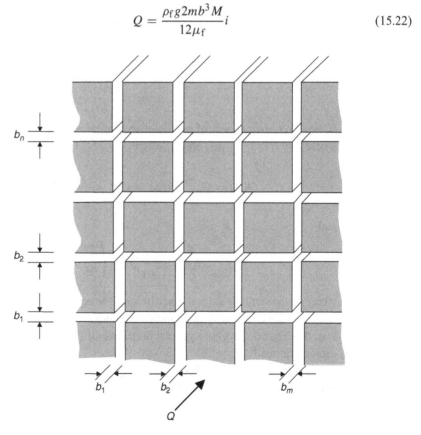

Fluid flow along two orthogonal sets of fractures, one set consisting of vertical fractures, the other of horizontal fractures or contacts. The fracture apertures are denoted by $b_1$, $b_2$, etc.

View west, two sets of orthogonal discontinuities in a normal-fault wall (the fault is seen in Fig. 3.9). The vertical discontinuities are columnar (cooling) joints. The horizontal discontinuities are contacts between flow units in the Holocene, pahoehoe lava flow. The fault wall is close to 30 m high.

and

$$K_f = \frac{\rho_f g 2 m b^3}{12 \mu_f L_1 L_2} \tag{15.23}$$

If, however, the two sets of parallel discontinuities have variable apertures, denoted by $b_i$, where $i = 1, 2, 3, \ldots, m_1$, and $b_j$, where $j = 1, 2, 3, \ldots, m_2$, then we get

$$Q_{\text{total}} = Q_{m_1} + Q_{m_2} = q L^2 = \frac{\rho_f g}{12 \mu_f} \left( \sum_{i=1}^{m_1} b_i^3 + \sum_{j=1}^{m_2} b_j^3 \right) i \tag{15.24}$$

and

$$K_f = \frac{\rho_f g}{12 \mu_f L_1 L_2} \left( \sum_{i=1}^{m_1} b_i^3 + \sum_{j=1}^{m_2} b_j^3 \right) \tag{15.25}$$

where all the symbols are as defined before. The equations are, however, only approximately correct because the flow through the discontinuity junctions is counted twice. Nevertheless, these equations give a general indication of fluid transport through orthogonal sets of discontinuities.

We can also define the fracture porosity $\phi_f$ of such a system of orthogonal fractures or discontinuities of variable aperture as

$$\varphi_f = \left( \sum_{i=1}^{m_1} b_i + \sum_{j=1}^{m_2} b_j \right) L_1^{-1} L_2^{-1} \tag{15.26}$$

## 15.7 Fractures and permeability

In this section we shall briefly comment on some aspects that relate rock fractures to the general permeability of the associated rock mass. All these topics are discussed in more detail in some of the suggested reading at the end of the chapter, such as de Marsily (1986), Bear (1993), Lee and Farmer (1993), and Singhal and Gupta (1999).

### 15.7.1 Percolation threshold

No flow takes place along a particular fracture network unless the fractures are interconnected. For fluid flow to occur from one site to another, there must be at least one interconnected cluster of fractures that links these sites. The condition that such a cluster exists is commonly referred to as a percolation threshold (Fig. 15.7).

### 15.7.2 Mechanical layering

Stratabound rock fractures in a reservoir contribute significantly less to its overall permeability than do non-stratabound fractures because stratabound fractures are less likely to

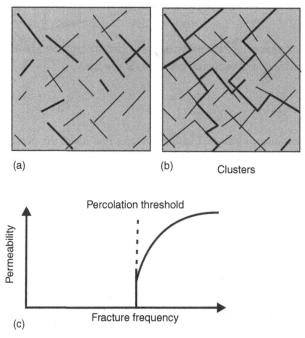

**Fig. 15.7**  For fluid transport to take place between two localities in a fractured rock mass, there must be at least one cluster of interconnected, fluid-conducting fractures connecting these localities. Original fractures (a) link up into a cluster (b) whereby the percolation threshold is reached (c).

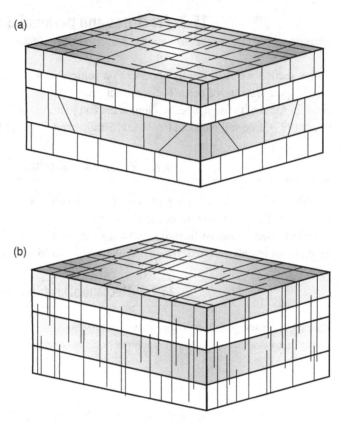

**Fig. 15.8** (a) Stratabound (layerbound) fractures are confined to a single stratum/layer, whereas (b) non-stratabound (non-layerbound) fractures cut through two or more layers.

form interconnected fracture systems that reach the percolation threshold (Figs. 15.8 and 15.9). Non-stratabound fractures may form extensive connected fracture systems, which can transport fluids from remote sources (Fig. 15.8).

Mechanical layering is a universal feature of many heterogeneous rock masses (Figs. 15.9 and 15.10). Layered reservoirs are well known in petroleum engineering. In that field, a practical distinction is often made between laminated and layered reservoirs in that a reservoir is referred to as layered if the layers are thick enough to be targeted by a horizontal well, whereas if the layers are too thin for this, the reservoir is referred to as laminated (Fig. 15.11). Many laminated reservoirs have a poor vertical permeability. In layered, fractured reservoirs, the permeability from layer to layer can vary considerably. In general, ignoring the variation in permeability between layers of a layered reservoir can lead to an overestimate of the overall permeability of the reservoir. If a layered rock mass has essentially the same Young's modulus throughout, and if the layers are welded together so that there are no weak or open contacts, the layers will function mechanically as a single unit.

**Fig. 15.9** View east, stratabound fractures (mineral veins) in limestone layers on the Somerset coast, Southwest England. The fractures were unable to propagate through the shale layers in between the limestone layers.

### 15.7.3  Effects of local fields

The local stress field also affects fluid flow in, and therefore the permeability of, fractured rocks and reservoirs. One reason for this is that fractures are sensitive to changes in the stress field and deform much more easily than circular pores. Another reason is that the stress field contributes to the fluid overpressure.

The single most important difference between the hydrogeological properties of porous media and fractured media is the sensitivity of the fractured media to changes in the associated local stress field. This sensitivity follows primarily from the geometrical difference between the near-spherical voids in the porous media and the crack voids in the fractured media. When the state of stress is changed the cracks deform and change their apertures, and hence the permeability, much more easily than the near-spherical voids (Fig. 15.12).

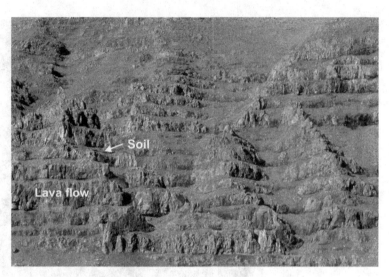

**Fig. 15.10**    View north, stiff basaltic lava flows alternate with soft soil (and scoria) layers in West Iceland.

The state of stress in a particular area has a great effect on the permeability in that area. The stress field controls the activity of existing faults and fractures and may also initiate new, highly water-conducting fractures. The geostatic stress (overburden pressure) affects average width (aperture) of fractures. Nevertheless, the general effect of normal and shear stresses on permeability of a rock mass is still not very well understood. Generally, the current stress field together with the geometry and connectivity of the fractures largely control fluid flow in fractured rocks.

### 15.7.4 Effects of fractures on in-situ permeability

Fractures have a great effect on rock permeability. In fact, where there are major fractures and faults in a rock mass, they usually control the fluid transport in that mass. Some of the main effects of fractures on permeability may be summarised as follows:

- In-situ permeability values range over 12 orders of magnitude.
- At a particular site a variation of 4–6 orders of magnitude is typical.
- A major active (water-conducting) fracture can easily increase the average permeability of a test area by several orders of magnitude over adjacent sections.
- Active (water-conducting) fractures may be only one fifth to one fourth of the total number at a particular site. Thus, permeability is not a simple function of the total fracture frequency.
- Rock-mass permeability is commonly significantly anisotropic. The ratio of horizontal to vertical permeability can range from 0.01 (for mainly vertical joints) to 1000 (for intact, layered sedimentary strata). Commonly used ratios are between 1 and 10.
- Conducting fractures may easily increase in-situ permeability by three orders of magnitude or more over laboratory values.

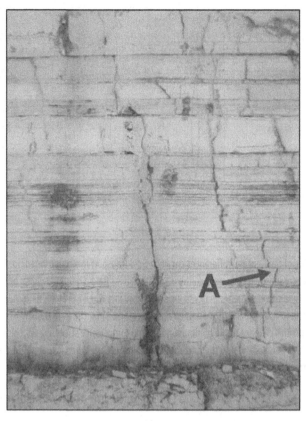

**Fig. 15.11** Laminated carbonate rocks in a quarry in southeast Italy (cf. Larsen *et al.*, 2009, 2010). Note that the fracture is offset (one offset indicated by A) at nearly all the contacts between the lamina, indicating that the fracture path is relatively long and partly mixed mode, both of which result in more energy being needed to propagate the fracture vertically (Chapter 10). Laminated reservoirs generally have a poor vertical permeability.

### 15.7.5  Crustal depth and fracture-related permeability

Fracture-related permeability changes with depth. The main effects may be summarised as follows (Figs. 15.13–15.15; cf. Lee and Farmer, 1993):

- Fracture **frequency** (intensity) normally decreases rapidly with depth in the uppermost few hundred metres of the crust and is then approximately constant in the uppermost 1–2 km (Fig. 15.13).
- Fracture **aperture** decreases rapidly in the uppermost few tens of metres and then decreases slightly with depth (Fig. 15.14).
- **Hydraulic conductivity** (or **permeability**) decreases somewhat in the uppermost few tens of metres of the crust, but is then roughly constant to depths of 2–3 km (Fig. 15.15).

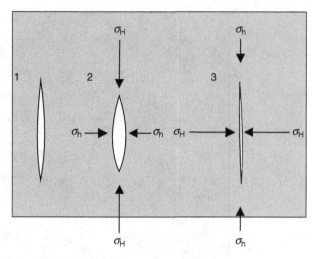

**Fig. 15.12** One of the principal differences between fluid transport in a porous medium and a fractured medium is that the latter is very sensitive to changes in the associated stress field. Equidimensional pores in a porous medium have comparatively little tendency to expand or contract when the stress field changes. By contrast, fractures such as in (1) tend to open (2) and increase their apertures (and, thus, the fluid transmissivity) when the stress perpendicular to the fractures changes to the minimum horizontal compressive stress, $\sigma_h$, but close (3) when the horizontal compressive stress perpendicular to the fractures is high, particularly when it changes to the maximum horizontal compressive stress, $\sigma_H$.

## 15.7.6 Permeability anisotropy

Dominant fracture trends may contribute to the highly anisotropic permeability of a rock mass or reservoir. Some of the main findings may be summarised as follows (Fig. 15.16):

- Permeability anisotropy is common, and often very important, in fractured rocks including reservoirs.
- If the rock mass contains only one set of fractures, then the permeability is likely to be highly anisotropic.
- If the rock mass contains three or more sets of fractures, of different attitudes, then the permeability anisotropy is less than if the rock contains only one set of fractures.
- Rocks are often highly permeable along bedding planes in layered sedimentary rocks and contacts between lava flows in a lava pile.

# 15.8 Summary

- Fluid transport in rocks is through pores, fractures, or both. Fluid flow in porous media is well understood and has a basis in Darcy's law. Fluid flow through fractured media, however, is still poorly understood and is a field of active research.

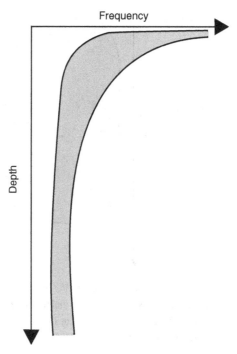

**Fig. 15.13**   Schematic illustration of how fracture frequency changes with depth in the uppermost 1000 m of the crust. It first decreases rapidly with depth, and then becomes essentially constant at about 3–4 fractures per metre (based on data in Lee and Farmer, 1993).

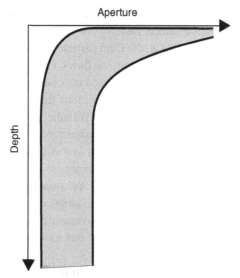

**Fig. 15.14**   Schematic illustration of how fracture aperture changes with depth in the uppermost 100 m of the crust. It first decreases rapidly with depth, and then becomes essentially constant at about 0.03–0.1 mm (based on data in Lee and Farmer, 1993). It should be noted, however, that much larger apertures (up to several millimetres) are common for fractures in surface rocks.

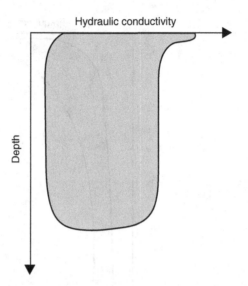

**Fig. 15.15** Schematic illustration of how hydraulic conductivity ($K$) changes with depth in crystalline rocks in the upper part of the crust. Both for crystalline and sedimentary rocks (not shown here) the average hydraulic conductivity changes little in the upper part of the crust, to a depth of about 3 km, but shows a very great variation (the horizontal width of the shaded area). Based on data in Lee and Farmer (1993).

- Darcy's law states that the fluid velocity (the discharge velocity or specific discharge) in a porous medium, such as sand, is proportional to the hydraulic gradient. The hydraulic gradient is the difference in water level (for ground water) in wells, that is, manometers over a certain distance. Darcy's law is generally valid for granular media so long as the flow is laminar, that is, the fluid particles move along straight, parallel paths in layers or lamina (also known as streamline flow). The law is widely used for flow in sediments and sedimentary rocks, and can be used as a crude approximation for flow in densely fractured rocks where the fractures have various trends and form an interconnected pore space. If $q$ is the flow velocity, $K$ the hydraulic conductivity (in m s$^{-1}$), and $i$ the hydraulic gradient, then Darcy's law may be written as $q = Ki$.
- The physical meaning of Darcy's law is that the fluid flow is driven towards regions of the crust where the potential energy is lower than in the surroundings. Only in the special case of horizontal flow is the fluid driven to regions of lower pressure. When the fluid is not ground water, for example oil or gas, permeability (rather than hydraulic conductivity) is used. Permeability measures the ease with which any fluid moves through the crust. Intrinsic permeability $k$ has units of m$^2$ and is measured in darcy, $d$, where $1d = 10^{-12}m^2$.
- The basic equation for fluid flow in fractures is the cubic law, which is based on the Navier–Stokes equation for the flow of a viscous fluid between parallel plates. The law states that the volumetric flow rate is proportional to the cube of the opening or aperture of the fracture, that is, the separation of the plates (the fracture walls). Because of the cubic law, the volumetric flow rate through a fracture may change greatly for

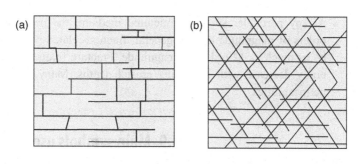

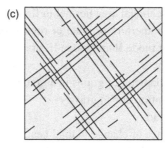

**Fig. 15.16** Different fracture orientations results in different fracture-related permeabilities (cf. EUR 17116, 1997). The set shown in (a) is very common in many rocks, particularly some sedimentary rocks (Larsen *et al.*, 2009, 2010). The vertical clusters may be well connected when the vertical fractures link through the horizontal contacts, giving rise to an anisotropic permeability, that is, dominating vertical permeability. The set in (b) consists of three subsets all of which are of similar intensity. All the sets are interconnected, resulting in close-to-isotropic permeability. Set (c) has varying fracture intensity (frequency) inside and between clusters. Sampling of a large rock volume is needed to obtain a realistic estimate of the anisotropic rock permeability. When the subsets consist of single fractures, the fractures may form a system similar to that in Fig. 15.17.

relatively small changes in the fracture aperture. The law also implies that, for a fracture with variation in aperture, much of the fluid transport may be focused on, or channelled to, the parts of the fracture that have the largest aperture. This is known as flow channelling.

- In a fractured reservoir, no fluid flow takes place from one locality to another in the reservoir unless the fractures between the localities form an interconnected, fluid-conducting cluster. When such a cluster forms, the percolation threshold is reached. Whether such a cluster forms depends on many factors, including mechanical layering. Fractures confined to individual layers are known as stratabound and are less likely to form an interconnected cluster of a significant size than non-stratabound fractures.

- The single most important difference between the fluid-transport properties of porous and fractured media is the sensitivity of fractured media to changes in the local stress field. When the stress field changes, the fractures of a fractured medium deform and change their apertures, and thus the associated rock permeability, much more readily than the near-spherical voids of a porous medium.

- At any particular site of a fractured medium, the permeability may range by 4–6 orders of magnitude and, in general, fracture-related permeability in the crust ranges by as much as 12 orders of magnitude. Fracture frequency and average aperture generally decrease with increasing crustal depths. Many rock masses have anisotropic permeabilities.

## 15.9  Main symbols used

| | |
|---|---|
| $A$ | cross-sectional area normal to fluid (water) flow |
| $b$ | fracture aperture or plate or rock-wall separation (for a mode I crack, equal to $\Delta u$; for mode II and mode III cracks, however, $b$ has little if any relation to $\Delta u$) |
| $C$ | constant |
| $d$ | mean diameter of grains in a granular material (sand in Darcy's experiments) |
| $g$ | acceleration due to gravity |
| $h$ | total head |
| $\Delta h$ | difference or change in total head |
| $\nabla h$ | gradient (the rate of change) of the scalar field $h$ ($\nabla$ is del or nabla) |
| $i$ | hydraulic gradient ($\Delta h / \Delta L$) |
| $K$ | hydraulic conductivity, coefficient of permeability |
| $K_f$ | hydraulic conductivity of a fractured rock body |
| $k$ | intrinsic permeability |
| $L$ | length; length of a fracture profile |
| $L_1$ | length of fracture profile 1 |
| $L_2$ | length of fracture profile 2 |
| $\Delta L$ | distance of fluid (water) flow |
| $m$ | number of fractures dissecting a profile |
| $N$ | dimensionless shape factor |
| $P$ | total pressure of a fluid (water) at the bottom of a water well (a piezometer) |
| $P_1$ | fluid total pressure at a certain point 1 |
| $P_2$ | fluid total pressure at a certain point 2 |
| $Q$ | volumetric flow rate, volume of fluid per unit time |
| $q$ | discharge velocity, specific discharge |
| $W$ | width of plates, or fracture dimension, perpendicular to fluid flow |
| $z$ | elevation head |
| $\mu_f$ | dynamic or absolute viscosity of a fluid |
| $\mu_w$ | dynamic or absolute viscosity of water |
| $\pi$ | 3.1416… |
| $\rho_f$ | fluid density |
| $\rho_r$ | rock or crustal density |
| $\rho_w$ | water density |
| $\phi_f$ | porosity of a fractured rock body |

# 15.10  Worked examples

---

### Example 15.1

Problem

A set of parallel, open joints is studied in a vertical section (cf. Fig. 15.4). The joints are confined to a single layer and all of essentially the same dip dimension. The apertures of the joints, however, show a variation by a factor of 5. That is to say, the narrowest joint has an aperture that is 1/5 of the joint with the greatest opening. If the aperture is the only parameter that is different among the joints, what would be the difference in volumetric flow through the joints with the smallest and the largest aperture?

Solution

From any of Eqs. (15.8)–(15.10) we see that the volumetric flow rate $Q$ depends on the cube of the aperture of the fracture, that is, on $b^3$. If the aperture $b$ is the only parameter that differs between these two fractures then a fivefold increase in aperture between the fractures would imply that the volumetric flow rate could increase by $5^3 = 5 \times 5 \times 5 = 125$ times.

This shows the great importance of apertures in general, and of the cubic law in particular, for the transport of fluids in fractures and fracture sets. The difference in apertures, and to a lesser degree the dimension of the fracture perpendicular to the flow ($W$ in the equations above), is the primary reason why one or a few fractures in a large fracture set may entirely dominate the fluid transport through that set.

---

### Example 15.2

Problem

A 4-m-long horizontal profile dissects 40 joints, giving the average frequency of 10 joints per metre. The height of the profile is 1 m. All the joints have the same aperture, 1 mm. Ground water at 4°C with a density of 1000 kg m$^{-3}$ and a dynamic viscosity of $1.55 \times 10^{-3}$ Pa s flows through the joint set. Find the hydraulic conductivity of the rock hosting the fracture set.

Solution

For a single set of fractures (Fig. 15.4), Eq. (15.18) gives the hydraulic conductivity $K_f$ as

$$K_f = \frac{\rho_f g m b^3}{12 \mu_f L}$$

Here the fluid density $\rho_f$ is 1000 kg m$^{-3}$, the acceleration due to gravity $g$ is 9.81 m s$^{-2}$, the number of fractures (here joints) $m$ is 40, the aperture of each fracture (all the apertures are equal) $b$ is $1 \times 10^{-3}$ m, the dynamic viscosity of the fluid (here water) $\mu_f$ is

$1.55 \times 10^{-3}$ Pa s, and the length of the profile $L$ is 4 m. The profile has a unit height, as is assumed in Eq. (15.18). Putting these values into Eq. (15.18) we get

$$K_f = \frac{1000 \, \text{kg m}^{-3} \times 9.81 \, \text{m s}^{-2} \times 40 \times (1 \times 10^{-3} \text{m})^3}{12 \times 1.55 \times 10^{-3} \, \text{Pa s} \times 4 \, \text{m}} = 5.27 \times 10^{-3} \text{m s}^{-1}$$

When compared with porous in-situ geomaterials, this conductivity is very similar to that of a typical sand (Lee and Farmer, 1993). No laboratory specimens, however, reach values so high: the highest conductivities of laboratory specimens are about $10^{-4}$ m s$^{-1}$. When compared with fractured rocks, these values are similar to those of highly fractured basalt (Singhal and Gupta, 1999), such as lava flows and sills with well-developed columnar joints (Figs. 4.6 and 15.6).

## Example 15.3

### Problem

An 8-m-long horizontal profile in a vertical section (a wall) dissects ten fractures. The fracture apertures are 0.1 mm, 0.2 mm, 0.3 mm, 0.4 mm, 0.5 mm, 0.6 mm, 0.7 mm, 0.8 mm, 0.9 mm, and 1.0 mm. The fractures occur in gneiss in a flat area. The estimated hydraulic gradient $i = \Delta h / \Delta L$ is 0.01. This means that the total head (the water level) in manometers (water wells) drops by 1 m for every 100 m of horizontal distance. The water density $\rho_f$ is 1000 kg m$^{-3}$, and its dynamic viscosity $\mu_f$ is $1.55 \times 10^{-3}$ Pa s. Calculate the following:

(a) the volumetric flow rate $Q$ through the fracture system
(b) the hydraulic conductivity $K_f$ of the fractured rock.

### Solution

(a) The volumetric flow rate $Q$ may be obtained from Eq. (15.20) thus:

$$Q = \frac{\rho_f g M}{12 \mu_f} \left( \sum_{i=1}^{m} b_i^3 \right) i$$

We start by calculating the sum of the apertures in an elementary way. For convenience we do the calculations in millimetres and then drop the units, but remember to change the final value from cubic millimetres to cubic metres before calculating the volumetric flow rate. This change into metres must be made because the results must be given in SI units. We have

$$\sum_{i=1}^{10} b_i^3 = (0.1)^3 + (0.2)^3 + (0.3)^3 + (0.4)^3 + (0.5)^3 + (0.6)^3 + (0.7)^3 + (0.8)^3$$
$$+ (0.9)^3 + (1.0)^3 = 0.001 + 0.008 + 0.027 + 0.064 + 0.125$$
$$+ 0.216 + 0.343 + 0.512 + 0.729 + 1.0 = 3.025 \, \text{mm}^3 = 3.025 \times 10^{-9} \, \text{m}^3$$

Using this value, and those given above (including the assumption of a unit height $M$ of the profile, a value that enters the calculations in the numerator as 1 m), we now calculate the volumetric flow rate $Q$ as follows:

$$Q = \frac{1000 \text{ kg m}^{-3} \times 9.81 \text{ m s}^{-2} \times 1\text{m} \times 3.025 \times 10^{-9}\text{m}^3 \times 0.01}{12 \times 1.55 \times 10^{-3}\text{Pa s}}$$

$$= 1.595 \times 10^{-5}\text{m}^3\text{s}^{-1}$$

On checking the dimensions of the equation we see that the dimension of length, namely 1 m, must be in the numerator as indicated by the assumption of unit height of the sections measured.

(b) From Eq. (15.21) the hydraulic conductivity of the rock hosting the fracture set is given by

$$K_f = \frac{\rho_f g}{12\mu_f L} \left( \sum_{i=1}^{m} b_i^3 \right)$$

We could do the calculations again along the lines of part (a). However, an inspection of Eqs. (15.20), (15.21) and (15.3) shows that the difference between these equations is that in Eq. (15.21) the hydraulic gradient $i$ and the length of the profile $L$ times its height (here a unit), that is, the cross-sectional area perpendicular to the flow (and the strike of the fractures), form a part of the denominator. Thus, we only need to divide the result in (a) by 0.01 (the hydraulic gradient) and by the cross-sectional area of the profile, $A = 8m^2$. Then we get

$$K_f = \frac{Q}{Ai} = \frac{1.595 \times 10^{-5}\text{m}^3\text{s}^{-1}}{8 \text{ m}^2 \times 0.01} = 1.994 \times 10^{-4}\text{m s}^{-1} \approx 2 \times 10^{-4}\text{m s}^{-1}$$

This result is very typical for many fractured rocks such as sandstones, limestones and basalt (Lee and Farmer, 1993).

---

## Example 15.4

### Problem

An orthogonal system of fractures/discontinuities, similar to those in Figs. 15.5, 15.6, and 15.18, is measured along a horizontal profile that is 20 m long and a vertical profile that is 5 m high. There are 40 vertical fractures meeting the horizontal profile of which seven have an aperture of 0.1 mm, six an aperture of 0.2 mm, four an aperture of 0.3 mm, eight an aperture of 0.4 mm, five an aperture of 0.5 mm, three an aperture of 0.7 mm, three an aperture of 0.8 mm, two an aperture of 0.9 mm, and two an aperture of 1.0 mm. There are ten fractures meeting the vertical profile of which four have an aperture of 0.1 mm, three an aperture of 0.2 mm, two an aperture of 0.4 mm, and one an aperture of 0.5 mm. The water density $\rho_f$ is 1000 kg m$^{-3}$, and its dynamic viscosity $\mu_f$ is $1.55 \times 10^{-3}$ Pa s. Calculate the hydraulic conductivity of the rock hosting the fracture sets.

## Solution

The hydraulic conductivity of the sets is obtained from Eq. (15.25), namely

$$K_f = \frac{\rho_f g}{12\mu_f L_1 L_2} \left( \sum_{i=1}^{m_1} b_i^3 + \sum_{j=1}^{m_2} b_j^3 \right)$$

As before, we start by calculating the sum of apertures in an elementary way. Again we do the calculations first in cubic millimetres and then drop the unit, later changing into cubic metres. We take set 1 as being the set of vertical fractures dissecting profile 1, which thus has the length $L_1 = 20$ m and fracture number $m_1 = 40$. For set 1 we have (where each aperture is multiplied by the number of fractures that have that particular aperture)

$$\sum_{i=1}^{40} b_i^3 = 7(0.1)^3 + 6(0.2)^3 + 4(0.3)^3 + 8(0.4)^3 + 5(0.5)^3 + 3(0.7)^3 + 3(0.8)^3$$
$$+ 2(0.9)^3 + 2(1.0)^3 = 0.007 + 0.048 + 0.108 + 0.512 + 0.625 + 1.029 + 1.536$$
$$+ 1.458 + 2.0 = 7.323 \text{ mm}^3 = 7.323 \times 10^{-9} \text{ m}^3$$

Then we calculate the sum of apertures for set 2, that is, the set of horizontal fractures. Thus, this set has the length $L_2 = 5$ m and fracture number $m_2 = 10$. Proceeding as before, the sum of apertures in set 2 becomes

$$\sum_{j=1}^{10} b_j^3 = 4(0.1)^3 + 3(0.2)^3 + 2(0.4)^3 + (0.5)^3 = 0.004 + 0.024 + 0.128 + 0.125$$
$$= 0.281 \text{ mm}^3 = 2.81 \times 10^{-10} \text{ m}^3$$

Using these values for the aperture sums as well as the values above for water viscosity, density and profile lengths, Eq. (15.25) gives the hydraulic conductivity of the rock hosting the sets as

$$K_f = \frac{1000 \text{ kg m}^3 \times 9.81 \text{ m s}^{-2}}{12 \times 1.55 \times 10^{-3} \text{Pa s} \times 20 \text{ m} \times 5 \text{ m}} \times \left( 7.323 \times 10^{-9} \text{m}^3 + 2.81 \times 10^{-10} \text{m}^3 \right)$$
$$= 4.01 \times 10^{-5} \text{ m s}^{-1}$$

Clearly, this is a high hydraulic conductivity, such as can be expected in lava flows. For example, the orthogonal systems of fractures are common in pahoehoe (basaltic) lava flows (Fig. 15.6). Then the vertical fractures or discontinuities are mostly columnar (cooling) joints, whereas the horizontal ones are partly cooling joints but mostly contacts between flow units. Thus, fractured basalt has commonly this or a higher hydraulic conductivity, as do many fractured sedimentary rocks such as limestones and sandstones, as well as sediments such as sand (Lee and Farmer, 1993; Singhal and Gupta, 1999). Such systems occur also in sedimentary rocks (Fig. 15.17) and in metamorphic rocks (Fig. 15.18). In the joint system in Fig. 15.18 the fractures consist of three mutually orthogonal sets, indicating a high permeability.

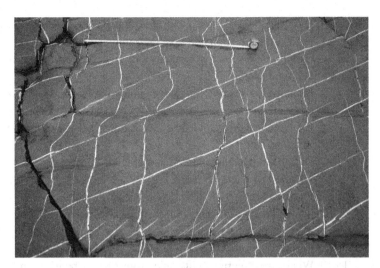

**Fig. 15.17**   Close-to-orthogonal sets of mineral veins in limestone on the Somerset coast of the Bristol Channel in Southwest England. The cross-cutting relationships between the veins indicate that most of them are pure extension fractures.

**Fig. 15.18**   Three subsets of mutually orthogonal joints/discontinuities in gneiss on an island offshore West Norway. These well-connected subsets, although of different orientation, are comparable with Fig. 15.16(b) and result in essentially isotropic permeability.

## 15.11 Exercises

15.1  Describe and provide an appropriate equation for Darcy's law. Explain all the symbols used in the equation and give their units.

15.2  Explain the two concepts of permeability and give their units. Which concept is used for the permeability of a petroleum reservoir?

15.3 Describe the cubic law and provide an appropriate equation. Explain all the symbols used in the equation and give their units.

15.4 Explain the following concepts and give their units: elevation head, pressure head, and total head.

15.5 Explain flow channelling. Give geological examples of processes that relate to flow channelling.

15.6 Explain the concept of a percolation threshold and its implications for fluid transport in fractured rocks.

15.7 Define stratabound and non-stratabound fractures and explain their effects on fluid transport in fractured rocks.

15.8 Explain briefly why fractured rocks are so much more sensitive as regards permeability to changes in local stresses than porous rocks.

15.9 How do fracture frequency (intensity) and aperture change with increasing crustal depth?

15.10 What is rock-permeability anisotropy? Describe its main characteristics.

15.11 During a rifting event, the aperture of a water-conducting tension fracture (an example of such a fracture is on the cover of the book) increases from 5 cm to 60 cm. If the aperture is the only hydrogeological factor that changes during the rifting event, what would be the difference in volumetric flow rate of the fracture before and after the rifting event.

15.12 A 5-m-long and 1-m-high horizontal profile is dissected by 50 vertical extension fractures, each with an aperture of 0.5 mm. Ground water with a dynamic viscosity of $1.55 \times 10^{-3}$ Pa s and a density of $1000 \, \text{kg m}^{-3}$ flows along the fractures. Calculate the associated hydraulic conductivity.

15.13 Repeat Exercise 15.12 but add a new set of horizontal extension fractures. That is to say, the rock is now dissected by an orthogonal system of fractures. The set of vertical fractures is as in Exercise 15.12, but there is now, in addition, a set of 20 horizontal fractures, each with an aperture of 0.2 mm and dissecting a vertical profile that is 4 m high. Calculate the associated hydraulic conductivity.

15.14 A 10-m-long and 1-m-high horizontal profile is dissected by 20 vertical, extension fractures of which five have an aperture of 0.2 mm, three an aperture of 0.3 mm, two an aperture of 0.4 mm, four an aperture of 0.6 mm, three an aperture of 0.7 mm, and three an aperture of 0.8 mm. The ground-water density and viscosity are as in Exercise 15.12 and the hydraulic gradient is 0.015. Calculate the volumetric flow rate and the hydraulic conductivity.

# References and suggested reading

Aguilera, R., 1995. *Naturally Fractured Reservoirs*. Tulsa, OK: PennWell Publishing Company.

Aguilera, R., 2000. Well test analysis of multi-layered naturally fractured reservoirs. *Journal of Canadian Petroleum Technology*, **39**, 31–37.

Al-Busaidi, A., Hazzard, J. F., and Young, R. P., 2005. Distinct element modeling of hydraulically fractured Lac du Bonnet granite. *Journal of Geophysical Research*, **110**, B06302, doi:10.1029/2004JB003297.

Amadei, B. and Stephansson, O., 1997. *Rock Stress and its Measurement*. London: Chapman & Hall.

Aydin, A., 2000. Fractures, faults, and hydrocarbon entrapment, migration and flow. *Marine Petroleum Geology*, **17**, 797–814.

Barton, C. A., Zoback, M. D., and Moos, D., 1995. Fluid flow along potentially active faults in crystalline rock. *Geology*, **23**, 683–686.

Bear, J., 1988. *Dynamics of Fluids in Porous Media*. New York: Dover.

Bear, J., 1993. Modelling flow and contaminant transport in fractured rocks. In: Bear, J., Tsang, C. F., and de Marsily, G. (eds.), *Flow and Contaminant Transport in Fractured Rock*. New York: Academic Press, pp. 1–37.

Berkowitz, B., 2002. Characterizing flow and transport in fractured geological media: a review. *Advances in Water Resources*, **25**, 861–884.

Bons, P. D., 2001. The formation of large quartz veins by rapid ascent of fluids in mobile hydrofractures. *Tectonophysics*, **336**, 1–17.

Brebbia, C. A. and Dominguez, J., 1992. *Boundary Elements: an Introductory Course*. Boston, MA: Computational Mechanics.

Brenner, S. L. and Gudmundsson, A., 2002. Permeability development during hydrofracture propagation in layered reservoirs. *Norges Geologiske Undersøkelse Bulletin*, **439**, 71–77.

Brenner, S. L. and Gudmundsson, A., 2004. Permeability in layered reservoirs: field examples and models on the effects of hydrofracture propagation. In: Stephansson, O., Hudson, J. A., and Jing, L. (eds.), *Coupled Thermo-Hydro-Mechanical-Chemical Processes in Geo-Systems, Geo-Engineering Series, Vol. 2*. Amsterdam: Elsevier, pp. 643–648.

Bruggeman, G. A., 1999. *Analytical Solutions of Geohydrological Problems*. Amsterdam: Elsevier.

Chapman, R. E., 1981. *Geology and Water: An Introduction to Fluid Mechanics for Geologists*. London: Nijhoff.

Chilingar, G. V., Serebryakov, V. A., and Robertson, J. O., 2002. *Origin and Prediction of Abnormal Formation Pressures*. London: Elsevier.

Coward, M. P., Daltaban, T. S., and Johnson, H. (eds.), 1998. *Structural Geology in Reservoir Characterization*. Geological Society of London Special Publication 127. London: Geological Society of London.

Dahlberg, E. C., 1994. *Applied Hydrodynamics in Petroleum Exploration*. New York: Springer-Verlag.

Daneshy, A. A., 1978. Hydraulic fracture propagation in layered formations. *AIME Society of Petroleum Engineers Journal*, 33–41.

de Marsily, G., 1986. *Quantitative Hydrogeology*. New York: Academic Press.

Deming, D., 2002. *Introduction to Hydrogeology*. Boston, MA: McGraw-Hill.

Domenico, P. A. and Schwartz, F. W., 1998. *Physical and Chemical Hydrogeology*, 2nd edn. New York: Wiley.

Dresen, G., Stephansson, O., and Zang, A., (eds.). 2006. Rock damage and fluid transport, part I. *Pure and Applied Geophysics*, **163**(5–6).

Economides, M. J. and Nolte, K. G. (eds.), 2000. *Reservoir Stimulation*. New York: Wiley.

EUR 17116, 1997. Interim Guide to Fracture Interpretation and Flow Modelling in Fractured Reservoirs. EC Report, Brussels.

Evans, J. P., Forster, C. B., and Goddard, J. V., 1997. Permeability of fault-related rocks, and implications for hydraulic structure of fault zones. *Journal of Structural Geology*, **19**, 1393–1404.

Faybishenko, B., Witherspoon, P. A., and Benson, S. M. (eds.), 2000. *Dynamics of Fluids in Fractured Rocks*. Washington, DC: American Geophysical Union.

Fetter, C. W., 1994. *Applied Hydrogeology*, 3rd edn. Upper Saddle River, NJ: Prentice-Hall.

Finkbeiner, T., Barton, C. A., and Zoback, M. D., 1997. Relationships among in-situ stress, fractures and faults, and fluid flow: Monterey formation, Santa Maria basin, California. *American Association of Petroleum Geologists Bulletin*, **81**, 1975–1999.

Fyfe, W. S., Price, N. J., and Thompson, A. B. 1978. *Fluids in the Earth's Crust*. Amsterdam: Elsevier.

Giles, R. V., 1977. *Fluid Mechanics and Hydraulics*, 2nd edn. New York: McGraw-Hill.

Gudmundsson, A., 1999. Fluid overpressure and stress drop in fault zones. *Geophysical Research Letters*, **26**, 115–118.

Gudmundsson, A., 2000. Active fault zones and groundwater flow. *Geophysical Research Letters*, **27**, 2993–2996.

Gudmundsson, A., Fjeldskaar, I., and Brenner, S.L., 2002. Propagation pathways and fluid transport of hydrofractures in jointed and layered rocks in geothermal fields. *Journal of Volcanology and Geothermal Research*, **116**, 257–278.

Haneberg, W. C., Mozley, P. S., Moore, J. C., and Goodwin, L. B. (eds.), 1999. *Faults and Subsurface Fluid Flow in the Shallow Crust*. Washington, DC: American Geophysical Union.

Hardebeck, J. L. and Hauksson, E., 1999. Role of fluids in faulting inferred from stress field signatures. *Science* **285**, 236–239.

Ingebritsen, S. E. and Sanford, W. E., 1998. *Groundwater in Geologic Processes*. Cambridge: Cambridge University Press.

Kümpel, H. J. (ed.), 2003. *Thermo-Hydro-Mechanical Coupling in Fractured Rock*. Berlin: Birkhäuser.

Labaume, P., Craw, D., Lespinasse, M., and Muchez, P. (eds.), 2002. Tectonic processes and the flow of mineralising fluids. *Tectonophysics*, **348**, 1–185.

Larsen, B., Grunnaleite, I., and Gudmundsson, A., 2009. How fracture systems affect permeability development in shallow-water carbonate rocks: an example from the Gargano Peninsula, Italy. *Journal of Structural Geology*, doi:10.1016/jsg.2009.05.009

Larsen, B., Gudmundsson, A., Grunnaleite, I., Sælen, G., Talbot, M. R., and Buckley, S., 2010. Effects of sedimentary interfaces on fracture pattern, linkage, and cluster formation in peritidal carbonate rocks. *Marine and Petroleum Geology*, doi: 10.1016/j.marpetgeo.2010.03.011.

Lee, C. H. and Farmer, I., 1993. *Fluid Flow in Discontinuous Rocks*. London: Chapman & Hall.

Nelson, R. A., 1985. *Geologic Analysis of Naturally Fractured Reservoirs*. Houston, TX: Gulf Publishing.

Philipp, S. L., 2008. Geometry and formation of gypsum veins in mudstones at Watchet, Somerset, SW-England. *Geological Magazine*, **145**, 831–844.

Secor, D. T., 1965. Role of fluid pressure in jointing. *American Journal of Science*, **263**, 633–646.

Selley, R. C., 1998. *Elements of Petroleum Geology*. London: Academic Press.

Sibson, R. H., 1996. Structural permeability of fluid-driven fault-fracture meshes. *Journal of Strucural Geology*, **18**, 1031–1042.

Simonson, E. R., Abou-Sayed, A. S., and Clifton, R. J., 1978. Containment of massive hydraulic fractures. *Society of Petroleum Engineers Journal*, **18**, 27–32.

Singhal, B. B. S. and Gupta, R. P., 1999. *Applied Hydrogeology of Fractured Rocks*. London: Kluwer.

Smits, A. J. 2000. *A Physical Introduction to Fluid Mechanics*. New York: Wiley.

Stauffer, D. and Aharony, A., 1994. *Introduction to Percolation Theory*. London: Taylor & Francis.

Stearns, D. W. and Friedman, M., 1972. Reservoirs in fractured rocks. *American Association of Petroleum Geologists Memoirs*, **16**, 82–106.

Sun, R. J., 1969. Theoretical size of hydraulically induced horizontal fractures and corresponding surface uplift in an idealized medium. *Journal of Geophysical Research*, **74**, 5995–6011.

Teufel, L. W. and Clark, J. A., 1984. Hydraulic fracture propagation in layered rock: Experimental studies of fracture containment. *Society of Petroleum Engineers Journal*, 19–32.

Tsang, C. F. and Neretnieks, I., 1998. Flow channeling in heterogeneous fractured rocks. *Reviews of Geophysics*, **36**, 275–298.

Turcotte, D. L., 1997. *Fractals and Chaos in Geology and Geophysics*, 2nd edn. Cambridge: Cambridge University Press.

Valko, P. and Economides, M. J., 1995. *Hydraulic Fracture Mechanics*. New York: Wiley.

Wang, J. S. Y., 1991. Flow and transport in fractured rocks. *Reviews of Geophysics*, **29**, 254–262.

Warpinski, N. R., 1985. Measurement of width and pressure in a propagating hydraulic fracture. *Society of Petroleum Engineers Journal*, 46–54.

Warpinski, N.R., Schmidt, R.A., and Northrop, D. A., 1982. In-situ stress: the predominant influence on hydraulic fracture containment. *Journal of Petroleum Technology*, 653–664.

Warpinski, N. R. and Teufel, L. W., 1987. Influence of geologic discontinuities on hydraulic fracture propagation. *Journal of Petroleum Technology*, 209–220.

Yew, C.H., 1997. *Mechanics of Hydraulic Fracturing*. Houston, TX: Gulf Publishing.

Zhang, X., Koutsabeloulis, N., and Heffer, K., 2007. Hydromechanical modeling of critically stressed and faulted reservoirs. *American Association of Petroleum Geologists Bulletin*, **91**, 31–50.

Zimmerman, R. W. and Yeo, I.-W., 2000. Fluid flow in rock fractures: from the Navier-Stokes equations to the cubic law. In: Faybishenko, B., Witherspoon, P. A., and Benson, S. M. (eds.), *Dynamics of Fluids in Fractured Rock*. Geophysical Monograph 122. Washington, DC: American Geophysical Union, pp. 213–224.

# Fluid transport in faults

## 16.1 Aims

The way active faults transport crustal fluids is important in many fields of earth sciences, including petroleum geology, geothermal research, volcanology, seismology, and hydrogeology. There is increasing evidence that active, or potentially active, faults largely control fluid flow in solid rocks. In order to understand the flow of ground water and other crustal fluids in a region, it is necessary to know the general permeability structure of the fault zones in that region. This follows because the general hydromechanical structure of a fault zone, in relation to that of the host rock, determines whether the fault zone acts as a conduit that transports fluids or as a barrier to fluid flow. Here the focus is on the effects of faults on ground-water flow, but the results are, with suitable modifications, applicable to the transport of other crustal fluids. The principal aims of this chapter are to:

- Explain how fluids migrate to fault zones, in particular active fault zones.
- Present a general model on fluid transport along a fault zone.
- Explain the effects of fluid pressure on the apertures of fractures in the damage zone.
- Provide a simple model of fluid transport along the fault damage zone.
- Provide a simple model of fluid transport along the fault core.
- Discuss the general implications of the results for fluid transport by faults.

## 16.2 Overview

The effects of fault zones on fluid flow can be complex. For example, part of a fault zone may act as a barrier to flow, while other parts act as conduits for the flow. Also, parts of an active fault zone that are highly conductive just prior to and during periods of seismogenic faulting may act as barriers during the greater part of the interseismic (non-seismic) period.

There exists considerable geological work on the permeability structure of major fault zones (e.g. Bruhn *et al.*, 1994; Caine *et al.*, 1996; Lin *et al.*, 2007; Anderson and Bakker, 2008; Cappa, 2009; Paul *et al.*, 2009). In-situ bulk hydraulic characteristics of fault zones and fracture zones have been measured in boreholes (e.g. Ahlbom and Smellie, 1991; Barton *et al.*, 1995; Fisher *et al.*, 1996; Braathen *et al.*, 1999; Nativ *et al.*, 1999) and modelled (Barton *et al.*, 1995; Bredehoeft, 1997; Cappa, 2009; Paul *et al.*, 2009).

In this chapter we present and discuss the implications of some general conceptual and numerical hydromechanical models of major active fault zones. The models are primarily based on field, analytical, and numerical studies of large fault zones, which are mostly normal and strike-slip fault zones. The analytical models are derived from those presented in Chapter 15 and related to fluid flow in porous and fractured media. The numerical studies are mainly concerned with the local stresses that develop around and inside fault zones when loaded to failure. As regards fluids, the focus is here on the inferred flow of ground water and geothermal water to and along the fault zones. However, with suitable modifications, the results can be applied to the flow of other fluids, such as oil and gas.

The permeability infrastructure is discussed with several field examples from fault zones, particularly well-exposed faults in Iceland. The numerical models are used primarily to indicate how stresses concentrate within, and around, fault zones and how the stress concentration affects the associated permeability. A typical fault zone has two main hydromechanical units, namely a fault core and a fault damage zone (Figs. 16.1 and 16.2; Chapter 14). Since the core is primarily composed of breccias, it is modelled as porous media, whereas the damage zone, which is primarily composed of fractures, is modelled as fractured media. Fluid flow in porous media is a well-established field, with many analytical solutions available (Chapter 15). By contrast, many aspects of fluid transport in fractured media are still poorly understood. This is partly because fracture apertures are very sensitive to stress, that is, tend to change when the associated local stresses or fluid pressure change. And the local stresses, in turn, depend on the fluid pressure and the mechanical interactions between the fractures (Chapters 13 and 14). The permeability development of and fluid transport in fractured rocks, in particular in fractured reservoirs and fault zones,

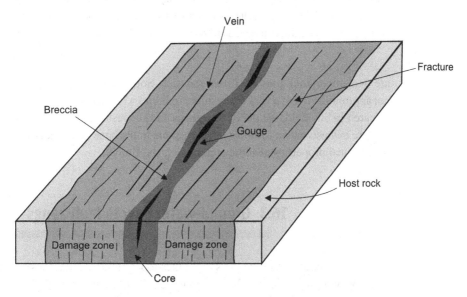

**Fig. 16.1**  Hydromechanical structure of a typical fault zone, here a strike-slip fault, focusing on the plan view. The zone consists of two main hydromechanical units: a fault core, mostly of breccias and gouge, and a fault damage zone, mostly of fractures (including veins) of various sizes and apertures/thicknesses.

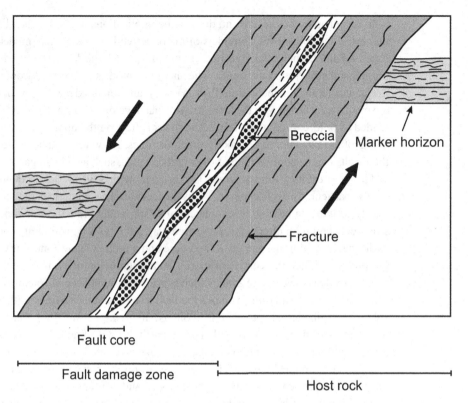

**Fault core**

**Fault damage zone**

**Host rock**

**Fig. 16.2** The hydromechanical structure of a typical fault zone has universal characteristics. Here the structure is shown for a dip-slip fault, focusing on the cross-sectional view. During interseismic (non-slipping) periods, the fractures in the damage zone are the main fluid conduits. However, during fault slip the temporary permeability of the core, and in particular the contact between the core and the damage zone, may increase by many orders of magnitude.

are thus important topics of current research. Some of these works are cited here, but it is not appropriate to have very many direct references in the text. To provide a guide to current research in this field, the list of references and suggested reading at the end of the chapter is more extensive than for most other chapters in the book (Appendix E.2 provides data on the physical properties of some common crustal fluids).

## 16.3 Ground-water flow to fault zones

Many fault zones occupy valleys and other topographic depressions and receive ground water by topography-driven flow. But fault zones occupying regions of gentle topography, or topographical highs, cannot collect ground water by means of this mechanism. Active faults, however, can collect ground water, rather than just acting as barriers to ground-water flow, by maintaining high permeability in the fault zone and its surroundings.

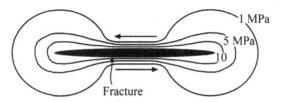

Fig. 16.3  Shear-stress concentration, in megapascals, around a sinistral strike-slip fault. The model is made using the boundary-element program Beasy (1991).

Concentration of stresses around fault zones, and associated permeability changes, is presumably one of the main reasons for changes in water table, water chemistry, and yield of springs prior to seismogenic fault slip, as are commonly observed (Roeloffs, 1996; Tsunogai and Wakita, 1995). Even more documented changes have occurred at and in the vicinity of seismogenic fault zones during the coseismic and postseismic periods (Muirwood and King, 1993; King *et al.*, 1999).

To explore the potential effects of loading a fault zone to failure on the surrounding permeability, consider a strike-slip fault (Fig. 16.3). In this model, a sinistral strike-slip fault zone is subject to fault-parallel loading of 6MPa, which is equal to the maximum in-situ tensile strength of most solid rocks, and also similar to the common stress drops in interplate earthquakes (Kanamori and Anderson, 1975; Amadei and Stephansson, 1997). The appropriate in-situ elastic properties of the uppermost part of a basaltic lava pile are used, namely 10 GPa for the Young's modulus and 0.25 for the Poisson's ratio (Chapter 9). Since the present fault is regarded as highly active, its core is assumed to consist of a weak, porous material, breccias and gouge, and accordingly it is modelled as an open, elliptic hole. This assumption in relaxed in many other fault models, where the core is modelled as, for example, an internal and soft elastic spring (Chapter 14; Gudmundsson *et al.*, 2009).

The fault-parallel loading of 6 MPa leads to symmetric shear-stress concentration around that fault zone (Fig. 16.3). However, the tensile-stress concentration (and the compressive-stress concentration) is asymmetric. That is, the tensile stresses concentrate in two quadrants around the fault zone (Fig. 16.4). The maximum tensile stress at the fault-zone tips reaches 50–60 MPa. This, however, is a theoretical value that depends on the geometry of the model and cannot be reached in nature because the maximum in-site tensile strength of most rocks is about 6 MPa. Thus, the tensile stress shown in Fig. 16.4 has a maximum value arbitrarily cut off at 10 MPa, as in many models in earlier chapters. Because the minimum in-situ tensile strength for most rocks is 0.5–1 MPa, the regions where the tensile stress exceeds 1 MPa may develop tensile fractures.

The contour showing one megapascal tensile stress is thus regarded as an outer limit for the effect of the local tensile stress around the fault close to or at slip on the permeability of its surroundings. This size estimate is conservative and the tensile-stress effects may extend to considerably larger areas. This follows because existing fractures concentrate stresses and, when favourably oriented, are likely to propagate at host-rock tensile stresses of a fraction of a megapascal (Chapter 6). At or near the surface these tensile

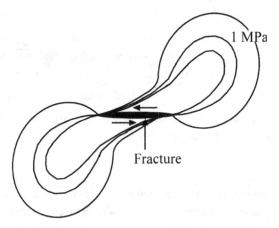

**Fig. 16.4** Tensile-stress concentration, in megapascals, around the sinistral strike-slip fault in Fig. 16.3. The model is made using the boundary-element program Beasy (1991).

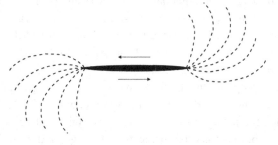

**Fig. 16.5** Stress trajectories (ticks) showing the trend of the maximum principal compressive stress, $\sigma_1$, around the ends of the sinistral strike-slip fault in Fig. 16.3. Tension and extension fractures tend to follow the $\sigma_1$-trajectories. The model is made using the boundary-element program Beasy (1991).

fractures would be pure mode I (opening mode) cracks, but at crustal depths exceeding a few hundred metres they would change into normal faults (Chapter 8). The formation of new, and propagation of existing, tensile fractures increases the porosity of the rocks in these quadrants, and if the fractures are interconnected they also increase the rock permeability.

Tensile fractures and normal faults develop parallel with the direction of the maximum principal compressive stress, $\sigma_1$ (Chapter 8). For the given loading conditions, the ticks in Fig. 16.5 indicate schematically the trend of $\sigma_1$. Clearly, the tensile fractures (and at greater depths, hydrofractures) following the $\sigma_1$-ticks or trajectories would curve towards the fault-zone tips. Some of these fractures are of the type commonly referred to as wing cracks (Fig. 16.6). When the general direction of ground-water flow coincides with the trend of the fault zone (Fig. 16.7), the permeability increase due to the loading (Figs. 16.4–16.6) results in ground water being channelled to the upstream tip of the fault zone and expelled from its downstream tip. As a consequence, the yield of springs and associated streams

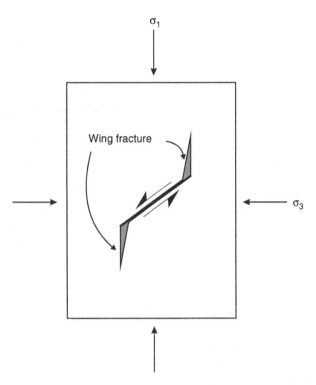

**Fig. 16.6**  Wing fractures (wing cracks) form in areas of tensile-stress concentration, transtension areas, around faults. They are very common around strike-slip as well as dip-slip faults in rocks of various types and environments, including sedimentary basins. Wing-fractures connected with faults may greatly increase the permeability of the rock (cf. Fig. 12.13).

should increase in the downstream region surrounding the tip, but decline in the upstream region surrounding the tip (Fig. 16.8).

Similar effects have been observed in numerous areas of active faulting (e.g. Muirwood and King, 1993). A particularly good example of such effects occurred during earthquakes in the South Iceland Seismic Zone in 2000. Two dextral strike-slip faults occurred in June 2000, each with a magnitude M6.5, and the early one triggering, within a few days, the second one (Arnadottir *et al.*, 2003; Jonsson *et al.*, 2003). During the earthquakes, the water level in geothermal wells in the areas of tensile-stress concentration (as in Fig. 16.4) fell, whereas the water level rose in the areas of transpression, located in the opposite quadrants.

The same kind of effects have been observed on a much larger scale, for example when the South Iceland Seismic Zone is treated as a single fault zone (Gudmundsson and Brenner, 2003). Then it can be shown, using similar models to those above, that the sinistral crustal movement across the zone as a whole generates transtension in two quadrants (Fig. 16.9). The transpression quadrants may be expected to concentrate compressive stresses, which results in crustal doming and increases the (micro)seismic activity before major earthquakes in the seismic zone. If the geothermal fields, magma chambers, and other

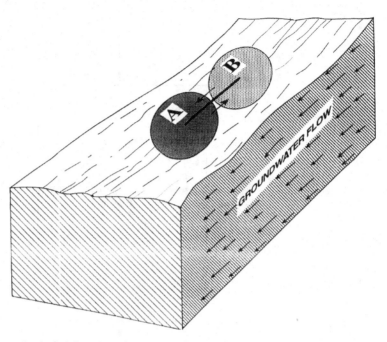

For an active strike-slip fault (here sinistral) that strikes parallel with the hydraulic gradient (the direction of ground-water flow), the upstream part (B) tends to collect water, channel it along the fault plane, and then expel it in the downstream part (A).

fluid-filled reservoirs are located within the transpression zones, then the 'squeezing' effect may change the geothermal activity, increase the seismicity, and result in crustal doming. Similarly, the transtension zones, if they coincide with central volcanoes, may result in unusually frequent eruptions prior to major earthquakes in the seismic zone. All these processes have been observed in three of the four quadrants around the South Iceland Seismic Zone that include active central volcanoes (Gudmundsson and Brenner, 2003). Thus the Hekla Volcano (in a transtension zone) erupted unusually frequently, and the Hengill Volcano and the Eyjafjallajökull Volcano (both in transpression areas) were subject to crustal doming and about 100 000 small earthquakes in the period up to the M6.5 earthquakes in the year 2000. Following the earthquakes in the year 2000, and the associated stress relaxation, doming in the Hengill Volcano ceased and the relaxation may have contributed to the triggering of an eruption in the Eyjafjallajökull Volcano in the year 2010.

## 16.4 Ground-water flow along fault zones

In order to analyse fluid transport along a fault zone, and in particular the changes in permeability of a zone following fault slip, the hydromechanical infrastructure of the fault zones must be considered. This follows because the two main structural units of a major fault

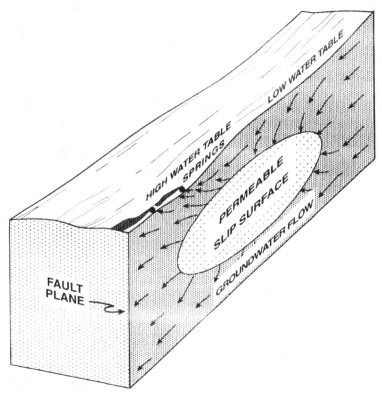

For an active strike-slip fault (here a sinistral fault), just prior to and following a slip, the water level in wells in the quadrant that concentrates the tensile stresses (transtension) in the upstream part falls, whereas the water level in the wells in the quadrant that concentrates compressive stresses (transcompression) rises.

zone, the core and the damage zone (Figs. 16.1 and 16.2), are very different hydromechanical units. Not only are their infrastructures, and therefore the associated flow patterns, different as regards local permeabilities and the scale of the fluid transport network, but the bulk properties are also different. Using the above results as a basis, the general conceptual hydromechanical model in Fig. 16.10 may be representative for major strike-slip fault zones. The model can, however, easily be extended to include dip-slip fault zones as well. In this model, the core is modelled as a porous medium during the interseismic period, but as a fractured medium during fault slip, whereas the damage zone is modelled at all times as a fractured medium. This is based on field observations of major strike-slip fault zones, where the core is occasionally fractured and conducts geothermal fluids, as indicated by vein networks.

For the fault core, Darcy's law for fluid transport in porous media may be written, using Eqs. (15.3) and (15.6) and omitting the minus sign indicating flow in the direction of decreasing head, in the form

$$Q = K_p A \nabla h \tag{16.1}$$

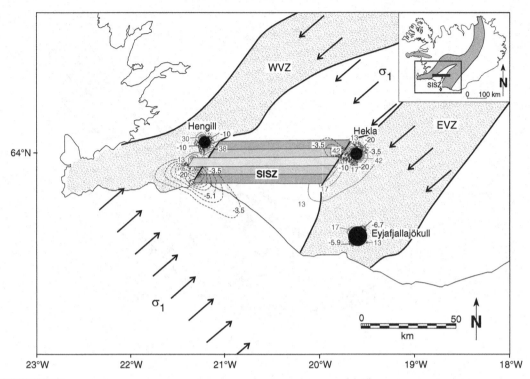

**Fig. 16.9**   South Iceland Seismic Zone (SISZ) is a major zone of general sinistral movement. That is, the crustal part north of it moves to the left and the crustal part south of it moves to the right. This is a finite-element model (see www.ansys.com) of the compressive and (relative) tensile stresses around the SISZ and its nearby composite volcanoes when the zone is loaded by a uniaxial compressive stress of 10 MPa at an angle of 40° to the zone. The contours show maximum principal compressive stress $\sigma_1$ and the minimum principal compressive (maximum tensile) stress $\sigma_3$ in megapascals. The maximum compressive stresses shown as contours are between 10 and 30 MPa for $\sigma_1$ and between 1 and 10 MPa for $\sigma_3$.

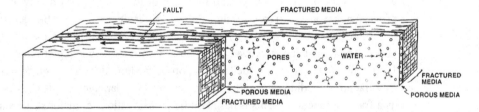

**Fig. 16.10**   Internal hydromechanical structure of the strike-slip (here dextral) fault. The core is composed of porous media and the damage zone of fractured media, and both (particularly the damage zone) may consist of several subzones with different mechanical properties (different rock types and different pore and fracture frequencies). When fault slip occurs, all the numerous small cracks and pores in the core that meet with the slip plane become interconnected along that plane. This is a major reason why the permeability of the core may temporarily increase by many orders of magnitude following the slip.

where $Q$ is the volumetric flow rate (volume of flow in unit time), $K_p$ is the hydraulic conductivity of the porous medium, $A$ is the cross-sectional area normal to the flow, and $\nabla h$ is the hydraulic gradient, where $\nabla$ is the operator nabla. The hydraulic conductivity of a porous medium with a mean (or effective) grain diameter $d$, from Eq. (15.4), is

$$K_p = \frac{N\rho_f g d^2}{\mu_f} \tag{16.2}$$

where $N$ is a dimensionless shape factor, $\rho_f$ the density of the fluid, $g$ the acceleration due to gravity, and $\mu_f$ the dynamic (or absolute) viscosity of the fluid. The property of the porous medium itself is given by the intrinsic permeability, $k = Nd^2$, which is related to $K_p$ through

$$K_p = \frac{k\rho_f g}{\mu_f} \tag{16.3}$$

The field data discussed above suggest that a damage zone normally behaves as a fractured medium. For a damage zone consisting of a set of smooth, parallel, planar vertical fractures striking in the direction of the fluid flow (Fig. 15.4) in the fault zone, the hydraulic conductivity, $K_f$, of the damage zone is, from Eq. (15.21), given by

$$K_f = \frac{\rho_f g}{12\mu_f L} \sum_{i=1}^{m} b_i^3 \tag{16.4}$$

where $\rho_f$ and $\mu_f$ are the density and the dynamic (or absolute) viscosity, respectively, of the fluid, $g$ is the acceleration due to gravity, and $L$ the length of the measured horizontal profile (of unit height) that dissects $m$ fractures, each with an aperture $b_i$ (Fig. 15.4). The corresponding equation for the volumetric flow rate $Q$ through a damage zone, from Eq. (15.20), is

$$Q = \frac{\rho_f g}{12\mu_f} \left( \sum_{i=1}^{m} b_i^3 \right) \nabla h \tag{16.5}$$

where all the symbols are as defined above. In an ideal case, all the fractures have the same aperture $b$, and Eq. (16.4) simplifies (cf. Eq. 15.18) to

$$K_f = \frac{\rho_f g m b^3}{12\mu_f L} \tag{16.6}$$

and the volumetric flow rate to

$$Q = \frac{\rho_f g m b^3}{12\mu_f} \nabla h \tag{16.7}$$

Some damage zones consist of two sets (rather than one) of roughly orthogonal fractures (Fig. 16.11). The hydraulic conductivity $K_f$ and the volumetric flow rate $Q$ for such zones can then be estimated using a model of two sets of smooth, parallel planar fractures meeting at right angles, with both sets striking in the direction of the associated fluid flow (Figs. 15.5

**Fig. 16.11**  Network of roughly orthogonal mineral veins in the damage zone of a transform fault (the Husavik-Flatey Fault) in North Iceland (cf. Gudmundsson *et al.*, 2002). Fracture networks of this type may be modelled crudely as in Fig. 15.5. The length of the measuring tape is 1 m.

and 16.10). From Eq. (15.25), the total hydraulic conductivity is then

$$K_\mathrm{f} = \frac{\rho_\mathrm{f} g}{12 \mu_\mathrm{f} L H} \left( \sum_{i=1}^{m} b_i^3 + \sum_{j=1}^{n} b_j^3 \right) \tag{16.8}$$

where the horizontal profile of length $L$ is dissected by $m$ fractures of varying apertures $b_i$, the vertical profile of height $H$ is dissected by $n$ fractures of varying apertures $b_j$, and the other symbols have the same meaning as above. From Eq. (16.1), with $A = L \times H$, and Eq. (16.8), the volumetric flow rate $Q$ through a damage zone consisting of these two orthogonal sets of fractures is

$$Q = \frac{\rho_f g}{12\mu_f} \left( \sum_{i=1}^{m} b_i^3 + \sum_{j=1}^{n} b_j^3 \right) \nabla h \tag{16.9}$$

Equations (16.8) and (16.9) can be simplified in a way analogous to Eqs. (16.6) and (16.7) in the ideal case of all the fractures having the same aperture (Bear, 1993). In their present forms, however, these equations are useful to model damage zones where the fracture apertures have the very commonly observed log-normal or power-law size distributions (or some other negative exponential aperture/thickness distribution), such as the veins in Fig. 16.11 and related veins in that fault zone (Gudmundsson et al., 2002).

## 16.5 Aperture and fluid pressure

The flow of ground water in porous media is commonly taken to be independent of the fluid pressure with respect to the stress field. This means that the potential effects of the fluid pressure on the permeability of the rock through which the fluid flows are assumed to be negligible. This assumption may often be justified in the case of fluid flow in porous media, particularly as regards ground-water flow at shallow depths, but can lead to errors in the case of fluid flow through fractured media, in particular for fluid flow in the damage zones of active fault zones. The assumption is also not normally valid for poroelastic reservoirs (Wang, 2000), such as oil and gas reservoirs at considerable depths (several kilometres). Here, however, the focus is on the effects of fluid pressure on fracture apertures.

Consider a fluid-filled extension (mode I) fracture (Figs. 8.2 and 9.16) with a strike dimension smaller than its dip dimension so that the strike dimension or length, $L$, controls the opening of the fracture and is its characteristic dimension. If the fracture is a through-crack, that is, extends through the rock layer hosting it, then its aperture $b$ is normally equal to its opening displacement, $\Delta u_I$, and related to the overpressure $p_0$ of the fluid through the plane-strain equation (Eq. 9.13)

$$\Delta u_I = b = \frac{2p_0(1 - v^2)L}{E} \tag{16.10}$$

where $E$ is Young's modulus of the rock and $v$ its Poisson's ratio. Note that while the aperture of a mode I crack is normally directly related to its opening displacement, no such relationship exists between $b$ and the mode II and mode III displacements. That is, $b \neq \Delta u_{II}$ and $b \neq \Delta u_{III}$.

If the fluid-filled extension fracture is a part-through (thumbnail) crack, that is, goes only partly into the rock layer hosting it (Chapter 9), and its strike dimension is greater than its dip dimension, the controlling (characteristic) dimension is the dip dimension $R$.

By analogy with Eq. (9.14) and substituting $b$ for $\Delta u_I$, a plane-stress mode I crack model gives the following equation for the fracture aperture:

$$b = \frac{4 p_0 V R}{E} \tag{16.11}$$

In this equation, the function $V\left(R/T\right)$, where $T$ is the total thickness of the rock layer hosting the fracture, is defined (using radians) as (Eq. 9.10)

$$V\left(\frac{R}{T}\right) = \frac{1.46 + 3.42\left[1 - \cos(\pi R/2T)\right]}{\left[\cos(\pi R/2T)\right]^2} \tag{16.12}$$

Equations (16.10) and (16.11) show that fluid overpressure can have a great effect on the aperture of water-conducting fractures in a fault zone. Because the volumetric flow rate, $Q$, depends on the third power of the aperture, as is indicated by the cubic law (Eqs. 15.13 and 16.5), any changes in the fracture aperture due to fluid-overpressure changes, or changes in the local stress field, can have a great effect on the volumetric flow rate through a single fracture or a fracture set. This latter is indicated when we consider fracture sets (Figs. 15.4 and 15.5) where the apertures may vary (Eqs. 16.5 and 16.9). There are, however, no analytical solutions available for the volumetric flow rate of fluids through complex fracture sets where the apertures change as the fluid overpressure and/or the local stresses change. The development of numerical models that can deal with complex fracture systems under these boundary conditions is currently a field of active research. In the absence of such analytical solutions, the fluid transport in the damage zone of a fault zone can only be estimated very crudely using some simplifying assumptions, such as assuming that the aperture stays constant during changes in fluid overpressure. Such a simple approach, with the aim of clarifying the main concepts, is made in the next sections. But it should be kept in mind that when fracture sets, such as those in the damage zones of major fault zones or in fractured reservoirs, are subject to fluid overpressure or local stress changes, the fault zone hydraulic conductivity and transmissivity may change substantially.

## 16.6  Fluid transport in the damage zone

The fracture systems in the damage zone can be complex (Figs. 16.11 and 16.12). Field studies indicate that most of the veins are pure extension fractures (Figs. 15.17 and 16.11). In fact, in the damage zone of the transform fault in North Iceland, part of which is seen in Figs. 16.11 and 16.12, some 80% of the veins are extension fractures (hydrofractures), the remaining 20% being shear fractures, that is, small faults (Gudmundsson *et al.*, 2002) as some of the mineral-filled fractures in Fig. 16.12.

For a typical vein, 60–70 cm long (strike dimension) (Fig. 9.16; Gudmundsson *et al.*, 2002), we make the following observations. An isolated, non-restricted vein of this length would have a typical maximum thickness (a former aperture) of some 2–3 mm. Based on common aspect ratios of extension fractures formed by external tensile stresses rather than internal fluid pressure (Chapter 9), such a fracture may, if formed by tensile stress, have

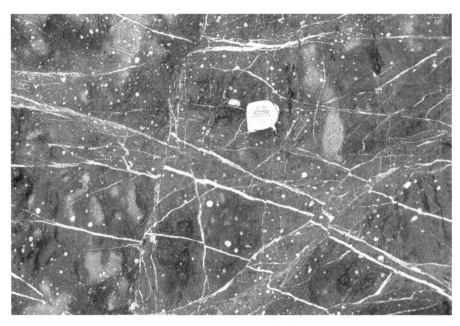

**Fig. 16.12**   Network of mineral veins and small faults (filled with minerals) in a basaltic lava flow of the damage zone of a transform fault (the Husavik-Flatey Fault) in North Iceland (cf. Gudmundsson *et al.*, 2002). View vertical, the conjugate strike-slip faults have displacements of 1–2 cm, whereas most of the mineral veins are extension fractures with thicknesses/apertures of a few millimetres or less. The measuring steel tape, 6 cm in diameter, provides a scale.

had a maximum aperture of 0.5–1 mm before injections of overpressured geothermal fluids increased it to its present value of, say, 3 mm. From Eqs. (16.6) and (16.7) it follows that the hydraulic conductivity of a single fracture, $K_c$, is given by

$$K_c = \frac{\rho_f g b^2}{12 \mu_f} \tag{16.13}$$

and, similarly, that the volumetric rate of water $Q_c$ through a crack of unit controlling (smaller) dimension is

$$Q_c = \frac{\rho_f g b^3}{12 \mu_f} \nabla h \tag{16.14}$$

From Eq. (16.13) it follows that, other things being equal, increasing the aperture from 1 mm to 3 mm would increase the hydraulic conductivity of the fracture/vein $K_c$ by a factor of 9, and, from Eq. (16.14), that the volumetric rate of flow, $Q_c$, would then increase by a factor of 27. Similarly, increasing the fracture aperture from 0.5 mm to 3 mm would increase $K_c$ by a factor of 36 and $Q_c$ by a factor of 216.

These results can easily be extended to fracture sets. Selecting 379 well-exposed, non-restrictive extension (mode I) veins from the on-land parts of the Husavik-Flatey Fault gives an average aspect (length/thickness) ratio of about 400 (Gudmundsson *et al.*, 2002). This indicates fluid overpressure, with reference to the minimum compressive principal stress, during vein emplacement of about 20 MPa (Chapter 17; Gudmundsson *et al.*, 2002).

For similar fractures generated by an external tensile stress, rather than by internal fluid overpressure, the fracture-forming tensile stress would commonly be similar to the in-situ tensile strength, 1–6 MPa, with an average of 3–4 MPa (Chapter 7).

Equations (16.10) and (16.11) show that, for a given fracture length, there is a linear relationship between fluid overpressure and fracture aperture. It follows that increasing the fracture-opening fluid overpressure from 4 MPa to 20 MPa increases the average fracture aperture by a factor of 5. Other things being equal, it follows from Eq. (16.7) that for a uniform aperture, the volumetric flow rate for the damage zone as a whole, as a consequence of its fracture network being subject to this fluid pressure increase, would be greater by a factor of 125.

A damage zone subject to fluid overpressure may thus easily transport 100 to 200 times as much water through a given fracture set as would be transmitted through that set during periods of no fluid overpressure in the fault zone. Large temporary changes in the ground-water transport along fault zones are common (King et al., 1999; Leonardi et al., 1998) and are, presumably, partly related to overpressured fluids in the fault zones. Seismogenic faulting is commonly associated with zones of fluid overpressure, although the origin of the fluid overpressure is not clear (Sibson, 1996; Melchiorre et al., 1999; Gudmundsson et al., 2002). Fault slip, however, affects not only the fluid flow in the damage zone, but also that in the fault core.

## 16.7  Fluid transport in the core

As indicated above, the fault core behaves as low-permeability porous media during the interseismic period, but as fractured media during fault slip along the core. Evans et al. (1997) and Seront et al. (1998) made laboratory tests of the permeability of fault-core materials, from a thrust fault and a normal fault, which should also be applicable to strike-slip faults. The results, based on small specimens, yield permeabilities, $k$, from $10^{-17}\,\mathrm{m}^2$ to $10^{-20}\,\mathrm{m}^2$. Using Eq. (16.3), with $g = 9.8\ \mathrm{m\ s}^{-2}$, $\rho_f = 1000\ \mathrm{kg\ m}^{-3}$, and, from Giles (1977), $\mu_f \approx (1.8 - 0.6) \times 10^{-4}$ Pa s (for water at 0°C and 50°C, respectively), we get $K_p = (0.54 - 1.6)Ck$. The constant $C = 10^7$ has units of $\mathrm{m}^{-1}\ \mathrm{s}^{-1}$. For water at 21°C, which is a rather typical temperature of water in the upper parts of many active fault zones, we get $\mu = 9.97 \times 10^{-4}$ Pa s and the approximate relation

$$K_p \cong Ck \qquad\qquad (16.15)$$

From this relationship between hydraulic conductivity $K_p$ and permeability $k$, the cited core tests give $K_p = 10^{-10}-10^{-13}\ \mathrm{m\ s}^{-1}$. These values, although based on centimetre-scale laboratory samples and therefore including few if any fractures larger than a few millimetres, are presumably not much lower than common in-situ values (Evans et al., 1997). This range of hydraulic conductivity overlaps with that of typical laboratory values for sediments such as silt and clay and sedimentary rocks such as siltstone and the lowest values of sandstone (Domenico and Schwartz, 1998). The reported fault-core laboratory values are similar to the lowest in-situ values for rock masses (Lee and Farmer, 1993).

Generally, the results lend further support to treating the fault core during the interseismic period as a porous medium of a very low hydraulic conductivity.

During fault slip, however, the hydraulic conductivity, and the associated transmissivity, of the core may temporarily increase dramatically. As an example, consider a seismogenic fault slip in the core that gives rise to a fault plane with a maximum aperture of 0.1 mm. This aperture is similar to the thinnest measurable mineral veins in the data set from the transform in North Iceland (Figs. 16.11 and 16.12), and in fault zones in sedimentary basins (Figs. 16.13–16.15). The fault slip could occur either somewhere inside the fault core or at its contact with the damage zone (Figs. 16.1, 16.2, and 16.10). The general transmissivity of the porous, interseismic fault core, $T_p$ is given by

$$T_p = K_p b_p \tag{16.16}$$

where $K_p$ is the average hydraulic conductivity of the porous fault core material (Eqs. 16.2 and 16.3) and $b_p$ is its thickness. From Eqs. (16.13) and (16.16) it can be inferred that the transmissivity $T_f$ of the fault-slip surface is

$$T_f = K_c b = \frac{\rho_f g b^3}{12\mu_f} \tag{16.17}$$

where $K_c$ is the hydraulic conductivity of a single fracture or a fault-slip surface, as defined in Eq. (16.13), and all the other symbols are as defined above.

To minimise the effects of fault slip on ground-water transmissivity, we not only assume the aperture of the resulting fault plane to be very small, but also use the maximum hydraulic conductivity values for the interseismic fault core. From Eq. (16.17), with $g = 9.8$ m s$^{-2}$, $\rho_f = 1000$ kg m$^{-3}$, $b = 0.0001$ m (0.1 mm) and $\mu_f = 9.97 \times 10^{-4}$ Pa s (for water at 21°C), we obtain the transmissivity of the fault plane as $T_f = 8.2 \times 10^{-7}$ m$^2$ s$^{-1}$. From Eq. (16.14) it follows that a porous fault core with the maximum hydraulic conductivity of $10^{-10}$ m s$^{-1}$ (the maximum value obtained above) would have to be 8200 m thick to transmit the same amount of water. This is of course a completely unrealistic core thickness, since most have thicknesses of less than a few tens of metres.

From these results we can make the following conclusions:

- The highest measured hydraulic conductivities of interseismic fault cores are very low, indicating that a fault core would normally act as a barrier to ground-water flow across the fault zone. This means that, during the interseismic period, the fault core tends to channel ground water along the damage zone.
- During fault slip, even if the resulting slip-surface aperture is a fraction of a millimetre, the ground-water transmissivity of a typical fault-zone core can increase, temporarily, by many orders of magnitude, probably commonly in the range $10^6$–$10^9$ times. While the transmissivity of the damage zone is also likely to increase prior to and during seismogenic faulting, an enormous increase in the transmissivity of the fault core may be one main reason for the commonly observed and inferred large-volume fluid transport associated with fault slip.

**Fig. 16.13**    View west, a dense network of calcite veins in a normal-fault zone in a sedimentary basin at Kilve, on the Somerset coast, Southwest England. The central part is the core, the outer parts the damage zone. The veins extend only for a few metres into the damage zone and are mostly confined to the stiff limestone layers (seen here), but rare in the soft shale layers in between the limestone layers. The vein density is highest in the hanging (right) wall of the fault zone (cf. Fig. 16.2).

## 16.8 Permeability development

Fault zones, or parts of them, can act either as barriers to or conduits for the flow of ground water and other crustal fluids. Observations and modelling indicate that inactive fault zones rapidly become sealed, through the precipitation of minerals as veins or amygdales, and healed, through chemical processes such as dissolution of minerals

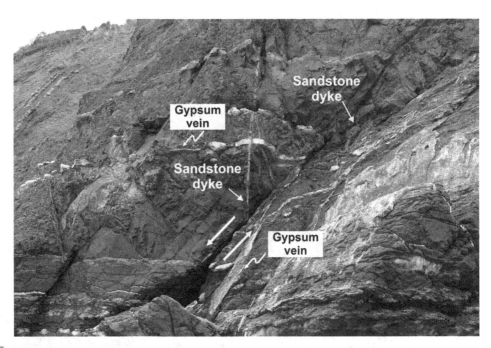

**Fig. 16.14** View southeast, mineral veins, of gypsum, and dip-slip fault (occupied with minerals) and a 3-cm-thick sandstone dyke in mudstones at Watchet, on the Somerset coast, Southwest England (Philipps, 2008). Mudstones commonly have a very low permeability, but well-interconnected networks of hydrofractures and faults may generate a high temporary permeability (cf. Fig. 16.15).

in high-energy environments in the vicinity of a fracture and precipitation of the material in low-energy environments within the fracture (Fisher and Brantley, 1992; Olsen *et al.*, 1998). Although there are other processes, such as creep and compaction (Sleep and Blanpied, 1992), that may reduce the hydraulic conductivity of a fault zone, healing and sealing are probably among the most significant. Fault zones that are inactive and subject to these processes would thus normally rapidly develop into barriers to the flow of ground water.

In hydrogeology, faults where the bulk hydraulic conductivity approaches zero are referred to as 'tight' (Shan *et al.*, 1995). By contrast, if the hydraulic conductivity approaches 'infinity', the fault is referred to as 'constant head'. It is referred to as 'leaky' if the conductivity is somewhere between these extremes. Most inactive faults develop through healing, sealing, creep, and compaction into tight faults. The barriers formed by tight faults may collect water and direct it to the surface, where it appears as springs. However, in that case the water is normally not conducted along the infrastructure of the fault zone itself but rather along its margins. By contrast, in many, and presumably most, active fault zones the hydromechanical units of the fault zone transmit the water.

Here we have divided the hydromechanical structure of fault zones into two contrasting units, the core and the damage zone, as is generally accepted. The fluid-transport analysis

Fig. 16.15 View east, a close-up of a gypsum-vein network at Watchet, Southwest England (cf. Fig. 16.14). Cross-cutting relationships and mineral orientation show that most of the veins are pure extension fractures and were thus formed as fluid-driven fractures. The measuring tape in the left-lower corner of the photograph is 1 m long.

in this chapter models the core as a porous medium and the damage zone as a fractured medium. This approach to modelling the main hydromechanical units of a fault zone, although schematic (for example, because the damage zone may also contain lenses of breccias, that is, porous media) is a useful conceptual presentation and allows approximate quantitative estimates of the volumetric flow rate of ground water (and other crustal fluids) in a fault zone. Here the model has been mainly applied to major strike-slip fault zones, but it is easily extended to major dip-slip fault zones. Some small-displacement fault zones lack porous-media fault cores. In such fault zones, the central parts consist of dense fracture systems rather than breccias or cataclastic rocks. For such small-displacement fault zones, the part of the model applicable to the damage zone may account for the hydraulic conductivity and transmissivity of the entire fault zone.

Many studies suggest that major fault zones may largely govern ground-water flow in bedrock on a regional scale (Mayer and Sharp, 1998; Ferill *et al.*, 1999). There is also evidence that, on a local scale, the hydraulically conductive fractures in a rock mass are primarily the active, or potentially active, small-scale faults (Barton *et al.*, 1995; Finkbeiner *et al.*, 1997). The effects of shear stress on fracture aperture are complex and depend on the associated normal stress and surface roughness of the fracture. In terms of the present model, however, the observation that potentially active faults appear to be the fractures most commonly hydraulically conductive may be partly attributed to the stress concentration around these faults. Active faults at any scale concentrate stresses, in a manner similar to that presented in Figs. 16.3–16.5, and 16.9, and thus collect ground water and other crustal fluids (such as geothermal water, gas and oil) from their surroundings. The permeability of an active fault zone is maintained by fault slip.

The most favourably oriented potentially active small-scale faults are also those that are most likely to link up into major fault zones (Chapter 14). Therefore, these fractures have the greatest probability of becoming interconnected and reaching the percolation threshold (Chapter 15; Stauffer, 1985). We know that for this to happen, the fractures must form interconnected clusters (Figs. 16.11–16.15). How this happens, that is, how interconnected fracture clusters form, is one of the main current research activities on fluid transport in fault zones and fractured reservoirs (Fig. 16.16; Chapter 15; Rawling et al., 2001; Lin et al., 2007; Larsen et al., 2009, 2010; Paul et al., 2009).

The major conclusions of this chapter include, first, that during the preseismic activity of strike-slip fault zones ground water is channelled into the fault zones, and, second, that temporary water transmissivity may increase by many orders of magnitude during the subsequent fault slip. The predicted preseismic effects on the surrounding permeability are supported by direct measurements of water-level changes in geothermal wells in South Iceland during earthquakes (Arnadottir et al., 2003; Jonsson et al., 2003) and by general observations of changes in ground water wells prior to and following earthquakes (Muirwood and King, 1993). The details of the transmissivity increase associated with fault slip, which part relates to the increased hydraulic conductivity of the damage zone and part to the increased conductivity of the core, are topics currently of great interest, particularly with regard to major, active fault zones, with thick cores and damage zones.

## 16.9 Summary

- Major fault zones consist of two main structural units: a fault core and a fault damage zone. The core, where most of the fault displacement is accommodated, consists mostly of breccia and gouge and, for major zones, is commonly up to a few tens of metres thick. On either side of the core is a damage zone, as much as several hundred metres thick. The damage zone includes numerous faults and fractures, many of which eventually become filled with secondary minerals, but lacks widespread breccias. During the interseismic period of an active fault zone, the fault core normally behaves as a porous medium with hydraulic conductivity which may be somewhere between $10^{-10}$ and $10^{-13}$ m s$^{-1}$. By contrast, the damage zone behaves as a parallel-fractured medium with a normal hydraulic conductivity many orders of magnitude higher than that of the interseismic core.
- During interseismic periods, stresses concentrate around active faults. In the chapter, the main examples used are active strike-slip fault zones, but the results are easily extended to include other types of faults. Fault-parallel loading prior to fault slip generates tensile stresses in large areas, that is, areas of transtension around the fault-zone tips. The resulting interconnected tensile fractures increase the rock permeabilities and porosities in the transtension areas. The tensile fractures curve towards,

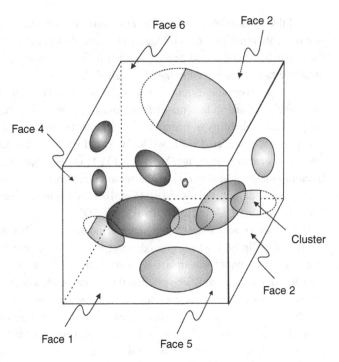

**Fig. 16.16** Modelling fractured rocks such as in Figs. 15.18, and 16.11–16.15 requires measurements of numerous fractures in sections with different trends (e.g. Figs. 15.18, 16.14, and 16.15). The measurements include fracture attitude (strike and dip), aperture, and height (dip dimension) and/or length (strike dimension). The measured fractures and their strike, dip, aperture, and size distributions are then extrapolated, using a numerical program, into the rock body to be modelled (a part of a fault zone or a reservoir, for example). That is to say, the hypothetical fractures inside the body are supposed to have the same attitudes and size distributions as the measured fractures. Commonly the fractures are assumed to be penny-shaped or somewhat elliptical (Chapter 9). Depending on their attitudes and size distributions, the extrapolated fractures may form one or many interconnected clusters where the percolation threshold is reached and which can thus conduct fluids. From the frequency and characteristics of these clusters, the overall fracture-related permeability of the rock body is calculated theoretically and the results compared, if possible, with in-situ measurements.

and channel ground water into the fault zone. As a consequence, the temporary fluid pressure in the fault zone may increase greatly, which in turn may increase the apertures of the associated fractures and thereby the hydraulic conductivity and fault-zone transmissivity.

- The aspect ratios (length/maximum thickness) of several hundred, non-restricted mineral-filled veins in the damage zone of a strike-slip fault zone, a transform fault, in North Iceland indicate an average fluid overpressure during their development, with reference to the minimum principal compressive stress, of about 20 MPa. Overpressured fluids increase fracture apertures so that the volumetric flow rate, $Q$, can become several hundred times greater than during fluid flow under hydrostatic pressure. Increased fluid pressure in a fault zone may also trigger fault slip along the fault core, which, in turn, may increase the core transmissivity (hydraulic conductivity times core thickness) by many orders of magnitude. For example, a single fault plane along the core with an aperture of only 0.1–1 mm can transmit many hundred to many hundred thousand times more water than a porous fault core, tens of metres thick and with a high hydraulic conductivity of $K = 10^{-10}$m s$^{-1}$. This partly explains the great volumes of water that are commonly inferred or observed to be transported during seismic activity in large-scale fault zones.

# 16.10 Main symbols used

| | |
|---|---|
| $A$ | cross-sectional area normal to the fluid (water) flow |
| $b$ | fracture aperture (for a mode I crack, equal to $\Delta u$) |
| $C$ | constant |
| $d$ | mean diameter of grains in granular media (sand in Darcy's experiments) |
| $E$ | Young's modulus |
| $g$ | acceleration due to gravity |
| $H$ | height of a vertical profile dissected by fractures |
| $\nabla h$ | gradient (the rate of change) of the scalar field $h$ ($\nabla$ is del or nabla) |
| $K_f$ | hydraulic conductivity of a fractured rock body |
| $K_p$ | hydraulic conductivity of porous rock (or medium) |
| $k$ | intrinsic permeability |
| $L$ | length; length of a fracture profile |
| $L$ | strike dimension (length) of a fracture in Eq. (16.10) |
| $m$ | number of fractures dissecting a profile |
| $N$ | dimensionless shape factor |
| $p_0$ | fluid overpressure (in a fracture) |
| $Q$ | volumetric flow rate, volume of fluid per unit time |
| $Q_f$ | volumetric flow rate through a fractured rock body |
| $R$ | dip dimension of a fracture |
| $T$ | total thickness of a crustal segment or a rock body hosting a part-through crack |
| $T_f$ | transmissivity of a fractured rock (or medium) |
| $T_p$ | transmissivity of a porous rock (or medium) |
| $V$ | function for a part-through crack |
| $\mu_f$ | dynamic or absolute viscosity of a fluid |
| $\mu_w$ | dynamic or absolute viscosity of water |
| $\nu$ | Poisson's ratio |
| $\rho_f$ | fluid density |

# 16.11  Worked examples

### Example 16.1

**Problem**

A 100-m-thick porous ground-water aquifer has an average hydraulic conductivity of $1 \times 10^{-6}$ m s$^{-1}$, which is a typical in-situ value for many sandstones, fine sands and tills (Lee and Farmer, 1993). The density of the ground water is 1000 kg m$^{-3}$ and its dynamic viscosity is $1.55 \times 10^{-3}$ Pa s. What must be the aperture (opening) of a fracture, of otherwise equal dimensions to that of the aquifer, in order for the fracture to transmit the same amount of water as the aquifer?

**Solution**

From Eq. (16.16) the transmissivity of a porous aquifer $T_p$ is

$$T_p = K_p b_p$$

where $K_p$ is the average hydraulic conductivity of the aquifer and $b_p$ its thickness. Similarly, from Eq. (16.17) the transmissivity of a fracture $T_f$ is

$$T_f = K_f b = \frac{\rho_f g b^3}{12 \mu_f}$$

where, as before, $\rho_f$ and $\mu_f$ are the density and the dynamic viscosity, respectively, of the ground water, $g$ is the acceleration due to gravity, $b$ is the aperture or total opening of the fracture, and $K_f$ is the hydraulic conductivity of the fracture. From Eq. (16.16) the transmissivity of the porous aquifer is

$$T_p = K_p b_p = 1 \times 10^{-6} \text{m s}^{-1} \times 100 \text{ m} = 1 \times 10^{-4} \text{m}^2 \text{s}^{-1}$$

Using the values for water viscosity and density, Eq. (16.17) gives the fracture transmissivity as

$$T_f = K_f b = \frac{\rho_f g b^3}{12 \mu_f} = \frac{1000 \text{ kg m}^{-3} \times 9.81 \text{ m s}^{-2} \times b^3}{12 \times 1.55 \times 10^{-3} \text{Pa s}} = 5.274 \times 10^5 \text{m}^{-1} \text{s}^{-1} \times b^3$$

Since the transmissivity of the fracture is supposed to be equal to that of the aquifer, we have

$$T_f = T_p = 1 \times 10^{-4} \text{m}^2 \text{s}^{-1}$$

So that

$$5.274 \times 10^5 \text{m}^{-1} \text{s}^{-1} \times b^3 = 1 \times 10^{-4} \text{m}^2 \text{s}^{-1}$$

Or

$$b^3 = \frac{1 \times 10^{-4} \text{m}^2 \text{s}^{-1}}{5.274 \times 10^5 \text{m}^{-1} \text{s}^{-1}} = 1.89 \times 10^{-10} \text{m}^3$$

Thus

$$b = (1.89 \times 10^{-10} \text{m}^3)^{1/3} = 0.000575 = 0.575 \text{ mm}$$

These calculations show that a fracture with an aperture of about 0.6 mm can conduct as much water as a 100-m-thick porous aquifer with a hydraulic conductivity of $1 \times 10^{-6}$ m s$^{-1}$. These results agree well with other studies and measurements.

## Example 16.2

### Problem

A fault core has normally a very low hydraulic conductivity. This follows because the core is partly composed of low-permeability breccias and gouge (a clay-like material). Laboratory estimates indicate that the hydraulic conductivity varies from $1 \times 10^{-10}$ m s$^{-1}$ to as low as $1 \times 10^{-13}$ m s$^{-1}$ (Section 16.7). While for some faults these hydraulic conductivity values may reflect the in-situ values, it is common for rocks that the in-situ values are 2–3 orders of magnitude greater than the laboratory values. For the present purpose, we take the in-situ core permeability during non-slipping periods as $1 \times 10^{-8}$ m s$^{-1}$, which is appropriate for many silts and clays. During seismogenic slip in the fault zone, the core-damage zone boundary is ruptured and a slip plane with an aperture of 0.1 mm is formed. The fault is transporting water at 21°C for which the dynamic viscosity is $\mu_f = 9.97 \times 10^{-4}$ Pa s (Giles, 1977) and the density is 1000 kg m$^{-3}$. How thick would the core have to be in order to have the same transmissivity as the slip plane?

### Solution

We first find the transmissivity of the fault plane. From Eq. (16.17) we have

$$T_f = K_f b = \frac{\rho_f g b^3}{12 \mu_f} = \frac{1000 \text{ kg m}^{-3} \times 9.81 \text{ m s}^{-2} \times (0.0001 \text{ m})^3}{12 \times 9.97 \times 10^{-4} \text{Pa s}} = 8.2 \times 10^{-7} \text{m}^2 \text{s}^{-1}$$

The estimated hydraulic conductivity of the intact (non-slipping) core is $1 \times 10^{-8}$ m s$^{-1}$. Using these values and rewriting Eq. (16.16) we get the thickness $b_p$ as

$$b_p = \frac{T_p}{K_p} = \frac{8.2 \times 10^{-7} \text{m}^2 \text{s}^{-1}}{1 \times 10^{-8} \text{m s}^{-1}} = 82 \text{ m}$$

So we would need an 82-m-thick core to provide a transmissivity equal to that of the slip plane. And of course the core would need to be much thicker if either the slip-plane aperture was larger (say 1–2 mm) or the core hydraulic conductivity lower, say $10^{-9}$ m s$^{-1}$. Cores, even in the largest transform faults, rarely exceed 10–20 m (Fig. 14.20), so that a core thickness of more than 80 m is generally unlikely to be reached. The results show the large effects that fault slip may have on the transmissivity of a fault in comparison with that of the non-slipping fault core and, therefore, why fault slip may have large effects on the associated ground water and geothermal fields (e.g. Gudmundsson, 2000).

## 16.12 Exercises

16.1 Describe the local stresses around a typical strike-slip fault just prior to slip. Where would fluids such as ground water or geothermal water be collected and where would they be expelled?

16.2 Describe the effects of fault trend in relation to the hydraulic gradient on ground-water transport.

16.3 For which parts of a typical fault zone would Darcy's law be appropriate and for which parts the cubic law?

16.4 Describe likely changes in the temporary permeability of a fault zone following a slip. Which part of the fault zone is likely to show the greatest changes? Why?

16.5 What are tight and leaky faults?

16.6 How does the permeability of a typical fault zone change during the interseismic period? Which processes are mainly responsible for these changes?

16.7 Mineral veins are very common in fault zones. What mechanical types of fractures are most of the veins and how do they form?

16.8 A strike-slip fault has a core that is 5 m thick and with an interseismic hydraulic conductivity of $1 \times 10^{-8}$ m s$^{-1}$. Calculate the transmissivity of the core.

16.9 If the fault zone in Exercise 16.8 strikes parallel with a hydraulic gradient of 0.01, what would be the volumetric flow rate of ground water through a 5 m high (dip dimension) part of the fault core?

16.10 A seismogenic slip occurs in the core of the fault zone in Exercise 16.8. If the core transmits ground water with a density of 1000 kg m$^{-3}$ and a viscosity of $1.55 \times 10^{-3}$ Pa s, what aperture must the resulting earthquake fracture have in order to increase the transmissivity of the core to $10^6$ times its pre-slip value?

## References and suggested reading

Ahlbom, K., and Smellie, J. A. T., 1991. Overview of the fracture zone project at Finnsjön, Sweden. *Journal of Hydrology*, **126**, 1–15.

Amadei, B. and Stephansson, O., 1997. *Rock Stress and its Measurement*. London: Chapman & Hall.

Anderson, E. I. and Bakker, M., 2008. Groundwater flow through anisotropic fault zones in multiaquifer systems. *Water Resources Research*, **44**(11): Art. No W11433.

Arnadottir, T., Jonsson, S., Pedersen, R., and Gudmundsson, G. B., 2003. Coulomb stress changes in the South Iceland Seismic Zone due to two large earthquakes in June 2000. *Geophysical Research Letters*, **30**(5): Art. No. 1205 March 5, 2003.

Babiker, M. and Gudmundsson, A., 2004. The effects of dykes and faults on ground-water flow in an arid land: the Red Sea Hills, Sudan. *Journal of Hydrology*, **297**, 256–273.

Barton, C. A., Zoback, M. D., and Moos, D., 1995. Fluid flow along potentially active faults in crystalline rock. *Geology*, **23**, 683–686.

Bear, J., 1988, *Dynamics of Fluids in Porous Media*. New York: Dover.

Bear, J., 1993. Modeling flow and contaminant transport in fractured rocks. In Bear, J., Tsang, C.F., and de Marsily, G. (eds.), *Flow and Contaminant Transport in Fractured Rock*. New York: Academic Press, pp. 1–37.

Beasy, 1991. *The Boundary Element Analysis System User Guide*. Boston, MA: Computational Mechanics.

Blanpied, M. L., Lockner, D. A., and Byerlee, J. D., 1992. An earthquake mechanism based on rapid sealing of faults. *Nature*, **358**, 574–576.

Braathen, A., Berg, S. S., Storro, G., Jaeger, O., Henriksen, H., and Gabrielsen, R., 1999. Fracture-zone geometry and ground water flow: results from fracture studies and drill tests in Sunnfjord. Geological Survey of Norway, report 99.017 (in Norwegian).

Bredehoeft, J. D., 1997. Fault permeability near Yucca Mountain. *Water Resources Research*, **33**, 2459–2463.

Bruhn, R. L., Parry, W. T., Yonkee, W. A., and Thompson, T., 1994. Fracturing and hydrothermal alteration in normal fault zones. *Pure and Applied Geophysics*, **142**, 609–644.

Byerlee, J., 1993. Model for episodic flow of high-pressure water in fault zones before earthquakes. *Geology*, **21**, 303–306.

Caine, J. S., Evans, J. P., and Forster, C. B., 1996. Fault zone architecture and permeability structure. *Geology*, **24**, 1025–1028.

Cappa, F., 2009. Modelling fluid transfer and slip in a fault zone when integrating heterogeneous hydromechanical characteristics in its internal structure. *Geophysical Journal International*, **178**, 1357–1362.

Domenico, P. A. and Schwartz, F. W., 1998, *Physical and Chemical Hydrogeology*, 2nd edn. New York: Wiley.

Evans, J. P., Forster, C. B., and Goddard, J. V., 1997. Permeability of fault-related rocks, and implications for hydraulic structure of fault zones. *Journal of Structural Geology*. **19**, 1393–1404.

Faulkner, D. R., Lewis, A. C., and Rutter, E. H., 2003. On the internal structure and mechanisms of large strike-slip fault zones: field observations of the Carboneras fault in southeastern Spain. *Tectonophysics*, **367**, 235–251.

Ferrill, D. A., Winterle, J., Wittmeyer, G. *et al.*, 1999. Stressed rock strains groundwater at Yucca Mountain, Nevada. *GSA Today*, **9**, 1–8.

Finkbeiner, T., Barton, C. A., and Zoback, M. D., 1997. Relationships among in-situ stress, fractures and faults, and fluid flow: Monterey Formation, Santa Maria Basin, California. *American Association of Petroleum Geologists Bulletin*, **81**, 1975–1999.

Fisher, D. M. and Brantley, S. L., 1992. Models of quartz overgrowth and vein formation: deformation and episodic fluid flow in an ancient subduction zone. *Journal of Geophysical Research*, **97**, 20 043–20 061.

Fisher, A. T., Zwart, G. and Ocean Drilling Program Leg 156 Scientific Party, 1996. Relation between permeability and effective stress along a plate-boundary fault, Barbados accretionary complex. *Geology*, **24**, 307–310.

Giles, R. V., 1977. *Fluid Mechanics and Hydraulics*. New York: McGraw-Hill.

Grecksch, G. F., Roth, F., and Kumpel, H. J., 1999. Coseismic well-level changes due to the 1992 Roermond earthquake compared to static deformation of half-space solutions. *Geophysical Journal International*, **138**, 470–478.

Gudmundsson, A., 1995. Stress fields associated with oceanic transform faults. *Earth and Planetary Science Letters*, **136**, 603–614.

Gudmundsson, A., 1999. Fluid overpressure and stress drop in fault zones. *Geophysical Research Letters*, **26**, 115–118.

Gudmundsson, A., 2000. Active fault zones and groundwater flow. *Geophysical Research Letters*, **27**, 2993–2996.

Gudmundsson, A. and Brenner, S.L., 2003. Loading of a seismic zone to failure deforms nearby volcanoes: a new earthquake precursor. *Terra Nova*, **15**, 187–193.

Gudmundsson, A., Fjeldskaar, I., and Brenner, S. L., 2002. Propagation pathways and fluid transport of hydrofractures in jointed and layered rocks in geothermal fields. *Journal of Volcanology and Geothermal Research*, **116**, 257–278.

Gudmundsson, A., Simmenes, T. H., Larsen, B., and Philipp, S. L., 2009. Effects of internal structure and local stresses on fracture propagation, deflection, and arrest in fault zones. *Journal of Structural Geology*, document doi:10.1016/j.jsg.2009.08.013.

Gutmanis, J. C., Lanyon, G. W., Wynn, T. J., and Watson, C. R., 1998. Fluid flow in faults: a study of fault hydrogeology in Triassic sandstone and Ordovician volcaniclastic rocks at Sellafield, north-west England. *Proceedings of the Yorkshire Geological Society*, **52**, 159–175.

Haimson, B. C. and Rummel, F., 1982. Hydrofracturing stress measurements in the Iceland research drilling project drill hole at Reydarfjordur, Iceland. *Journal of Geophysical Research*, **87**, 6631–6649.

Jonsson, S., Segall, P., Pedersen, R., and Bjornsson, G., 2003. Post-earthquake ground movements correlated to pore-pressure transients. *Nature*, **424**, 179–183.

Kanamori, H. and Anderson, D. L., 1975. Theoretical basis of some empirical relations in seismology. *Bulletin of the Seismological Society of America*, **65**, 1073–1095.

King, C. Y., Azuma, S., Igarashi, G., Ohno, M., Saito, H., and Wakita, H., 1999. Earthquake-related water-level changes at 16 closely clustered wells in Tono, central Japan. *Journal of Geophysical Research*, **104**, 13 073–13 082.

Knipe, R. J., 1993. The influence of fault-zone processes and diagenesis on fluid flow. In: Horbury, A. D. and Robinson, A. (eds.), *Diagenesis and Basin Development*. Tulsa, OK: American Association of Petroleum Geologists, pp. 135–151.

Larsen, B., Grunnaleite, I., and Gudmundsson, A., 2009. How fracture systems affect permeability development in shallow-water carbonate rocks: an example from the Gargano Peninsula, Italy. *Journal of Structural Geology*, doi:10.1016/jsg.2009.05.009

Larsen, B., Gudmundsson, A., Grunnaleite, I., Sælen, G., Talbot, M. R., and Buckley, S., 2010. Effects of sedimentary interfaces on fracture pattern, linkage, and cluster formation in peritidal carbonate rocks. *Marine and Petroleum Geology*, doi: 10.1016/j.marpetgeo.2010.03.011.

Lee, C. H. and Farmer, I., 1993. *Fluid Flow in Discontinuous Rocks*. New York: Chapman & Hall.

Leonardi, V., Arthaud, F., Tovmassian, A., and Krakhanian, A., 1998. Tectonic and seismic conditions for changes in spring discharge along the Garni right lateral strike slip fault (Armenian Upland). *Geodinamica Acta*, **11**, 85–103.

Lin, A., Maruyama, T., and Kobayashi, K., 2007. Tectonic implications of damage zone-related fault-fracture networks revealed in drill core through the Nojima fault, Japan. *Tectonophysics*, **443**, 161–173.

Lopez, D. L. and Smith, L., 1995. Fluid flow in fault zones: Analysis of the interplay between convective circulation and topographically driven groundwater flow. *Water Resources Research*, **31**, 1489–1503.

Mayer, J. R. and Sharp, J. M., 1998. Fracture control of regional ground-water flow in a carbonate aquifer in a semi-arid region. *Geological Society of America Bulletin*, **110**, 269–283.

Melchiorre, E. B., Criss, R. E., and Davisson, M. L., 1999. Relationship between seismicity and subsurface fluids, central Coast Ranges, California. *Journal of Geophysical Research*, **104**, 921–939.

Muirwood, R. and King, G. C. P., 1993. Hydrologic signatures of earthquake strain. *Journal of Geophysical Research*, **98**, 22 035–22 068.

Nativ, R., Adar, E. M., and Becker, A., 1999. Designing a monitoring network for contaminated ground water in fractured chalk. *Ground Water*, **37**, 38–47.

Ohno, M., Sato, T., Notsu, K., Wakita, H., and Ozawa, K., 1999. Groundwater-level changes in response to bursts of seismic activity off the Izu Peninsula, Japan. *Geophysical Research Letters*, **26**, 2501–2504.

Olsen, M. P., Scholz, C.H., and Leger, A., 1998. Healing and sealing of simulated fault gouge under hydrothermal conditions: Implications for fault healing. *Journal of Geophysical Research*, **103**, 7421–7430.

Paul, P., Zoback, M., and Hennings, P., 2009. Fluid flow in fractured reservoir using a geomechanically constrained fault-zone-damage model for reservoir simulation. *SPE Reservoir Evaluation & Engineering*, **12**, 562–575.

Philipp, S. L., 2008. Geometry and formation of gypsum veins in mudstones at Watchet, Somerset, SW England. *Geological Magazine*, **145**, 831–844.

Rawling, G. C., Goodwin, L. B., and Wilson, J. L., 2001. Internal architecture, permeability structure, and hydrologic significance of contrasting fault-zone types. *Geology*, **29**, 43–46.

Roberts, S. J., Nunn, J. A., Cathles, L., and Cipriani, F. D., 1996. Expulsion of abnormally pressured fluids along faults: *Journal of Geophysical Research*, **101**, 28 231–28 252.

Roeloffs, E. A., 1988. Hydrologic precursors to earthquakes: a review. *Pure and Applied Geophysics*, **126**, 177–209.

Roeloffs, E. A., 1996. Poroelastic techniques in the study of earthquake-related hydrologic phenomena. *Advances in Geophysics*, **37**, 135–195.

Rognvaldsson, S. Th., Gudmundsson, A., and Slunga, R., 1998. Seismotectonic analysis of the Tjornes Fracture Zone, an active transform fault in north Iceland. *Journal of Geophysical Research*, **103**, 30 117–30 129.

Rojskczer, S., Wolf, S., and Michel, R., 1995. Permeability enhancement in the shallow crust as a cause of earthquake-induced hydrological changes. *Nature*, **373**, 237–239.

Schultz, R. A., 1995. Limits on strength and deformation properties of jointed basaltic rock masses. *Rock Mechanics and Rock Engineering*, **28**, 1–15.

Seront, B., Wong, T. F., Caine, J. S., Forster, C. B., and Bruhn, R. L., 1998. Laboratory characterisation of hydromechanical properties of a seismogenic normal fault system. *Journal of Structural Geology*, **20**, 865–881.

Shan, C., Javandel, I., and Witherspoon, P.A., 1995. Characterization of leaky faults: study of water flow in aquifer-fault-aquifer systems. *Water Resources Research*, **31**, 2897–2904.

Sibson, R. H., 1996. Structural permeability of fluid-driven fault-fracture meshes. *Journal of Structural Geology*, **18**, 1031–1042.

Sibson, R. H., McMoore, J., and Rankine, A. H., 1975. Seismic pumping – a hydrothermal fluid transport mechanism. *Journal of the Geological Society of London*, **131**, 653–659.

Sleep, N. H. and Blanpied, M. L., 1992. Creep, compaction and the weak rheology of major faults. *Nature*, **359**, 687–692.

Sneddon, I. N. and Lowengrub, M., 1969. *Crack Problems in the Classical Theory of Elasticity*. New York: Wiley.

Stauffer, D., 1985. *Introduction to Percolation Theory*. Philadelphia, PA: Taylor & Francis.

Tada, H., Paris, P. C., and Irwin, G. R., 2000. *The Stress Analysis of Cracks Handbook*. New York: American Society of Mechanical Engineers.

Tsunogai, U. and Wakita, H., 1995. Precursory chemical changes in ground water: Kobe earthquake, Japan. *Science*, **269**, 61–63.

Wang, H. F., 2000. *Theory of Linear Poroelasticity with Applications to Geomechanics and Hydrogeology*. Princeton, NJ: Princeton University Press.

# 17 Fluid transport in hydrofractures

## 17.1 Aims

How hydrofractures transport crustal fluids is important in many fields of earth sciences, including petroleum geology, geothermal research, volcanology, seismology, and hydrogeology. In the previous chapters the focus has been on the transport of ground water, geothermal water, and gas and oil. Here the main examples come from two fields: man-made hydraulic fractures, as are used to increase the permeability of reservoirs (petroleum and geothermal); and volcanology, that is, the transport of magma through vertical (dykes), inclined (sheets), and horizontal (sills) intrusions. It should be stressed, however, that the basic principles of fluid transport in hydrofractures are the same irrespective of the type of fluid. Thus, the results obtained here apply equally well, with suitable modifications, to hydrofracture transport of ground water, oil, and gas. The principal aims of this chapter are to explain and discuss:

- The source of the driving pressure (overpressure) in hydrofractures, including the effects of buoyancy.
- The initiation, propagation, and fluid transport in man-made (artificial) hydraulic fractures.
- Fluid transport in magma-driven fractures, primarily dykes and inclined sheets.
- Fluid transport in fractures driven by hydrothermal fluids, that is, fractures that subsequently become mineral veins.

## 17.2 Driving pressure and volumetric flow rate

When a hydrofracture has generated its path through the linking up of discontinuities in the host rock, it forms a continuous fracture that conducts fluid. The volumetric rate of flow of the fluid through the fracture can be estimated using equations from fluid mechanics.

Consider a hydrofracture injected from a source at a certain depth below the surface (Fig. 17.1). Here the source of the hydrofracture is taken to be sill-like, that is, an oblate ellipsoid. This is not a necessary geometry for the analysis that follows. The source may be shaped like any general ellipsoid or a sphere, or, in two dimensions, an ellipse or a circle. When the host rock (and the source) is assumed to be rigid, there will be no change in either the geometry of the hydrofracture or the source as the fluid is driven from the source through

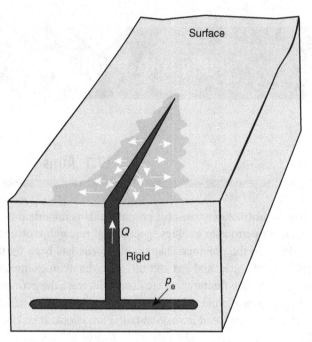

Schematic presentation of a model of a self-supporting or rigid vertical hydrofracture. In this model, neither the hydrofracture nor its sill-like source change their size or shape when the fluid overpressure changes. $Q$ is the volumetric flow rate and $p_e$ is the excess pressure in the sill-like reservoir before rupture and hydrofracture formation. A rigid model is commonly used for modelling ground-water transport, particularly that associated with comparatively small aquifers in the uppermost part of the crust.

the fracture, that is, when the fluid excess pressure in the source decreases. However, when the rock and the hydrofracture are regarded as elastic, then it is assumed that the source responds elastically, that is, contracts or shrinks, when excess pressure decreases as fluid flows out of the source (Fig. 17.2). If the internal excess pressure at the time of rupture of the upper wall of the source, the sill, is $p_e$ (which is thus the fluid pressure in excess of the lithostatic pressure in the upper wall of the sill), then the conditions for rupture and hydrofracture initiation is

$$p_l + p_e = \sigma_3 + T_0 \tag{17.1}$$

where $p_l$ is the lithostatic stress or pressure at the depth of the sill-like source; $p_e = p_t - p_l$ is the excess fluid pressure, that is, the difference between the total fluid pressure, $p_t$, in the source at the time of its rupture and the lithostatic pressure, $p_l$; and $\sigma_3$ and $T_0$ are the minimum principal compressive stress and the in-situ tensile strength, respectively, in the upper wall or surface of the sill-like source.

When the walls of the source become ruptured and the hydrofracture forms, the associated volumetric flow rate up the fracture can be calculated. A hydrofracture formed in this way is generally assumed to propagate through crustal layers which behave, to a first approximation, as elastic. As indicated above, when the fluid from the source flows up the fracture, the excess pressure in the source decreases and the source shrinks or contracts. The assumption of an elastic crustal behaviour is a common approach in petroleum geology and volcanology. In many models used in hydrogeology, however, the crust is assumed to be rigid, particularly for flow at very shallow depths, say at less than a few hundred metres. This assumption is, for example, used in the fracture-set models in Chapter 15. Since rigid and elastic assumptions are both used in theoretical studies of fluid transport in fractures, we shall here discuss and compare the results of both modelling approaches.

For a rigid host rock, we refer to the resulting hydrofracture and fluid source as self-supporting and use the superscript s for the volumetric flow through the fracture. By contrast, if the hydrofracture and its source deform elastically as the excess pressure changes, the fracture is referred to as elastic, using the superscript e for the volumetric flow through the fracture.

Consider first a vertical, self-supporting hydrofracture that extends from its sill-like source to the Earth's surface (Fig. 17.1). From the Navier–Stokes equations (Lamb, 1932; Milne-Thompson, 1996) it follows that the volumetric flow rate, $Q_z^s$, up through the fracture (along the vertical coordinate, $z$) is

$$Q_z^s = \frac{\Delta u^3 W}{12\mu_f} \left[ \rho_f g - \frac{\partial p_e}{\partial z} \right] \qquad (17.2)$$

where $\Delta u$ is the opening of the fracture and thus equal to the aperture as used in hydrogeology, that is, $\Delta u = b$, $W$ is its width in a direction that is perpendicular to the flow direction (so that the fracture cross-sectional area perpendicular to the flow is $A = \Delta u W$) and it is assumed that $W \gg \Delta u$. Also, $\mu_f$ is the dynamic (absolute) viscosity and $\rho_f$ the density of the fluid (both assumed to be constant in the examples in this book), $g$ is the acceleration due to gravity, and $\partial p_e / \partial z$ is the vertical pressure gradient in the direction $z$ of the flow. Appendix E.2 gives data on the densities and viscosities of some common crustal fluids.

When the walls of the hydrofracture and its source are free to deform elastically as fluid is transported from the sill-like source (Fig. 17.2), the weight of the rock above the source must be supported by its internal fluid pressure. Since the host-rock density $\rho_r$ is different from the fluid density $\rho_f$, a buoyancy term is added to the excess-pressure gradient. Thus, for a vertical, elastic hydrofracture, Eq. (17.1) is rewritten in the form

$$Q_z^e = \frac{\Delta u^3 W}{12\mu_f} \left[ (\rho_r - \rho_f)g - \frac{\partial p_e}{\partial z} \right] \qquad (17.3)$$

where $Q_z^e$ is the volumetric rate of fluid flow through a hydrofracture in an elastic host rock.

Equations (17.2) and (17.3) apply for fluid transport in vertical hydrofractures, such as many mineral veins, joints, and dykes. Vertical hydrofractures are most common in regions undergoing extension such as active rift zones and many sedimentary basins where the fluid source is sill-like. This follows from the considerations of the local stresses around such fluid sources. For example, numerical models indicate that, as long as the crustal segment

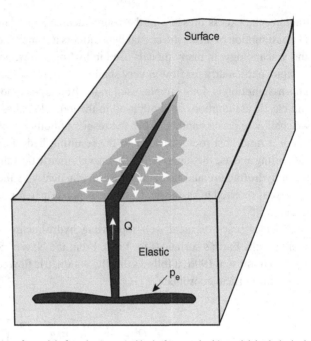

Surface

Q

Elastic

$p_e$

Fig. 17.2 Schematic presentation of a model of an elastic vertical hydrofracture. In this model, both the hydrofracture and its sill-like source change their shape and size when the fluid overpressure changes. $Q$ is the volumetric flow rate and $p_e$ is the excess pressure in the sill-like reservoir before rupture and hydrofracture formation. An elastic model is commonly used to model fluid transport in hydrofractures, such as magma-driven fractures (dykes, sheets) and gas- and oil-driven fractures, injected from large reservoirs at considerable depths and (often) temperatures.

hosting the source is homogeneous and isotropic, the trajectories of $\sigma_1$ tend to be vertical above the roof of a sill-like source (Chapters 6 and 13). Since hydrofractures follow the $\sigma_1$-trajectories, they would also normally be vertical.

As indicated above, many fluid sources are likely to be of a shape different from that of a sill. Here we shall consider the example of a shallow magma chamber of a circular cross-sectional area. We use a magma chamber, but the formulas apply to any fluids. Many shallow magma chambers inject not only dykes but also inclined sheets (Chapter 11). For inclined sheets and other non-vertical hydrofractures, Eqs. (17.2) and (17.3) must be modified to take the fracture dip $\alpha$ into account (Fig. 17.3). The dip dimension of the sheet, $L$, is here the linear distance from the point of rupture and sheet initiation, at the boundary of the magma chamber, and to the surface.

Consider first the case of a rigid crustal segment. For an inclined, self-supporting hydrofracture, the volumetric rate $Q_L^s$ of fluid flow along the fracture is, by analogy with Eq. (17.2), given by

$$Q_L^s = \frac{\Delta u^3 W}{12\mu_f}\left[\rho_f g \sin\alpha - \frac{\partial p_e}{\partial L}\right] \qquad (17.4)$$

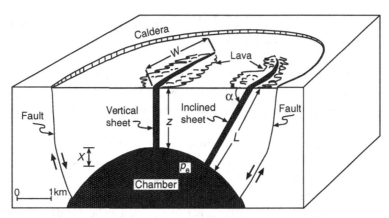

**Fig. 17.3**  Dyke and inclined sheet injected from a magma chamber with a circular cross-section (only the upper part is seen here) and reaching the surface to supply magma to fissure eruptions. In this cross-section, the distance that the magma travels to the surface along the vertical dyke is denoted by $z$, whereas the distance travelled by the magma along the inclined sheet to the surface is denoted by $L$ ($x$ is explained in Example 17.3). For a flat volcano, such as a collapse caldera, commonly $L > z$, in which case the excess pressure gradient for the inclined sheet becomes smaller than that for the vertical dyke. The length (strike dimension) of the volcanic fissure at the surface is denoted by $W$ and the caldera or ring fault is indicated (modified from Gudmundsson and Brenner, 2005).

Consider next an inclined sheet located in an elastic crustal segment. For an inclined, elastic hydrofracture the volumetric flow rate $Q_L^e$ is, by analogy with Eq. (17.3), given by

$$Q_L^e = \frac{\Delta u^3 W}{12\mu_f}\left[(\rho_r - \rho_f)g\sin\alpha - \frac{\partial p_e}{\partial L}\right] \qquad (17.5)$$

We see that Eqs. (17.2) and (17.3) may be regarded as special cases of Eqs. (17.4) and (17.5), namely when the dip of the sheet is vertical, or $\alpha = 90°$, so that $\sin\alpha = 1$. Similarly, fluid flow along a horizontal fracture, a sill, is a special case of Eqs. (17.4) and (17.5). For the sill, the fracture dip is zero, or $\alpha = 0°$, so that $\sin\alpha = 0$ and the first term in the brackets drops out. For a hydrofracture in the horizontal $xy$-plane, a sill, whose width $W$ is measured along the $y$-axis and the length $L$ measured along the $x$-axis, we may substitute $x$ for $L$ and obtain, from Eqs. (17.4) and (17.5), the volumetric flow rate $Q_x$ as

$$Q_x = -\frac{\Delta u^3 W}{12\mu_f}\frac{\partial p_e}{\partial x} \qquad (17.6)$$

In Eq. (17.6), the only pressure driving the flow along the sill, that is, through the horizontal fracture, is the excess pressure $p_e$ (Fig. 13.9). In the case where the horizontal hydrofracture is supplied with fluid from a vertical hydrofracture, such as when a vertical dyke supplies magma to a horizontal sill (Fig. 13.9; Chapter 11), then the excess pressure in the sill is equal to the overpressure that develops in the dyke on its path towards the contact with the sill.

# 17.3 Man-made hydraulic fractures

Man-made hydrofractures, also known as hydraulic fractures, are generated when fluid is injected under overpressure from a drill hole into the host rock (Chapter 11). The aim is normally to increase the permeability of a reservoir, commonly an oil or geothermal reservoir.

Hydraulic fractures form by the same mechanism as natural hydrofractures. They propagate in a direction that is perpendicular to the local minimum principal compressive stress, $\sigma_3$ (Chapter 13; Valko and Economides, 1995). Like a natural hydrofracture, a hydraulic fracture may use weaknesses such as joints or segments of steeply dipping normal faults for a part of their paths, but the general path is perpendicular to $\sigma_3$. A hydraulic fracture, injected from a drill hole to increase or generate permeability in a certain target layer, is supposed to stay within that layer (Figs. 13.28 and 17.4). Whether the propagating hydraulic fracture stays within the target layer depends on several factors. Factors that favour the arrest of the tips of the hydraulic fracture at the contact of the target layer and the adjacent layers (above and below the target layer) are the same as encouraging the arrest of extension fractures in general (Chapter 13), and include the following:

- The adjacent layers are stress barriers, that is, have a local stress field that is unfavourable to the propagation of a vertical hydraulic fracture. Such barriers may, for example, be generated by local high horizontal compressive stresses (Fig. 17.5) in a stiff layer or by unfavourable contact properties, such as material toughness (Fig. 17.6).
- The fracture-parallel tensile stresses induced by, and ahead of the tip of, the hydraulic fracture may also open up the contact and generate an arrested T-shaped fracture (Figs. 13.11–13.13). Some or all these mechanisms may operate together to contribute to the arrest of the hydraulic fracture in the same way as they contribute to the arrest of natural hydrofractures (Fig. 17.7).

A hydraulic fracture may be horizontal like a sill (Fig. 11.20). More commonly, a hydraulic fracture is injected horizontally as a 'blade-like fracture' from the drill hole, but the fracture

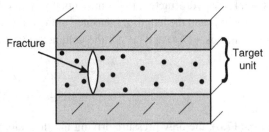

**Fig. 17.4**    For a hydraulic fracture, injected into a target unit from a drill hole, the aim is that the fracture stays within the target unit, as indicated here. The chance that the hydraulic fracture becomes confined to the target unit is great when there is an abrupt change in the mechanical properties, particularly in stiffness, between the target unit and the units above and below. In particular, if the adjacent units are stiffer than the target unit, the hydraulic fracture tends to be arrested at the contacts between the units and thus be confined to the target unit (Chapter 14).

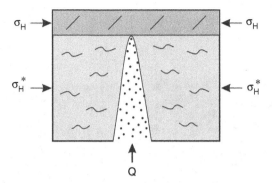

**Fig. 17.5** One way by which a fracture may become arrested on meeting a contact between mechanically dissimilar layers is through unfavourable local stresses, that is, stress barriers. For a vertical hydrofracture, a common stress barrier is a layer or unit above the fracture tip where the horizontal compressive stress perpendicular to the fracture strike, here $\sigma_H$, is much greater than the horizontal stress in the layer hosting the fracture, here $\sigma_H^*$. Here it is assumed that $\sigma_H = \sigma_h = \sigma_3$.

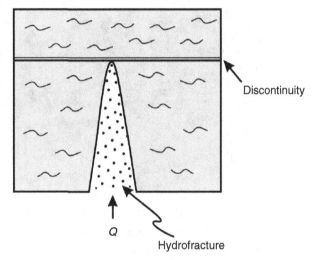

**Fig. 17.6** Many hydrofractures become arrested at discontinuities such as faults, fractures, and contacts. This commonly happens when there is an abrupt change in the mechanical properties across the contact and/or the material toughness changes across the contact are unfavourable for hydrofracture penetration (Chapter 14).

itself is vertical or steeply inclined (Fig. 13.28). Such a fracture may extend, to either side of the drill hole, by as much as 500 m and is sometimes referred to as a 'wing crack' in the petroleum industry (Valko and Economides, 1995), a term that has a somewhat different meaning in structural geology (Fig. 16.6). The exact attitude of the hydraulic fracture, however, depends on the local stress field. It is well known that a hydraulic fracture may change its orientation, sometimes curving or rotating along its path away from the source drill hole (Valko and Economides, 1995).

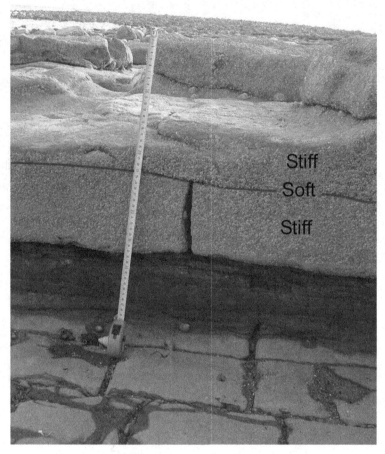

**Fig. 17.7** T-shaped fracture forms when a vertically propagating hydrofracture (here generating a joint) becomes deflected into the contact between a soft shale layer (millimetres in thickness) and a stiffer limestone layer (Chapter 13). In volcanology such a contact is referred to as a dyke–sill contact (Figs. 10.4, 10.5, and 13.8).

Unless the local stress field and the rock heterogeneities and discontinuities, including fractures and faults, are known in detail, which is rare, it is difficult to forecast with accuracy the exact propagation path of a hydraulic fracture. Because of heterogeneities and variation in local stresses and strengths, the hydraulic-fracture path may be somewhat irregular even if the overall path is perpendicular to the minimum principal stress, $\sigma_3$, an observation that is used in stress measurements. If a horizontally propagating hydraulic fracture meets a major fault zone trending obliquely or at right angles to its propagation path, the hydraulic fracture may either penetrate the fault zone, become deflected into the fault zone, or become arrested (Fig. 17.8). The factors that control whether the hydraulic fracture becomes deflected or

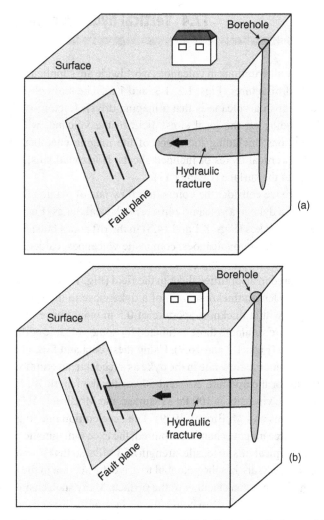

**Fig. 17.8** Hydraulic fractures are formed by injection (commonly as lateral, blade-like or wing-like fractures) into a target layer (Fig. 17.4). The aim is for the fracture to follow a certain, predicted path. That path, however, may be somewhat irregular because of heterogeneities in the rock. In addition, the hydrofracture may become deflected or stop altogether on meeting a major discontinuity such as a transverse fault zone. (a) The hydraulic fracture propagates horizontally until (b) it meets an inclined fault plane where the fracture becomes arrested.

arrested are the same as for a vertically propagating hydrofracture, discussed above and in Chapter 13. Generally, to reach the maximum possible length (strike dimension) of a hydraulic fracture, an attempt should be made to select its potential path so that it avoids major oblique or transverse fault zones.

For horizontal, sill-like hydraulic fractures, the volumetric flow rate may be calculated from Eq. (17.6). For vertical hydraulic fractures, injected laterally to form blade-like fractures or wing cracks on either side of the drill hole (Fig. 13.28), more complex formulas are needed (Valko and Economides, 1995).

# 17.4 Vertical hydrofractures (dykes)

Nearly all eruptions in volcanoes worldwide are supplied with magma-driven fractures, that is, hydrofractures (Figs. 1.2, 1.3, and 1.6). The main physical requirement for an eruption to occur in a volcano is that a magma-driven fracture is able to propagate to its surface. Depending on the local stress field in the volcano, which is a function of the loading conditions, including the shape of the magma chamber, the magma-filled fractures are either vertical dykes or inclined sheets. Horizontal sills, while common, normally do not erupt at the surface (Chapter 11).

Here we consider the volumetric flow rate of magma through a typical vertical fracture, that is, a dyke in a volcanic edifice. Such local dykes (Fig. 11.18) tend to be thinner than the regional dykes (Figs. 8.1 and 14.8) in the rift zones outside major volcanic edifices (central volcanoes, stratovolcanoes, composite volcanoes, calderas).

We take the opening or aperture of the dyke, which is roughly equal to the thickness as measured in a solidified dyke in the field (Fig. 17.3; Chapters 9 and 11), as 0.5 m. A typical outcrop length/thickness ratio of a dyke close to the surface is about 1000, in which case a dyke with a thickness/aperture of 0.5 m would have an outcrop length of 500 m. These values are similar to those estimated for some volcanic edifices in Iceland, such as the Hekla Volcano (Figs. 1.2 and 16.9). Using these data and Eqs. (17.2) and (17.3), we can estimate the volumetric flow rate in the dyke as a feeder to an eruption. In the equations, we substitute 0.5 m for the aperture $\Delta u$, and 500 m for the length $W$. For a typical tholeiite basalt, the dynamic viscosity is 100 Pa s (Murase and McBirney, 1973) and the density 2650 kg m$^{-3}$ (Williams and McBirney, 1979). The acceleration due to gravity is $g = 9.8$ m s$^{-2}$. Before the source magma chamber ruptured, the excess magmatic pressure is assumed to be similar to the typical in-situ tensile strength of 3 MPa, so that $p_e = T_0 = 3$ MPa. We assume that this excess pressure has the potential to drive the magma in the dyke from the uppermost part of a shallow magma chamber to the surface. Many such chambers, for example in Iceland, are at depths of 1.5–2.5 km below the surface of the associated volcanoes. We use the average depth of 2 km, which thereby becomes the vertical distance that the magma in the dyke has to flow to reach the surface of the volcano. For a vertical depth of 2000 m, it follows that the vertical pressure gradient in Eqs. (17.2) and (17.3) is $\partial p_e / \partial z = -1500$ Pa m$^{-1}$.

Consider first the case where the crustal segment, including the volcano, behaves as rigid. Substituting the above values into Eq. (17.3), the volumetric flow rate through a dyke of the above dimensions is $Q_z^s \approx 1.4 \times 10^3$ m$^3$s$^{-1}$. Next consider the case when the crustal segment behaves elastically. For the same values of magma density and viscosity, and dyke opening or thickness and outcrop length, and taking the average host-density for the uppermost 2 km of the crust as 2540 kg m$^{-3}$ (an appropriate value for Iceland), then from Eq. (17.4) we get $Q_z^e \approx 22$ m$^3$s$^{-1}$. Thus, a dyke in a rigid crustal segment yields a volumetric flow rate that is more than 60 times greater than that for the same dyke in an elastic crust. This difference is largely because of the negative buoyancy effects (reduction in overpressure) on the overpressure (driving pressure) of the dense basaltic magma as it moves upwards through a lower-density, elastic host crust.

# 17.5 Inclined hydrofractures (cone sheets)

Now we consider the effect of hydrofracture dip on fluid transport, using the example of an inclined sheet (Chapter 11). If the volcano to which the inclined sheet supplies magma is essentially flat, as is common in collapse calderas (Fig. 17.3), the linear distance of magma transport along an inclined sheet, $L$, is normally considerably longer than the shortest vertical distance, $z$, to the surface. By contrast, if the volcano is an edifice, that is, forms a large topographic high, such as most stratovolcanoes (Fig. 1.2), so that the elevation difference is several hundred metres or more, $L$ may be similar to or even smaller than $z$. Clearly, the difference between these distances depends on the topography of the volcano, the geometry of the associated shallow magma chamber, and the dip of the inclined sheet transporting the magma to the surface of the volcano. Presumably, the condition $L = z$ is appropriate for many stratovolcanoes and other large edifices.

Here we focus on magma transporting in an inclined sheet hosted by an elastic crust. We use the same magma density and viscosity and host-rock density and size of fracture as for the dyke above. The mean dip of inclined sheets in Iceland is around $\alpha = 60°$. Then $\sin 60° \approx 0.87$ and Eq. (17.5) gives $Q_z^e \approx 19 \text{ m}^3\text{s}^{-1}$. This value differs so little from the earlier value for a vertical dyke ($22 \text{ m}^3\text{s}^{-1}$) that the assumption that steeply dipping feeder sheets are vertical is reasonable when estimating magmatic flow (effusion) rates. Many shallow dipping sheets, however, have dips as low as $\alpha = 30°$ (Chapter 11). Then $\sin 30° = 0.5$ and Eq. (17.5) yields $Q_z^e \approx 11 \text{ m}^3\text{s}^{-1}$. This is half the value obtained for the vertical sheet or dyke. For shallow dipping feeder sheets, therefore, assuming that they are vertical can lead to significant errors in estimating their potential for transporting magma to the surface of a volcano.

For a flat volcano such as a caldera (Fig. 17.3), both the sheet dip and the geometry of the shallow magma chamber contribute to make the length $L > z$. If a sheet with a dip $\alpha$ is injected from a roughly spherical magma chamber at a depth of $x$ metres below the chamber top (Fig. 17.3), then, with $L$ and $z$ in metres, from trigonometry

$$L = \frac{z + x}{\sin \alpha} \tag{17.7}$$

Consider a sheet dipping at $\alpha = 60°$, for example, and injected at a depth $x = 500$ m below the top of the shallow magma chamber. Both these values would be expected to be common in local sheet swarms injected from roughly spherical chambers with a radius of 1–2 km (Gudmundsson, 2002). If the distance $z$ is 2 km, as in the calculations above, Eq. (17.7) gives $L \sim 2.9$ km. The average host-rock density for the uppermost 2.5 km of the crust in Iceland is about $2560 \text{ kg m}^{-3}$ (Gudmundsson, 1988). Using this value for $\rho_r$, 2900 m for $L$, so that the gradient $\partial p_e / \partial L = -1030 \text{ Pa m}^{-1}$, and other values as in the example above, Eq. (17.5) yields $Q_z^e \approx 7.9 \text{ m}^3\text{s}^{-1}$. This volumetric flow rate is about 40% of the value of $Q_z^e \approx 19 \text{ m}^3\text{s}^{-1}$ obtained when the effect of sheet dip on $L$ and the associated gradient $\partial p_e / \partial L$ was ignored.

## 17.6 Mineral veins

Typical springs in the low-temperature areas of Iceland yield 5 l/s, whereas the maximum yield of single springs is 180 l/s (Gudmundsson *et al.*, 2002). We shall now estimate the dimensions of the hydrofracture networks, conceptualised as single hydrofractures or as fracture clusters, that are necessary to supply these springs with the measured volumetric rates of hydrothermal water. Many, perhaps most, geothermal springs worldwide are associated with faults, particularly normal faults (Figs. 16.13 and 16.14). Many springs in the low-temperature areas of Iceland are associated with active faults, some with steep normal faults, others with near-vertical strike-slip faults. Many hydrofracture paths are also vertical or steeply dipping (Figs. 16.11 and 16.13), whereas others are gently dipping or subhorizontal (Figs. 16.11, 16.14, and 16.15). The dip of the hydrofracture path depends much on the local stress field during the hydrofracture formation, in the same way as the dips of dykes and inclined sheets depend on the local stress field around the source magma chamber (Chapter 11). Here we first model the hydrofractures as vertical and then later explore the effects of changing the dip.

Consider first the case of fluid transport up a vertical hydrofracture, using Eq. (17.5) as the appropriate model. If the hydrofracture propagates from a sill-like source, the excess pressure $p_e$ before rupture of the source and vertical propagation of the hydrofracture may be taken as equal to the typical in-situ tensile strength of the basaltic lava flows in Iceland, 3 MPa. The surface-water temperature in low-temperature areas of Iceland is most commonly 20°–100°C. In the calculations we use a surface-water temperature of 90°C, equal to that of Geysir in South Iceland and Old Faithful in Yellowstone Park in the United States (Ingebritsen and Sanford, 1998).

Water at 90°C has a dynamic viscosity of about $3 \times 10^{-4}$ Pa s and a density of about 965 kg m$^{-3}$ (Smits, 2000). In detail, these properties depend on the composition of the hydrothermal water. For the purpose of the present calculations, however, these values are sufficiently accurate. The excess pressure changes from 3 MPa in the source to zero at the Earth's surface (where the fracture ends and the hot spring is located). It follows that the excess-pressure gradient $\partial p_e/\partial y$ is equal to the pressure difference, 3 MPa – 0 MPa = 3 MPa = $3 \times 10^6$ Pa, divided by the dip dimension (height) of the fracture in metres. For hydrothermal water originating at a depth of 2 km, the gradient is thus $\partial p_e/\partial y = 3 \times 10^6$ Pa/2000 m = 1500 Pa m$^{-1}$. Since the excess pressure is decreasing, its sign is negative and thus $-1500$ Pa m$^{-1}$. For the uppermost 2 km of the crust in Iceland the average density is about 2540 kg m$^{-3}$ (Gudmundsson, 1988). Substituting these values and the acceleration due to gravity, 9.81 m s$^{-2}$, into Eq. (17.3) we get

$$Q_y = CW\,\Delta u^3 \tag{17.8}$$

where $W$ is here the strike dimension or length and $\Delta u$ the aperture of the hydrofracture that supplies water to the hot spring. It should be noted that this hydrofracture may propagate to the surface along a fault, such as the damage zone of a fault zone, in which case only the part of the fault zone occupied by the hydrofracture at the surface is regarded as the length $W$. For the values given above, the factor $C$ is about $4.7 \times 10^6$ m$^{-1}$s$^{-1}$. Using the relationship

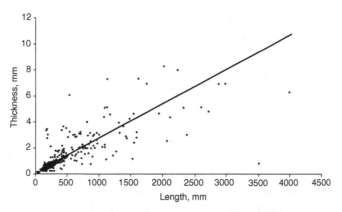

Fig. 17.9 Thickness (palaeo-apertures) versus length (strike dimension) of 382 mineral veins in a transform fault in North Iceland (Fig. 12.16). The veins are exposed in the damage zone of the fault zone, mostly in basaltic lava flows, at a crustal depth of about 1500 m below the original top of the lava pile. The average length/thickness ratio is about 400, the coefficient of determination of the regression line (between vein thickness and length) is 0.66, and the linear correlation coefficient is 0.81 (Gudmundsson *et al.*, 2002). This linear relationship agrees well with theoretical models indicating that, for constant fluid pressure, the aperture varies positively as the controlling dimension, here the length (Chapter 9).

$W/\Delta u \approx 400$, as was established for the veins supplying water to the palaeo-geothermal fields of the transform fault in North Iceland (Fig. 17.9; Gudmundsson *et al.*, 2002), then Eq. (17.8) can be rewritten approximately as

$$Q_y \approx N \Delta u^4 \qquad (17.9)$$

where the factor $N$ has a value of about $1.9 \times 10^9$ m$^{-1}$s$^{-1}$.

For a hot spring yielding 5 1/s, or $5 \times 10^{-3}$ m$^3$s$^{-1}$, Eq. (17.9) gives a $\Delta u$ of about 0.0013 m or 1.3 mm. From the ratio $W/\Delta u \approx 400$ it follows that the water-conducting strike dimension of the hydrofracture need be only about 0.5 m for supplying this volumetric flow rate of water to the hot spring. These fracture dimensions are very similar to those of the most common mineral veins in the palaeo-geothermal fields of the transform fault in North Iceland (Fig. 17.9). By contrast, in the highest-yield hot spring in Iceland, Deildartunguhver in West Iceland, the yield is 180 1/s, or 0.18 m$^3$s$^{-1}$. For such a high yield, Eq. (17.9) gives a $\Delta u$ of about 0.003 m or 3 mm. From the ratio $W/\Delta u \approx 400$, the water-conducting length of this hydrofracture would be around 1.2 m, which, again, is similar to the common strike dimensions of mineral veins in palaeo-geothermal fields in Iceland (Fig. 17.9).

If the geothermal water has no specific, well-defined source, for example it does not accumulate in a sill-like source as in Figs. 17.1 and 17.2, then its only driving pressure arises from the fluid buoyancy. For such a geothermal fluid, the gradient $\partial p_e/\partial y$ in Eq. (17.3) is equal to zero so that, in Eq. (17.9), the factor $N$ becomes about $1.7 \times 10^9$ m$^{-1}$s$^{-1}$. The other parameters in the example above, however, remain the same. Calculations show that the minimum strike dimension and aperture of a hydrofracture that is sufficiently large to sustain a typical spring yield of 5 1/s, or the maximum yield of 180 1/s, does not change much from the dimensions obtained when the excess pressure is included. These results indicate that the pressure that drives geothermal fluids to the surface to supply water to a hot spring is primarily the result of fluid buoyancy. Thus, the excess fluid pressure in a specific sill-like source, or sources of other geometries, has little effect. The main reason for this is that the density difference between the hydrothermal fluid and the host rock is so large that, for an elastic host rock, buoyancy dominates the driving pressure. These results thus differ from those for magma flow to the surface in volcanic eruptions. Because the density of a basaltic magma is similar to, and commonly somewhat greater than, the density of the upper part of the crust, buoyancy has little positive, and sometimes a negative, effect on the pressure that drives basaltic magma to the surface, particularly when the magma originates in a shallow magma chamber (Fig. 17.3).

As indicated above, many hot springs are associated with normal faults. It follows that many hydrofractures supplying geothermal fluids propagate to the surface along dipping normal faults (Figs. 16.2, 16.13, and 16.14). The hydrofractures thus propagate along inclined paths (Fig. 17.10), the dip of which should be taken into account in calculations.

Consider, for example, mineral veins along the normal-fault planes in the lava pile of Iceland. The Tertiary and Pleistocene normal faults in Iceland dip between $42°$ and $89°$, with an average of around $73°$ (Chapters 12 and 14). For example, the spring with the highest yield in Iceland, Deildartunguhver in West Iceland, is associated with a normal fault. If the dip of that fault is equal to the average dip of $73°$, and all the other parameters for the fluid transport are the same as above, then the gradient $\partial p_e / \partial L$ in Eq. (17.4) is 1435 Pa m$^{-1}$. Furthermore, the buoyancy term decreases by a factor of $\sin 73° = 0.956$. Carrying out the

**Fig. 17.10**  Core (central, mineralised part) and damage zone of a normal fault at Kilve, Somerset coast, Southwest England (cf. Fig. 16.13). The damage zone contains many mineral veins, but these are mostly confined to the stiff limestone layers. Most fractures are unable to propagate through the soft shale layers, and those that do are normally shear fractures.

calculations as before, we obtain an aperture of about 0.003 m or 3 mm. This is the same value as obtained above for the vertical hydrofracture. Thus, for a hot spring of a given yield associated with a fault, the dimensions of the fluid-transporting part of the fault are likely to be similar, regardless of whether it is a typical (for Iceland) normal fault or a vertical strike-slip fault.

We can summarise some of the main conclusions of these calculations as follows:

- The yield of a hot spring associated with a fault may be similar regardless of whether the fault is a vertical strike-slip fault or a steeply dipping normal fault. However, if the fluid overpressure, and thus the aperture, of the hydrofracture associated with the spring depends on the stress field associated with the fault, the transmissivity for a hot spring associated with a normal fault and a strike slip fault may differ. Also, if the normal fault is shallow-dipping, as in many sedimentary basins (Fig. 17.10), the dip effect, when compared with a vertical strike-slip fault, may be expected to be more significant.
- A fluid-filled fracture following a fault is normally not in a principal stress plane, so its overpressure is no longer with reference to the minimum compressive principal stress but rather with reference to the normal stress $\sigma_n$ on that part of the fault plane where the fluid transport occurs. The normal stress may be different for normal faults and strike-slip faults, in which case the fluid overpressure and aperture, and thus the fluid transport, may differ between these types of faults. It follows that, for such a situation, the aperture $\Delta u$ does not correspond to the total opening displacement $\Delta u_I$ of a hydrofracture formed in the same area and propagating in a direction perpendicular to the minimum principal compressive stress, $\sigma_3$, but rather that of a fracture propagating in a direction perpendicular to the normal stress $\sigma_n$.

## 17.7 Summary

- Hydrofractures initiate when the excess fluid pressure in the fluid source (a reservoir, a magma chamber, an aquifer) reaches the host-rock tensile strength so that the walls or roof of the source become ruptured. The source and the resulting hydrofracture may be modelled either as self-supporting and rigid, or as elastic. In the rigid model, neither the hydrofracture nor its source deform as the fluid pressure changes; this means that the host rock is assumed to have an infinite Young's modulus. By contrast, in the elastic model, the source and the hydrofracture deform elastically as the fluid pressure changes. The rigid model is commonly used for fluid-transporting fractures at shallow crustal depths in hydrogeology. By contrast, the elastic model is more generally used for fluid-transporting fractures in volcanology, geothermal studies, and petroleum geology.
- Man-made hydrofractures, hydraulic fractures, are generated when a fluid (normally water) is injected under overpressure from a drill hole into the host rock. There are two main reasons for making hydraulic fractures: the first, using small hydraulic fractures, is to measure the in-situ state of stress in a certain layer at a certain depth; the second, using much larger hydraulic fractures, is to increase the permeability of a certain target layer.

For this second aim, the focus is mostly on increasing the permeability of a petroleum reservoir (for oil or gas) or to generate a reservoir of interconnected fractures and pores to form a man-made (artificial) geothermal reservoir.

- Dykes, sills, and inclined sheets are natural magma-driven hydrofractures. Nearly all volcanic eruptions in the world are supplied with magma from dykes and inclined (cone) sheets. The attitude (strike and dip) of the fractures depends on the local stress field around the source magma chamber or reservoir. The overpressure that drives the magma through the dyke or sheet and, in the case of a volcanic eruption, to the surface is related to two main factors: the excess pressure in the source chamber before rupture and dyke formation; and the difference in density between the magma in the dyke and the host rock.

- The excess pressure is normally equal to the tensile strength of the host rock, and thus commonly several megapascals. The density difference between the magma and the host rock contributes to buoyancy, which may be negative (if the average host-rock density is less than that of the magma) or positive (if the average host-rock density is greater than that of the magma). For a positive buoyancy, the overpressure of the magma in a dyke increases on its path to the surface and may reach a maximum of some tens of megapascals. The density of the uppermost several hundred metres, or in places the uppermost one or two kilometres, of the crust (continental and oceanic) is less than that of a typical basaltic magma (density 2650–2700 kg m$^{-3}$), so that the buoyancy is negative. Nevertheless, when the stress conditions are suitable, basaltic dykes and sheets are able to propagate through these low-density crustal layers because of (i) the excess pressure, and (ii) the average density of the crustal segment through which the magma propagates may be higher than that of the magma even if the uppermost crustal layers are less dense than the magma.

- Because of the cubic law, the volumetric flow rate of magma through a dyke or sheet during an eruption depends greatly on the aperture or opening of the dyke-fracture or volcanic fissure. The cubic law means that the volume of magma (or other fluid) per unit time is a function of the third power of the aperture. Slight changes in feeder-dyke aperture during an eruption may thus have strong effects on the volumetric flow rate. By contrast, changes in the length of the eruptive fissure, as are common, during an eruption generally have much less effects on the volumetric flow rate.

- The overpressure gradient that drives magma to the surface through an inclined sheet, and thus the volumetric flow rate, depends on the dip of the sheet. For a flat volcano, such as a collapse caldera, the shallower the dip of the sheet, the longer the distance that the magma has to travel to the surface and, other things being equal, the smaller the overpressure gradient. This relation, however, does not necessarily hold if the volcano forms a topographic high, such as many stratovolcanoes do. Then the dip dimension of a vertical feeder dyke, that is, the distance that the magma must flow along the dyke to reach the surface, may be equal to or less than the dip dimension of an inclined sheet in the same volcano.

- The dimensions of fractures associated with typical hot springs in geothermal fields may be quite small. The strike dimension of a water-conducting fracture need only be 0.5–1 m and its aperture 1–3 mm to provide a yield typical of the large hot springs in Iceland. These dimensions are similar to commonly observed dimensions of mineral veins in palaeo-geothermal fields.

# 17.8  Main symbols used

$g$      acceleration due to gravity

$L$      length of fluid transport along a fracture, for example, the dip dimension of an inclined hydrofracture (such as a sheet) or the strike dimension of a laterally emplaced dyke

$p_e$    excess pressure in a fluid source (such as a reservoir, a water sill, or a magma chamber)

$p_l$    lithostatic stress or pressure; in rock it is stress, in fluid it is pressure

$Q_z^e$   volumetric flow rate through a fracture located in an elastic crust

$Q_z^s$   volumetric flow rate through a fracture located in a rigid (self-supporting) crust

$T_0$    tensile strength

$W$      length (strike dimension) of fracture in a direction perpendicular to the fluid flow direction

$\Delta u$   total opening (aperture) of a hydrofracture (an extension fracture)

$x$      horizontal distance, horizontal coordinate

$z$      vertical distance, vertical coordinate

$\alpha$      dip of hydrofracture

$\mu_f$     dynamic or absolute viscosity of a fluid

$\rho_f$     fluid density

$\rho_r$     rock or crustal density

$\sigma_3$     minimum compressive (maximum tensile) principal stress

# 17.9  Worked examples

---

### Example 17.1

Problem

A hydrofracture in a fault zone transports water to the surface of the zone. The fracture consists of interconnected segments with a total height of 1500 m (Fig. 17.11). The length (strike dimension or width) of the fracture perpendicular to the flow is 50 cm, its aperture is 0.1 cm, the viscosity of the water is $5.5 \times 10^{-4}$ Pa s at $50\,^{\circ}$C and its density is 990 kg m$^{-3}$ (Giles, 1977). The vertical network of fracture segments originates in a water sill (at 1500 m depth) in the damage zone of the fault zone. The fault rock hosting the water sill has a tensile strength of 3 MPa. Calculate:

(a)  the volumetric flow rate $Q$ up through the fracture network if the fault rock hosting the fracture segments is modelled as rigid

(b)  the volumetric flow rate $Q$ if the fault rock is modelled as elastic and has an average density of 2500 kg m$^3$.

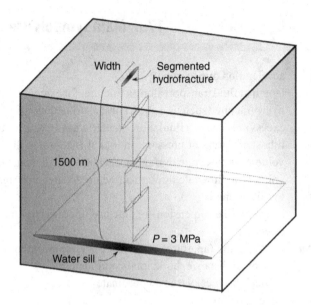

Example 17.1 An interconnected cluster of hydrofractures transports water from a source at 1500 m depth to the surface.

Solution

(a) When the fault rock hosting the fracture network is supposed to behave as rigid or self-supporting, then the volumetric flow rate can be calculated using Eq. (17.2). All the values in Eq. (17.2) are given except the fluid-pressure gradient $\partial p_e/\partial z$. To evaluate this gradient we proceed as follows. The water sill ruptures and forms the hydrofracture network when the excess pressure in it reaches the tensile strength of the host rock, namely 3 MPa. Since the pressure at the Earth's surface is atmospheric and, in comparison with pressures (including fluid pressure) in the crust is regarded as zero, we may assume that the excess pressure varies from 3 MPa in the water sill to 0 MPa at the Earth's surface. (There may be exceptions to this assumption of zero fluid pressure in a fracture at the Earth's surface, such as during some powerful volcanic eruptions, but as a first approximation it is valid.) Then the excess pressure gradient becomes

$$\frac{\partial p_e}{\partial z} = -\frac{3 \times 10^6 \text{ Pa}}{1500 \text{ m}} = -2000 \text{ Pa m}^{-1}$$

The gradient is negative because the $z$-coordinate axis is positive downwards, whereas the flow of the water is upwards and towards a decreasing pressure, namely the Earth's surface. That is, the gradient is doubly negative and thus, in the calculations, becomes positive. Using this result and the values of the parameters above, Eq. (17.2) gives

$$Q_z^s = \frac{b^3 W}{12\mu_f} \left[ \rho_f g - \frac{\partial p_e}{\partial z} \right] = \frac{(0.001 \text{ m})^3 \times 0.5 \text{ m}}{12 \times 5.5 \times 10^{-4} \text{Pa s}}$$

$$\left[ 990 \text{ kg m}^{-3} \times 9.81 \text{ m s}^{-2} - (-2000 \text{ Pa m}^{-1}) \right]$$

$$= 8.88 \times 10^{-4} \text{m}^3\text{s}^{-1}$$

(b) Here the only change from (a) is that the density difference between the rocks of the damage zone of the fault zone and the water comes into play and contributes to buoyancy effects on the water overpressure as it flows upwards through the fault zone and to the Earth's surface. From Eq. (17.3) the volumetric flow rate is

$$
\begin{aligned}
Q_z^e &= \frac{\Delta u^3 W}{12 \mu_f} \left[ (\rho_r - \rho_f) g - \frac{\partial p_e}{\partial z} \right] \\
&= \frac{(0.001\ \text{m})^3 \times 0.5\ \text{m}}{12 \times 5.5 \times 10^{-4}\,\text{Pas}} \\
&\quad \times [(2500\ \text{kg m}^{-3} - 990\ \text{kg m}^{-3}) \times 9.81\ \text{m s}^2 + 2000\ \text{Pa m}^{-1}] \\
&= 1.27 \times 10^{-3}\,\text{m}^3\,\text{s}^{-1}
\end{aligned}
$$

Clearly, therefore, the volumetric rate is considerably higher for a fracture in a damage zone that behaves elastically than for a fracture in a damage zone that behaves as self-supporting or rigid. This is because when the rock behaviour is elastic, buoyancy increases the pressure that drives the water up through the fracture and towards the surface. By contrast, when the rock behaviour is rigid, the only driving pressure is the excess pressure in the water sill. Buoyancy, however, can both increase and decrease the driving pressure. If the fluid has a density that is greater than that of the host rock through which the fluid migrates, then buoyancy has negative effects on the driving pressure. This latter is not uncommon in basaltic eruptions of composite volcanoes (stratovolcanoes, central volcanoes), as is discussed in Example 17.3 below.

## Example 17.2

### Problem

A basaltic dyke reaches the surface to form a volcanic-fissure segment that is 1 km long at the surface (Fig. 17.12) and with an aperture of 1 m. The viscosity of the magma, tholeiite basalt, is 100 Pa s (Murase and McBirney, 1973) and its density is 2650 kg m$^{-3}$ (Williams and McBirney, 1979). The source magma chamber is at a depth of 2 km below the rift zone and the average density of the crust down to that depth is 2550 kg m$^{-3}$. The excess magma pressure in the chamber before rupture and dyke injection is 4 MPa. Calculate:

(a) the volumetric flow rate $Q$ up through the dyke if the crustal segment hosting the dyke and the magma chamber is modelled as rigid
(b) the volumetric flow rate $Q$ up through the dyke if the host rock behaves elastically.

### Solution

(a) As in Example 17.2 we assume that the fluid (here magma) overpressure becomes zero when it reaches the surface of the volcano. Proceeding as in Example 17.1, the excess pressure gradient is

$$
\frac{\partial p_e}{\partial z} = -\frac{4 \times 10^6 \text{Pa}}{2000\ \text{m}} = -2000\ \text{Pa m}^{-1}
$$

**Fig. 17.12**    Example 17.2. The beginning stages of the fissure eruption of July 1980 in the Krafla Volcanic System, North Iceland (photo: Aevar Johannsson; located in Figs. 1.3–1.4). The volcanic fissure consists of offset segments, each eventually reaching several hundred metres to about 1 km in length.

From Eq. (17.2) we then have

$$Q_z^s = \frac{b^3 W}{12\mu_f}\left[\rho_f g - \frac{\partial p_e}{\partial z}\right] = \frac{(1\text{ m})^3 \times 1000\text{ m}}{12 \times 100\text{ Pa s}}$$

$$\left[2650\text{ kg m}^{-3} \times 9.81\text{ m s}^{-2} + 2000\text{ Pa m}^{-1})\right]$$

$$= 2.33 \times 10^4 \text{m}^3\text{s}^{-1}$$

(b) For the same values as in (a) but now using Eq. (17.3) we have

$$Q_z^e = \frac{\Delta u^3 W}{12\mu_f}\left[(\rho_r - \rho_f)g - \frac{\partial p_e}{\partial z}\right]$$

$$= \frac{(1\text{ m})^3 \times 1000\text{ m}}{12 \times 100\text{ Pa s}} \times$$

$$\left[(2550\text{ kg m}^{-3} - 2650\text{ kg m}^{-3}) \times 9.81\text{ m s}^2 + 2000\text{ Pa m}^{-1}\right]$$

$$= 850\text{ m}^3\text{s}^{-1}$$

Thus, the volumetric flow rate is about 27 times greater for the rigid or self-supporting magma chamber and dyke than for the same dyke and magma chamber in an elastic crustal segment. The reason for this great difference is that negative buoyancy of the dense magma propagating through the less dense, elastic crust in (b) decreases the volumetric flow rate. The presentation in (b), however, is much more realistic and the volumetric flow rates obtained in (a) are unrealistic.

### Example 17.3

#### Problem

Consider an inclined sheet injected from a shallow magma chamber associated with a collapse caldera (Fig. 17.3). We use collapse caldera because then the surface is comparatively flat and the topography of the volcano need not be considered. However, the results are of course equally applicable to cone-shape volcanoes such as major stratovolcanoes. An inclined basaltic sheet (a cone sheet) reaches the surface to form a fissure that is 1 km long at the surface with an aperture of 1 m. As in Example 17.2, the viscosity of the magma is 100 Pa s, its density is 2650 kg m$^{-3}$, and the source magma chamber is at a minimum depth of 2 km below the caldera floor. Although the minimum depth to the chamber, of a circular cross-sectional area, is 2000 m, the sheet itself is injected from a point on the boundary of the chamber that is located 100 m deeper in the crust ($x$ in Fig. 17.3). The host rock above the chamber has an average density of 2550 kg m$^{-3}$. The excess magma pressure in the chamber before rupture and sheet injection is 4 MPa. Calculate the volumetric flow rate $Q$ up through the inclined sheet if the host rock behaves elastically and the sheet dip is (a) 60° and (b) 30°, these values being common for sheets in Iceland (Gudmundsson, 1995, 2002).

#### Solution

Since the crustal segment hosting the inclined sheet is assumed to behave elastically, we used Eq. (17.5), namely

$$Q_L^e = \frac{\Delta u^3 W}{12\mu_f} \left[ (\rho_r - \rho_f)g \sin\alpha - \frac{\partial p_e}{\partial L} \right]$$

where $\alpha$ is the dip of the sheet and its dip dimension $L$ is here the linear distance from the point of rupture in the roof of the magma chamber to the surface (Fig. 17.3).

(a) To use Eq. (17.5) for an inclined sheet dipping 60°, we must first find the excess-pressure gradient $\partial p_e/\partial L$. From Eq. (17.7) we obtain the dip dimension $L$ as

$$L = \frac{z+x}{\sin\alpha} = \frac{2000\ \text{m} + 100\ \text{m}}{\sin 60} = \frac{2100\ \text{m}}{0.866} = 2425\ \text{m}$$

Then the excess-pressure gradient becomes

$$\frac{\partial p_e}{\partial L} = -\frac{4 \times 10^6\ \text{Pa}}{2425\ \text{m}} = -1650\ \text{Pa m}^{-1}.$$

Using this result, the volumetric magma flow rate becomes

$$Q_L^e = \frac{(1\text{m})^3 \times 1000\ \text{m}}{12 \times 100\ \text{Pa s}}$$

$$\left[ (2550\ \text{kg m}^{-3} - 2650\ \text{kg m}^{-3}) \times 9.81\ \text{m s}^{-2} \times 0.866 + 1650\ \text{Pa m}^{-1} \right]$$

$$= 667\ \text{m}^3\ \text{s}^{-1}$$

We see that this is less than half the volumetric flow rate for Example 17.2(b), where all the parameters are the same except that the present sheet is inclined and originates from the boundary of the chamber that is 100 m deeper in the crust than in that example.

(b) For the sheet that dips at 30°, we proceed as in (a) to get the dip dimension $L$ as

$$L = \frac{z+x}{\sin\alpha} = \frac{2000\ \text{m} + 100\ \text{m}}{\sin 30} = \frac{2100\ \text{m}}{0.5} = 4200\ \text{m}$$

The excess-pressure gradient is then

$$\frac{\partial p_e}{\partial L} = -\frac{4 \times 10^6\ \text{Pa}}{4200\ \text{m}} = -952\ \text{Pa m}^{-1}$$

The volumetric magma flow rate up the inclined sheet is then

$$Q_L^e = \frac{(1\ \text{m})^3 \times 1000\ \text{m}}{12 \times 100\ \text{Pa s}}$$

$$[(2550\ \text{kg m}^{-3} - 2650\ \text{kg m}^{-3}) \times 9.81\ \text{m s}^{-2} \times 0.5 + 952\ \text{Pa m}^{-1}]$$

$$= 385\ \text{m}^3\text{s}^{-1}$$

If the negative density difference was somewhat greater, the negative buoyancy of the magma would imply that it became unable to reach the surface. Thus, for the given

**Fig. 17.13**    Dyke ending vertically inside a thick pyroclastic layer in Tenerife, Canary Islands. View west, the dyke gradually decreases in thickness from 0.25 m at the foot of the wall to about 0.02 m (2 cm) at its tip.

conditions, an inclined sheet dipping at 30° would not have high enough pressure gradient to drive the magma to the surface. This is, in fact, in agreement with observations. Numerous gently dipping sheets (and many steeply dipping sheets as well) are unable to reach the surface; they become arrested at certain crustal depths. Sheets and dykes

**Fig. 17.14**    Dyke ending vertically at a contact between a stiff lava flow (on top) and a soft pyroclastic unit (cf. Fig. 6.20).

**Fig. 17.15**    Part of the sheet swarm associated with the extinct, late Tertiary central volcano Geitafell in Southeast Iceland. The wall marked by 'sheets' consists almost entirely of 0.5–1-m thick, inclined and mostly basaltic sheets. The 'gabbro/sheet' contact, a depression, marks the boundary between the extinct shallow magma chamber (presently a gabbro body) and the inclined sheets. This contact is also seen in Fig. 6.1.

that end because their driving pressure or overpressure decreases gradually to zero may end anywhere within a layer (Fig. 17.13), whereas dykes and sheets that end because of abrupt changes in mechanical or stress conditions tend to end at layer contacts or other discontinuities (Fig. 17.14). A typical swarm of inclined/cone sheets contains tens of thousands of sheets (Fig. 17.15), and the majority of these never reach the surface to feed eruptions.

## Example 17.4

### Problem

It is well known that ground-water transport in the crust is normally along the maximum hydraulic gradient or, for a pressure-driven flow, along the maximum pressure gradient (Chapter 15). In many volcanoes some of the dykes are driven essentially laterally from the magma chamber, or from a central cylindrical conduit, to the flanks of the volcano where they supply magma to flank eruptions (Acocella and Neri, 2003).

Consider again a collapse caldera with little or no topography in a rift zone, similar to that of the Krafla Caldera (Fig. 17.3). Consider a basaltic dyke injected from its magma chamber at a depth of 2 km, where the average density of the crust above the chamber is 2550 kg m$^{-3}$ and the density of the basaltic magma in the dyke is 2650 kg m$^{-3}$. Assume the excess pressure in the chamber at the time of rupture and dyke formation to be 4 MPa and that the dyke is free to propagate vertically or laterally. That is, the potential effects of local stresses on the direction of the dyke propagation are not considered. Suppose the lateral dyke propagation is along layers with density equal to that of its magma (the level of 'neutral buoyancy'). What maximum lateral distance could such a dyke propagate to the surface in an elastic crust before it becomes more favourable for the dyke to propagate vertically to the surface?

### Solution

Since a laterally emplaced dyke is vertical (similar to the hydraulic fracture in Figs. 13.28 and 17.8), the dip is 90° and Eq. (17.5) can be used in the modified form

$$Q_z^e = \frac{\Delta u^3 W}{12\mu_f}\left[(\rho_r - \rho_f)g - \frac{\partial p_e}{\partial L}\right] \tag{17.10}$$

where $L$ is the distance that the magma is transported to the surface (as in Eqs. (17.4) and (17.5)), which is the strike dimension for a laterally propagating dyke. The source magma chamber, and the lateral dyke, are both assumed to be located at a level of neutral buoyancy for the simple reason that this is the normal assumption made by authors analysing lateral dyke emplacement. This means that the density of the host rock and the magma is the same so that for the laterally emplaced dyke we have

$$(\rho_r - \rho_m)_L = 0 \tag{17.11}$$

where the subscript L stands for laterally emplaced. Clearly, therefore, from Eq. (17.10) the only driving pressure for the laterally emplaced dyke is the excess-pressure gradient. For the vertical dyke, and assuming that the source magma chamber is at the level of neutral buoyancy, we have

$$(\rho_r - \rho_m)_V \leq 0 \tag{17.12}$$

where the subscript V stands for vertically emplaced. Inequality (17.12) follows from the assumption that the source chamber is at a level of neutral buoyancy and, for a homogeneous, isotropic crust, the density normally stays the same or decreases from that level and to the Earth's surface.

Clearly, if $(\rho_r - \rho_m)_V = 0$ (Eq. 17.12) then the distance that a dyke may propagate laterally, for the given conditions, before it becomes more favourable to propagate vertically, is equal to the depth to the dyke initiation in the roof of the chamber – here 2 km. The maximum length that the lateral dyke can propagate is when the vertical and horizontal gradients are equal. Using Eq. (17.3) for the vertically propagating dyke and Eqs. (17.10) and (17.11) for the laterally propagating dyke, the maximum distance is obtained from equating these two, namely when

$$-\frac{\partial p_e}{\partial L} = (\rho_r - \rho_m)g - \frac{\partial p_e}{\partial z} \tag{17.13}$$

which can be rewritten as (because the excess-pressure gradient is negative but we want the distance $L$ to be positive)

$$\frac{\partial p_e}{\partial L} = \frac{\partial p_e}{\partial z} - (\rho_r - \rho_m)g \tag{17.14}$$

For the case when the average crustal density above the chamber is 2500 kg m$^{-3}$, and the other values as given, we have from Eq. (17.14):

$$\frac{\partial p_e}{\partial L} = \frac{4 \times 10^6 \text{Pa}}{2000 \text{ m}} - (2500 \text{ kg m}^{-3} - 2650 \text{ kg m}^{-3}) \times 9.81 \text{ m s}^{-2} = 528.5 \text{ Pa m}^{-1}$$

It follows that

$$\partial L = \frac{4 \times 10^6 \text{ Pa}}{528.5 \text{ Pa m}^{-1}} = 7568.6 \text{ m} \approx 7.6 \text{ km}$$

This is similar to the inferred lateral propagation distances of many dykes that erupted during the Krafla Fires in North Iceland (1975–1984) and in many other volcanoes (Acocella and Neri, 2003).

Clearly, as the maximum distance for lateral dyke propagation to the surface increases the density of the host rock above the magma chamber, at a given depth, decreases. However, if the chamber also becomes shallower, the maximum lateral propagation distance may, in fact, decrease. For example, very shallow magma chambers are at a depth of only 1 km. In Iceland, the average density of the uppermost 1 km of the crust in the rift zone is about 2450 kg m$^{-3}$ (Gudmundsson, 1988). For the same excess pressure and other values as above, the corresponding maximum distance for a laterally propagating dyke would be 1963 m or less than 2 km.

By contrast, if the magma chamber was at the same depth but the crustal density above the chamber was reduced to 2450 kg m$^{-3}$, which would be unrealistic for basaltic rift zones, then the maximum lateral distance would reach about 100 km. However, in that case the excess pressure gradient is so small that the velocity of the magma transport will be very low (centimetres per second or less), so that, depending on the dyke aperture, cooling and viscosity increase of the magma may become a limiting factor for how long distance a lateral dyke could propagate for the given conditions.

## Example 17.5

### Problem

During the 1991 eruption of the Hekla Volcano in South Iceland (Fig. 17.16), the maximum volumetric flow rate during the first hours of the eruption reached 800 $m^3$ $s^{-1}$ (Gudmundsson et al., 1992). During these early stages of the eruption, the active volcanic fissure was about 3 km long. The depth of the magma chamber or reservoir supplying magma to the Hekla eruptions is not known in detail, but is generally thought to be about 10 km for the basaltic magma (Gudmundsson et al., 1992). The crustal segment hosting the volcano and the reservoir is known, from geodetic measurements, to behave as approximately elastic during eruptions (Gudmundsson et al., 1992). The average crustal density of the uppermost 10 km of the crust of the volcanic zones in Iceland is about 2800 kg $m^{-3}$ (Gudmundsson, 1988). For a basaltic magma we may take the density as 2650 kg $m^{-3}$ and the viscosity as 100 Pa s, both values that could vary somewhat, and the excess pressure in the reservoir before rupture and dyke emplacement as 3 MPa. Given this information,

**Fig. 17.16**  Hekla Volcano in South Iceland during its 1991 eruption (cf. Fig. 1.2). View northeast, the volcano mainly erupted through a 3-km-long fissure, partly located at its top (reopening of the top fissure referred to as the Hekla Fissure) and partly located on the southeast sloping flanks of the volcano where, subsequently, the main crater cone formed.

calculate the aperture or opening of the 3-km-long fissure during the beginning stages of the eruption.

Solution

Since the feeder dyke is assumed to be vertical and the crustal segment behaves elastically, we use Eq. (17.3), namely

$$Q_z^e = \frac{\Delta u^3 W}{12\mu_f} \left[ (\rho_r - \rho_f)g - \frac{\partial p_e}{\partial z} \right]$$

From the values given in the problem, we have

$$800 \text{ m}^3 = \frac{\Delta u^3 \times 3000 \text{ m}}{12 \times 100 \text{ Pa s}} \left[ (2800 \text{ kg m}^{-3} - 2650 \text{ kg m}^{-3}) \times 9.81 \text{ m s}^{-2} + \frac{3 \times 10^6 \text{Pa}}{10\,000 \text{ m}} \right]$$

So that

$$\Delta u^3 \times 4428.75 = 800 \Rightarrow \Delta u = \left( \frac{800}{4428.75} \right)^{1/3} = 0.56 \text{ m} \approx 0.6 \text{ m}$$

Thus, during the early stages of the 1991 Hekla eruption the feeder dyke may have had an aperture or opening of about 0.6 m. While these results are quite crude, since none of the estimated values, such as the viscosity, fissure length, and maximum volumetric flow rate in the early stages are very accurate, they agree well with previous estimates indicating that the feeder dykes formed in the Hekla eruptions are of thickness of the order of 0.5–1 m (Gudmundsson and Brenner, 2005).

# 17.10 Exercises

17.1 Define the concept of a hydrofracture (a fluid-driven fracture) and list some fields of earth sciences where the transport of fluids by hydrofractures plays an important role.

17.2 What is the difference between the total fluid pressure and the excess fluid pressure in a reservoir?

17.3 How does the behaviour of a rigid (self-supporting) fluid source differ from that of an elastic fluid source during fluid transport out of the source? What does rigid mean in terms of the assumed Young's modulus of the rock?

17.4 Man-made hydraulic fractures injected from drill holes sometimes propagate along curved paths next to the drill hole and then, at a greater distance, along essentially planar paths. What do such curved paths indicate as regards the local stress field?

17.5 What are hydraulic wing cracks?

17.6 In comparison with an elastic model, a self-supporting (rigid) model of a shallow magma chamber normally overestimates, whereas self-supporting model of a deep-seated reservoir underestimates, the volumetric magma flow through a dyke (and an associated volcanic fissure). Why is there this difference, in comparison with elastic models, between shallow and deep-seated rigid reservoirs?

17.7 In early stages of the 2010 eruption of Eyjafjallajökull Volcano in South Iceland (Fig. 16.9), the volcanic fissure was about 700 m long. The magma chamber is at a depth of about 5 km and the average crustal density above the chamber is about 2670 kg m$^{-3}$. The magma density is 2650 kg m$^{-3}$ and its viscosity 100 Pa s. If the feeder-dyke is vertical, the excess magma pressure in the chamber before rupture 1 MPa, and the fissure aperture 0.5 m, calculate the volumetric flow rate assuming a self-supporting crustal model.

17.8 Repeat Exercise 17.7 assuming an elastic crustal model.

17.9 A vertical hydrofracture in a fault zone transports water to the surface of the zone. The fracture consists of interconnected segments with a total height of 500 m. The length (strike dimension) or width of the fracture perpendicular to the flow is 1 m, its aperture is 0.6 mm, the viscosity of the water is 5.5 × 10$^{-4}$ Pa s at 50° C and its density is 990 kg m$^{-3}$. The vertical network of fracture segments originates in a water sill (at 500 m depth) in the damage zone of the fault zone. The fault rock hosting the water sill has a tensile strength of 1 MPa. Calculate the volumetric flow rate $Q$ up through the fracture network if the fault rock hosting the fractures is modelled as rigid.

17.10 Repeat exercise 17.9 assuming the fault rock hosting the fractures behaves elastically and has an average density of 2400 kg m$^3$.

17.11 An inclined basaltic sheet (a cone sheet) reaches the flat floor of a collapse caldera to form a fissure that is 0.5 km long at the surface with an aperture of 0.4 m. The magma is 100 Pa s, its density is 2650 kg m$^{-3}$, and the source sill-like magma chamber at the location of rupture and sheet injection is at depth of 4 km below the caldera floor. The average host-rock density is 2650 kg m$^{-3}$. The excess magma pressure in the chamber before rupture and sheet injection is 2 MPa. Calculate the volumetric flow rate $Q$ up through the inclined sheet if the host rock behaves elastically and the sheet dip is (a) 70° and (b) 25°.

# References and suggested reading

Acocella, V. and Neri, M., 2003. What makes flank eruptions? The 2001 Etna eruption and its possible triggering mechanism. *Bulletin of Volcanology*, **65**, 517–529.

Acocella, V. and Neri, M., 2009. Dike propagation in volcanic edifices: overview and possible developments. *Tectonophysics*, **471**, 67–77.

Advani, S. H., Lee, T. S., Dean, R. H., Pak, C. K., and Avasthi, J. M., 1997. Consequences of fluid lag in three-dimensional hydraulic fractures. *International Journal of Numerical and Analytical Methods in Geomechanics*, **21**, 229–240.

Berkowitz, B., 2002. Characterizing flow and transport in fractured geological media: a review. *Advances in Water Resources*, **25**, 861–884.

Bons, P. D., 2001. The formation of large quartz veins by rapid ascent of fluids in mobile hydrofractures. *Tectonophysics*, **336**, 1–17.

Brebbia, C. A. and Dominguez, J., 1992. *Boundary Elements: an Introductory Course.* Boston, MA: Computational Mechanics.

Brenner, S. L. and Gudmundsson, A., 2002. Permeability development during hydrofrac-
    ture propagation in layered reservoirs. *Norges Geologiske Undersøkelse Bulletin*, **439**,
    71–77.
Brenner, S. L. and Gudmundsson, A. 2004. Permeability in layered reservoirs: field exam-
    ples and models on the effects of hydrofracture propagation. In: Stephansson, O.,
    Hudson, J. A., and Jing, L. (eds.), *Coupled Thermo-Hydro-Mechanical-Chemical Pro-
    cesses in Geo-Systems, Geo-Engineering Series, Vol. 2*. Amsterdam: Elsevier, pp.
    643–648.
Chilingar, G. V., Serebryakov, V. A., and Robertson, J. O., 2002. *Origin and Prediction of
    Abnormal Formation Pressures*. London: Elsevier.
Coward, M. P., Daltaban, T. S., and Johnson, H. (eds.), 1998. *Structural Geology in Reservoir
    Characterization*. Geological Society of London Special Publication 127. London:
    Geological Society of London.
Dahlberg, E. C., 1994. *Applied Hydrodynamics in Petroleum Exploration*. New York:
    Springer-Verlag.
Daneshy, A. A., 1978. Hydraulic fracture propagation in layered formations. *AIME Society
    of Petroleum Engineers J*, 33–41.
Davis, P. M., 1983. Surface deformation associated with a dipping hydrofracture. *Journal
    of Geophysical Research*, **88**, 5826–5834.
de Marsily, G., 1986. *Quantitative Hydrogeology*. New York: Academic Press.
Deming, D., 2002. *Introduction to Hydrogeology*. McGraw-Hill, Boston.
Economides, M. J. and Boney, C., 2000. Reservoir stimulation in petroleum production. In:
    Economides, M. J. and Nolte, K. G. (eds.), *Reservoir Stimulation*. New York: Wiley,
    pp. 1–1 to 1–30.
Economides, M. J. and Nolte, K. G. (eds.), 2000. *Reservoir Stimulation*. New York: Wiley.
Faybishenko, B., Witherspoon, P. A., and Benson, S. M. (eds.), 2000. *Dynamics of Fluids
    in Fractured Rocks*. Washington, DC: American Geophysical Union.
Finkbeiner, T., Barton, C. A., and Zoback, M. D., 1997. Relationships among in-situ stress,
    fractures and faults, and fluid flow: Monterey formation, Santa Maria basin, California.
    *American Association of Petroleum Geologists Bulletin*, **81**, 1975–1999.
Flekkoy, E. G., Malthe-Sorenssen, A., and Jamtveit, B., 2002. Modeling hydrofracture.
    *Journal of Geophysical Research*, **107**(B8), 2151, doi:10.1029/2000JB000132.
Garagash, D. and Detournay, E., 2000. The tip region of a fluid-driven fracture in an elastic
    medium. *Journal of Applied Mechanics*, **67**, 183–192.
Geshi, N., Kusumoto, S., and Gudmundsson, A., 2010. Geometric difference between non-
    feeders and feeder dikes. *Geology*, **38**, 195–198.
Giles, R. V., 1977. *Fluid Mechanics and Hydraulics*, 2nd edn. New York: McGraw-Hill.
Grecksch, G., Roth, F., and Kümpel, H. J., 1999. Coseismic well-level changes due to the
    1992 Roermond earthquake compared to static deformation of half-space solutions.
    *Geophysical Journal International*, **138**, 470–478.
Gudmundsson, A., 1988. Effect of tensile stress concentration around magma chambers on
    intrusion and extrusion frequencies. *Journal of Volcanology and Geothermal Research*,
    **35**, 179–194.
Gudmundsson, A., 1995. The geometry and growth of dykes. In: G. Baer and A. Heimann
    (eds.), *Physics and Chemistry of Dykes*. Rotterdam: Balkema, pp. 23–34.
Gudmundsson, A., 2002. Emplacement and arrest of sheets and dykes in central volcanoes.
    *Journal of Volcanology and Geothermal Research*, **116**, 279–298.
Gudmundsson, A. and Brenner, S. L., 2005. On the conditions of sheet injections and
    eruptions in stratovolcanoes. *Bulletin of Volcanology*, **67**, 768–782.
Gudmundsson, A. and Philipp, S. L., 2006. How local stresses prevent volcanic eruptions.
    *Journal of Volcanology and Geothermal Research*, **158**, 257–268.

Gudmundsson, A., Fjeldskaar, I., and Brenner, S. L., 2002. Propagation pathways and fluid transport of hydrofractures in jointed and layered rocks in geothermal fields. *Journal of Volcanology and Geothermal Research*, **116**, 257–278.

Gudmundsson, A., Oskarsson, N., Gronvold, K. *et al.*, 1992. The 1991 eruption of Hekla, Iceland. *Bulletin of Volcanology*, **54**, 238–246.

Gulrajani, S. N. and Nolte, K. G., 2000. Fracture evaluation using pressure diagnostics. In: Economides, M. J. and Nolte, K. G. (eds.), *Reservoir Stimulation*. New York: Wiley, pp. 9–1 to 9–63.

Haimson, B. C. and Rummel, F., 1982. Hydrofracturing stress measurements in the Iceland research drilling project drill hole at Reydarfjordur, Iceland. *Journal of Geophysical Research*, **87**, 6631–6649.

Haneberg, W. C., Mozley, P. S., Moore, J. C., and Goodwin, L. B. (eds.), 1999. *Faults and Subsurface Fluid Flow in the Shallow Crust*. Washington, DC: American Geophysical Union.

Hardebeck, J. L. and Hauksson, E., 1999. Role of fluids in faulting inferred from stress field signatures. *Science*, **285**, 236–239.

Heimpel, M. and Olson, P., 1994. Buoyancy-driven fracture and magma transport through the lithosphere: models and experiments. In: Ryan, M. P. (ed.), *Magmatic Systems*. London: Academic Press, pp. 223–240.

Hubbert, M. K. and Rubey, W. W., 1959. Role of fluid pressure in mechanics of overthrust faulting Part I. *Geological Society of America Bulletin*, **70**, 115–166.

Hubbert, M. K. and Willis, D. G., 1957. Mechanics of hydraulic fracturing. *Transactions of the American Institute of Mining, Metallurgical and Petroleum Engineers*, **210**, 153–168.

Ingebritsen, S. E. and Sanford, W. E., 1998. *Groundwater in Geologic Processes*. Cambridge: Cambridge University Press.

King, C. Y., Azuma, S., Igarashi, G., Ohno, M., Saito, H., and Wakita, H., 1999. Earthquake-related water-level changes at 16 closely clustered wells in Tono, central Japan. *Journal of Geophysical Research*, **104**, 13 073–13 082.

Lamb, H., 1932. *Hydrodynamics*, 6th edn. Cambridge: Cambridge University Press.

Larsen, B., Grunnaleite, I., and Gudmundsson, A., 2009. How fracture systems affect permeability development in shallow-water carbonate rocks: an example from the Gargano Peninsula, Italy. *Journal of Structural Geology*, document doi:10.1016/j.jsg.2009.05.009.

Lister, J. R. and Kerr, R. C., 1991. Fluid-mechanical models of crack propagation and their application to magma transport in dykes. *Journal of Geophysical Research*, **96**, 10 049–10 077.

Mahrer, K. D., 1999. A review and perspective on far-field hydraulic fracture geometry studies. *Journal of Petroleum Science and Engineering*, **24**, 13–28.

Mandl, G. and Harkness, R. M., 1987. Hydrocarbon migration by hydraulic fracturing. In: Jones, M. E. and Preston, R. M. E. (eds.), *Deformation of Sediments and Sedimentary Rocks*. Geological Society of London Special Publication 29. London: Geological Society of London, pp. 39–53.

Milne-Thompson, L. M., 1996. *Theoretical Hydrodynamics*, 5th edn. New York: Dover.

Muirwood, R. and King, G. C. P., 1993. Hydrological signatures of earthquake strain. *Journal of Geophysical Research*, **98**, 22 035–22 068.

Murase, T. and McBirney, A. R., 1973. Properties of some common igneous rocks and their melts at high temperatures. *Bulletin of the Geological Society of America*, **84**, 3563–3592.

Nelson, R. A., 1985. *Geologic Analysis of Naturally Fractured Reservoirs*. Houston, TX: Gulf Publishing.

Neuzil, C. E., 2003. Hydromechanical coupling in geologic processes. *Hydrogeological Journal*, **11**, 41–83.

Nunn, J. A. and Meulbroek, P., 2002. Kilometer-scale upward migration of hydrocarbons in geopressured sediments by buoyancy-driven propagation of methane filled fractures. *American Association of Petroleum Geologists Bulletin*, **86**, 907–918.

Odling, N. E., Gillespie, P., Bourgine, B. *et al.*, 1999. Variations in fracture system geometry and their implications for fluid flow in fractured hydrocarbon reservoirs. *Petroleum Geoscience*, **5**, 373–384.

Philipp, S. L., 2008. Geometry and formation of gypsum veins in mudstones at Watchet, Somerset, SW-England. *Geological Magazine*, **145**, 831–844.

Pollard, D. D. and Segall, P., 1987. Theoretical displacement and stresses near fractures in rocks: with application to faults, points, veins, dikes, and solution surfaces. In: Atkinson, B. (ed.), *Fracture Mechanics of Rocks*. London: Academic Press, pp. 277–349.

Ray, R., Sheth, H. C., and Mallik, J. 2007. Structure and emplacement of the Nandurbar–Dhule mafic dyke swarm, Deccan Traps, and the tectonomagmatic evolution of flood basalts. *Bulletin of Volcanology*, **69**, 537–551.

Rijsdijk, K. F., Owen, G., Warren, W. P., McCarroll, D., and van der Meer, J. J. M., 1999. Clastic dykes in over-consolidated tills: evidence for subglacial hydrofracturing at Killiney Bay, eastern Ireland. *Sedimentary Geology*, **129**, 111–126.

Roeloffs, E. A., 1988. Hydrologic precursors to earthquakes – a review. *Pure and Applied Geophysics*, **126**, 177–209.

Rojstaczer, S., Wolf, S., and Michel, R., 1995. Permeability enhancement in the shallow crust as a cause of earthquake-induced hydrological changes. *Nature*, **373**, 237–239.

Rubin, A. M., 1995. Propagation of magma-filled cracks. *Annual Reviews of Earth and Planetary Sciences*, **23**, 287–336.

Rummel, F., 1987. Fracture mechanics approach to hydraulic fracturing stress measurements. In: Atkinson, B. (ed.), *Fracture Mechanics of Rock*. London: Academic Press, pp. 217–239.

Schön, 2004, *Physical Properties of Rocks: Fundamentals and Principles of Petrophysics*, 2nd edn. Amsterdam: Elsevier.

Schultz, R. A., 1995. Limits on strength and deformation properties of jointed basaltic rock masses. *Rock Mechanics and Rock Engineering*, **28**, 1–15.

Secor, D. T., 1965. Role of fluid pressure in jointing. *American Journal of Science*, **263**, 633–646.

Secor, D. T. and Pollard, D. D., 1975. On the stability of open hydraulic fractures in the earth's crust. *Geophysical Research Letters*, **2**, 510–513.

Selley, R. C., 1998. *Elements of Petroleum Geology*. London: Academic Press.

Simonson, E. R., Abou-Sayed, A. S., and Clifton, R. J., 1978. Containment of massive hydraulic fractures. *Society of Petroleum Engineers Journal*, **18**, 27–32.

Singhal, B. B. S. and Gupta, R. P., 1999. *Applied Hydrogeology of Fractured Rocks*. London: Kluwer.

Smart, B. G. D., Somerville, J. M., Edlman, K., and Jones, C., 2001. Stress sensitivity of fractured reservoirs. *Journal of Petroleum Science and Engineering*, **29**, 29–37.

Smith, M. B. and Shlyapobersky, J. W., 2000. Basics of hydraulic fracturing. In: Economides, M. J. and Nolte, K. G. (eds.), *Reservoir Stimulation*. New York: Wiley, pp. 5–1 to 5–28.

Smits, A. J., 2000. *A Physical Introduction to Fluid Mechanics*. New York: Wiley.

Sneddon, I. N. and Lowengrub, M., 1969. *Crack Problems in the Classical Theory of Elasticity*. New York: Wiley.

Spence, D. A. and Turcotte, D. L. 1985. Magma-driven propagation of cracks. *Journal of Geophysical Research*, **90**, 575–580.

Spence, D. A., Sharp, P. W., and Turcotte, D. L. 1987. Buoyancy-driven crack-propagation – a mechanism for magma migration. *Journal of Fluid Mechanics*, **174**, 135–153.

Stearns, D. W. and Friedman, M., 1972. Reservoirs in fractured rocks. *American Association of Petroleum Geologists Memoirs*, **16**, 82–106.

Stewart, M. A., Klein, E. M., Karson, J. A., and Brophy, J. G., 2003. Geochemical relationships between dikes and lavas at the Hess Deep Rift: implications for magma eruptibility. *Journal of Geophysical Research*, **108**(B4), 2184, doi:10.1029/2001JB001622.

Sun, R. J., 1969. Theoretical size of hydraulically induced horizontal fractures and corresponding surface uplift in an idealized medium. *Journal of Geophysical Research*, **74**, 5995–6011.

Teufel, L. W. and Clark, J. A., 1984. Hydraulic fracture propagation in layered rock: Experimental studies of fracture containment. *Society of Petroleum Engineers Journal*, **24**, 19–32.

Tsang, C. F. and Neretnieks, I., 1998. Flow channeling in heterogeneous fractured rocks. *Reviews of Geophysics*, **36**, 275–298.

Valko, P. and Economides, M. J., 1995. *Hydraulic Fracture Mechanics*. New York: Wiley.

van Eekelen, H. E. M., 1982. Hydraulic fracture geometry: fracture containment in layered formations. *Society of Petroleum Engineers Journal*, **22**, 341–349.

Vermilye, J. M. and Scholz, C. H., 1995. Relation between vein length and aperture. *Journal of Structural Geology*, **17**, 423–434.

Wang, J. S. Y., 1991. Flow and transport in fractured rocks. *Reviews of Geophysics*, **29**, 254–262.

Warpinski, N. R., 1985. Measurement of width and pressure in a propagating hydraulic fracture. *Society of Petroleum Engineers Journal*, **25**, 46–54.

Warpinski, N. R., Schmidt, R. A., and Northrop, D. A., 1982. In-situ stress: the predominant influence on hydraulic fracture containment. *Journal of Petroleum Technology*, **34**, 653–664.

Warpinski, N. R. and Teufel, L. W., 1987. Influence of geologic discontinuities on hydraulic fracture propagation. *Journal of Petroleum Technology*, **39**, 209–220.

Weertman, J., 1971. Theory of water-filled crevasses in glaciers applied to vertical magma transport beneath oceanic ridges. *Journal of Geophysical Research*, **76**, 1171–1183.

Weertman, J., 1996. *Dislocation Based Fracture Mechanics*. London: World Scientific.

Williams, H. and McBirney, A. R., 1979. *Volcanology*. San Francisco, CA: W.H. Freeman.

Yew, C. H., 1997. *Mechanics of Hydraulic Fracturing*. Houston, TX: Gulf Publishing.

Zhang, X., Jeffrey, R. G., and Thiercelin, M., 2007a. Deflection and propagation of fluid-driven fractures at frictional bedding interfaces: a numerical investigation. *Journal of Structural Geology*, **29**, 396–410.

Zhang, X., Koutsabeloulis, N., and Heffer, K., 2007b. Hydromechanical modeling of critically stressed and faulted reservoirs. *American Association of Petroleum Geologists Bulletin*, **91**, 31–50.

Zienkiewicz, O. C., 1977. *The Finite Element Method*. London: McGraw-Hill.

# Appendix A: Units, dimensions, and prefixes

Scientists and engineers use the International System of Units (in French: Système Internationale d'Unités – hence the acronym SI units), which has **seven base units**, relying on simple physical effects, and many **derived units**, only some of which (such as force and stress) are listed below. Also provided below are the dimensions of some of the quantities, where L denotes length, M mass, T time, and F force. More details are given by Huntley (1967), Gottfried (1979), Emiliani (1995), Benenson *et al.* (2002), Woan (2003), and Deeson (2007). The reference list for all the appendices is at the end of Appendix E.

## A.1 SI base units

| Physical quantity | Unit name | Symbol | Dimensions |
|---|---|---|---|
| Length | metre | m | L |
| Mass | kilogram | kg | M |
| Time (interval) | second | s | T |
| Electric current | ampere | A | I |
| Thermodynamic temperature | kelvin | K | $\Theta$ |
| Amount of substance | mole | mol | N |
| Luminous intensity | candela | cd | J |

## A.2 Derived SI units of some quantities

| Physical quantity | Unit name | Symbol | Dimensions | SI units |
|---|---|---|---|---|
| Area | | | $L^2$ | $m^2$ |
| Energy | joule | J | $L^2 M T^{-2}$ | N m |
| Force | newton | N | $L M T^{-2}$ | $kg\ m\ s^{-2}$ |
| Heat | joule | J | $L^2 M T^{-2}$ | N m |
| Power | watt | W | $L^2 M T^{-3}$ | $J\ s^{-1}$ |
| Pressure | pascal | Pa | $L^{-1} M T^{-2}$ | $N\ m^{-2}$ |
| Shear modulus | pascal | Pa | $L^{-1} M T^{-2}$ | $N\ m^{-2}$ |

| | | | | |
|---|---|---|---|---|
| Speed | | | $LT^{-1}$ | $m\ s^{-1}$ |
| Stress | pascal | Pa | $L^{-1}\ M\ T^{-2}$ | $N\ m^{-2}\ (kg\ m^{-1}\ s^{-2})$ |
| Velocity | | | $LT^{-1}$ | $m\ s^{-1}$ |
| Viscosity (dynamic) | | | $L^{-1}\ M\ T^{-1}$ | $Pa\ s\ (kg\ m^{-1}s^{-1})$ |
| Viscosity (kinematic) | | | $L^2T^{-1}$ | $m^2\ s^{-1}$ |
| Volume | | | $L^3$ | $m^3$ |
| Weight | newton | N | $L\ M\ T^{-2}$ | $kg\ m\ s^{-2}$ |
| Work | joule | J | $L^2M\ T^{-2}$ | $N\ m$ |
| Young's modulus | pascal | Pa | $L^{-1}\ M\ T^{-2}$ | $N\ m^{-2}$ |

# A.3 SI prefixes

Here the EU name means the name of the factor as commonly used in continental European countries and many others. The UK and US names are specified only when they differ from the EU names. Note that SI multipliers and submultipliers go up and down in steps of thousand ($10^3$). It follows that hecto, deca, deci, and centi are not strictly SI prefixes. Also, widely used abbreviations for centimetres and millimetres are cm and mm, respectively.

| Value/factor | Prefix | Symbol | EU name | UK name | US name |
|---|---|---|---|---|---|
| $10^{24}$ | yotta | Y | quadrillion | | septillion |
| $10^{21}$ | zetta | Z | trilliard | thousand trillion | septillion |
| $10^{18}$ | exa | E | trillion | | quintillion |
| $10^{15}$ | peta | P | billiard | thousand billion | quadrillion |
| $10^{12}$ | tera | T | billion | | trillion |
| $10^{9}$ | giga | G | milliard | thousand million | billion |
| $10^{6}$ | mega | M | million | | |
| $10^{3}$ | kilo | k | thousand | | |
| $10^{2}$ | hecto | h | hundred | | |
| $10^{1}$ | deca/deka | da | ten | | |
| $10^{0}$ | none | none | none | | |
| $10^{-1}$ | deci | d | tenth | | |
| $10^{-2}$ | centi | c | hundredth | | |
| $10^{-3}$ | milli | m | thousandth | | |
| $10^{-6}$ | micro | $\mu$ | millionth | | |
| $10^{-9}$ | nano | n | milliardth | thousand millionth | billionth |
| $10^{-12}$ | pico | p | billionth | | trillionth |
| $10^{-15}$ | femto | f | billiardth | thousand billionth | quadrillionth |
| $10^{-18}$ | atto | a | trillionth | | quintillionth |
| $10^{-21}$ | zepto | z | trilliardth | thousand trillionth | sextillionth |
| $10^{-24}$ | yocto | y | quadrillionth | | septillionth |

# Appendix B: The Greek alphabet

| Lower case | Upper case | Name |
|---|---|---|
| $\alpha$ | A | alpha |
| $\beta$ | B | beta |
| $\gamma$ | $\Gamma$ | gamma |
| $\delta$ | $\Delta$ | delta |
| $\varepsilon$ | E | epsilon |
| $\zeta$ | Z | zeta |
| $\eta$ | H | eta |
| $\theta$ | $\Theta$ | theta |
| $\iota$ | I | iota |
| $\kappa$ | K | kappa |
| $\lambda$ | $\Lambda$ | lambda |
| $\mu$ | M | mu |
| $\nu$ | N | nu |
| $\xi$ | $\Xi$ | xi |
| $o$ | O | omicron |
| $\pi \, \varpi$ | $\Pi$ | pi |
| $\rho$ | P | rho |
| $\sigma \, \varsigma$ | $\Sigma$ | sigma |
| $\tau$ | T | tau |
| $\upsilon$ | $\Upsilon$ | upsilon |
| $\phi, \varphi$ | $\Phi$ | phi |
| $\chi$ | X | chi |
| $\psi$ | $\Psi$ | psi |
| $\omega$ | $\Omega$ | omega |

# Appendix C: Some mathematical and physical constants

| Quantity | Symbol | Value |
|---|---|---|
| Ratio of the circumference to the diameter of a circle | $\pi$ | 3.1416 |
| Exponential constant; base of natural logarithms | e | 2.7183 |
| Typical acceleration due to gravity at the Earth's surface | $g$ | 9.81 m s$^{-2}$ |
| Acceleration due to gravity at the surface at the equator | $g_e$ | 9.78 m s$^{-2}$ |
| Acceleration due to gravity at the surface at the poles | $G_p$ | 9.83 m s$^{-2}$ |
| Gravitational constant | $G$ | $6.673 \times 10^{-11}$ N m$^2$ kg$^{-2}$ |
| One calendar year, 365 days | yr | $3.1536 \times 10^7$ s |
| Earth's equatorial radius | $R_e$ | $6.3781 \times 10^6$ m |
| Earth's polar radius | $R_p$ | $6.3567 \times 10^6$ m |
| Earth's mean radius | $R$ | $6.37 \times 10^6$ m |
| Earth's volume | $V$ | $1.083 \times 10^{21}$ m$^3$ |
| Earth's mass | $M$ | $5.974 \times 10^{24}$ kg |
| Earth's mean density | $\rho$ | $5.515 \times 10^3$ kg m$^{-3}$ |
| Earth's surface area | $A$ | $5.10 \times 10^{14}$ m$^2$ |

Below are values of Young's modulus and Poisson's ratios of some common rocks (Appendix D.1). All the other elastic moduli or constants can be derived from these two using the relations in Appendix D.2. These values are derived from (mostly static) laboratory measurements, and these are normally different from the in-situ values. All rocks show a great variety in elastic properties, in particular when measured in situ. The 'typical values' shown below are rounded and are meant as a rough guide. These data on rock properties as measured in the laboratory are derived from tables and information in many books including Jaeger and Cook (1979), Jumikis (1979), Carmichael (1989), Jeremic (1994, for rock salt), Waltham (1994), Hansen (1998), Nilsen and Palmström (2000), Myrvang (2001), Schön (2004), Paterson and Wong (2005), Fjaer *et al.* (2008), and Mavko *et al.* (2009). When data are not known, or not measurable (such as the tensile strength of unconsolidated sand or gravel), there is no number in the appropriate box in the tables in Appendices D and E.

## D.1 Typical Young's moduli and Poisson's ratios

| Rock type | Young's modulus: normal range, GPa | Young's modulus: typical values, GPa | Poisson's ratio: normal range | Poisson's ratio: typical values |
|---|---|---|---|---|
| **Sedimentary rocks** | | | | |
| Limestone | 7.8–150.0 | 20–70 | 0.10–0.44 | 0.22–0.34 |
| Dolomite | 19.6–93.0 | 30–80 | 0.08–0.37 | 0.10–0.30 |
| Sandstone | 0.4–84.3 | 20–50 | 0.05–0.45 | 0.10–0.30 |
| Shale | 0.14–44.0 | 1–20 | 0.04–0.30 | 0.10–0.25 |
| Marl/mudstone | 0.002–0.25 | 0.025–0.10 | | |
| Chalk | 0.5–30.0 | 1–15 | 0.05–0.37 | 0.15–0.30 |
| Anhydrate | 48.8–86.4 | 70–80 | 0.20–0.34 | 0.25–0.30 |
| Gypsum | 0.74–36.0 | 15–35 | 0.14–0.47 | 0.18–0.30 |
| Rock salt | 5.0–42.3 | 15–30 | 0.09–0.50 | 0.29–0.38 |
| Conglomerate | 24.0–51.0 | 30–40 | 0.22–0.23 | 0.22–0.23 |

| | | | | |
|---|---|---|---|---|
| Unconsolidated sand | 0.01–0.1 | 0.05 | 0.45 | 0.45 |
| Clay | 0.003–0.5 | 0.05–0.1 | 0.40 | 0.40 |
| **Igneous rocks** | | | | |
| Basalt | 20.0–128.0 | 50–90 | 0.14–0.30 | 0.23–0.27 |
| Diabase | 29.4–114.0 | 55–100 | 0.10–0.33 | 0.24–0.28 |
| Gabbro | 58.4–108.0 | 60–80 | 0.12–0.48 | 0.15–0.25 |
| Granite | 17.2–70.5 | 20–50 | 0.12–0.34 | 0.15–0.24 |
| Diorite | 43.0–92.5 | 50–80 | 0.13–0.26 | 0.18–0.22 |
| Syenite | 21.3–86.3 | 30–50 | 0.15–0.34 | 0.20–0.30 |
| Tuff | 0.05–5 | 0.5–1.0 | 0.21–0.33 | 0.25–0.30 |
| Rhyolite | 15.5–37.6 | 20–30 | 0.10–0.28 | 0.14–0.19 |
| **Metamorphic rocks** | | | | |
| Gneiss | 9.5–147.0 | 20–60 | 0.03–0.346 | 0.1–0.2 |
| Marble | 28.0–100.0 | 40–90 | 0.11–0.38 | 0.24–0.30 |
| Schist | 13.7–98.1 | 20–50 | 0.01–0.31 | 0.1–0.2 |
| Quartzite | 25.5–97.5 | 40–60 | 0.09–0.23 | 0.15–0.20 |
| **Water ice** | 8.6–12.0 | 9–10 | 0.33–0.35 | 0.33 |

## D.2 Relations among the elastic constants for isotropic rock

Each elastic constant is here presented in terms of two other constants. For example, Young's modulus $E$ is presented in terms of Poisson's ratio $v$ and shear modulus $G$. The other two constants used here are Lamé's constant $\lambda$ and the bulk modulus $K$. Some of the more complex expressions are omitted (cf. Slaughter, 2002).

| | $E$ | $v$ | $G$ | $K$ | $\lambda$ |
|---|---|---|---|---|---|
| $\lambda, G$ | $\dfrac{G(3\lambda + 2G)}{\lambda + G}$ | $\dfrac{\lambda}{2(\lambda + G)}$ | $G$ | $\lambda + \dfrac{2}{3}G$ | $\lambda$ |
| $\lambda, E$ | $E$ | | | | $\lambda$ |
| $\lambda, v$ | $\dfrac{\lambda(1+v)(1-2v)}{v}$ | $v$ | $\dfrac{\lambda(1-2v)}{2v}$ | $\dfrac{\lambda(1+v)}{3v}$ | $\lambda$ |
| $\lambda, K$ | $\dfrac{9K(K-\lambda)}{3K-\lambda}$ | $\dfrac{\lambda}{2(\lambda + G)}$ | $\dfrac{3(K-\lambda)}{2}$ | $K$ | $\lambda$ |
| $G, E$ | $E$ | $\dfrac{E-2G}{2G}$ | $G$ | $\dfrac{GE}{3(3G-E)}$ | $\dfrac{G(E-2G)}{3G-E}$ |
| $G, v$ | $2G(1+v)$ | $v$ | $G$ | $\dfrac{2G(1+v)}{3(1-2v)}$ | $\dfrac{2Gv}{1-2v}$ |
| $G, K$ | $\dfrac{9KG}{3K+G}$ | $\dfrac{3K-2G}{6K+2G}$ | $G$ | $K$ | $K - \dfrac{2}{3}G$ |

| | | | | | |
|---|---|---|---|---|---|
| $E, \nu$ | $E$ | $\nu$ | $\dfrac{E}{2(1+\nu)}$ | $\dfrac{E}{3(1-2\nu)}$ | $\dfrac{E\nu}{(1+\nu)(1-2\nu)}$ |
| $E, K$ | $E$ | $\dfrac{3K-E}{6K}$ | $\dfrac{3KE}{9K-E}$ | $K$ | $\dfrac{3K(3K-E)}{9K-E}$ |
| $K, \nu$ | $3K(1-2\nu)$ | $\nu$ | $\dfrac{3K(1-2\nu)}{2(1+\nu)}$ | $K$ | $\dfrac{3K\nu}{1+\nu}$ |

# E Appendix E: Properties of some crustal materials

## E.1 Rock densities, strengths, and internal friction

All the values given below refer to dry conditions except that the upper values for gravel-sand and for clay refer to saturated specimens. The sources of the data are mostly the same as in Appendix D.

| Rock type | Dry density, $\rho_r$, kg m$^{-3}$ | Compressive strength, $C_0$ MPa | Tensile strength, $T_0$ MPa | Coefficient of internal friction, $\mu$ |
|---|---|---|---|---|
| **Sedimentary rocks** | | | | |
| Limestone | 2300–2900 | 4–250 | 1–25 | 0.70–1.20 |
| Dolomite | 2400–2900 | 80–250 | 3–25 | 0.40 |
| Sandstone | 2000–2800 | 6–170 | 0.4–25 | 0.50–0.70 |
| Shale | 2300–2800 | 10–160 | 2–10 | 0.27–0.58 |
| Marl/mudstone | 2300–2700 | 26–70 | 1 | 0.78–1.07 |
| Chalk | 1400–2300 | 5–30 | 0.3 | 0.47 |
| Anhydrate | 2700–3000 | 70–120 | 5–12 | |
| Gypsum | 2100–2300 | 4–40 | 0.8–4 | 0.58 |
| Rock salt | 1900–2200 | 9–23 | 0.2–3 | |
| Sand, gravel | 1400–2300 | | 0 | 0.58–1.20 |
| Clay | 1300–2300 | 0.2–0.5 | 0.2 | 0.35–0.52 |
| **Igneous rocks** | | | | |
| Basalt | 2600–3000 | 80–410 | 6–29 | 1.11–1.19 |
| Diabase | 2600–2900 | 120–250 | 5–13 | 1.19–1.43 |
| Gabbro | 2850–3100 | 150–290 | 5–29 | 0.18–0.66 |
| Granite | 2500–2700 | 120–290 | 4–25 | 0.60–1.73 |
| Diorite | 2650–2900 | 70–180 | 10–14 | |
| Andesite | 2500–2800 | 70–200 | | |
| Syenite | 2550–2700 | 100–340 | 10–13 | |
| Tuff | 1600 | 0.4–44 | 0.1–0.8 | 0.21–0.36 |
| Rhyolite | 2150–2500 | 180–260 | 16–21 | |

**Metamorphic rocks**

| | | | | |
|---|---|---|---|---|
| Gneiss | 2500–2900 | 80–250 | 4–20 | 0.60–0.71 |
| Marble | 2600–2800 | 50–200 | 5–20 | 0.62–1.20 |
| Schist | 2500–2900 | 20–100 | 2 | 1.90 |
| Quartzite | 2600–2700 | 90–300 | 3–5 | 0.48–1.73 |
| **Water ice,0°C** | 917 (–934) | | | |

# E.2 General rock and fluid properties

The main data presented in the first part of this Appendix are from Byerlee (1978), Atkinson (1987), Schultz (1995), Amadei and Stephansson (1997), Bell (2000), Schön (2004), Paterson and Wong (2005), and Gudmundsson (2009). Other references are cited where appropriate.

1. The **in-situ tensile strengths** of most rocks are between 0.5 and 6 MPa (Haimson and Rummel, 1982; Schultz, 1995; Amadei and Stephansson, 1997). The highest value, 9 MPa, was obtained at a depth of about 9 km in the KTB drill hole in Germany (Amadei and Stephansson, 1997).

2. Experimental **rock-friction relationships** for different normal stresses, $\sigma_n$, were obtained by Byerlee (1978). These relationships are sometimes referred to as **Byerlee's law** and, using Eq. (7.1), may be stated as follows:

For normal stress in the range 10 MPa $\leq \sigma_n \leq$ 200 MPa, Eq. (7.1) becomes

$$\tau = 0.85\sigma_n \tag{E.1}$$

For normal stress in the range 200 MPa $\leq \sigma_n \leq$ 1500 MPa, Eq. (7.1) becomes

$$\tau = 50 + 0.6\sigma_n \tag{E.2}$$

In Eq. (E.1) the inherent shear strength, $\tau_0$, in Eq. (7.1) would be zero and the coefficient of internal friction, $\mu$, 0.85. In Eq. (E.2) the inherent shear strength, $\tau_0$, in Eq. (7.1) would be 50 MPa and the coefficient of internal friction, $\mu$, 0.6. These laws hold to normal stresses that would correspond to crustal depths of some 50–60 km and are largely independent of roughness on the fault surface and the type of rock being faulted (that is, they apply to a variety of rock types). The coefficients of internal friction, 0.6–0.85, are similar to those of many rocks in Appendix E.1. However, fluid pressure (effective stress) does not enter these laws (Chapters 7 and 8) and they are, as yet, purely empirical since they have no clear theoretical explanation.

3. **Laboratory fracture and material toughness**. Compilations of laboratory data on fracture and material toughness are provided by Fourney (1983), Atkinson and Meredith (1987), Paterson and Wong (2005), Nasseri *et al.* (2006), and Nasseri and Mohanty (2008).

Typical values for the **fracture toughness** of rock, $K_c$, for mode I (extension) fractures are 0.5–3 MPa m$^{1/2}$. Few rock specimens have values of fracture toughness as high as 5–20 MPa m$^{1/2}$ (Fourney, 1983). These values do not change much with increasing temperature and pressure. For example, basalt specimens from Iceland and Italy subject to temperatures of

20–600°C and pressures of as much as 30 MPa yield values for fracture toughness of 1.4–3.8 MPa m$^{1/2}$ (Balme $et\ al.$, 2004). At pressures of 60–100 MPa, however, some limestone and sandstone specimens yield values of fracture toughness of 5 MPa m$^{1/2}$ or higher.

Typical values for the **material toughness** of rock, $G_c$, for mode I fractures are 20–400 J m$^{-2}$. The highest and lowest specimen values given by Atkinson and Meredith (1987) are 1580 J m$^{-2}$ (for sandstone normal to bedding) and 15 J m$^{-2}$ (for limestone), respectively. The highest and lowest values given by Fourney (1983) are 2298 J m$^{-2}$ (for basalt) and 8 J m$^{-2}$ (for limestone). The material toughness of shear fractures is normally much higher. For example, for mode II fractures in various rock specimens, most material toughness values are of the order of 0.01 MJ m$^{-2}$ (Li, 1987).

4. **In-situ fracture and material toughness**. Many values for in-situ fracture and material toughness for various crustal segments and fracture types have been provided, as summarised by Gudmundsson (2009). First consider the estimates for faults (modes II and III). Estimates of material toughness for strike-slip faults range from 260 J m$^{-2}$ to 100 MJ m$^{-2}$, but the most common values are in the range 0.15–17 MJ m$^{-2}$ (Li, 1987; Rice, 2006; Rice and Cocco, 2007). For dip-slip faults, the material toughness is estimated at 2.3 MJ m$^{-2}$ and the corresponding fracture toughness (stress intensity) at 150 MPa m$^{1/2}$ (Gudmundsson, 2009).

Many estimates have been made of the in-situ toughness of mode I fractures, primarily using dykes. Most values of fracture toughness (stress-intensity) are in the range 30–150 MPa m$^{1/2}$ (Delaney and Pollard, 1981; Rubin and Pollard, 1987; Parfitt, 1991; Rivalta and Dahm, 2006), although much higher values have been suggested (Jin and Johnson, 2008). For typical dykes, Gudmundsson (2009) obtained values for material toughness of 1.3–47 MJ m$^{-2}$ and corresponding fracture toughness (stress intensities) of 114–690 MPa m$^{1/2}$.

5. **Yield strengths and the brittle–ductile transition stresses of some rocks**. Bell (2000) and Paterson and Wong (2005) provide the yield strengths and stresses at the brittle–ductile transition for some rocks. For gypsum, the yield strengths, that is, the stresses at which plastic yield (or flow) starts are generally between 7 and 29 MPa (Bell, 2000). Paterson and Wong (2005) give the following brittle–ductile transition stresses:

| Rock | Stress at brittle–ductile transition, MPa |
|---|---|
| Limestone/marble | 10–100 |
| Chalk | < 10 |
| Dolomite | 100–200 |
| Gypsum | 40 |
| Anhydrate | 100 |
| Rock salt | < 20 |
| Quartzite | 600 |
| Sandstone | 200–300 |
| Shale/siltstone | < 100 |
| Basalt | 300 |
| Porous lava flows | 30–100 |

Non-porous igneous and metamorphic rocks at **room temperature** (25 °C) are normally brittle up to confining pressures ($\sigma_3$) of at least 1000 MPa, corresponding to a depth of roughly 30 km. Some granites behave as brittle up to confining pressures of 3000 MPa. Increasing temperature generally increases ductility, that is, decreases the depth to the brittle–ductile transition. However, the effects of temperature have much more limited effects in increasing ductility if the confining pressure does not also increase. Generally, for non-porous igneous and metamorphic rocks a temperature of at least 300–500 °C (600–800 K) is needed to make them behave as ductile. For example, limited ductility occurred in specimens of basalt and granite at temperatures of 600–900 °C under compressive loading of 500 MPa. This pressure corresponds to crustal depths of roughly 18 km. Strain rates, however, also affect the depth of the brittle–ductile transition; lower strain rates decrease the depth.

6. **Densities and viscosities of some crustal fluids**. The data on lavas and magmas are from Murase and McBirney (1973), Kilburn (2000) and Spera (2000); those on other fluids from Smits (2000) and Middleton and Wilcock (2001). The values for the (flowing) lavas refer to their eruption sites (at the crater cones). The values refer to atmospheric pressure. Note that the viscosities of the magmas vary over several orders of magnitude depending on their composition, temperature, volatile content, and other factors.

| Fluid | Density, kg m$^{-3}$ | Dynamic viscosity, Pa s |
|---|---|---|
| Water at 0 °C | 999.9 | $1.787 \times 10^{-3}$ |
| Water at 5 °C | 1000 | $1.519 \times 10^{-3}$ |
| Water at 40 °C | 992.2 | $0.653 \times 10^{-3}$ |
| Water at 80 °C | 971.8 | $0.355 \times 10^{-3}$ |
| Water at 100 °C | 968.4 | $0.282 \times 10^{-3}$ |
| Seawater at 20 °C | 1025 | $1.070 \times 10^{-3}$ |
| Crude oils | 850–950 | 0.1–0.01 |
| Glacier ice | 920 | $1 \times 10^{13}$ |
| Rock salt | 1900–2200 | $3 \times 10^{15}$ |
| Basaltic lava, 1050–1200 °C | 2500–2800 | $1 \times 10^{2-3}$ |
| Andesitic lava, 950–1170 °C | 2400–2500 | $1 \times 10^{4-7}$ |
| Rhyolite lava, 700–900 °C | 2100–2300 | $1 \times 10^{9-13}$ |
| Basaltic magma, 1200 °C | 2650 | $1 \times 10^{2}$ |
| Rhyolitic magma, 900 °C | 2250 | $5 \times 10^{5}$ |

# References

Amadei, B. and Stephansson, O., 1997. *Rock Stress and its Measurement*. London: Chapman & Hall.

Atkinson, B. K. (ed.), 1987. *Fracture Mechanics of Rock*. London: Academic Press.

Atkinson, B.K. and Meredith, P.G., 1987. Experimental fracture mechanics data for rocks and minerals. In: Atkinson, B. K. (ed.), *Fracture Mechanics of Rock*. London: Academic Press, pp. 477–525.

Balme, M. R., Rocchi, V., Jones, C., Sammonds, P. R., Meredith, P. G., and Boon, S., 2004. Fracture toughness measurements on igneous rocks using a high-pressure, high-temperature rock fracture mechanics cell. *Journal of Volcanology and Geothermal Research*, **132**, 159–172.

Bell, F. G., 2000. *Engineering Properties of Rocks*, 4th edn. Oxford: Blackwell.

Benenson, W., Harris, J.W., Stocker, H., and Lutz, H. (eds.), 2002. *Handbook of Physics*. Berlin: Springer-Verlag.

Byerlee, J. D., 1978. Friction of rocks. *Pure and Applied Geophysics*, **116**, 615–626.

Carmichael, R. S., 1989. *Practical Handbook of Physical Properties of Rocks and Minerals*. Boca Raton, Boston, MA: CRC Press.

Deeson, E., 2007. *Internet-linked Dictionary of Physics*. London: Collins.

Delaney, P. and Pollard, D. D., 1981. Deformation of host rocks and flow of magma during growth of minette dikes and breccia-bearing intrusions near Ship Rock, New Mexico. *US Geological Survey Professional Paper*, **1202**.

Emiliani, C., 1995. *The Scientific Companion*, 2nd edn. New York: Wiley.

Fjaer, E., Holt, R. M., Horsrud, P., Raaen, A. M., and Risnes, R., 2008. *Petroleum Related Rock Mechanics*, 2nd edn. Amsterdam: Elsevier.

Fourney, W. L., 1983. Fracture control blasting. In: Rossmanith, H. P. (ed.), *Rock Fracture Mechanics*. New York: Springer-Verlag, pp. 301–319.

Gottfried, B. S., 1979. *Introduction to Engineering Calculations*. New York: McGraw-Hill.

Gudmundsson, A., 2009. Toughness and failure of volcanic edifices. *Tectonophysics*, **471**, 27–35.

Haimson, B. C. and Rummel, F., 1982. Hydrofracturing stress measurements in the Iceland research drilling project drill hole at Reydarfjordur, Iceland. *Journal of Geophysical Research*, **87**, 6631–6649.

Hansen, S. E., 1998. *Mechanical Properties of Rocks*. Report STF22 A98034. Trondheim: Sintef (in Norwegian).

Huntley, H. E., 1967. *Dimensional Analysis*. New York: Dover.

Jaeger, J. C. and Cook, N. G. W., 1979. *Fundamentals of Rock Mechanics*, 3rd edn. London: Chapman & Hall.

Jeremic, M. L., 1994. *Rock Mechanics in Salt Mining*. Rotterdam: Balkema.

Jin, Z. H. and Johnson, S. E., 2008. Magma-driven multiple dike propagation and fracture toughness of crustal rocks. *Journal of Geophysical Research*, **113**, B03206.

Jumikis, A. R., 1979. *Rock Mechanics*. Clausthal, Germany: Trans Tech Publications.

Kilburn, C. J., 2000. Lava flows and flow fields. In: Sigurdsson, H. (ed.), *Encyclopedia of Volcanoes*. New York: Academic Press, pp. 291–305.

Li, V. C., 1987. Mechanics of shear rupture applied to earthquake zones. In: Atkinson, B. K. (ed.), *Fracture Mechanics of Rock*. London: Academic Press, pp. 351–428.

Mavko, G., Mukerji, T., and Dvorkin, J., 2009. *The Rock Physics Handbook*, 2nd edn. Cambridge: Cambridge University Press.

Middleton, G. V. and Wilcock, P. R., 2001. *Mechanics in the Earth and Environmental Sciences*. Cambridge: Cambridge University Press.

Murase, T. and McBirney, A. R., 1973. Properties of some common igneous rocks and their melts at high temperatures. *Geological Society of America Bulletin*, **84**, 3563–3592.

Myrvang, A., 2001. *Rock Mechanics*. Trondheim: Norway University of Technology (NTNU) (in Norwegian).

Nasseri, M. H. B. and Mohanty, B., 2008. Fracture toughness anisotropy in granitic rocks. *International Journal of Rock Mechanics and Mining Science*, **45**, 167–193.

Nasseri, M. H. B., Mohanty, B., and Young, R. P., 2006. Fracture toughness measurements and acoustic activity in brittle rocks. *Pure and Applied Geophysics*, **163**, 917–945.

Nilsen, B. and Palmström, A., 2000. *Engineering Geology and Rock Engineering*. Oslo: Norwegian Soil and Rock Engineering Association (NJFF).

Parfitt, E. A., 1991. The role of rift zone storage in controlling the site and timing of eruptions and intrusions at Kilauea Volcano, Hawaii. *Journal of Geophysical Research*, **96**, 10101–10112.

Paterson, M. S. and Wong, T. W., 2005. *Experimental Rock Deformation –the Brittle Field*, 2nd edn. Berlin: Springer-Verlag.

Pinel, V. and Jaupart, C., 2005. Some consequences of volcanic edifice destruction for eruption conditions. *Journal of Volcanology and Geothermal Research*, **145**, 68–80.

Rice, J. R., 2006. Heating and weakening of faults during earthquake slip. *Journal of Geophysical Research*, **111**, B05311.

Rice, J. R. and Cocco, M., 2007. Seismic fault rheology and earthquake dynamics. In: Handy, M. R., Hirth, G., and Horius, N. (eds.), *Tectonic Faults: Agents of Chance on a Dynamic Earth*. Cambridge, MA: The MIT Press, pp. 99–137.

Rivalta E. and Dahm, T., 2006. Acceleration of buoyancy-driven fractures and magmatic dikes beneath the free surface. *Geophysical Journal International*, **166**, 1424–1439.

Rubin, A. M. and Pollard, D. D., 1987. Origins of blade-like dikes in volcanic rift zones. *US Geological Survey Professional Paper*, **1350**, pp. 1449–1470.

Schön, J. H., 2004. *Physical Properties of Rocks: Fundamentals and Principles of Petrophysics*. Oxford: Elsevier.

Schultz, R. A., 1995. Displacement-length scaling for terrestrial and Martian faults: Implications for Valles Marineris and shallow planetary grabens. *Journal of Geophysical Research*, **102**, 12 009–12 015.

Slaughter, W. S., 2002. *The Linearized Theory of Elasticity*. Berlin: Birkhauser.

Smits, A. J., 2000. *A Physical Introduction to Fluid Mechanics*. New York: Wiley.

Spera, F. J., 2000. Physical properties of magmas. In: Sigurdsson, H. (ed.), *Encyclopedia of Volcanoes*. New York: Academic Press, pp. 171–190.

Waltham, A. C., 1994. *Foundations of Engineering Geology*. London: Spon.

Woan, G., 2003. *The Cambridge Handbook of Physics Formulas*. Cambridge: Cambridge University Press.

# Index

Printed in the United States
By Bookmasters